Statistical
Foundations for
Econometric
Techniques

Statistical Foundations for Econometric Techniques

Asad Zaman

Bilkent University
Ankara, Turkey

Academic Press

San Diego   New York   Boston   London   Sydney   Tokyo   Toronto

This book is printed on acid-free paper. ∞

Copyright © 1996 by ACADEMIC PRESS, INC.

All Rights Reserved.
No part of this publication may be reproduced or transmitted in any form or by any
means, electronic or mechanical, including photocopy, recording, or any information
storage and retrieval system, without permission in writing from the publisher.

Academic Press, Inc.
A Division of Harcourt Brace & Company
525 B Street, Suite 1900, San Diego, California 92101-4495

United Kingdom Edition published by
Academic Press Limited
24-28 Oval Road, London NW1 7DX

Library of Congress Cataloging-in-Publication Data

Zaman, Asad.
 Statistical foundations for econometric techniques / by Asad
Zaman.
        p.   cm. -- (Economic theory, econometrics, and mathematical
economics)
 Includes index.
 ISBN 0-12-775415-6 (pbk. : alk paper)
 1. Econometrics. I. Title. II. Series.
HB139.Z36 1996
330'.01'5195--dc20   95-44359
                                                              CIP

PRINTED IN THE UNITED STATES OF AMERICA
95 96 97 98 99 00 MM 9 8 7 6 5 4 3 2 1

In the name of Allah, the Merciful, the Benificient

Preface

When I started this book, it was with the intention of expositing the theory of Stein estimates and related paraphernalia, a neglected topic at that time. After spending several years gathering the material for this book (during which several expositions have appeared), I find that what I thought was relatively ripe theory, with answers to all important questions, appears to be virgin territory, largely unexplored. In the process of writing the book, the number of questions that have arisen is greater than the number of answers I have learned. Some of these questions are indicated in the text, but many fall outside the intended scope of the text. In this sense the book has an unfinished feeling; I have had to leave many questions open and to put pointers to various discussions that I had meant to include. Nonetheless, the search for perfection appears not to be a convergent process, and one must truncate at some arbitrary iteration.

In the process of writing, I have accumulated debts too numerous to list, both intellectual and personal. The classes of students at University of Pennsylvania, Columbia University, and Johns Hopkins, who sat through early formulations of various topics covered by these notes, undoubtedly suffered heavily. A large number of topics presented in this text were not available except in advanced texts and articles. Students frequently asked for references and were disappointed to learn that other 'elementary' presentations were unavailable. I understand that my course was referred to as Econometrics 2001 by some of my classes. Numerous colleagues have commented on material directly or indirectly included in the text. Indirect debts are too numerous to list, but I would like to acknowledge my first mentors, Professors Charles Stein, T. W. Anderson and Takeshi Amemiya, for arousing my interest in the subject and challenging me intellectually. Professors Johann Pfanzagl, Peter Rousseeuw, and Sam Weerahandi were kind enough to provide comments on Chapters 16, 5, and 10, respectively; of course, only I am responsible for any remaining errors.

Institutional support provided by Columbia University, Johns Hopkins University, and Bilkent University is gratefully acknowledged. Special thanks to Bilkent librarians for searching out odd articles from all over the world. Thanks are also due to developers and supporters of LATEX, who have substantially simplified the task of creating and revising mathematical manuscripts. I hereby acknowledge, with thanks, permissions from

- John Wiley & Co. to reproduce the data in Table 5.1 and Figure 5.2.

- Springer-Verlag to use data on which Figure 5.3 is based.

- American Statistical Association to reproduce the Toxomplasmosis data in Table 16.4.

My greatest personal debt is to my father, Mohammed Masihuzzaman. The support and encouragement of all members of my extended family were important to this project. I dedicate this book to my wife, Iffat, and my children, Meryem, Talha, and Zainab, who all made great sacrifices for this book.

Contents

I   Estimators for Regression Models

II Hypothesis Tests for Regression Models

IV   Empirical Bayes: Applications

V   Appendices

List of Symbols and Notation

$\mathbb{E},(\mathbb{E}^{X|Y})$	Expected Value
$\mathbb{P}(\mathrm{E})$	Probability of Event E
$\mathbb{V}ar(X)$	Variance of real-valued RV (Random Variable) X
$\mathbb{C}ov(X)$	Covariance matrix of vector-valued RV X
Qlim	limit in quadratic mean
plim	limit in probability
$\mathbf{tr}(M)$	trace of matrix M
$\mathbb{1}$	A vector of ones
$X \sim F$	Random Variable X has distribution F
iid	independent and identically distributed
inid	independent non-identically distributed
$X \sim \mathcal{B}i(n,\theta)$	X is Binomial: $\mathbb{P}(X=x) = \binom{n}{x}\theta^x(1-\theta)^{n-x}$
$X \sim N(\mu,\sigma^2)$	X is scalar Normal with mean μ, variance σ^2.
$X \sim N_k(\mu,\Sigma)$	X is k-variate Normal; $\mathbb{E}X = \mu$ and $\mathbb{C}ov(X) = \Sigma$
$X \sim \mathcal{G}(p,\lambda)$	X is Gamma with parameters p and λ:
	$f^X(x) = \lambda^p x^{p-1} \exp(-\lambda x)/\Gamma(p)$
$X \sim \sigma^2 \chi_k^2$	X is Chi-Square: $X \sim \mathcal{G}(k/2, 1/(2\sigma^2))$

Learning is from cradle to grave ...
Seek knowledge even if it is in China ...
—Sayings of Prophet Muhammad, peace be upon him

Introduction

This book focuses on the statistical and theoretical background as well as optimality properties of econometric procedures (almost exclusively in the context of linear regression models). I have used it as a second year text for several years. Typical first year econometrics sequences need minor tailoring to provide an adequate background; some of the supplements needed are covered in the appendices to this text.

I have attempted, not always successfully, to provide intuition, theoretical background, as well as detailed recipes needed for applications. My hope is that by skipping over the occasionally difficult theory, less well-prepared students as well as applied econometricians will be able to benefit from the detailed presentation of several new developments in econometrics. While it is widely realized that the structure of the traditional econometrics course (which has remained more or less the same since the 1960s) is inadequate for current needs, no acceptable alternative has emerged. Several pedagogical innovations have been made in this text in the hope of providing a more efficient route to the frontiers. Novel aspects of presentation and contents are discussed in the first section below.

Since views on methodology have influenced my choice of topics, these are discussed in the second section. Briefly, although alternative methodologies developed by Hendry (1991), Spanos (1986), Leamer (1978, 1982, 1983), Pratt and Schlaifer (1984), and others have pointed out important defects in conventional econometrics, my view is that conventional methodology can be patched up to meet the challenges. This is roughly what is undertaken in the text. However, the text focuses primarily on technique, and methodological issues are not treated in depth. See Darnell and Evans (1990) for an illuminating review of the critiques and a defense of conventional methodology. The third section below discusses the Bayesian vs. Frequentist debate, since this is of importance in selection of estimators and test statistics.

Features

What is the best econometric tool for use in a given application? This text focuses on how to make such judgements. This distinguishes it and makes it a complement to rather than an alternative for several excellent modern texts with different goals:

- Judge *et al.* (1985) provide a thorough survey of contemporary econometric theory.

- Berndt (1991) teaches the craft of econometric practice by studying chosen applications in some depth.

- Hendry (1991) and Spanos (1986) are concerned with appropriate methodology for use in econometric practice.

- Conventional texts, such as Greene (1990), exposit the existing corpus of econometric theory and practice.

- Advanced texts, such as Amemiya (1985), focus on the numerous complex and specialized applications of econometric techniques in different areas.

In answer to our question, we conclude, not surprisingly, that for the most part, we should use Maximum Likelihood (ML) to estimate and Likelihood Ratio (LR) to test. Nearly 600 pages provide details, support, and occasionally, counterexamples to this blanket statement.In situations amenable to Empirical Bayes, ML can be substantially inferior. Also, inference based on asymptotic distributions of ML and LR is frequently misleading, and more accurate approximations based on Bootstrap or Edgeworth Expansions should be used more often. This forest-level perspective of the text is followed by a more tree-level description below.

Part I describes, in five chapters, five different classes of estimators for linear regression. Chapter 1 discusses projection theory, the modern version of least squares (or nonstatistical) curve fitting. Chapter 2 describes an ML-based unified approach to single equation, simultaneous equations, and dynamic models, though the latter two topics are treated sketchily. Chapter 3 develops empirical/hierarchical Bayes as attempts to address certain difficulties with more classical Bayesian estimators. Chapter 4 develops the theory of minimax and Stein-type estimators. Chapter 5 tells the tangled tale of robust regression estimators, describing the many blind alleys and a few promising avenues.

Part II deals with hypothesis testing for linear regression models. It is well known that LR, Wald, and LM tests are locally asymptotically equivalent. It is less well known that this equivalence breaks down at the third order term in the asymptotic expansions, and LR is superior to the other two. Whether or not the third order term is significant depends on the statistical 'curvature' of the testing problem, introduced by Efron (1975). This is documented via an example and some intuition for the phenomenon is provided in Chapter 6. Chapter 7 discusses one approach to finite sample optimality of tests, based on invariance and the Hunt-Stein Theorem. Chapters 8, 9, and 10 discuss different types of testing situations, and what is known and not known about properties of various tests. Numerous widely used tests (such as Chow, Goldfeld-Quandt, Eicker-White, Durbin-Watson) are assessed to be deficient in some respect, and alternatives are suggested.

Part III of the text develops asymptotic theory. An effort is made to match assumptions to requirements of practice. For example, the routinely made assumption that $\lim_T (X'X)/T = Q$ is routinely violated by time series data. Such assumptions are avoided wherever possible. Chapters 11 and 12 discuss consistency of estimators and tests. Chapter 13 develops the asymptotic distribution theory based on the Central Limit Theorem (CLT). Since CLT-based approximations are frequently inadequate in applications, two tools for getting more accurate approximations are introduced in Chapter 14. These are Edgeworth approximations and the Bootstrap. Chapter 15 discusses asymptotic optimality of tests and estimators.

Part IV is devoted to applications of empirical Bayes estimators. This sophisticated but valuable tool is explained via a series of examples of increasing complexity.

Methodological Issues

The rise and fall of empiricism and logical positivism as a modern philosophy has a parallel in the fortunes of econometrics as a subdiscipline of economics. In both cases, too much reliance was placed on observations as a means of learning about the world. This led to disappointment as it was learned that data alone are not enough to lead us to the truth. More or less, this is for the simple reason that any set of observations is finite and compatible with an infinite number of alternative theories.

Some economic theorists go to the opposite extreme of ignoring all empirical evidence. It would be fair to say that we are still in the process of learning what we can and cannot discover about economic phenomena by the means of observations. Nonetheless, strategies which insulate theories from having to explain observations can only be damaging in the long run. Most fruitful theories come from the need to explain recalcitrant observations. While econometrics is not the solution to all our problems, we cannot get by without it either. After these preliminary comments, we turn to a more detailed description of the process of formulating and estimating econometric models.

Selection of a Family of Models

Econometric and statistical analysis begins *after* a suitable initial family of models has been selected. The rather broad and difficult issues involved in making a suitable initial choice are discussed by Cox and Hinkley (1974, Introduction), Cox (1990), and also by Lehmann (1990) who provides further references. Some issues especially relevant to econometric practice are discussed briefly below. It is convenient to break down this topic into three subtopics:

A How do we select a suitable family of possible models?

B How do we decide whether this family is adequate?

C What should we do if we find the family to be inadequate?

Selection of a suitable family of models is guided by available theory, past experience with similar data sets, and exploratory or preliminary analysis of the data set. The family of models should satisfy all theoretical restrictions. If it is desired to test the theory, it may be useful to enlarge the family to include some models which do not satisfy the theoretical restrictions. It is desirable to select a family in which the parameters correspond to theoretical entities so as to facilitate interpretation of the model. Ease of statistical analysis is also of importance in selecting a family of models. Parsimoniously parameterized models are also useful in extracting maximum information out of a data set. The *Encompassing Principle* is potentially in conflict with this last suggestion for selection of a suitable family of models, and is discussed in greater detail in a separate section below. In practice, it frequently happens that the initial family selected proves unsuitable in some way and has to be replaced by some modification. Thus an iterative process is used in selection of a suitable family.

In judging adequacy of a family of models, we use two types of criteria: *internal* and *comparative* (or *external*). By internal criteria, we refer to the consistency of the data with the econometric and theoretical assumptions of the family of models. Tests based on comparing different models will be referred to as comparative. Internal adequacy of a family of models can be judged by assessing if the estimates satisfy theoretical restrictions and also the statistical assumptions made by the model. For example, consider a class of models suggesting that current aggregate consumption is explained by current disposable income. Estimation of such models typically leads to autocorrelated error terms, which shows that there are unmodeled dynamic effects at work. Thus we should reject the initial class of models as inadequate and consider a larger family which allows for the possibility of effects across time. Similarly if signs, magnitudes, etc. of estimated coefficients and error variances are out of line with those suggested by theory, then we may question the adequacy of the initial family of models (and also possibly, that of the theory). Internal inadequacy may sometimes (but not always) furnish clues regarding issue C, namely the construction of an adequate model, given that the initial specification is deemed inadequate.

A model can also be judged by comparing it to other models. A given model may be in accord with theory and apparently consistent with the data, but if it gives significantly poorer forecasts than a simple atheoretical model, we may well question its validity. Similarly, comparisons across models can give some guidance as to relative adequacy of the models. If one of the models is outright superior, then issue C is simple: we just use the better model. It can happen that two models each have superior performance over different ranges of observations. In such cases, hybrid models may be required in response to external inadequacy.

The Encompassing Principle

In discussing issue A listed above, it is important to take note of the *Encompassing Principle* formulated and vigorously defended by Hendry (1991). According to this principle, we should make the initial family of models as large as possible. How large is large? More or less, the goal is to make it large enough to minimize problem B, and avoid C altogether. Clearly, if the family is large enough, it will never be found inadequate, and testing for adequacy will not be a major concern. In fact, it is quite awkward from several points of view to have to re-specify the initial family of models, after finding it inadequate. From the frequentist point of view, the meaning of all classical statistics and significance levels become suspect if such a procedure is followed. The Bayesian perspective also provides strong support for the encompassing principle; the need to respecify the model corresponds to a situation where an event occurs which is impossible according to the prior density (technically: outside the support of the prior). In such situations classical Bayesian updating formulae are invalid, and the performance of Bayesian procedures does not have the usual optimality properties. To avoid these difficulties, a classical Bayesian prescription is to *build gigantic models*.

Effectively, the encompassing principle asks us to avoid being surprised by the data; we should build sufficient flexibility into our models so that all possible data twists are potentially explainable by some setting of the parameters. In fact, unexpected patterns in the data form an important part of the process of learning from data. It appears impossible to anticipate all possible deviations from a given model. Indeed, one can find violations of the encompassing methodology in published works of proponents of the methodology (see Darnell and Evans (1990) for some examples). It is illuminating to consider this issue in the context of the distinction between 'Exploratory' versus 'Confirmatory' data analysis introduced by Tukey. Encompassing is a priceless tool for confirmatory data analysis but not of much value in the exploratory mode. Unfortunately, in econometrics as elsewhere, the two categories can never be neatly separated. Section 9.9.1 contains additional discussion of contexts where the principle is and is not suitable.

Inference within a Family of Models

While the issue of selection of a family of models is of great significance and colors all subsequent analysis, it will only be studied incidentally in this book. This is primarily because this issue belongs largely to the creative realm of scientific analysis and is not easily subjected to mathematical study. Most of the present text is concerned with inference under the assumption that the initial selection of a family of models is appropriate. Having made an initial selection, one can to some extent study whether statistical assumptions incorporated in the family of models are compatible with the data. One can also study the relative performance of different families of models and thereby obtain measures of comparative validity, as discussed in an earlier section.

Once we have selected a family of models, a vast paraphernalia of statistical theory and methods become available to us. It is then an issue of some importance to select, among a class of estimators, those which are efficient, or optimal, relative to suitable criteria. Overemphasis on optimality can also prove dangerous, as it is frequently the case that a method which is optimal for a given model is not good for a nearby model. In such cases, it may worth sacrificing optimality to obtain reasonable performance for a wide range of models. While the issue of optimal inference within a family has been widely studied, the issue of robust inference is not so well understood. Some recently developed tools are of great value in analysis of some aspects of robustness.

Bayesian vs. Frequentist Views

This section discusses the Bayesian philosophy with a view to introducing the student to the foundations controversy and also to clarify the author's point of view. This tale is too tangled for a detailed discussion; our goal is merely to provide an adequate background for justifying the procedures proposed in the text.

What is the meaning of the sentence 'The probability of the coin coming up heads is one-half'? There are two major schools of thought and many minor ones. The objectivist views hold that the sentence describes a property of the coin which is best demonstrated by the fact that in repeated trials the heads will occur about one-half of the time (in fact, exactly half of the time in the limit as trials increase to infinity). The subjectivist view is that the sentence describes an attitude on the part of the person who says it, namely that he considers heads and tails to be equally likely. This attitude need not be associated with any particular characteristic of the coin.

Given that subjectivists consider probability theory to be a study of quantifiable beliefs of people and how they change in response to data, while frequentists consider probablity theory to be the study of behavior of certain types of physical objects, or events of certain types, it would seem that there is no room for conflict between the two. The theories have different domains as their object of study, and one might reasonably expect that experts in these domains would have little to say to each other. In fact, they are *not* on speaking terms, but this is because the theories overlap, and appear to conflict, when it comes to studying decision making behavior. It seems clear to the author that decision making necessarily involves assessing the facts and making subjective judgements. Thus both theories are indispensable — frequentist theory provides us the tools to assess the facts and Bayesian theory provides convenient tools to incorporate subjective judgement into the analysis. Unnecessary conflict is created by the determination of certain subjectivists and 'objectivists' to deny legitimacy to each other.

The objectivists say that the attitudes of people about probability are irrelevant to a scientific approach. This is correct to the extent that the facts of the matter do not depend on opinions (even this is denied by extremist Bayesians). Thus if we flip a coin and it comes up heads 10 times in a row, we can objectively state that if the coin is fair, this is an event of probability $1/2^{10}$. Similarly, in any model we can make objective statements of the type that if the data-generating mechanism is of such and such type, the observed events have such and such probability. To go from the facts to an analysis, or an inference, we must introduce some judgements into the picture. We can take different rational decisions based on identical facts because of subjective elements. For example, Copernicus decided that the facts were sufficiently at variance with the Ptolemaic theory to justify working on alternatives. For a less talented man, the same decision would have been wrong. As another example, if a coin comes up heads 10 times in a row, we would be inclined to reject the null hypothesis that the coin is fair. However, if a man claiming ESP guesses heads or tails correctly 10 times in a row, we would be more inclined to suspect cheating or luck rather than the null hypothesis. Given that most data analysis is done with a view to make decisions of some type, and decisions necessarily involve subjective elements, Bayesian analysis provides convenient tools for adding subjective judgements to an analysis. Wholesale rejection of Bayesian methods seems unjustified.

Extremist subjectivists deny that there are any 'facts.' They consider both the statements 'The probability of rain today is 30%' and 'The probability of a balanced coin coming up heads is 50%' to be equally subjective judgements. According to classical frequentist theory, the second statement is a fact verifiable by evaluating the limiting frequency of the event in repeated trials. The subjectivists rightly attacked this definition as being (a) impossible to verify and (b) severely limiting (events which could not be repeated could not have probabilities). A more sensible view of the objectivist theory, suggested to me by Professor Charles Stein, is the following. Just as physicists introduce the term *mass* as a primitive concept, a quality of objects not subject to further explication, so we assume that probability is a quality of events not subject to further explication. We know certain rules about mass by virtue of which it becomes possible to measure mass in certain circumstances. Similarly the laws of probability permit us to measure the probability of certain types of events (approximately) in the situation where repeated trials are possible. This view avoids the usual subjectivist objections to the objectivist theory.

It seems to us rather peculiar to insist (as de Finetti and certain other leading subjectivists do) that there is no such thing as objective probability but there is subjective probability. In saying this, we are saying that random events themselves do not satisfy the laws of objective probability, but our beliefs about these events do satisfy the laws of subjective probability. How can it be rational to believe that random events display probabilistic regularity when the events themselves do not necessarily possess such regularity? This seems a contradiction in terms. See Zaman (1995) for a discussion of flaws in the Bayesian argument that rational behavior 'must' be Bayesian.

The main reason for this detour into the foundations controversy is because we will discuss various Bayesian estimators and evaluate them on the basis of their objective characteristics. Also, we will evaluate prior probability distributions and suggest that some are better than others. Both of these procedures are distasteful to true blue Bayesians who reject the validity of objective probability. They also find it a non-sequitur to say that a certain prior belief about a probability is wrong, since the prior belief is all we have to define probability (we cannot test the prior against an objective probability). Since we hold both subjective and objective views to be valid, the first as an approximate description of thought processes about probability and the second as an approximate description of an objective characteristic of random events, so we can both evaluate Bayesian procedures by objective methods and also say that a given subjective probability distribution is wrong, because it does not accord with the objective probability distribution.

Part I

Estimators for Regression Models

Returning home in the evening, Hoja Nasruddin saw a figure moving in his garden and thought it was a burglar. Quickly fetching his bow and arrow, he put an arrow through the figure. Closer examination revealed that the suspected burglar was just his robe hanging out to dry on the clothesline. The arrow had passed through and ruined his expensive robe. 'God be praised that I was not in the robe, when it was shot,' Hoja said.

Chapter 1

Least Squares and Projections

1.1 Introduction

The theory of projection is mathematically elegant, easy to grasp intuitively, and illuminates numerous otherwise obscure matrix equalities. It is surprising how far you can get with a glorified version of Pythagoras' theorem. Time-constrained readers can nonetheless skip the chapter and take on faith references to it made elsewhere.

Basic finite dimensional vector space concepts are assumed known. Elementary matrix theory is also assumed known. In particular, the concepts of vector space, subspace, the span of a set of vectors, distance between two vectors, and the orthogonality of two vectors should be familiar to the reader. Chapter contents can be summarized as follows.

Given a vector $y \in \mathbb{R}^T$, suppose we want to find a vector $\hat{y}_V$ in some vector subspace V of $\mathbb{R}^T$ which is the closest to y. The projection theorem, a close kin of the Pythagorean theorem, characterizes the projection $\hat{y}_V$ by the property that the error $e = y - \hat{y}$ is orthogonal to vector space V. This simple insight has numerous ramifications, explored at book length in Luenberger (1969). We explore a few consequences which are useful in understanding various econometric formulae of least squares theory. An easy consequence is the decomposition theorem, according to which each vector y can be uniquely expressed as the sum of two projections $\hat{y}_V$ and $\hat{y}_W$ when V and W are orthogonal vector spaces. The Gram-Schmidt procedure uses projections to construct an orthogonal sequences of vectors starting from a linearly independent set. The projection theorem holds without change, after suitable redefinitions, with respect to different geometries on $\mathbb{R}^T$. These geometries, generated by nonidentity inner products, are explored. Kruskal's theorem characterizes conditions under which projections relative to different inner products lead to the same vector.

1.2   Projection: The Simplest Case

A number of properties of and formulas related to least squares estimates in linear regression become apparent once the geometry of projection is understood. Our goal in this chapter is to clarify this geometry.

A series of observations on a variable can be regarded as a vector. Say we observe y_t for $t = 1, 2, \ldots, T$. Then we can take $y = [y_1, y_2, \ldots, y_2]'$ as a $T \times 1$ vector in $\mathbb{R}^T$. Let $x \in \mathbb{R}^T$ be some other vector and $\alpha \in \mathbb{R}$ a scalar. The set of vectors αx is the one-dimensional vector space generated by x. This vector space contains all vectors which can be 'explained' by x. The projection of y on x is a vector of the type αx which comes as close as possible to y. In other words, it is the best explanation of y which is possible using x. It is easy to calculate the projection of y on x using calculus. We want to find α which minimizes the distance between αx and y. This distance is $d(\alpha) = \|y - \alpha x\|^2 = \|y\|^2 - 2\alpha y'x + \alpha^2\|x\|^2$. Minimizing this quadratic with respect to α yields

$$\hat{\alpha} = \frac{y'x}{\|x\|^2}.$$

Thus $\hat{\alpha}x$ is the projection of y on x. A geometric approach sheds more

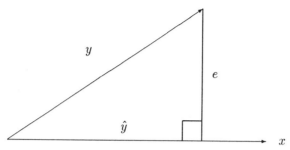

Figure 1.1: Projection of y on x

light on this situation.

Let $\hat{y} = \hat{\alpha}x$ be the projection of y on x and let $e = y - \hat{y}$ be the 'error' or discrepancy between the projection and y itself. As shown in Figure 1, the error e is orthogonal to x, so that $\hat{y}$ and e form two sides of a right angle triangle, with y being the hypotenuse. By invoking Pythagoras, we get

$$\|\hat{y}\|^2 + \|e\|^2 = \|y\|^2.$$

This shows immediately that the projection is always smaller in length than the original vector. As we will soon show, the fact that the error must be orthogonal to x can be established geometrically, without recourse to calculus. Furthermore, this orthogonality alone is enough to yield the formula for the projection.

1.3 The Projection Theorem

We now consider a more general formulation of the problem of the previous section. Given a vector $y \in \mathbb{R}^T$, we wish to 'explain' it using vectors $x_1, x_2, \ldots, x_K$. It will be convenient to let X be a $T \times K$ matrix with i-th column equal to x_i. These K vectors can explain all vectors in the K-dimensional vector space formed by the linear combinations of x_i. Given any $T \times K$ matrix X, the vector space spanned by the columns of X will be denoted by $\mathbf{X}$. This vector space can also be described as the set of all vectors of the type $X\alpha$, where $\alpha = [\alpha_1, \alpha_2, \ldots, \alpha_K]'$ is any vector in $\mathbb{R}^K$, since $X\alpha = \sum_{i=1}^{K} \alpha_i x_i$.

We wish to find a vector in the space $\mathbf{X}$ which comes as close as possible to y. This approximation problem can be solved using the theorem below:

Theorem 1.1 (The Projection Theorem) *For any vector $y \in \mathbb{R}^T$ there exists a unique vector $\hat{y} \in \mathbf{X}$ closest to y, in the sense that for all $w' \neq \hat{y}, w' \in \mathbf{X}$ we have $\|w'-y\| > \|\hat{y}-y\|$. The vector $\hat{y}$ is characterized by the property that the 'error of approximation' $e = y - \hat{y}$ is orthogonal to all vectors in $\mathbf{X}$.*

Proof: The proof of existence is somewhat complex and will not be given here; see for example Luenberger(1969). Suppose $y-\hat{y}$ is orthogonal to all vectors $w \in \mathbf{X}$. A direct argument shows that for any $w_1 \in \mathbf{X}$, the distance $\|y - w_1\|^2$ must be bigger than the distance $\|y - \hat{y}\|$:

$$
\begin{aligned}
\|y - w_1\|^2 &= \|y - \hat{y} + \hat{y} - w_1\|^2 \\
&= \|y - \hat{y}\|^2 + \|\hat{y} - w_1\|^2 + 2(y - \hat{y})'(\hat{y} - w_1) \\
&= \|y - \hat{y}\|^2 + \|\hat{y} - w_1\|^2.
\end{aligned}
$$

The cross-product term must be zero because $(y - \hat{y})$ is orthogonal to $\hat{y}$ and also to w_1. This proves the desired inequality and shows that $\hat{y}$ minimizes distance to y. Note that uniqueness of $\hat{y}$ also follows from this computation: for any $w_1 \neq \hat{y}$, we must have $\|w_1 - \hat{y}\|^2 > 0$ so that $\|y - w_1\|$ must be strictly larger than $\|y - \hat{y}\|$.

Using the projection theorem, one can also obtain an explicit formula for the projection $\hat{y}$. Note that any vector subspace of $\mathbb{R}^T$ is generated by a finite collection of vectors, say $x_1, \ldots, x_k$, and hence is of the form $\mathbf{X}$ for some matrix X. The space orthogonal to $\mathbf{X}$ is denoted $\mathbf{X}^\perp$ (pronounced X perp). This is the $T - K$ dimensional space of vectors v satisfying $v \perp x$ (equivalently $v'x = 0$) for all $x \in \mathbf{X}$. Note that $X'v = 0$ if and only if $x_i'v = 0$ for $i = 1, 2, \ldots, K$. From this, it follows that $v \in \mathbf{X}^\perp$ if and only if $X'v = 0$. Now, according to the projection theorem, $\hat{y}$ is characterized by the property that $e = y - \hat{y} \in \mathbf{X}^\perp$. This is true if and only if $0 = X'e = X'(y - \hat{y})$. Since $\hat{y} \in \mathbf{X}$, $\hat{y} = X\hat{\alpha}$ for some $\hat{\alpha} \in \mathbb{R}^K$. Substituting, we get $X'y - X'X\hat{\alpha} = 0$ so that $\hat{\alpha} = (X'X)^{-1}X'y$. We will henceforth denote the projection of y on the vector space $\mathbf{X}$ by $\Pi_{\mathbf{X}}(y)$. We have just proven that $\hat{y} = X\hat{\alpha}$ so that an explicit formula for the projection is

$$\Pi_{\mathbf{X}}(y) = X(X'X)^{-1}X'y.$$

A matrix P is called *idempotent* if it is symmetric and satisfies $P^2 = P$. There is an important connection between idempotent matrices and projections. Note that since $\hat{y} \in \mathbf{X}$, $\Pi_{\mathbf{X}}(\hat{y}) = \hat{y}$ for any y. Letting $P = X(X'X)^{-1}X'$, we see that $P^2 y = Py$ for any $y \in \mathbb{R}^T$ so that $P^2 = P$. This matrix equality can also be directly verified from the form of P. It follows that projection of y onto any vector subspace is equivalent to premultiplying y by an idempotent matrix. The converse is also true. Every idempotent matrix defines a projection mapping. The details are left to exercises.

1.4   Decomposition

The mapping from y to the projection $\Pi_{\mathbf{X}}(y) = X(X'X)^{-1}X'y = \hat{y}$ breaks up y into two parts, the 'fitted value' $\hat{y}$ and the error $e = y - \hat{y}$. It is of importance to note that the error is also a projection of y: in fact, $e = \Pi_{\mathbf{X}^{\perp}}(y)$. With reference to a vector space $\mathbf{X}$ every vector y can be uniquely decomposed into two parts as $y = \Pi_{\mathbf{X}}(y) + \Pi_{\mathbf{X}^{\perp}}(y)$. This is the content of our next result.

Lemma 1.1 (Decomposition) *Suppose $\mathbf{X}$ is a vector subspace of $\mathbb{R}^T$. Every $y \in \mathbb{R}^T$ has a unique decomposition $y = y_0 + y_1$ with $y_0 \in \mathbf{X}$ and $y_1 \in \mathbf{X}^{\perp}$. Furthermore, y_0 is the projection of y into $\mathbf{X}$ and y_1 is the projection of y into $\mathbf{X}^{\perp}$.*

Proof: Let $y_0 = \Pi_{\mathbf{X}}(y)$ and $y_1 = \Pi_{\mathbf{X}^{\perp}}(y)$ be the projections of y into $\mathbf{X}$ and $\mathbf{X}^{\perp}$, respectively. We claim that $y = y_0 + y_1$. Let $e = y - y_0$. By the projection theorem, e is orthogonal to all vectors in $\mathbf{X}$ and hence belongs to $\mathbf{X}^{\perp}$. Next note that $y - e = y_0 \in \mathbf{X}$ is orthogonal to all vectors in $\mathbf{X}^{\perp}$. But this is precisely the property which characterizes the projection of y into $\mathbf{X}^{\perp}$. It follows that e must be the projection of y into $\mathbf{X}^{\perp}$ so that $e = y_1$. This shows the existence of a decomposition $y = y_0 + y_1$ with $y_0 \in \mathbf{X}$ and $y_1 \in \mathbf{X}^{\perp}$. To prove uniqueness, suppose

$y = y_0' + y_1'$ with $y_0' \in \mathbf{X}$ and $y_1' \in \mathbf{X}^\perp$. Note that $y_0' \in \mathbf{X}$ and $y - y_0' = y_1' \in \mathbf{X}^\perp$ is orthogonal to all vectors in $\mathbf{X}$. It follows from the projection theorem that y_0' is the projection of y into $\mathbf{X}$ and hence, by uniqueness of the projection, $y_0' = y_0$. The proof that $y_1' = y_1$ is similar.

We can also give an explicit formula for the projection of y into the space $\mathbf{X}^\perp$. By the decomposition theorem, we have

$$\Pi_{\mathbf{X}^\perp}(y) = y - \hat{y} = y - X(X'X)^{-1}X'y = \left(\mathbf{I} - X(X'X)^{-1}X'\right)y.$$

Defining $M = \mathbf{I} - X(X'X)^{-1}X'$, we see that M is a projection matrix and hence must be idempotent. Furthermore note that for any $x \in \mathbf{X}$, the projection of x onto $\mathbf{X}^\perp$ must be zero. In matrix terms, for any x of the type $x = X\alpha$, we must have $Mx = MX\alpha = 0$. Since this holds for all α, we must have $MX = 0$. These facts about M will be useful in the sequel.

1.5   The Gram-Schmidt Procedure

The Gram-Schmidt procedure is used to take a collection of vectors and convert them into an orthonormal set of vectors which spans the same space. This is done by projections. Let $x_1, \ldots, x_k$ be a linearly independent set of vectors in R^T. Define $v_1 = x_1/\|x_1\|$. This makes v_1 a unit vector. We define the remaining sequence by induction.

Let $\mathbf{X_j}$ for $j = 1, 2 \ldots, k$ be the vector space spanned by $x_1, \ldots, x_j$. Note that v_1 spans the space $\mathbf{X_1}$. Suppose we have defined an orthonormal set of vectors $v_1, \ldots, v_n$ which spans $\mathbf{X_n}$. We then define v_{n+1} as follows. Let $u_{n+1} = x_{n+1} - \Pi_{\mathbf{X_n}}(x_{n+1})$. It is easily checked that u_{n+1} added to any collection of vectors which spans $\mathbf{X_n}$ will span $\mathbf{X_{n+1}}$. It follows that $v_{n+1} = u_{n+1}/\|u_{n+1}\|$ added to the collection $v_1, \ldots, v_n$ forms an orthonormal set of vectors spanning $\mathbf{X_{n+1}}$ as desired.

It is also profitable to look at the Gram-Schmidt procedure in terms of matrix operations. Let X be the $T \times K$ matrix with columns $x_1 \ldots, x_k$. The first step forms the matrix $U_1 = XD_1$ where $D_1 = \mathrm{diag}(1/\|x_1\|, 1, 1, \ldots, 1)$. Multiplication by this diagonal matrix rescales the first column of X to have unit length, while all other columns remain the same. The $(n+1)$-th step has two parts. Consider the projection of x_{n+1} on the first n columns of U_n. This will be of the form $\Pi_{\mathbf{X_n}}(x_{n+1}) = \beta_1 u_1 + \cdots + \beta_n u_n$. Define the matrix $U_{n+1}^* = U_n E_{n+1}$ where E_{n+1} is the same as the identity matrix except that in the upper half of the $n+1$ column, put the entries $-\beta_1, -\beta_2, \ldots, -\beta_n$. It is easily verified that all columns of U_{n+1} are the same as columns of U_n except the $(n+1)$ column, which is now orthogonal to the first n columns since the projection on the first n columns has been subtracted off. Defining $U_{n+1} = U_{n+1}^* D_{n+1}$ where D_{n+1} is identity except for the $n+1, n+1$ entry which is the reciprocal of the length of the $(n+1)$ column of U^*. This completes the induction by rescaling the $(n+1)$ column to have unit length. This process is compactly described in the following theorem.

Theorem 1.2 (Gram-Schmidt) *Let X be any $T \times K$ matrix of rank K. There exists a $K \times K$ nonsingular upper triangular matrix Δ such that the columns of U defined by $U = X\Delta$ are orthonormal.*

Proof: This follows from the process described just prior to the theorem. X is transformed to U via a sequence of multiplications by nonsingular upper triangular matrices on the right. This can be written as follows: $U = X E_1 D_2 E_2 D_3 E_3 \cdots D_K E_K = X\Delta$. Here we have defined Δ to be the product of all the matrices used to transform X to U. Then Δ is nonsingular and upper triangular since each E_i and D_i is, and it accomplishes the required transformation of X.

Corollary 1.2.1 *Let $x_1, x_2, \ldots, x_k$ be a linearly independent collection of vectors in $\mathbb{R}^T$. Then we can find orthonormal vectors $u_1, \ldots, u_T$ which span $\mathbb{R}^T$ such that for $j \leq k$ $u_1, \ldots, u_j$ span the same vector space as $x_1, \ldots, x_j$.*

This follows from the previous result. First add vectors x_{k+1} to x_T to the initial k such that the T vectors span $\mathbb{R}^T$. Then apply the Gram-Schmidt procedure to this set.

Using the Gram-Schmidt procedure it is possible to give alternative formulas for the projections $\Pi_{\mathbf{X}}$ and $\Pi_{\mathbf{X}^\perp}$. Let Δ be a $K \times K$ matrix such that $U = X\Delta$ is orthonormal. Since the columns of U and X span the same space, we have

$$\Pi_{\mathbf{X}}(y) = X(X'X)^{-1}X'y = \Pi_{\mathbf{U}}(y) = U(U'U)^{-1}U'y = UU'y.$$

Thus by choosing a suitable basis for the space spanned by the columns of X, we can simplify the projection formula considerably. The formula for projection onto $\mathbf{X}^{\perp}$ can also be simplified. Choose $T - K$ vectors $g_1, g_2, \ldots, g_{T-K}$ which form an orthonormal set and span the space $\mathbf{X}^{\perp}$. Let G be the $T \times (T - K)$ matrix with i-th column g_i. Then it is clear that

$$\Pi_{\mathbf{X}^{\perp}}(y) = (\mathbf{I} - X(X'X)^{-1}X')y = G(G'G)^{-1}G'y = GG'y.$$

Since each vector in $\mathbf{U} = \mathbf{X}$ is orthogonal to each vector in $\mathbf{G}$, we must have $U'G = 0$ and also $G'U = 0$. The columns of U and G together form an orthonormal basis for the full space $\mathbb{R}^T$, and hence the matrix $W = (H|G)$ will be an orthogonal matrix. It follows that $\mathbf{I}_T = WW' = UU' + GG'$. This gives us an explicit form for the decomposition of y into its projections on $\mathbf{X}$ and $\mathbf{X}^{\perp}$:

$$y = \mathbf{I}y = UU'y + GG'y = \Pi_{\mathbf{X}}(y) + \Pi_{\mathbf{X}^{\perp}}(y).$$

These formulas clarify an important aspect of regression. We can regard the columns of U as an alternative set of regressors to the original regressors in X. As long as these span the same vector space, the projection $\hat{y}$ and the effectiveness of the regression will be exactly the same in the two cases. The value of the coefficients could vary considerably across such reformulations, however. For example in situations where the original set of regressors is multicollinear, one could easily construct an orthogonal basis for the vector space spanned by the regressors and get an equivalent regression without any multicollinearity. From the point of view of goodness of fit or improved R^2 (discussed in a Section 1.8.1) there is no advantage to removing multicollinearity. However, if the coefficients of the particular regressors are desired then multicollinearity does make a big difference.

1.6   GLS and Nonidentity Covariance

The method of least squares finds coefficients β which minimize the sum of squares $\sum_{t=1}^{T}(y_t - x_t'\beta)^2 = \|y - X\beta\|^2$. We have seen that this problem can be solved by projections. Least squares treats all coordinates of y equally. In cases where different coordinates are measured with different degrees of precision, this is not appropriate. The more precisely measured components should receive greater weight. We will show in Chapter 2 that it is optimal to minimize the weighted sum $\sum_{t=1}^{T} |y_t - x_t'\beta|^2/\sigma_t^2$. Similarly, if there are correlations between different coordinates, this must be taken into account by any procedure for estimating β. If two coordinates are highly correlated, least squares in effect is double counting what amounts to one observation.

The method of least squares works best when the covariance matrix of the observations is a scalar multiple of identity. If, on the other hand, $\mathbb{C}ov(y) = \Sigma$, some nonidentity matrix, other methods of estimation are appropriate. In general, the GLS estimator is defined to be an estimate $\tilde{\beta}$ which minimizes $(y - X\beta)'\Sigma^{-1}(y - X\beta)$. This corresponds to the maximum likelihood estimate under certain assumptions and has other optimality properties as well. There are two equivalent ways to solve this problem. The first is to transform the variables so as to make the covariance identity. After the transformation, the problem can be solved using techniques already discussed. The second method is to use a different geometry on the original space. The two are more or less equivalent, but both are convenient in different contexts. We discuss both methods below.

1.6.1   Transformation to Identity Covariance

It is easily checked that if $\mathbb{C}ov(y) = \Sigma$ then $\mathbb{C}ov(Qy) = Q\Sigma Q'$. Given any covariance matrix Σ, we can find a (nonunique) matrix Q such that $Q\Sigma Q' = \mathbf{I}$. Defining $v = Qy$, we see that the covariance matrix of v is $Q\Sigma Q' = \mathbf{I}$. Let $W = QX$. We can find the projection of v on the space $\mathbf{W}$ by the projection theorem. This must be $\hat{v} = W(W'W)^{-1}W'v = QX(X'Q'QX)^{-1}X'Q'v$. Now we can transform this projection back to the original space by applying the inverse transformation Q^{-1}. The desired vector $\tilde{y}$, which is the best explanation of y by X after adjusting for nonidentity covariance, is therefore

$$\tilde{y} = Q^{-1}\hat{v} = Q^{-1}QX\left\{X'Q'QX\right\}^{-1}X'Q'Qy = X(X'Q'QX)^{-1}X'Q'Qy.$$

Now note that $Q\Sigma Q' = \mathbf{I}$ implies $\Sigma = Q^{-1}(Q')^{-1}$ and hence $\Sigma^{-1} = Q'Q$. Substituting, we get the usual formula for the projection under nonidentity covariance: $\tilde{y} = X(X'\Sigma^{-1}X)^{-1}X'\Sigma^{-1}y$.

The coefficient of the projection is the value of $\tilde{\beta}$ such that $\tilde{y} = X\tilde{\beta}$. For the nonidentity covariance case, the coefficient is $\tilde{\beta} = (X'\Sigma^{-1}X)^{-1}X'\Sigma^{-1}y$.

1.6.2   Nonidentity Inner Product

Instead of transforming y and X, doing the projection in the transformed space, and then undoing the transformation, it is possible to carry out the projection in the original space. The device used to do this is to change the geometry of R^T to match the geometry after the transformation. Geometry is governed by distances and angles. Every positive definite $T \times T$ matrix S defines an *inner product* on R^T. Geometry on R^T relative to the inner product S is defined as follows. For any vector y, we define the (squared) length of the vector to be $\|y\|_S^2 = y'Sy$. Similarly, the angle θ between two vectors x and y relative to the inner product S is defined by $\cos_S(\theta) = (x'Sy)/\sqrt{(x'Sx)(y'Sy)}$. In particular the two vectors are orthogonal if and only if $x'Sy = 0$. This notion of distance and angle produces a Euclidean geometry on R^T which satisfies all the usual laws.

If we let $S = \Sigma^{-1}$ in the above definition, it is easily seen that $\|y\|_S^2 = \|Qy\|^2$. That is, the length of y relative to the inner product Σ^{-1} is the same as the length of the transformed vector Qy (where $Q\Sigma Q' = \mathbf{I}$ as before). Similarly, the angle between x and y relative to the inner product S is the same as the angle between Qx and Qy. Thus the geometry on R^T relative to the inner product Σ^{-1} is the same as the geometry produced by transforming y to Qy. By using this geometry, we avoid the need to transform and retransform, and can do the desired calculations directly in the original variables.

It is easily verified that the projection theorem remains valid with any inner product (see exercise 2). The projection of y on the space $\mathbf{X}$ spanned by the columns of X relative to the inner product S will be denoted by $\Pi_{\mathbf{X}}^S(y)$. According to the projection theorem, the difference between y and its projection is orthogonal to the space $\mathbf{X}$ *relative to the inner product S*. That is, the projection must satisfy the equation:

$$X'S(\Pi_{\mathbf{X}}^S(y) - y) = 0. \qquad (1.1)$$

Since the projection $\Pi_{\mathbf{X}}^S(y)$ belongs to the space $\mathbf{X}$, it must be a linear combination of the columns of X. Thus $\Pi_{\mathbf{X}}^S(y) = X\tilde{\beta}$ for some vector of coefficients $\tilde{\beta}$. Substituting in Eq. (1.1) and solving, we get $\tilde{\beta} = (X'SX)^{-1}X'Sy$. Thus the formula for projection relative to inner product S is

$$\Pi_{\mathbf{X}}^S(y) = X\tilde{\beta} = X(X'SX)^{-1}X'Sy.$$

If we take the inner product $S = \Sigma^{-1}$ we get the same formula as we do by using the transformation method done earlier.

For later reference, we note that the decomposition theorem is also valid in the geometry defined by any inner product S. The relevant formulas can be summarized as

$$\Pi_{\mathbf{X}}^S(y) = X(X'SX)^{-1}X'Sy, \tag{1.2}$$
$$\Pi_{\mathbf{X}^\perp}^S(y) = \left(\mathbf{I} - X(X'SX)^{-1}X'S\right)y. \tag{1.3}$$

The second equation follows from the decomposition of y into its projection on $\mathbf{X}$ and $\mathbf{X}^\perp$. It is important to note that $\mathbf{X}^\perp$ is *not* the set of vectors v satisfying $X'v = 0$ but rather the set w satisfying $X'Sw = 0$.

1.7 Kruskal's Theorem

In the previous section, we defined the projection of y on X relative to the inner product Σ^{-1}. In general this is not the same as the projection of y on X. However, the two coincide in some special cases. Kruskal's theorem gives necessary and sufficient conditions for the two to be the same. We develop this result in this section. See Section 6.1.3 of Amemiya (1985) for alternative formulations of this result and applications.

An important tool is the definition of *invariant subspaces*:

Definition 1.1 *A subspace V of $\mathbb{R}^n$ is said to be invariant under a transformation $T : \mathbb{R}^n \to \mathbf{R}^n$ if for every $v \in V$, $T(v) \in V$.*

An algebraic description of invariant subspaces is also useful. Let S be a $T \times T$ matrix. Let $\mathbf{X}$ be the K-dimensional vector space spanned by the columns of the $T \times K$ matrix X.

Lemma 1.2 *The vector space* **X** *is invariant under the transformation* $S : \mathbb{R}^T \to \mathbb{R}^T$ *if and only if there exists a* $K \times K$ *matrix* Δ *such that* $SX = X\Delta$. *Furthermore, if* S *is nonsingular and the columns of* X *are linearly independent,* Δ *must also be nonsingular.*

Proof: Let $x_1, \ldots, x_k$ be the columns of X. Suppose **X** is invariant under S. Then $Sx_j \in \mathbf{X}$, so that $Sx_j = X\delta_j$ for some $K \times K$ vector δ_j. Letting δ_j form the columns of the $K \times K$ matrix Δ, we get $SX = X\Delta$. Conversely, if $SX = X\Delta$, then $Sx_j = X\delta_j$, which means that the vector Sx_j can be expressed as a linear combination of columns of X. Thus $Sx_j \in \mathbf{X}$ for $j = 1, 2, \ldots, K$. But this implies that for all $x \in \mathbf{X}$, $Sx \in \mathbf{X}$.

To prove the last statement, assume that Δ is singular. Then there exists a nonzero $K \times 1$ vector u such that $\Delta u = 0$. It follows that $SXu = 0$. If the columns of X are linearly independent and $u \neq 0$ then $v = Xu \neq 0$. It follows that $Sv = 0$ so that S is also singular.

We can now prove that the GLS and OLS estimates coincide whenever the space **X** is invariant under Σ. As shown in exercise 5, this condition is equivalent to requiring **X** to be invariant under the transformation Σ^{-1}, which is the hypothesis of the following theorem. The converse also holds and is stated separately below. This result was presented in Kruskal (1968), though without a claim of originality.

Theorem 1.3 (Kruskal) *Let* **X** *be the vector subspace of* $\mathbf{R}^T$ *spanned by the columns of* X. *Suppose* **X** *is invariant under the transformation* $\Sigma^{-1} : \mathbb{R}^T \to \mathbb{R}^T$. *Then* $\hat{y} \equiv \Pi_{\mathbf{X}}(y) = \Pi_{\mathbf{X}}^{\Sigma^{-1}}(y) \equiv \tilde{y}$; *that is, the projection of* y *on* **X** *coincides with the same projection under the inner product* Σ^{-1}.

Proof: The projection $\hat{y}$ of y on X is characterized by the orthogonality condition (*) $(\hat{y} - y)'X = 0$. The projection $\tilde{y}$ of y on X relative to inner product Σ^{-1} is characterized by the orthogonality condition (**) $(\tilde{y}-y)'\Sigma^{-1}X = 0$. If X is invariant under Σ^{-1}, the space spanned by the columns of $\Sigma^{-1}X$ is the same as the space spanned by X and hence the two conditions are equivalent. Algebraically, if X has linearly independent columns, we can write $\Sigma^{-1}X = X\Delta$ where Δ is nonsingular. Then the second condition (**) is $(\tilde{y}-y)'\Sigma^{-1}X = (\tilde{y}-y)'X\Delta = 0$. If Δ is non-singular, postmultiplying both sides by Δ^{-1} shows that $(\tilde{y}-y)'X = 0$. Since projections are unique, we must have $\tilde{y} = \hat{y}$.

Theorem 1.4 (Kruskal:Converse) *If least squares and generalized least squares coincide (i.e. $\hat{y} = \tilde{y}$ for all $y \in \mathbb{R}^T$) then the space $\mathbf{X}$ spanned by columns of X is invariant under the transformation Σ : $\mathbb{R}^T \to \mathbb{R}^T$.*

Remark: $\hat{y} = \tilde{y}$ may hold for some values of y without the invariance conditions. The converse holds only if this equality is valid for all y.

Proof: Note that $\hat{y}$ is characterized by the conditions that $(y - \hat{y})$ is orthogonal $\mathbf{X}$. If $\hat{y} = \tilde{y}$ for all y then it must also satisfy the condition that $(\hat{y}-y)'\Sigma^{-1}X = 0$; that is, it is orthogonal to X relative to the inner product defined by Σ^{-1}. Letting $W = \Sigma^{-1}X$, the second condition is equivalent to $(\hat{y} - y)'W = 0$, so that $(\hat{y} - y)$ is orthogonal to the vector space $\mathbf{W}$ spanned by the columns of W. Consider the function $f(y) = y - \hat{y} = y - \tilde{y}$. Then we have $y = f(y) + \hat{y}$, where $\hat{y} \in \mathbf{X}$ and $f(y) \in \mathbf{X}^{\perp}$. It follows from the decomposition theorem that $f(y)$ is the projection of y into $\mathbf{X}^{\perp}$. We have also established that $f(y) \in \mathbf{W}^{\perp}$ for all y. This can only hold if $\mathbf{X}^{\perp} = \mathbf{W}^{\perp}$ which implies that $\mathbf{W} = \mathbf{X}$. Thus the space spanned by columns of $\Sigma^{-1}X$ is the same as the space spanned by the columns of X. This is the same as saying the $\mathbf{X}$ is invariant under Σ^{-1}.

The fact that the columns of X span a vector subspace that is invariant under the transformation Σ^{-1} can also be expressed in alternative forms. For example, this property holds if and only if each column of X is a linear combination of some K eigenvectors of Σ. See exercises 5 and 6.

1.8   Simplifying Models

Let W be a $T \times J$ matrix consisting of the first J columns of the $T \times K$ matrix X. Assuming $1 \leq J < K$, the vector space $\mathbf{W}$ is a proper subspace of $\mathbf{X}$. The J columns of W correspond to the first J

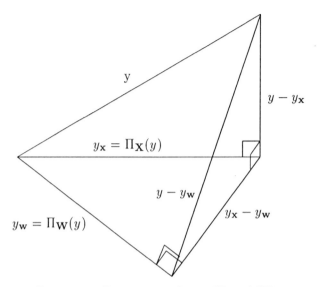

Figure 1.2: Projection of y on $\mathbf{X}$ and $\mathbf{W}$

regressors. The question of interest is whether the last $K - J$ regressors make a significant contribution to explaining y. If they do not, we can simplify our model for y by dropping the additional regressors. We must compare the projections $y_{\mathbf{X}} = \Pi_{\mathbf{X}}(y)$ and $y_{\mathbf{W}} = \Pi_{\mathbf{W}}(y)$. We can approximate y by either $\Pi_{\mathbf{X}}(y)$ or $\Pi_{\mathbf{W}}(y)$. Obviously $\Pi_{\mathbf{X}}(y)$ will always be a better approximation, since $\mathbf{X}$ is a larger space then $\mathbf{W}$. There are many situations where if $\Pi_{\mathbf{W}}(y)$ is only slightly worse than $\Pi_{\mathbf{X}}(y)$, we would prefer to use $\Pi_{\mathbf{W}}$ because of the fact that $\mathbf{W}$ is of lower dimension (corresponding to a simpler model with fewer regressors). This situation has a very neat geometry which leads to many matrix equalities that are hard to understand without reference to Figure 1.2 above.

It is important to note that $y_{\mathbf{W}} = \Pi_{\mathbf{W}}(y) = \Pi_{\mathbf{W}}(y_{\mathbf{X}}) = \Pi_{\mathbf{W}}(\Pi_{\mathbf{X}}(y))$. That is, since $\mathbf{W}$ is a subspace of $\mathbf{X}$, doing projections in stages, first on $\mathbf{X}$ and then on $\mathbf{W}$ leads to the same result as directly projecting y on $\mathbf{W}$; see exercise (8). Let $\mathbf{W}^{\perp\mathbf{X}}$ stand for the set of vectors in $\mathbf{X}$ which are orthogonal to all vectors in $\mathbf{W}$, so that $\mathbf{W}^{\perp\mathbf{X}} = \mathbf{W}^{\perp} \cap \mathbf{X}$. The vector space $\mathbf{W}^{\perp\mathbf{X}}$ is called the orthogonal complement of $\mathbf{W}$ in the space $\mathbf{X}$. By projecting y on $\mathbf{X}$, and $\mathbf{W}$ directly, and by projecting $\Pi_{\mathbf{X}}(y)$ on $\mathbf{W}$, we have three decompositions; note that in each case, the two vectors on the right hand side are orthogonal.

$$
\begin{aligned}
y &= \Pi_{\mathbf{X}}(y) + \Pi_{\mathbf{X}^{\perp}}(y) \equiv y_{\mathbf{x}} + (y - y_{\mathbf{x}}). \\
y &= \Pi_{\mathbf{W}}(y) + \Pi_{\mathbf{W}^{\perp}}(y) \equiv y_{\mathbf{w}} + (y - y_{\mathbf{w}}). \\
\Pi_{\mathbf{X}}(y) &= \Pi_{\mathbf{W}}(\Pi_{\mathbf{X}}(y)) + \Pi_{\mathbf{W}^{\perp}\mathbf{X}}(\Pi_{\mathbf{X}}(y)) \Leftrightarrow y_{\mathbf{x}} = y_{\mathbf{w}} + (y_{\mathbf{x}} - y_{\mathbf{w}}).
\end{aligned}
$$

Of crucial importance is the relation between the three errors:

$$
\|y - y_{\mathbf{w}}\|^2 = \|y - y_{\mathbf{x}}\|^2 + \|y_{\mathbf{x}} - y_{\mathbf{w}}\|^2. \tag{1.4}
$$

This is obvious from the geometry, and easily established algebraically using the orthogonality relations between the projections and the errors. Assessment of the validity of the linear restriction is based on comparisons of these three quantities.

The additional explanation provided by the last $K - J$ regressors can be measured by $\|y_{\mathbf{x}} - y_{\mathbf{w}}\|^2$. To eliminate dependence on units, it is useful to divide this by some measure of the scale. While other choices are possible, dividing by $\|y - y_{\mathbf{x}}\|^2$ leads to the usual statistic for testing the significance of the last $K - J$ regressors. Explicitly

$$
F = \frac{\|y_{\mathbf{x}} - y_{\mathbf{w}}\|^2 / (K - J)}{\|y - y_{\mathbf{x}}\|^2 / (T - K)} = \frac{T - K}{K - J} \frac{\|y_{\mathbf{x}} - y_{\mathbf{w}}\|^2}{\|y - y_{\mathbf{x}}\|^2}.
$$

The virtue of this choice is that it has a known distribution under some additional assumptions, as we will show in a later chapter. For now, we note that large values of F show that the additional regressors provide substantial additional explanation of y and should be retained in the model. Conversely, small values of F suggest the simpler model obtained by dropping the last $K - J$ regressors provides an adequate explanation of y.

The advantage of the F statistic is that it has a known distribution. However, an alternative measure is more intuitive and easily understood and interpreted. This is $R^2 = \|y_{\mathbf{x}} - y_{\mathbf{w}}\|^2 / \|y - y_{\mathbf{w}}\|^2$. The error made by using the simpler model W is $e_{\mathbf{w}} = y - y_{\mathbf{w}}$. This is what remains in need of explanation after using the simpler model. From the projection geometry, we get the relation that

$$
\|y - y_{\mathbf{w}}\|^2 = \|y - y_{\mathbf{x}}\|^2 + \|y_{\mathbf{x}} - y_{\mathbf{w}}\|^2. \tag{1.5}
$$

Thus the error $e_{\mathbf{w}}$ made by the simple model is partitioned into two parts: $\|e_{\mathbf{x}}\|^2 = \|y - y_{\mathbf{x}}\|^2$ is the portion left unexplained by the more complex model, while $\|y_{\mathbf{x}} - y_{\mathbf{w}}\|^2$ measures the additional explanation made possible by including the $K - J$ regressors. The ratio R^2 thus measures the proportion of the error made by the simpler model $e_{\mathbf{w}}$ which can be explained by the more complex model.

1.8.1 Coefficient of Determination: R^2

In order to assess the validity of the regression, we need to evaluate how good $\hat{y}$ is as an approximation of y. A natural measure of the accuracy of approximation is simply the ratio $\|\hat{y}\|^2/\|y\|^2$, which is also the squared cosine of the angle between the two vectors. This can yield misleading results, because in typical regressions one of the regressors is a constant. If the mean of y is large (relative to its variation), the ratio $\|\hat{y}\|^2/\|y\|^2$ can be large even though none of the regressors are related to y, simply because the constant term will do an adequate job of explaining y. So the question is to assess whether a large value of the ratio is essentially due to the constant, or whether the remaining regressors also provide additional explanatory power. The technique discussed above provides an answer to this question.

Let W be the first column of the X matrix, which we will assume is the column of ones. We want to assess whether the projection of y on $\mathbf{X}$ or $y_{\mathbf{x}}$ is a significant improvement over the projection of y on $\mathbf{W}$ or $y_{\mathbf{w}}$. According to Eq. (1.4), we have

$$\|y - y_{\mathbf{w}}\|^2 = \|y - y_{\mathbf{x}}\|^2 + \|y_{\mathbf{x}} - y_{\mathbf{w}}\|^2. \tag{1.6}$$

This equality is given the following interpretation in the present context. Since W is just a column of ones, the projection $y_{\mathbf{w}}$ is easily calculated: $y_{\mathbf{w}} = \overline{y}W$, where $\overline{y} = (1/T)\sum_{t=1}^{T} y_t$. It follows that $\|y - y_{\mathbf{w}}\|^2 = \sum_{t=1}^{T}(y_t - \overline{y})^2$. This quantity is called the *Total Sum of Squares* or SST. We are trying to explain the variations in y by means of a regression model. The SST is a measure of the variations in y. If this is small, y is nearly constant and there is nothing to explain. If SST is large, y fluctuates a lot, and we have to assess how well these fluctuations are explained by the regressors $x_2, \ldots, x_k$. On the right hand side of Eq. (1.6), the first term is $\|y - y_\mathbf{x}\|^2 = \sum_{t=1}^{t}(y_t - \hat{y}_t)^2$. This is called the *Residual Sum of Squares* or SSR. This is the distance between the closest vector in $\mathbf{X}$ and y and measures the variation in y that is *not* explained by our regressors. The second term is $\|y_\mathbf{x} - y_\mathbf{w}\|^2 = \sum_{t=1}^{T}(\hat{y}_t - \bar{y})^2$. This is called the *Explained Sum of Squares* or SSE. This is the portion of the variation of y explained by our model. Note that $\bar{y} = (1/T)\sum_{t=1}^{T} y_t = (1/T)\sum_{t=1}^{T} \hat{y}_t$. The *coefficient of determination* or R^2 is the proportion of the Explained Sum of Squares to Total Sum of Squares or SSE/SST. It is clear that this must vary between 0 and 1, and measures the percentage of the fluctuations in y explained by the nonconstant regressors.

The distribution of the R^2 is hard to calculate. For this reason, an alternative measure of the goodness of the model is also used. This is

$$F = \frac{SSE/(K-1)}{SSR/(T-K)} = \frac{T-K}{K-1}\frac{SSE}{SSR}.$$

If there is in fact no relationship between the (nonconstant) regressors and y, this statistic has an F distribution with $K-1$ and $T-K$ degrees of freedom. Large values of F show that the SSE is large relative to SSR and therefore the regressors are significant in explaining the variations in y.

1.9 Exercises

1. Show that regression residuals sum to zero if a constant term is included among the regressors. Also show that the average value of y is the same as the average value of the projection $\hat{y}$; equivalently, $\sum_{t=1}^{t} y_t = \sum_{t=1}^{T} \hat{y}_t$ Hint: use orthogonality of the residuals to the space spanned by columns of X.

2. Define $\tilde{y} = X\beta$ to be the vector in $\mathbf{X}$ which is closest to y relative to the distance defined by the inner product S. Show that $y - \tilde{y}$ must be orthogonal to X relative to the inner product S: that is, $X'S(\tilde{y} - y) = 0$.

3. Suppose that X and Z are $T \times K$ matrices of rank K and that the columns of X and Z span the same K-dimensional vector

subspace of $\mathbb{R}^T$. Show that $X(X'X)^{-1}X' = Z(Z'Z)^{-1}Z'$ using two different methods: (I) by the projection theorem, and (II) by matrix algebra. Hint: I is easy. For II show that if every column of Z is a linear combination of the columns of X then $Z = XA$ for some $K \times K$ matrix. Furthermore, A must be nonsingular if Z has rank K.

4. Note that a projection map must satisfy $P^2 = P$. Py is the closest vector in $\mathbf{X}$ to y, and P^2y is the closest to Py. But since $Py \in \mathbf{X}$, the closest vector to Py is Py itself so that $PPy = Py$. A symmetric matrix $P : \mathbb{R}^T \to \mathbb{R}^T$ is called idempotent if it satisfies $P^2 = P$. Let $\mathbf{X} = \{x \in \mathbb{R}^T : Px = x\}$. Show that if $P \neq 0$ then $\mathbf{X}$ is a vector subspace of $\mathbb{R}^T$ of dimension bigger than 0. Show that P projects all vectors to $\mathbf{X}$ by proving that $Py \in \mathbf{X}$ and $(y - Py)'v = 0$ for all $v \in \mathbf{X}$.

5. Suppose that $\mathbf{X}$ is an invariant subspace under the transformation Σ. Show that $\mathbf{X}^{\perp}$ is also an invariant subspace. Show that both are also invariant subspaces of Σ^{-1}.

6. Let $v_1, \ldots, v_T$ be the eigenvectors of symmetric positive definite matrix Σ. Show that a $T \times K$ matrix X generates an invariant subspace $\mathbf{X}$ under the transformation Σ if and only if its columns are linear combinations of some fixed subset of K eigenvectors of Σ.

7. Define $\tilde{\beta} = (X'\Sigma^{-1}X)^{-1}X'\Sigma^{-1}y$. Prove Kruskal's theorem without using projection theory by showing that if $\Sigma^{-1}X = X\Delta$ for some nonsingular matrix Δ then $\tilde{\beta} = (X'X)^{-1}X'y$.

8. Let V be a vector subspace of W and W a vector subspace of $\mathbb{R}^T$. For any $y \in \mathbb{R}^T$ show that
$$\Pi_V(y) = \Pi_V(\Pi_W(y)).$$

9. Show that the overall F statistic for a regression is a monotonic function of the coefficient of determination R^2. Hint: first show that
$$F = \frac{T-K}{K-1}\frac{R^2}{1-R^2}.$$

While walking back home with a case of eggs on his head, Ibne Zubeyde started speculating about the likelihoods. "These eggs will hatch into chicks, which will grow to chickens, which will lay more eggs. Soon I will have a huge poultry farm. I will sell some chickens and buy some cows which will give milk and produce more cows. In no time, I will be the wealthiest man in the country, and the King will marry his daughter to me. When the princess brings me my breakfast, I will note that the toast is slightly burnt, and say 'Humph!'... At this point, Ibne shook his head to match deed to word, and the case of eggs fell down and all the eggs were broken.

Chapter 2

Maximum Likelihood

2.1 Introduction

This chapter obtains the maximum likelihood and some related estimates for linear regression models. The theory is somewhat different for three types of regressors considered below. The simplest case is that of nonstochastic regressors. Here both the derivation and the distribution theory for the ML estimate is quite easy. The second case is that where the regressors are stochastic but ancillary. In this case one can condition on the observed values of the regressors. Except for minor differences, one can then proceed as in the case of nonstochastic regressors. Finally, the case where the regressors are not ancillary is discussed. This breaks down into two main cases. The first is that of dynamic equations, where lagged values of the dependent variable enter the right hand side. It turns out that this case is similar to the first two in many ways. The second case is that of simultaneous equations. The theory of ML estimation is quite complex in this case, and some simpler approximations are described. The distribution theory is also quite complex. We merely provide an overview and references, and do not go into details of this case.

Several facts about the multivariate normal and related distributions are assumed known. These are reviewed in Appendix A on the multivariate normal distribution. Although there are links to the previous chapter on projection theory, the results have been derived differently, so the chapter can be read independently.

In the standard linear regression model $y = X\beta + \epsilon$, with $\epsilon \sim N(0, \sigma^2 \mathbf{I}_T)$, we derive the ML estimators under the assumption that the regressors are nonstochastic. The estimator $\hat{\beta}_{ML}$ is MVU (minimum variance unbiased), while the variance estimator $\hat{\sigma}^2_{ML}$ needs a slight adjustment to become MVU. More typically, regressors are stochastic. This breaks down into three cases. The simplest case is where the regressors are ancillary — in this case, we can just condition on the regressors and pretend they are constant. A few differences between ancillary regressors and nonstochastic ones are also discussed briefly. The second case is where some of the regressors are lagged values of endogenous variables, also known as *predetermined* variables. This is the case of dynamic model, and simple conditional arguments are no longer possible. The third and most complex case is that of simultaneous equations, where the values of the stochastic regressors effect the errors of the relationship we wish to estimate. This simultaneous equations case is also discussed briefly.

2.2   Nonstochastic Regressors

Suppose that a variable y is largely determined by variables $\mathbf{x} = (x_1, x_2, \ldots, x_n)$. We posit a relationship of the form $y = f(x_1, \ldots, x_n) = f(\mathbf{x})$. A major portion of econometric theory is concerned with determining suitable approximations to the function f. Assuming the function is smooth, it can be expanded in a Taylor series around some suitable value as follows:

$$y = f(\mathbf{x}^0) + \sum_{i=1}^{k}(\mathbf{x} - \mathbf{x}^0)\frac{\partial}{\partial x_i}f(\mathbf{x}^0) + \sum_{i=k+1}^{n}(\mathbf{x} - \mathbf{x}^0)\frac{\partial}{\partial x_i}f(\mathbf{x}^0).$$

The variables x_i are called explanatory variables, or regressors, while y is called the dependent variable. In the above equation we have separated the regressors from $1, \ldots, k$ and $k+1, \ldots, n$ to account for the possibility that x_j may not be observed for $j > k$. Suppose that we have data $y_1, y_2, \ldots, y_T$ on y and corresponding observations $(\mathbf{x}_1, \ldots, \mathbf{x}_T)$ on the *observed* regressors. Form the $T \times 1$ matrix y by stacking the y observations and the $T \times k$ matrix with j-th row $\mathbf{x}_j$. Then we can write $y = X\beta + \epsilon$, where the parameters β are the unknown partial derivatives of f, and ϵ represents the contribution of the unknown regressor. If the error is caused by a lot of small effects added together, the central limit theorem allows us to assume that it is approximately normal. It is worth noting that approximate linearity of f is not required for the theory which follows. We can easily take second or higher order Taylor expansions if desired. We can also take nonlinear functions of x_j as regressors. As long as the parameters β enter linearly into the equation, the theory remains essentially the same.

The standard linear regression model, briefly motivated above, can be described as follows. We observe a $T \times 1$ vector y satisfying $y = X\beta + \epsilon$, where X is the $T \times k$ matrix of explanatory variables, and $\epsilon \sim N(0, \sigma^2\mathbf{I}_T)$. The error term ϵ is not observed. The parameters β and σ^2 are unknown and must be estimated. The matrix X consists of nonstochastic variables. The assumption that all diagonal elements of the covariance matrix of the errors are σ^2 is referred to as *homoskedasticity*. The case of *heteroskedasticity*, where errors may have different variances, will be studied later. The assumption that off-diagonal elements of the covariance matrix are zero means that the errors must be independent. All of these conditions together are sometimes called the full ideal conditions. We derive below the maximum likelihood estimates of the parameters β and σ^2 in a regression model under these conditions. We will make frequent use of the fact that if $z \sim N(\mu, \Sigma)$ and $y = Az + b$, then $y \sim N(A\mu + b, A\Sigma A')$. That is, linear transforms of multivariate normals are multivariate normal (see

Appendix A for details).

It is convenient to take the log of the likelihood function of y before maximizing over β and σ^2. Because y is a linear transform of $\epsilon \sim N(0, \sigma^2 I_T)$, we know that it must have density $y \sim N(X\beta, \sigma^2 I_T)$. Thus,

$$\log l(y) = -\frac{T}{2}\log(2\pi) - \frac{T}{2}\log(\sigma^2) - \frac{1}{2\sigma^2}(y - X\beta)'(y - X\beta).$$

To take the derivative with respect to the vector parameter β, it is convenient to introduce the some notation. Define ∇_θ to be the $k \times 1$ gradient vector:

$$\nabla_\theta f = \begin{pmatrix} \frac{\partial}{\partial\theta_1}f \\ \frac{\partial}{\partial\theta_2}f \\ \vdots \\ \frac{\partial}{\partial\theta_k}f \end{pmatrix}.$$

We will also use the notation $\nabla_\theta \nabla'_\theta f$ for the $k \times k$ matrix of second partial derivatives of f, with (i, j) entry $\partial^2 f/\partial\theta_i\partial\theta_j$. The first order conditions for a maximum require that $\nabla_\beta \log l(y, \beta, \sigma^2) = 0$ and also $(\partial/\partial\sigma^2)\log l(y, \beta, \sigma^2) = 0$. Also note that $(y - X\beta)'(y - X\beta) = y'y - (X\beta)'y - y'(X\beta) + (X\beta)'(X\beta) = y'y - 2\beta'Xy + \beta'(X'X)\beta$. We can now write the first order condition for a maximum with respect to the parameter β as follows (see exercise 7 for some rules of differentiation with respect to a vector):

$$\begin{aligned} \nabla_\beta \log l(y, \beta, \sigma^2) &= -\frac{1}{2\sigma^2}\nabla_\beta[y'y - 2y'X\beta + \beta'(X'X)\beta] \\ &= -2X'y + 2X'X\beta = 0. \end{aligned}$$

Since the matrix of second partial derivatives $X'X = \nabla_\beta \nabla'_\beta \log l(\cdot)$ is always positive semidefinite, any solution of the above equations must be a maximum (see exercise 8). Assuming, as we will throughout this discussion, that $X'X$ is of full rank (and hence positive definite), the unique solution for β is $\hat{\beta} = (X'X)^{-1}X'y$.

Next, we differentiate the log likelihood with respect to σ^2:

$$\frac{\partial}{\partial(\sigma^2)}\log l(y, \beta, \sigma^2) = -\frac{T}{2\sigma^2} + \frac{1}{2\sigma^4}\|y - X\hat{\beta}\|^2 = 0.$$

Solving this for the $\hat{\sigma}^2$, we get $\hat{\sigma}^2_{ML} = \|y - X\hat{\beta}\|^2/T = SSR/T$.

That the ML estimate for β is unbiased follows from $\mathbb{E}\hat{\beta} = \mathbb{E}(X'X)^{-1}X'y = (X'X)^{-1}X'\mathbb{E}y$. Now, since $y = X\beta + \epsilon$, it follows that $\mathbb{E}y = \mathbb{E}(X\beta + \epsilon) = X\beta + \mathbb{E}\epsilon = X\beta$. Substituting into the previous formula gives the result that $\mathbb{E}\hat{\beta} = \beta$ for all values of β, proving the desired result.

As we soon show, the ML estimate $\hat{\sigma}^2_{ML}$ is biased. By modifying the T to $T - K$, one gets the commonly used MVU estimate which is $\hat{\sigma}^2_{MVU} = \|y - X\hat{\beta}\|^2/(T - K) = SSR/(T - K)$.

2.3 Distribution of the MLE

In this section we derive the joint density of the maximum likelihood estimators $\hat{\beta}, \hat{\sigma}^2$. This is given by the theorem below. We will see later that this same distribution must be approximately valid in large samples under substantially weaker assumptions; see Theorem 1.5. In situations where we expect the error distributions to be skewed or otherwise nonnormal, improved approximations to the distribution of OLS can be obtained using Edgeworth expansions or bootstrapping, as discussed in Chapter 14.

Theorem 2.1 *The maximum likelihood estimators $\hat{\beta}$ and $\hat{\sigma}^2$ are independent with densities $\hat{\beta} \sim N(\beta, \sigma^2(X'X)^{-1})$ and $T\hat{\sigma}^2/\sigma^2 \sim \chi^2_{T-k}$.*

Proof: We know that $y \sim N(X\beta, \sigma^2 I_T)$ and hence we can calculate the density of $\hat{\beta} = (X'X)^{-1}X'y$ using the theorem on linear transforms of multinormals given in Appendix A. This says that $\hat{\beta}$ is multinormal with mean vector $(X'X)^{-1}X'(X\beta) = \beta$ and covariance matrix

$$(X'X)^{-1}X'(\sigma^2 I_T)[(X'X)^{-1}X']' = \sigma^2(X'X)^{-1}X'X(X'X)^{-1} = \sigma^2(X'X)^{-1}.$$

This proves that the density of $\hat{\beta}$ is as stated.

We will show that $T\hat{\sigma}^2/\sigma^2 = Z'MZ$, where $Z \sim N(0, \mathbf{I})$ and M is idempotent. It then follows from standard results (see Appendix A) that the density of $T\hat{\sigma}^2/\sigma^2$ is χ^2_f where f is the trace (i.e. sum of diagonal entries) of M. Define $M = I_T - X(X'X)^{-1}X'$ and note that

$$T\hat{\sigma}^2 = \|y - X\hat{\beta}\|^2 = \|y - X(X'X)^{-1}X'y\|^2 = \|My\|^2.$$

Now note two important properties of M: (i) $M^2 = M$, and (ii) $MX = 0$. This is easily proven using matrix algebra; see exercise 1 for the projection geometry explanation of these formulae. Note that $y = X\beta + \epsilon$, so that $T\hat{\sigma}^2 = \|M(X\beta + \epsilon)\|^2 = \|M\epsilon\|^2$, using property (ii). Set $Z = \epsilon/\sigma$ and note that Z has the standard normal density. Then we have

$$T\hat{\sigma}^2/\sigma^2 = \|M\epsilon/\sigma\|^2 = \|MZ\|^2 = Z'M'MZ = Z'MZ.$$

To complete the proof we must show that the trace of M is $T - K$. Recall two simple but important properties of the trace: $\text{tr}(A + B) = \text{tr}(A) + \text{tr}(B)$, and $\text{tr}(AB) = \text{tr}(BA)$, for any conformable matrices A and B. Thus we have:

$$\begin{aligned} f &= \mathbf{tr}\{I_T - X(X'X)^{-1}X'\} &= \mathbf{tr}\{I_T\} - \mathbf{tr}\{X(X'X)^{-1}X'\} \\ &= T - \mathbf{tr}\{X'X(X'X)^{-1}\} &= T - \mathbf{tr}\,I_k &= T - k. \end{aligned}$$

This calculation (in which one must pay careful attention to the order of the various matrices in question) establishes that the marginal densities of $\hat{\beta}$ and $\hat{\sigma}^2$ are as stated in the theorem. It only remains to be proven that the two are independent. This follows from Lemma 2.1 below.

Lemma 2.1 *Let $Z_1 = M\epsilon$ and $Z_2 = X'\epsilon$. Then Z_1 and Z_2 are independent.*

To prove the lemma define $Z = \begin{pmatrix} Z_1 \\ Z_2' \end{pmatrix} = \begin{pmatrix} M\epsilon \\ X'\epsilon \end{pmatrix}$. Since Z is a linear transform of ϵ it must also be normal. Now $\mathbb{C}ov(Z_1, Z_2) = \mathbb{E}Z_1 Z_2' = \mathbb{E}M\epsilon\epsilon'X = M\,\mathbb{C}ov(\epsilon)X = \sigma^2 MX = 0$. It follows that Z_1 and Z_2 are independent, since zero covariance implies independence when the joint distribution of Z_1, Z_2 is multivariate normal (see Appendix A).

Recall from probability theory that if Z_1 and Z_2 are independent, then any function $h(Z_1)$ is independent of any function $g(Z_2)$. Thus if we could write $\hat{\sigma}^2$ as a function of $Z_1 = M\epsilon$ and $\hat{\beta}$ as a function of $Z_2 = X'\epsilon$, we could establish their independence. But we have $\hat{\sigma}^2 = \|Z_1\|^2$ and $\hat{\beta} = \beta + (X'X)^{-1}Z_2$. This completes the proof of the theorem.

If $Z \sim \chi_m^2$ then $\mathbb{E}Z = m$. Since $T\hat{\sigma}_{ML}^2/\sigma^2 \sim \chi_{T-K}^2$, it follows that $\hat{\sigma}_{MVU}^2 = \|y - X\hat{\beta}\|^2/(T - K)$ is unbiased. That it is also MVU is discussed in the following section.

2.4 MVU Estimates for Regression

A vector of estimates $\hat{\theta}$ of θ is *minimum variance unbiased* (MVU) if $\mathbb{E}\hat{\theta} = \theta$ so that $\hat{\theta}$ is unbiased and if for any other unbiased estimator $\tilde{\theta}$, $\mathbb{C}ov(\tilde{\theta}) \geq \mathbb{C}ov(\hat{\theta})$. Here the inequality between the covariance matrices is understood in the sense that $A \geq B$ if and only if $A - B$ is positive semidefinite. For a one-dimensional parameter, this just means that $\hat{\theta}$ has less variance than $\tilde{\theta}$. In the general k-dimensional case, let $\lambda \in \mathbb{R}^k$ be any vector and consider the estimators $\lambda'\hat{\theta}$ and $\lambda'\tilde{\theta}$ of $\lambda'\theta$. Both are unbiased and their variances satisfy

$$\mathbb{V}ar(\lambda'\hat{\theta}) = \lambda' \, \mathbb{C}ov(\hat{\theta})\lambda \leq \lambda' \, \mathbb{C}ov(\tilde{\theta})\lambda = \mathbb{V}ar(\lambda'\tilde{\theta}).$$

Thus every linear functional of θ can be estimated unbiased with less variance by the corresponding linear functional of $\hat{\theta}$. We now show that the usual regression estimates have this optimality property.

Theorem 2.2 *The estimates* $\hat{\beta} = (X'X)^{-1}X'y$ *and* $\hat{\sigma}^2 = \|y - X\hat{\beta}\|^2/(T - K)$ *are MVU.*

Partial Proof: There are two routes to proving this optimality property. One way is to use the Cramer-Rao Lower Bound (CRLB, see Theorem 15.1). This works for $\hat{\beta}$, but the bound for $\hat{\sigma}^2$ is slightly smaller than the variance of $\hat{\sigma}^2$. Thus the CRLB does not provide a proof of the fact that $\hat{\sigma}^2$ is MVU. Instead, we have to use the Lehmann-Schéffe Theorem, for which the reader is referred to any standard statistics text.

According to the CRLB, if we observe y with density $f(y, \theta)$ and $\hat{\theta}$ is an unbiased estimator of θ then $\mathbb{C}ov(\hat{\theta}) \geq \mathcal{I}^{-1}(\theta)$ where the information matrix $\mathcal{I}(\theta)$ is the covariance of the score function $\mathcal{I} = \mathbb{C}ov \, S(y, \theta)$ and the score function is defined by $S(y, \theta) = \nabla_\theta \log f(y, \theta)$. According to Lemma 1.1, an equivalent expression for the information matrix is $\mathcal{I} = \mathbb{E} - \nabla_\theta \nabla'_\theta \log f(y, \theta) = \nabla_\theta S'(y, \theta)$ (with S' being the transpose of the vector S). This second form is easier to calculate.

Noting that $y \sim N(X\beta, \sigma^2\mathbf{I}_T)$, and letting $\theta = (\beta, \sigma^2)$, we have

$$\log f(y, \theta) = -\frac{T}{2}\log 2\pi - \frac{T}{2}\log(\sigma^2) - \frac{\|y - X\beta\|^2}{2\sigma^2}.$$

The score function $S(y, \theta)$ is the $(K + 1) \times 1$ vector with the first K coordinates $\nabla_\beta \log f = (y'X - \beta'X'X)/\sigma^2$ and the last coordinate $\nabla_{\sigma^2} f = -T/(2\sigma^2) + \|y - X\beta\|^2/(2\sigma^4)$. It follows that the information matrix is

$$
\begin{aligned}
\mathcal{I} &= \mathbb{E} \begin{pmatrix} \nabla_\beta(\nabla_\beta \log f)' & \nabla_\beta(\nabla_{\sigma^2} \log f) \\ \nabla_{\sigma^2}(\nabla_\beta \log f)' & \nabla_{\sigma^2}(\nabla_{\sigma^2} \log f) \end{pmatrix} \\
&= \begin{pmatrix} (X'X)/\sigma^2 & \mathbb{E} - (y'X - \beta'X'X)'/\sigma^4 \\ \mathbb{E} - (y'X - \beta'X'X)/\sigma^4 & +T/(2\sigma^4) - \mathbb{E}\|y - X\beta\|^2/\sigma^6 \end{pmatrix}.
\end{aligned}
$$

The top $K \times K$ block of the information matrix is $X'X/\sigma^2$. Since the off-diagonals are zero because $\mathbb{E}(y - X\beta) = \mathbb{E}\epsilon = 0$, the top $K \times K$ block of the inverse of the information matrix will be $\sigma^2(X'X)^{-1}$, which is precisely the covariance of the ML estimator. It follows that $\hat\beta$ is MVU. As for the bottom $(K + 1), (K + 1)$ entry, since $\mathbb{E}\|y - X\beta\|^2 = \mathbb{E}\|\epsilon\|^2 = T\sigma^2$, it follows that $\mathcal{I}_{K+1,K+1} = -T/(2\sigma^4) + T/\sigma^4 = T/(2\sigma^4)$. Since the information matrix is block diagonal, the CRLB for variance of unbiased estimators of σ^2 is $2\sigma^4/T$. Now $SSR = \|y - X\hat\beta\|^2$ is distributed as $\sigma^2\chi^2_{T-K}$ so that $\mathbb{V}ar(SSR) = 2\sigma^4(T - K)$. It follows that

$$
\mathbb{V}ar(\hat\sigma^2_{MVU}) = \mathbb{V}ar\left(\frac{\sigma^2\chi^2_{T-K}}{T - K}\right) = \frac{2\sigma^4(T - K)}{(T - K)^2} = \frac{2\sigma^4}{T - K}.
$$

This is strictly larger than the CRLB $2\sigma^4/T$, although for large T the two will be very close. Thus it is not clear from this line of argument that the estimate $\hat\sigma^2_{MVU}$ is really minimum variance unbiased.

As discussed in detail in Chapter 4, it is possible to dominate $\hat\beta_{ML}$ in terms of mean square error by using a biased estimator. Similarly, the biased estimator $\tilde\sigma^2_* = \|y - X\hat\beta\|^2/(T - K + 2)$ performs better than the MVUE in terms of mean squared error. This is easily established as follows. Consider the class of estimators $\tilde\sigma^2_c = c(SSR)$ where $SSR \sim \sigma^2\chi^2_{T-K}$. The mean squared error of these estimators is

$$
\begin{aligned}
\mathbb{E}(cSSR - \sigma^2)^2 &= c^2\mathbb{E}(SSR)^2 - 2c\sigma^2\mathbb{E}SSR + \sigma^4 \\
&= c^2\left(\mathbb{V}ar(SSR) + (\mathbb{E}SSR)^2\right) - 2c(T - K)\sigma^4 + \sigma^4 \\
&= c^2\sigma^4(2(T - K) + (T - K)^2) - 2(T - K)\sigma^4 c + \sigma^4.
\end{aligned}
$$

Minimizing this quadratic in c, we find that the estimator with smallest mean square error is $\tilde{\sigma}_c^2 = SSR/(T - K + 2)$. In particular, this estimator dominates the usual MVU estimator $\hat{\sigma}_{MVU}^2$ in terms of mean squared error. Which estimator should we use? Neither the mean squared error criterion nor the unbiasedness criterion is compelling, making it hard to choose. Furthermore, the difference between the two estimators is quite small in practice, so the question is not of great practical significance. However, biased versions of the ML $\hat{\beta}_{ML}$ can provide significant gains over the MVU in suitable circumstances. The background theory for these biased estimators is developed in the next two chapters, and several applications are discussed in part IV of the text.

2.5 Stochastic Regressors: Ancillarity and Exogeneity

In the previous sections we have assumed that the regressors X are nonstochastic. This is possible in an experimental setting where values of various factors are determined by the experimenter. In economics, most regressors are stochastic. Sometimes it is possible to condition on the observed values of the regressors. In this case, one can effectively treat the observed values as constants and the analysis of the previous sections is applicable. This case is discussed in the present section. Cases where this cannot be done are dealt with in later sections.

An essential element in being able to reduce the case of stochastic regressors to constants is the concept of ancillarity. We will first illustrate this concept by a simple example. Suppose a student walks into a lab, picks up a voltmeter from the shelf and measures the voltage of a battery to be, say $v = 12.3$. Suppose that if the true voltage is μ, the measured voltage satisfies $v \sim N(\mu, \sigma^2)$. Half the voltmeters in the lab are Hi-Grade and have $\sigma^2 = 0.0001$. The other half are Lo-Grade and have $\sigma^2 = 0.01$. Knowing this, the student can claim that the variance of his reading $v = 12.3$ is $0.00505 = (1/2)0.01 + (1/2)0.0001$. However, this would be silly if the voltmeters are labeled Lo-Grade or Hi-Grade. Even if he randomly picked a meter off the shelf and so had equal probabilities of getting either kind, in reporting his results he should condition on the type that he actually had. The fact that he could have picked another type should not enter his reported variance.

If we let X be the type of the voltmeter (X equals Lo or Hi with equal probability), then X is called an *ancillary* random variable. Given observations $X_1, \ldots, X_n$ with density $f^X(x_1, \ldots, x_n, \theta)$, a statistic $S(X_1, \ldots, X_n)$ is said to be ancillary (for the parameter θ) if its distribution does not depend on θ. An ancillary statistic is, in a sense, the opposite of a sufficient statistic – the distribution of a sufficient statistic contains all available information in the sample about the parameter, while the distribution of an ancillary statistic contains no information whatsoever about the parameter. A generally accepted principle of inference is that one should condition on the observed value of the ancillary statistic. This is because the distribution of ancillary statistic, being independent of the parameter of interest, only adds noise to the experiment. One eliminates the noise by conditioning on the observed value of the ancillary statistic. In the example above, the unconditional variance measure of 0.00505 is not very useful — it is too high for the case of the Hi-Grade meter and too low for the Lo-Grade. Conditioning on the grade of the meter gives more accurate measures.

A similar situation frequently obtains in regressions with stochastic regressors. If the marginal distribution of the regressors does not depend on the parameters of interest, we can condition on the regressors, and thereby effectively treat them as constants. Conditionally on the observed values, all of the theory developed in the previous sections is valid. This is an important extension since nonstochastic regressors are rare in econometrics. However, cases where we can condition on the observed values of regressors and thereby treat them as constants are quite common.

We can give a somewhat more formal definition of ancillarity, and also relate this to exogeneity. How do we decide whether or not a given set of stochastic regressors is ancillary? Suppose that the joint distribution of y_t, x_t depends on the parameters ψ_1, ψ_2. The joint distribution can always be factorized as a product of the conditional distribution of y given x and the marginal distribution of x. Suppose the factorization takes the following form:

$$f(y_t, x_t, \psi_1, \psi_2) = f(y_t | x_t, \psi_1) \times f(x_t, \psi_2).$$

In this case, if the parameters of interest are ψ_1, and there is no relationship between ψ_2 and ψ_1, we can say that x_t is ancillary (for the parameters of interest ψ_2). This definition is given in Cox and Hinkley (1974), and was labeled *weak exogeneity* in an important paper by Engle *et al.* (1983) which generalized this idea to dynamic models. An essential aspect of ancillarity is that it depends entirely on the parameters of interest. If one is interested in predictions of y conditional on observed values of x, then x is usually ancillary. Exceptions occur where the marginal density of x provides additional information on the parameters of interest.

Even though the case of ancillary regressors is close to the case of constant regressors, there are some important differences between the two models which we now discuss. The case where regressors are not ancillary for the parameters of interest is discussed in a later section.

2.6   Misspecification Analysis

Doctors were sent out to fight plague in a remote African area. After a while the natives noticed that death casualties from plague were highest in the villages where the doctors were. They remedied the problem by killing all the doctors.

The object of this section is to study the consequences of estimating a 'wrong' regression model. In order to do this properly, it is essential to distinguish between models of correlation and models of causation — a task made all the more difficult by popular misconceptions which confound causation with constant conjunction. The language of econometrics is that of causation, where we speak of dependent variables being (caused or) controlled by independent variables. Our tools are, however, those of statistics and correlation, and there is a tension between the two. Our topic is closely connected to methodology, but we will not be able to develop the links in any detail. See Hendry (1992) for a detailed and insightful book-length treatment of the subject, expositing one unified approach to methodology. See also Darnell and Evans (1990) for a comparative discussion of several approaches to methodology.

Consider the simple Keynesian consumption function $C_t = \alpha + \beta Y_t + \epsilon_t$. If we interpret this as a 'correlation' model, the model states that the conditional distribution of C given Y is $N(\alpha + \beta Y, \sigma^2)$. The correlation model is symmetric — one could work equally well with the conditional distribution of Y given C, and write $Y = a + bC + \nu$. When we write down a consumption function, we have in mind more than simply specifying a conditional distribution. We think that there is a mechanism whereby consumers make their consumption decision on the basis of their income. In terms of the causal interpretation, the model $Y = a + bC + \nu$ has an entirely different meaning (which may separately have, or fail to have, validity). To discuss specification analysis, it is necessary to be clear on the fact that a regression equation by itself does not constitute a model. We must add to the equation one of the following clarifications:

1. (C, Y) are jointly random. This is a correlation model.

2. Y is one of the causes of C.

3. Y causes C and no other variable causes C.

4. Y causes C and C does not cause Y.

Depending on the circumstances, any of the four statements above could be considered as the proper interpretation of the Keynesian consumption function (indeed even some combinations of the three causal statements are possible). However, all four have substantially different implications statistically. The last case is one of simultaneity, and is discussed in the following sections. We discuss some of the differences in the first three cases below.

In the correlation model, the problem of omitted variables has quite a different meaning from the same problem in the case of causal variables. Let W stand for wealth, and suppose that C is actually determined by Y and W. If the triple (C, Y, W) has a stable joint distribution — which is a premise of correlation models — then the marginal distribution of C given Y will also be stable. Running the regression of C on Y will produce estimates of the conditional expectation of C given Y, say $\hat{\alpha} + \hat{\beta} Y$. Running the regression of C on Y and W will produce the conditional expectation of C given Y and W, say $\hat{a} + \hat{b}Y + \hat{c}W$. Now $\hat{\beta}$ is the coefficient of the Y in the conditional distribution of C given Y while $\hat{b}$ is the corresponding coefficient in the conditional distribution of C given Y and W. Obviously the two parameters have different meanings as well as different values. OLS estimates are consistent in *both* models. That is, regressing C on Y will produce a good estimate of the conditional expectation of C given Y and regressing C on Y and W will produce a good estimate of the conditional expectation of C given Y and W. The two will be different (except when W has no effect on C) because the models being estimated refer to different distributions. In the correlation model it does not make sense to say that the regression of C on Y produces biased estimates because of misspecification caused by the exclusion of W.

Classical misspecification analysis implicitly presupposes a causation model. In this case the analysis of misspecification and its effects is more complex. Suppose for example that Wealth and Income determine consumption and there are no feedback (simultaneity) effects — that is, consumption does not influence income. This could be true in some micro data sets, though it will typically not hold at the macro level. Suppose however that data on wealth is not available and/or the econometrician is unaware that wealth enters the consumption function. In some sense, the regression model $C = \alpha + \beta Y$ is misspecified because it neglects an important determinant of consumption. To be more precise, if we interpret the regression model in the sense 3 discussed above, which specifies that Y is the sole causal determinant of consumption, then we have a misspecified model. The regression of C on Y estimates a well-defined and meaningful quantity, as we will now show.

To simplify calculations, let c, y, and w be differences from means, so the $\sum_{t=1}^{T} c_t = \sum_{t=1}^{T} y_t = \sum_{t=1}^{T} w_t = 0$. The regression coefficient $\hat{\beta}$ of c on $\alpha + \beta y$ is $\hat{\beta} = \sum_{t=1}^{T} x_t y_t / \sum_{t=1}^{T} y_t^2$, as can easily be verified. Assuming that $c_t = \gamma y_t + \delta w_t + \epsilon_t$, we see that

$$\hat{\beta} = \frac{\sum_{t=1}^{T} \gamma y_t^2 + \delta y_t w_t}{\sum_{t=1}^{T} y_t^2} = \gamma + \delta \frac{(1/T) \sum_{t=1}^{t} w_t y_t}{(1/T) \sum_{t=1}^{T} y_t^2} = \gamma + \delta \frac{\widehat{Cov}(y, w)}{\widehat{Var}(y)}.$$

This estimate has a clear interpretation. In calculating the effect of a unit change in y on c, γ is the direct effect of the change. In addition, the OLS estimates take into account the fact that y is correlated with w. A change in y is linked to a change in w and the change in w effects c via the coefficient δ. Thus the OLS estimates compensate for the misspecification introduced by excluding w, and adjusts the coefficient of y to optimally take into account the relationship of y with the omitted variable w. It is easily shown that the OLS forecasts from misspecified models are unbiased, a reflection of this property just discussed.

In classical specification analysis, γ is called the true parameter, and since OLS estimates $\gamma + \delta\, \mathcal{C}ov(y,w)/\, \mathbb{V}ar(y)$ it is said that the OLS is biased. Such a statement needs to be made with greater care. Having estimated β by $\hat{\beta}$ from a regression of c on y, if we observe a person from this or a similar population with income y^*, we can safely and correctly forecast his income using our OLS estimates of the misspecified model. The same kind of property holds for forecasts. However suppose we wish to run a social experiment, say of negative income tax. If we artificially increase incomes of some segment of the population, then the correlation between wealth and income of this population will be different from the prevailing correlation in the population. Our OLS estimates will give misleading results regarding the outcome of such experiments.

2.7   Endogenous Variables and Identification

Cases where there is feedback from the dependent variable to other variables complicate analysis substantially, but are very important in econometrics. There are two main types of regressors which require special treatment. One type is called a *predetermined* variable. For example, in the regression $y_t = a y_{t-1} + b x_t + \epsilon_t$, y_{t-1} is a predetermined variable; it is random, but its value is fixed before period t. For such variables, it is always possible to redefine the error term so that $\mathbb{E} y_{t-1}\epsilon_t = 0$; see Section 8.3. Such variables, which are uncorrelated with the current error term, are called *weakly exogenous.* Models with lagged variables are called dynamic models and are analyzed later in the text (see Section 11.7 for results on consistency, and Section 1.5 for asymptotic distribution theory). Roughly speaking, the main result is that in large samples such models are essentially similar to the case of ancillary regressors analyzed earlier. In small samples, the ML estimates differ somewhat from the OLS and their distributions are not the ones derived above. In the remaining chapter we analyze the somewhat deeper problem which arises when exogeneity also fails to hold, and the regressors are correlated with the errors.

A common case where weak exogeneity fails to hold in econometric applications is the following. Suppose the pair of variables (x, y) is jointly determined by market forces. A classic example is the equilibrium price and the equilibrium quantity in a demand and supply model. Suppose we wish to estimate a demand equation:

$$Q_t^D = a + bp_t + \epsilon_t. \tag{2.1}$$

We have seen that OLS will estimate the *total* effect of p_t on Q^D. If we observe equilibrium quantities, the total effect will involve both supply and demand relationships. It can be shown that the OLS estimates an equation which is a mix of both supply and demand. It follows that the parameters of the demand equation cannot be estimated by OLS.

Even worse, no technique can estimate the parameters of the demand equation without further information. This problem, known as the identification problem, arises whenever it is not possible to assume $\mathit{Cov}(p_t, \epsilon_t) = 0$ in equations like Eq. (2.1). To see why, define $\epsilon_t(c) = \epsilon_t - cp_t$ and note that

$$Q_t^D = a + (b + c)p_t + \epsilon_t(c).$$

This equation is mathematically equivalent to Eq. (2.1) for all values of c. When we do not assume the error to be uncorrelated with p_t, then the error can be arbitrarily modified by adding or subtracting multiples of p_t, so that the coefficient of p_t is arbitrary. This is not possible when we assume $\mathit{Cov}(p_t, \epsilon_t(c)) = 0$, since there is a unique value of c for which this equality holds.

As discussed earlier, a variable is called exogenous for a given equation if it is uncorrelated with the error term of the equation.

This terminology is somewhat ambiguous since in the presence of endogenous variables, the error term can be redefined. Therefore, we will say that z is *exogenous* for parameters (a^*, b^*) in Eq. (2.1) if $\mathbb{C}ov(z, Q_t^D - (a^* + b^* p_t)) = 0$. This formulation makes it clear the exogeneity depends on the parameter values. We will say the z is *endogenous* for parameters (a^*, b^*) if the zero covariance condition cannot be asserted; we do not know whether or not the covariance is zero. The argument outlined in the previous paragraph shows that if one of the variables on the right hand side of the equation is endogenous, then its coefficient is not identified. We now discuss how to estimate coefficients when endogenous variables are present.

2.7.1   Rank and Order Conditions

Consider a linear equation

$$y_t = \beta_1 x_{1t} + \beta_2 x_{2t} + \cdots + \beta_k x_{kt} + \epsilon_t. \tag{2.2}$$

Suppose the parameters of interest are $\beta_1^*, \beta_2^*, \ldots, \beta_k^*$ and it is known in advance that $x_1, \ldots, x_m$ are endogenous for these parameters, while $x_{m+1}, \ldots, x_k$ are exogenous. The following $k - m$ equations express the condition of exogeneity; for $j = m + 1, \ldots, k$, β^* must satisfy

$$\mathbb{C}ov(x_j, y) = \beta_1^* \, \mathbb{C}ov(x_j, x_1) + \cdots + \beta_k^* \, \mathbb{C}ov(x_j, x_k). \tag{2.3}$$

While the covariances are unknown, they can be estimated by the sample covariance $\mathbb{C}ov(x_j, z) \approx (1/T) \sum_{t=1}^{T} (x_j - \overline{x_j})(z - \overline{z})$. If $m = 0$ and there are no endogenous variables, then replacing the covariances by their estimates leads to k linear equations in k unknowns. If the matrix $X'X$ is invertible, these equations have a unique solution $\hat{\beta}$ which serves as an estimate of β^*. It is easily verified that this solution is in fact identical to the least squares solution.

When $m > 0$ then we only have $k-m$ linear equations in k unknowns and hence we cannot solve Eq. (2.3) for β^*. Suppose we could find m variables z_i which are exogenous for the parameters β^*. Then for $j = 1, \ldots, m$,

$$\mathbb{C}ov(z_j, y) = \beta_1^* \, \mathbb{C}ov(z_j, x_1) + \cdots + \beta_k^* \, \mathbb{C}ov(z_j, x_k). \tag{2.4}$$

This supplies us with an additional m equations, so that it is possible to solve for β^*. Note that when we have m endogenous variables, this leads to a deficit of m equations and hence we need at least m exogenous variables to compensate; with less than m the number of equations would be less than the number of variables so that a unique solution would be impossible. This is known as the *order condition* for identification of β^*: the number of excluded exogenous variables (i.e. variables not included in the equation) must be at least as large as the number of included endogenous variables.

If the order condition holds then we have enough equations to solve for β^*, but it is always possible that one of these equations is a linear combination of the others. This will make it impossible to solve for β^*. The set of Eqs. (2.4) and (2.3) can be written as a linear system $c = A\beta^*$; if the A matrix is invertible, then we say that the *rank condition* for identification holds. The parameter β^* is identified if and only if the rank condition for identification holds. The rank condition can fail to hold if there is some relationship of the type $y_1 = \gamma_1 x_1 + \cdots + \gamma_k x_k + \nu$ for parameters γ_i different from β^*. In this case it can be checked that

$$
\begin{aligned}
0 &= \mathbb{C}ov(z_1, y) - \mathbb{C}ov(z_1, y) \\
&= (\gamma_1 - \beta_1^*)\, \mathbb{C}ov(z_1, x_1) + \cdots + (\gamma_k - \beta_k^*)\, \mathbb{C}ov(z_1, x_k).
\end{aligned}
$$

This shows that there is a linear dependence among the equations and hence the k equations cannot be solved for k unknowns.

It is important to note that the rank condition is a population condition. It holds (or fails to hold) for the true but unknown covariances. It can be shown that whether or not the rank condition holds for the true covariances, the estimates of the covariances will, with probability one, yield an invertible matrix and a solution for the estimate of β^*. Thus if the order condition is fulfilled we will always find an estimate for β^*, even if in fact the parameter is not identified and hence impossible to estimate. In such situations, as the sample size increases the estimated covariance matrix will come closer and closer to singularity and hence the solution will fluctuate substantially and will have large variances in large samples (contrary to the usual case).

2.8   Exogenous, Endogenous, and Instrumental Variables

In this section we discuss how variables can be classified into the three categories mentioned in the title. Again it must be stressed that this depends on the parameter it is desired to estimate. For every equation of the type Eq. (2.2), it is possible to find a set of parameter values for which all variables on the right hand side are exogenous. For $i = 1, \ldots, k$, the set of k equations,f $Cov(x_i, y) = \beta_1 Cov(x_i, x_1) + \cdots + \beta_k Cov(x_i, x_k) = 0$, can always be solved for a unique β, say $\beta = \beta_0$. Then for the parameter β_0 all variables on the RHS will be exogenous. Furthermore, the equation parametrized by β_0 is equivalent to the structural equation parametrized by some other parameter value, since one can be obtained from the other by redefining the error term. Nonetheless, the parameters for which all the regressors are exogenous may not be the ones of interest to the econometrician.

Given a structural relationship described by the parameters of interest β^*, how can we decide whether or not the x_j are exogenous for this set of parameters? The key to making this decision is the presence or absence of feedback relationships from the dependent variable y to the x. For example, in a demand relationship, the price effects the quantity demanded; if there is excess demand or supply, this will in turn affect the price. There are also effects through the supply relationship. Due to these feedback effects, errors in the demand relationship will affect the price and we cannot expect the two to be independent. Of course in situations where price is fixed (by government, or by the world market) independently of the quantity demanded, it can be taken as exogenous. Typical textbook expositions classify variables as endogenous or exogenous according to whether or not they appear on the left hand side of other equations in systems of demand equations. This can be misleading because any equation can be rewritten to put any variables on the left hand side. Also, it may be that price is fixed by the government and hence is exogenous in the demand equation. We may nonetheless be able to build a model for how the government determines prices and thereby include an equation explaining the price in our model. This would make price endogenous by traditional criteria but statistically we can validly treat prices as exogenous for the demand equation. This point is discussed in detail by Klein (1990).

It was argued by Wold (1954) that feedbacks could not naturally occur since everything acts with lags. Systems in which exogenous variables determine one endogenous variable which in turn determines another, and so on, are called *recursive systems*. Since there are no feedback loops in such systems, there is no difficulty in estimating them by OLS. The generally accepted view on this issue is that if the period of observation is long relative to the lag with which feedback occurs, then the data must be treated as having feedbacks.[1]

To estimate an equation with endogenous regressors, we need instrumental variables (also called excluded exogenous variables) at least equal in number to the endogenous variables occurring on the right hand side of the equation (these are called included endogenous variables). According to the definition, an instrumental variable should be uncorrelated with the error term of the equation to be estimated. This is more or less equivalent to saying that it should not affect the dependent variable. If a variable is correlated with the errors of an equation, its inclusion in the equation will reduce the error and improve the fit of the equation. At the same time, good instrumental variables must have high correlation with the endogenous explanatory variables. If instruments have small correlations with the endogenous explanatory variables then the resulting estimates are very imprecise. The traditional approach is to take variables occurring as explanatory variables in other relationships in the system of equations. For example, rainfall may be an important determinant of the supply of a given crop. Thus it may be highly correlated with the price. We can reasonably expect it not to directly affect the demand; hence, this would be an ideal instrument.

An instrumental variable must have strong effects on variables which influence the dependent variable, but have no direct effects on the dependent variable. Such a delicately balanced variable can be found only

---

[1]Suppose today's price affects tomorrow's demand and tomorrow's excess demand leads to higher prices the day after tomorrow which leads to increased supply on the third day and a fall in prices on the fourth. If we have price and quantity data on a daily basis we could avoid the simultaneity problem, but we must assume that monthly averaged prices and quantities interact with each other.

when we have strong theories about the dependent variable. For example, utility theory tells us that determinants of demand are income and prices. Thus it would be safe to take variables which influence prices as instruments. In cases where we don't have strong theories, it is much harder to exclude the possibility of direct effects from the instrument to the dependent variable. Also, general equilibrium theory suggests that everything influences everything else. For these reasons finding suitable instruments can be a difficult task.

2.9   Estimating Equations with Endogenous Regressors

Assuming that enough suitable instruments can be found, how can we estimate structural equations? This section is devoted to this issue. We first consider the case where the number of instruments equals the numbers of included endogenous variables. This is called the just-identified case.

Let $z = (z_1, \ldots, z_m, x_{m+1}, \ldots, x_k)$ be the matrix of all the exogenous variables — m instrumental variables and $K - m$ included exogenous variables. Then Eqs. (2.3) and (2.4) can be written together as

$$\mathbb{C}ov(z, y) = \mathbb{C}ov(z, x\beta).$$

Assuming that z has been centered around its mean, we can estimate the covariance between z and y by $(1/T)Z'y$ and similarly $(1/T)Z'X\beta$ estimates $\mathbb{C}ov(z, X\beta)$. Using these estimated covariances, we can estimate β from the equation $(1/T)Z'y = (1/T)Z'X\beta$, which solves for $\hat{\beta} = (Z'X)^{-1}Z'y$. In the just-identified case, $Z'X$ is a square matrix and hence can be inverted (assuming it is nonsingular). This estimator is, as we shall soon see, the maximum likelihood estimator for β.

What about the case of more than m instruments? All existing estimation procedures handle this by reducing it to the case of m instruments as follows. Suppose $z_1, \ldots, z_n$ are n instruments. Then $w_1 = d_{11}z_1 + \cdots + d_{1n}z_n$ is also an instrument. By using m linear combination of the original n instruments we can cut them down to m and then apply the procedure discussed in the previous paragraph.

Algebraically, let D be an $m \times n$ matrix. Then $W = ZD$ is $T \times m$ matrix of instruments when Z is an $T \times n$ matrix. For any choice of D we get an instrumental variable estimator

$$\hat{\beta}_{IV} = (W'X)^{-1}W'y = (D'Z'X)^{-1}D'Z'y.$$

The simplest choice for D is known as a selection matrix. This simply chooses to include some of the instruments and ignore the others. This can be achieved by making rows of D have zeros everywhere and a 1 in the column for the variable selected. m such rows will select m instruments and disregard the rest.

The simple procedure is evidently unsatisfactory as it is arbitrary. A substantial improvement is the Two-Stage Least Squares procedure often abbreviated to 2SLS or TSLS. It can be shown that the best instruments are those which are most closely correlated with the endogenous variables $x_1, \ldots, x_m$. Given that we have more than m instruments, we can attempt to approximate $x_1, \ldots, x_m$ via the first stage regression:

$$\hat{x}_i = Z(Z'Z)^{-1}Z'x_i = Z\hat{\pi}_i.$$

The second stage involves regressing y on $(\hat{x}_1, \ldots, \hat{x}_m, x_{m+1}, \ldots, x_k)$. Let $\hat{\Pi} = (\hat{\pi}_1, \ldots, \hat{\pi}_m)$ be the $n \times m$ matrix formed by the regression coefficients from these first stage regressions. Then setting $D = \hat{\Pi}$ in the previous paragraph gives the formula for the two-stage least squares estimator.

The TSLS uses the instruments $Z\hat{\Pi}$, where $\hat{\pi}$ is an estimate of the matrix of *reduced form* coefficients Π which satisfy $\mathbb{E}(x_1, \ldots, x_m) = Z\Pi$. If we write

$$Y\beta = X\Pi\beta + E,$$

we note that $\Pi\beta$ must have zeros in certain positions. Thus the reduced form coefficients satisfy certain restrictions which the TSLS procedure does not account for properly in estimation. The Limited Information Maximum Likelihood estimator estimates Π subject to the constraint that $\Pi\beta$ has zeros in suitable places. This makes it more efficient in finite samples (assuming the overidentifying restrictions are correct). Asymptotically, the two estimators have the same distribution. See Judge *et al.* (1985) for a detailed exposition and further references to the literature on simultaneous equation models.

2.9.1   Distributions of Estimators for Stochastic Regressors

We discussed the distributions of estimators for the case of nonstochastic regressors in an earlier section. This section analyzes the same problem for the case of stochastic regressors.

First consider the situation where the regressors are ancillary. In this case if we condition on the observed values of the regressors, the distribution of $\hat{\beta}$ and $\hat{\sigma}^2$ is exactly the same as that in the case of constant regressors. Some authors have attempted to obtain the marginal (or unconditional) distribution of $\hat{\beta}$. This is quite complex as we have to integrate the matrix $(X'X)^{-1}$ with respect to the density of the x_j. This can just barely be managed; see for example, Ullah and Nagar (1988) and Wegge (1971). However, as we have noted earlier, for the case of ancillary regressors the conditional distribution is the one that should be used for inference. The marginal distribution of $\hat{\beta}$ is based on considering how the estimate would have varied had you in fact observed values of x_i different from the one you did. Taking this variation into consideration adds unnecessary impreciseness to the observed density of $\hat{\beta}$.

Both in dynamic models and in simultaneous equations models, exact distributions of estimators are extremely messy, while the asymptotic theory can be substantially misleading. Some progress towards obtaining computationally tractable expressions for dynamic models has been made recently by Abadir (1992, 1993), who also provides references to earlier literature. Greenberg and Webster (1983) give a good introduction to the exact distribution theory of estimators for simultaneous equation models and references to the literature. Higher order asymptotic approximations are also valuable tools in studying distributions of estimators. These are less sensitive to the assumptions made regarding normality of errors, and more accurate than standard asymptotics. Higher order approximations come in two flavors. The plain vanilla flavor is Edgeworth and allied expansions, which are also discussed by Greenberg and Webster (1983). The chocolate flavor is bootstrap approximations which provide a less brain-intensive and more computer-intensive method of achieving higher order accuracy. Both are discussed in Chapter 14. Some discussion and references to the literature on bootstrap approximations for simultaneous equations estimators are given by Jeong and Maddala (1992).

2.10 Exercises

1. Given $T \times K$ matrix X of full rank, define $M(X) = \mathbf{I} - X(X'X)^{-1}X'$. Show that for any y, My is the projection of y into the space orthogonal to the columns of X. Conclude, without matrix algebra, that $M^2 = M$ and also $MX = 0$.

2. The L_p estimate of the regression equation $y_t = \beta' x_t + \epsilon_t$ is defined to be the value of β which minimizes the following function:

$$\left(\sum_{t=1}^{T} |y_t - \beta' x_t|^p \right)^{1/p}.$$

Here β is $k \times 1$ and x_t is $k \times 1$, the t-th row of the $T \times k$ matrix X. Show that the L_2 estimate coincides with the maximum likelihood estimate computed in this section. In the case where $k = 1$ and $T = 5$, suppose we observe $y' = (0, 1, 3, 6, 7)$ and $X' = (0, 1, 2, 3, 4)$.

   (a) Calculate the L_p estimate for $p = 1/2, 1, 2, 4$. Can you define an estimate for $p = 0$ and/or $p = \infty$ by taking limits?

   (b) In the standard linear model, if we assume the density of the errors is the double exponential:

$$f(\epsilon_t) = \frac{1}{2\sigma} \exp\left(-\frac{|\epsilon_t|}{\sigma} \right).$$

   Show that the ML estimate is an L_p estimate. What is the value of p in this case?

3. Suppose $y = (aX_1^\alpha + bX_2^\beta)^\gamma$. Find the best linear approximation $(y = c_0 + c_1 X_1 + c_2 X_2)$ to this function in the neighborhood of $X_1 = X_2 = 1$.

4. In the linear regression model, suppose that $\epsilon \sim N(0, \sigma^2 \Sigma)$ where Σ is a known $T \times T$ positive definite matrix (instead of the identity). Calculate the maximum likelihood estimates $\tilde{\beta}$ and $\tilde{\sigma}$ of β and σ in this case. Are these estimates independently distributed?

5. Suppose $y = X\beta + \epsilon$, where $\mathbb{C}\text{ov}\,\epsilon = \Sigma$. Show that every linear unbiased estimator Ay of β can be written as

$$Ay = \left((X'\Sigma^{-1}X)^{-1}X' + Z' \right) \Sigma^{-1}y,$$

where Z satisfies $Z'\Sigma^{-1}X = 0$. Calculate the covariance matrix for all estimators of this form, and show that the smallest covariance matrix is obtained by setting $Z = 0$.

6. Obtain the formulas for R^2 and F in the case where $\mathbb{C}\text{ov}\,\epsilon = \Sigma$.

7. Let $f : R^n \to R$ be a function. Recall that the gradient $\nabla_x f$ is defined to be the $n \times 1$ vector of partial derivatives of x: Let b be an $n \times 1$ vector and Q an $n \times n$ matrix. Show that

   (a) If $f(x) = v'x$ then $\nabla_x f = v$.
   (b) If $f(x) = x'Qx$ than $\nabla_x f = (Q + Q')x$.

8. Let $f(x) : R^n \to R$ be a smooth function. Show that if $\|x - x_0\|^2$ is small,

$$f(x) \approx f(x_0) + (x - x_0)'\nabla_x f(x_0) + (x - x_0)'\nabla_x \nabla'_x f(x_0)(x - x_0),$$

where $\nabla_x \nabla'_x f$ is the Jacobian matrix of f with (i, j) entry $\partial^2 f / \partial x_i \partial x_j$. From this approximation, show that if $\nabla_x f \neq 0$ then there exist points close to x_0 such that $f(x) > f(x_0)$ so that x_0 cannot be a local maximum. If $\nabla_x f(x_0) = 0$ then x_0 satisfies the necessary condition to be a local maximum. If $\nabla_x \nabla'_x f(x_0)$ is negative definite, then x_0 is a local maximum.

9. Calculate $\nabla_x F$ for the following vector functions:

   (a) $F(x) = \|x\|^2$
   (b) $F(x) = \log \|x\|^2$

(c) $F(x) = \left(\sum_{j=1}^{k} x_j^{\alpha}\right)^{\beta}$.

10. Let $Z \sim N(0, I_p)$. Calculate explicitly and directly the densities of the following: (i) Z_1^2, (ii) $\|Z\|^2$, (iii) $(Z'AZ)/(Z'BZ)$, where A and B are symmetric idempotent matrices such that $AB' = 0$.

11. As discussed in the text, if Z_1 and Z_2 are independent of each other, $h(Z_1)$ and $g(Z_2)$ are also. Suppose A and B are matrices and X and Y vector valued random variables such that AX and BY are independent. Explain how $X'AX$ may fail to be independent of $Y'BY$. Show that the two must be independent if either (i) A and B are nonsingular, or (ii) A and B are idempotent.

12. Show that $\mathbb{E}\hat{\beta}'X'X\hat{\beta} = \beta'X'X\beta + K\sigma^2$. Hint: there are several ways of doing this. One way uses the following facts. For any random variable z,

$$\mathbb{E}z'z = \mathbb{E} \, \mathbf{tr} \, zz' = \mathbf{tr} \, \mathbb{E}zz',$$

where the *trace* (denoted '$\mathbf{tr}$') of a square matrix is the sum of its diagonal entries. Also note that $\mathbb{C}ov(z) = \mathbb{E}zz' - (\mathbb{E}z)(\mathbb{E}z)'$.

13. Find the CRLB for estimating for σ in a linear regression model. The Lehmann-Scheffé theorem permits us to conclude that if a function of the sufficient statistics $\hat{\beta}$, $\|y - X\hat{\beta}\|^2$ is an unbiased estimator of some parameter (such as σ), it is automatically the MVU estimate of this parameter. Use this result to find an MVUE of σ. Compare the variance with the CRLB.

One of the disciples of Sheikh Abdul Qadir Jeelani asked to be shown an apple of Heaven. After putting off the request for a while, the Sheikh finally produced an apple. The disciple examined the apple carefully and exclaimed, 'But Sheikh, this apple has a worm in it !.' 'Exactly,' replied the Sheikh. 'Defects in our eyes cause the apples from Heaven to appear to have worms.'

Chapter 3

Bayesian Estimators for Regression

3.1 Introduction

In this chapter we will study some Bayesian estimators for the linear regression model. As we will see, mathematically convenient natural conjugate priors are typically not sufficiently flexible to adequately represent prior information. However, they form the basis for empirical and hierarchical Bayes estimators which are substantial improvements over ML-type estimators in certain applications.

According to the Bayesian view, information about unknown parameters must be represented in the form of a density. Before observing the data, our information is summarized by the prior density. After observing the data, Bayes formula is used to update the prior and get the posterior density. This represents the sum of the prior information and the data information. Formulas for the prior-to-posterior transform for the normal distributions have easy interpretation in these terms.

The posterior distribution contains all our information about the parameter after observing the data. The mean of the posterior represents a good one-point summary of this information, and is an optimal Bayesian estimator for a class of loss functions including quadratic loss. Thus the prior-to-posterior transformation formulae immediately yield formulae for Bayesian estimators of regression parameters. These formulae are easiest when the prior information is in the form of a normal density. We refer to estimators based on normal priors as 'classical Bayesian estimators' since they were developed first. Specifying prior information in the form of a normal density leads to estimators which have certain undesirable properties, which are discussed below in Section 3.4. Empirical Bayes estimators can be viewed as attempts to remove or reduce some of these undesirable features of classical Bayes estimators.

Empirical Bayes estimators require us to estimate the parameters of the prior density together with the parameters of the regression model. It is rather surprising and unexpected that such an odd procedure works at all. There are nonetheless certain difficulties introduced by estimating the prior in the usual empirical Bayes way. It turns out that if the prior is estimated using a second stage Bayesian procedure, these difficulties can be removed. This leads to the concept of hierarchical Bayes estimators. Even though the theory is not complex, it takes some time to grasp the different stages of sophistication of these procedures. It helps matters to have concrete examples, and Part IV of the text is devoted to several such examples.

3.2 Priors and Posteriors for the Multivariate Normal

The general method for Bayesian calculations is as follows. If y is the vector of observations (or data) and θ the unknown parameters, the data density $f(y, \theta)$ is considered to be the conditional density $f(y|\theta)$. We assume the existence of prior information about θ in the form of a prior density $\pi(\theta)$. Then the joint density of y and θ can be written as the product

$$f(y, \theta) = f(y|\theta)\pi(\theta) = \pi(\theta|y)m(y).$$

The first factorization of the joint density is available to us from the functions $\pi(\theta)$ and $f(y|\theta)$. For Bayesian calculations we need $\pi(\theta|y)$, the conditional density of θ given the data y. This is frequently called the posterior density and interpreted to mean the updated information available about θ after observing the data y. The other factor, $m(y)$, is, of course, the marginal density of y. It is usually possible to simply rearrange the factors in the joint density to obtain the desired second factorization into the posterior and marginal. For the case of normal densities, the prior-to-posterior transformation is given below. The rather lengthy proof of the theorem is given in the exercises.

Theorem 3.1 *Let y and θ be k-variate normal vectors such that the conditional density of y given θ is $N(\theta, \Sigma_{y|\theta})$, while the marginal density of θ is $N(\mu, \Sigma_\theta)$. Then the marginal density of y is $N(\mu, \Sigma_{y|\theta} + \Sigma_\theta)$. The conditional density of θ given y is also normal with mean vector $\mathbf{E}[\theta|y] = P^{-1}(\Sigma_{y|\theta}^{-1}y + \Sigma_\theta^{-1}\mu)$ and covariance matrix $\mathbb{C}ov(\theta|y) = P^{-1}$, where $P = \Sigma_{y|\theta}^{-1} + \Sigma_\theta^{-1}$.*

Remarks: This result has a simple heuristic explanation (as opposed to a tedious proof) which also makes it easy to remember. Since conditional on θ, y is symmetrically distributed around θ, while θ is symmetric around μ, it is obvious that the marginal density of y will be symmetric around μ. Similarly, there are two components to the variance of the marginal density of y: the conditional variance of y given θ and the variance of θ itself. We just add the two to get the marginal variance of y. See exercise 2.

The formula for the (posterior) density of θ given y is best understood in Bayesian terminology. Think of θ as the parameter about which we have some prior information represented by the marginal density of θ. The inverse of the covariance matrix Σ_θ^{-1} is a measure of 'sharpness' or 'accuracy' of this information, and is called the precision matrix in Bayesian calculations. Additional information about θ is provided by the data y which has the precision $\Sigma_{y|\theta}^{-1}$. Posterior to the observation of y, our knowledge about θ is based on both the prior and the data. A useful property of the normal distribution is that the precision of our information about θ given the data y is simply the sum of the prior precision Σ_θ^{-1} and the data precision $\Sigma_{y|\theta}^{-1}$, which is exactly the quantity labeled P in the statement of the theorem above. Since the precision is the inverse of the covariance, the covariance matrix of θ given y is just P^{-1}. Finally, we come to the formula for the mean of the posterior. A priori, we believe the mean of θ is μ. Since y is symmetric around θ, after observing y we would expect (in the absence of prior information) θ to have mean y. The posterior mean is just a weighted average of the prior mean μ and the data mean y. Furthermore, the weights are just the precisions of the two measures: each mean is multiplied by its precision and the sum is 'divided by' (premultiplied by the inverse of) the total precision. Thus the posterior mean is $\mathbb{E}\theta|y = \left(\Sigma_{y|\theta}^{-1} + \Sigma_\theta^{-1}\right)^{-1}\left(\Sigma_{y|\theta}^{-1}y + \Sigma_\theta^{-1}\mu\right)$.

3.3 Conjugate Priors

We will be discussing many kinds of Bayesian estimates for the linear regression model. The first kind we will label *classical* Bayes, simply because they were introduced first, are the easiest to compute, and are based on the so-called *natural conjugate prior*. We attach this label only for convenience of reference to this type of estimate.

Using the theorem of the previous sections, it is easy to obtain Bayes estimates for the coefficients in a linear regression model, assuming the error variance σ^2 is known. The case of unknown σ^2 is technically more complicated but gives essentially the same qualitative results. This case is studied in conjunction with Empirical Bayes estimators in Part IV of the text. Note that when σ^2 is known, $\hat{\beta}$ is a sufficient statistic for the data, and we can base estimation on its distribution. We first apply the theorem to get the posterior density for the coefficients β, conditional on the sufficient statistic $\hat{\beta}$.

Corollary 3.1.1 *If the data density is $\hat{\beta}|\beta \sim N(\beta, \sigma^2(X^tX)^{-1})$, and the prior is $\beta \sim N(\mu, \Sigma_\beta)$, then the posterior density of β is multivariate* *normal with mean*

$$\mathbb{E}[\beta|\hat{\beta}] = \left(\frac{1}{\sigma^2}(X^tX) + \Sigma_\beta^{-1}\right)^{-1} \left(\frac{1}{\sigma^2}(X^tX)\hat{\beta} + \Sigma_\beta^{-1}\mu\right)$$

and covariance matrix

$$\mathbb{C}ov(\beta|\hat{\beta}) = \left(\frac{1}{\sigma^2}(X^tX) + \Sigma_\beta^{-1}\right)^{-1}.$$

See also Theorem 18.1 for alternative formulae for the posterior density in the regression case. Now with quadratic loss, the Bayes estimate $\hat{\beta}_B$ of β is simply the mean of the posterior, which is a matrix weighted average of the prior mean μ and the data mean $\hat{\beta}$. There is an important special case which yields an estimator sometimes used in applications. Suppose the prior mean μ is zero, and the prior covariance matrix is $\Sigma_\beta \equiv \nu^2\mathbf{I}_k$. Letting $r = \sigma^2/\nu^2$, we can write the posterior mean as

$$\mathbb{E}[\beta|\hat{\beta}] = (X^tX + r\mathbf{I}_k)^{-1}(X^tX)\hat{\beta} = (X^tX + r\mathbf{I}_k)^{-1}(X^tX)y.$$

This last expression is called the ridge estimator for β. It does not involve the unknown σ^2 and only requires us to specify (what we believe to be) the ratio of the data variance to the prior variance. Note that if r is very small, the prior is very imprecise relative to the data, and the estimate is close to the MLE $\hat{\beta}$. If r is very large, on the other hand, the data are very imprecise, and the MLE is pulled strongly towards 0, the mean of the prior, which is more precisely known. Normally r is assumed to be small, so as to not bias the estimator strongly towards zero. When the matrix X^tX is close to singular, addition of a small quantity to each of the diagonal elements makes the computation of the inverse more stable numerically. This fact contributed to the proposal of this estimator and its recommendation for the case of nearly singular design matrices. The link with Bayes estimates was discovered later.

To analyze the properties of the ridge regression estimators, it is very convenient to make a change of variables. Write $X'X = P'\Lambda P$, where P is the orthogonal matrix of eigenvectors and Λ the diagonal matrix of eigenvalues of $X'X$. Assume $\Lambda = \text{diag}(\lambda_1, \lambda_2, \ldots, \lambda_k)$, where the eigenvalues are arranged in decreasing order. Define $\hat{\alpha} = P\hat{\beta}$, $\alpha = P\beta$

and note that $\hat{\alpha} \sim N(\alpha, \sigma^2 \Lambda^{-1})$. Note that the prior $\beta \sim N(0, \tau^2 \mathbf{I})$ transforms to the prior $\alpha \sim N(0, \tau^2 \mathbf{I})$. Let $\alpha^* = (\Lambda + r\mathbf{I})^{-1}\Lambda y$ stand for the ridge estimates for α. Expressed in terms of coordinates, this yields

$$\alpha_i^* = \frac{\lambda_i}{\lambda_i + r}\hat{\alpha}_i \equiv (1 - B_i)\hat{\alpha}_i.$$

The *Bayes shrinkage factor* $B_i = 1/(\lambda_i + r) \in (0,1)$ is 0 for no shrinkage and 1 for maximum shrinkage of $\hat{\alpha}_i$ towards the prior mean 0. Since $\lambda_1 \geq \lambda_2 \geq \cdots \geq \lambda_k$, it follows that $B_1 \leq B_2 \leq \cdots \leq B_k$. Note that λ_i/σ^2 is the precision (reciprocal of the variance) of $\hat{\alpha}_i$. This ordering of the Bayes shrinkage factor makes sense since the coordinates with higher data variance are more influenced by the prior, while those about which the data is very informative (low variance) remain relatively unaffected by the prior information. As we will see later, this natural and intuitively plausible property of Bayes estimates causes them to fail to be minimax when the eigenvalues λ_i are 'too unequal.'

3.4   Three Difficulties with Conjugate Priors

The classical Bayes estimators derived in the previous section cannot seriously be recommended for use in practical settings. There are three main related difficulties which are discussed below. We will show that the empirical Bayes procedure, discussed in the next section offers solutions to all three. In this section we consider the simplest case of the classical Bayesian estimators; the nature of the difficulties remains exactly the same in the more complex cases.

Consider a one-dimensional special case of the regression model so that $\hat{\beta} \sim N(\beta, \sigma^2/(x'x))$. We will also assume that $x'x = 1$. The prior density will be $\beta \sim N(0, \nu^2)$. In this case the Bayes estimator is

$$\hat{\beta}_B = \frac{\nu^2}{\nu^2 + \sigma^2}\hat{\beta}.$$

We now discuss the properties of this estimator. Qualitatively, the properties of the classical Bayesian estimator for linear regression discussed in the previous section are identical.

3.4.1 Unbounded Risk

Since the Bayes estimates are biased towards zero (the prior mean), we must use the mean squared error (and not the variance) to evaluate them. Define $\beta_c^* = c\hat\beta$. Then, setting $c = \nu^2/(\nu^2 + \sigma^2)$ yields the Bayesian estimator discussed above. The risk, or mean squared error of this Bayes estimate can be calculated as follows:

$$
\begin{aligned}
R(\beta_c^*, \beta) &= \mathbb{E}(c\hat\beta - \beta)^2 = \mathbb{E}\{c(\hat\beta - \beta) + (c-1)\beta)\}^2 \\
&= c^2\mathbb{E}(\hat\beta - \beta)^2 + 2c(c-1)\beta\,\mathbb{E}(\hat\beta - \beta) + (c-1)^2\beta^2 \\
&= c^2\sigma^2 + (1-c)^2\beta^2.
\end{aligned}
$$

The risk of the Bayes estimator is lowest at $\beta = 0$ the prior mean, where it is $c^2 = \nu^4\sigma^2/(\nu^2 + \sigma^2)^2 < \sigma^2$. Since the risk of the ML is σ^2, it is easily calculated that the risk of the Bayes rule is less than that of ML for all β satisfying $\beta^2 < 2\nu^2 + \sigma^2$. Thus by choosing ν^2 large enough we can make the set on which the Bayes rule dominates ML arbitrarily large. On the other hand, the risk of the Bayes rule goes to infinity for large β. Regardless of how we choose the prior parameters the risk of the Bayes rule is arbitrarily worse than that of the ML for β large enough. This creates a problem with the use of the Bayes rule since cases where certain and definite prior information is available are rare. If the prior is wrong, there is always a chance of tremendously poor performance by the Bayes rule.

3.4.2 Choice of Hyperparameters

Parameters entering the prior density are referred to as hyperparameters to distinguish them from parameters of the regression model itself. From the formulas for the risk, the mean of the prior governs the point at which improvement occurs. Both the range of values and the amount of improvement in risk over OLS depend on the prior variance. Increasing the prior variance increases the range of parameter values over which the Bayes rule is superior but also decreases the amount of gain. Thus, the performance of the Bayes rule depends very much on the hyperparameters (the prior mean and variance). Unfortunately, no rule is available which permits us to make a good choice for these hyperparameters in applications.

The classical Bayesian technique for choosing the hyperparameters requires subjectively assessing a prior distribution. It can be shown that in certain decision making environments, rational choices can be made by acting as if one has a prior (see exercise 12 and also Zaman (1995) for a critique of the Bayesian rationality argument). Non-Bayesians maintain that the usual statistical decision making problem is quite different from the types of problems for which this result holds. Nonetheless, even if we accept the Bayesian arguments at face value, this leaves us with the following problem. The class of all Bayes procedures is very large, and all of these are *rational* since all of these are Bayes for some prior. Now the Bayesian rationality arguments do not exhaust the limits of reason, and it may be possible to prefer some of the Bayesian procedures over others on rational grounds. This is especially true since different Bayesian procedures (corresponding to different choices for hyperparameters) perform very differently from each other. For example, a recommendation frequently made by Bayesians is that if there is a large amount of a priori uncertainty about the parameters then the prior variance ν^2 should be made very large. For large ν^2 the range of values of β on which the Bayes estimator is superior to ML is large. On the other hand, the amount of the improvement over ML is negligible. At the same time, outside a finite neighborhood of the prior mean, the Bayes rule is arbitrarily worse than the ML. Thus very large values of ν^2 lead to a very undesirable estimation rule: in case of gain, the gain is negligible and for a large set of parameter values there is arbitrarily large loss relative to the ML. For smaller values of ν^2, the Bayes procedure will produce measurable gains over the ML on some neighborhood of the prior mean. The smaller the ν^2 the larger the gain; at the same time, the interval of good performance is also smaller for smaller ν^2.

As we have discussed, the choice of hyperparameters corresponds to the choice of a set of parameters for which we get improved performance; the amount of the improvement and the size of the set are both determined by this choice. Unfortunately, all the classical Bayes rule described above share the property that they are arbitrarily worse than the ML outside a compact set of parameter values. Since it is possible to find procedures which achieve the gains of the classical Bayesian procedures without incurring similar losses, classical Bayesian procedures cannot be recommended for applications. Some insight into the reason why Bayesian procedures perform so poorly outside a finite interval is provided in Section 3.4.3.

3.4.3   Conflicting Prior and Data

Related to the previous problem is the difficulty of conflicting prior and data information. This can always be a source of difficulties in the Bayesian technique. Empirical Bayes techniques to be discussed can be viewed as an attempt to fix this difficulty with classical Bayesian estimates. It is convenient to illustrate this difficulty for the special case that $\sigma^2 = \nu^2 = 1$. In this case we have $\hat{\beta}|\beta \sim N(\beta, 1)$ and the prior distribution is $\beta \sim N(0,1)$, so the Bayes estimator is $\beta_B^* = \mathbb{E}(\beta|\hat{\beta}) = \hat{\beta}/2$. This estimator is problematic for the following reason. Suppose we observe $\hat{\beta} = 15$. Then the Bayes estimate is $\beta_B^* = 7.5$ the midpoint of the prior mean 0 and the data mean 15. This is more than 7 standard deviations away from the MLE for β (which is, of course, 15). One could argue that this is not a defect since the Bayes estimate uses two sources of information, prior and data, and therefore could differ substantially from an estimate based on only one of the sources. However, there appears to be a conflict between the prior information and data information. According to both the prior and the data, $|\hat{\beta}-\beta|$ should be within -3 and 3 with very high probability. From the prior density $\beta \sim N(0,1)$ so that β is within -3 and 3 with very high probability and, therefore, observations $\hat{\beta}$ outside the range -6 to 6 are exceedingly improbable. According to the data $\hat{\beta} = 15$, so values of β outside the range 12 to 18 are exceedingly unlikely.[1] In such a situation, it does not make sense to aggregate the two sources of information. If one trusts the prior, one should discard the data, or alternatively, if the prior is deemed less trustworthy one should discard the prior information. To put it more graphically, if we have information about a person suggesting that either he is from California or he is from New York, it does not make sense to average and say that he is probably from Kansas. It is mainly due to this implausible estimate of β for

---

[1] Note that this problem is not resolved by looking at the posterior distribution instead of just the point estimate for β. The posterior distribution is also centered at 7.5 and hence incompatible with both the data and the prior.

large values of $\hat{\beta}$ that the Bayes estimator has unbounded risk.

3.5   Empirical Bayes Estimates

We will now discuss a type of estimate which alleviates all three of the difficulties associated with classical Bayesian estimates discussed above. One source of the difficulty is the choice of the prior variance ν^2. If this is chosen small, the risk of the Bayes rule is good relative to ML for β near the prior mean, and very poor other wise. If it is large, the range of values of β for which the Bayes rule is superior is much enlarged, but the amount of gain is also substantially reduced. Thus the performance of the Bayes rule depends critically on the parameter ν but we have no guidance as to how to select it. A question arises as to whether it is possible to choose the value of ν after looking at the data to avoid this problem. This is known as the 'Empirical' Bayes approach, where the data are used to estimate not only the parameter β but also the prior parameter ν. Since the prior is estimated from the data, the data are automatically in conformity with the prior and the case where the data essentially contradicts the prior is avoided.

There are three different ways to implement the empirical Bayes approach. In all cases, the marginal density of the observations is used to provide estimates for the hyperparameters. Note that the data density depends on the parameters, and the prior density of the parameters depends on the hyperparameters so that the marginal density of the observation (after integrating out the parameters) will depend directly on the hyperparameters. The simplest empirical Bayes approach is to directly estimate the hyperparameters. Once these are in hand, one can proceed in the classical Bayesian fashion, using the estimated prior density as a real prior. One difficulty (which does not matter much for larger sample sizes) with this approach is that the estimation of the hyperparameters introduces some uncertainty about the prior which is not accounted for correctly in the estimation. This is because we take the hyperparameters to be equal to their estimates with certainty and therefore ignore the variance of the estimates. There are two approaches which avoid this problem. One approach directly estimates the decision rule. A second approach estimates the parameters of the posterior distribution. We will illustrate all three approaches in the sequel.

3.5.1 Estimating the Prior

We continue to consider the case of known σ^2. This makes it easier to see the essential concepts.[2] In this case, the sufficient statistic is $\hat{\beta} \sim N(\beta, \sigma^2(X'X)^{-1})$. Consider a prior of the form $\beta \sim N(\mu, \nu^2\Omega)$, so that the posterior is $\beta|\hat{\beta} \sim N(m, V)$, where

$$V^{-1} = \frac{1}{\sigma^2}X'X + \frac{1}{\nu^2}\Omega^{-1}$$

$$m = V\left(\frac{1}{\sigma^2}X'X\hat{\beta} + \frac{1}{\nu^2}\Omega^{-1}\mu\right). \tag{3.1}$$

If the hyperparameters μ, ν^2, and Ω are specified a priori then m, the mean of the posterior density, is a classical Bayes estimator for β. Empirical Bayes methods require us to estimate the hyperparameters. To do this, we must look at how the distribution of the observed data depends on the hyperparameters. The marginal density of $\hat{\beta}$ is $\hat{\beta} \sim N(\mu, \sigma^2(X'X)^{-1} + \nu^2\Omega)$. Let us assume that μ and Ω are known or specified a priori, so that ν^2 is the only unknown hyperparameter. More complex cases will be considered later.

The simplest type of empirical Bayes procedure is based on directly estimating the hyperparameter ν^2. This may be done by the method of maximum likelihood, by the method of moments, or by some other suitable method. There is some consensus that the method of estimation used does not significantly affect the properties of the resulting estimator. For example, to estimate ν by ML, we would look at the log likelihood function

$$l(\mu) \propto -\frac{1}{2}\log(\boldsymbol{det}(\sigma^2(X'X)^{-1} + \nu^2\Omega))$$

$$-\frac{1}{2}(\hat{\beta} - \mu)'\left(\sigma^2(X'X)^{-1} + \nu^2\Omega\right)^{-1}(\hat{\beta} - \mu). \tag{3.2}$$

---

[2]Empirical experience shows that estimation of σ^2 does not make a big difference. If we take the formulas for known σ^2 and replace by the estimate $\hat{\sigma}^2$, this gives satisfactory results for most practical purposes. The case of unknown σ^2 will also be treated in Part IV of the text.

This can be maximized with respect to ν using numerical methods but not analytically. It is frequently convenient to have an explicit formula for ν. This can be achieved by looking at a method of moments estimator. Let a_i, b_i be the (i, i) diagonal entry of the matrix $(X'X)^{-1}$ and Ω^{-1}. Then it is easily checked that $\mathbb{E}(\hat{\beta}_i - \mu_i)^2 = \sigma^2 a_i + \nu^2 b_i$. It follows that

$$\mathbb{E}\frac{1}{K}\sum_{i=1}^{K}\frac{1}{b_i}\left(\{\hat{\beta}_i - \mu_i\}^2 - \sigma^2 a_i\right) = \nu^2.$$

The quantity on the left hand side is therefore an unbiased estimator for ν^2 and estimates of this kind have been used in applications. Each of the terms in the sum is an unbiased estimate of ν^2 which is necessarily positive, though the term may be negative. In practice it is advantageous to take the positive part of each term, even though this introduces bias, so as to avoid negative estimates for the variance ν^2.

Let $\hat{\nu}$ be an estimate of ν, whether ML or MOM or some other kind. The empirical Bayes procedure involves proceeding with a classical Bayes analysis using this estimate as if it was the prior parameter. In particular, the empirical Bayes estimator of β is

$$\hat{\beta}_{EB} = \hat{V}_{EB}\left(\frac{1}{\sigma^2}X'X\hat{\beta} + \frac{1}{\hat{\nu}^2}\Omega^{-1}\mu\right),$$

where $\hat{V}_{EB}^{-1}$ is the empirical Bayes estimate of the (posterior) covariance matrix of β which is

$$\hat{V}_{EB}^{-1} = \left(\frac{1}{\sigma^2}(X'X) + \frac{1}{\hat{\nu}^2}\Omega^{-1}\right)^{-1}.$$

Note that using the posterior covariance matrix we can get estimates for the standard errors of each of the coefficient estimates, as well as other useful information about the estimated parameter.

A special case of some importance arises if $\Omega = (X'X)^{-1}$. This situation arises if the prior covariances between the different β_i display exactly the pattern as the data covariances. While this is a crassly opportunistic assumption in that it simplifies all computations considerably, it appears to do reasonably well in applications, justifying its use. Following Zellner (1986) and Ghosh et al. (1988), who present some justification for the use of this prior, we will call it the *g-Prior*. In this case the marginal likelihood function of $\hat{\beta}$ given in Eq. (3.2) simplifies to

$$l(\mu) \propto -\frac{T}{2}\log(\sigma^2 + \nu^2) - \frac{Q}{2(\sigma^2 + \nu^2)},$$

where $Q = (\hat{\beta} - \mu)'X'X(\hat{\beta} - \mu)$. This is easily maximized with respect to ν, yielding the ML estimate

$$\hat{\nu}_{ML} = \frac{Q}{T} - \sigma^2 = \frac{1}{T}\left(\hat{\beta} - \mu\right)' X'X \left(\hat{\beta} - \mu\right) - \sigma^2.$$

One can check that the expected value of the RHS is ν^2, so that this can also be interpreted as a MOM estimator.

3.5.2 Estimating the Decision Rule

In the empirical Bayes technique described above, we estimate the hyperparameters, but then pretend that our estimates equal the hyperparameters with certainty, and proceed to use classical Bayes formulas. Obviously this technique ignores the uncertainty involved in the estimates of the hyperparameter. All further developments attempt to account for the estimation and therefore the uncertainty about the hyperparameters in some way or the other. Instead of directly estimating the hyperparameter ν^2, a second approach to empirical Bayes attempts to estimate the decision rule to be used. Uncertainty about hyperparameters is factored into the way the decision rule is estimated on the basis of available information about the hyperparameters. This is an indirect way of addressing the problem.

By some algebraic manipulation of Eq. (3.1), the classical Bayes estimate for β can be written as

$$\hat{\beta}_B = \mu + \left\{ \mathbf{I}_k - \left(\mathbf{I}_k + \frac{\nu^2}{\sigma^2}\Omega X'X \right)^{-1} \right\} \left\{ \hat{\beta} - \mu \right\}.$$

The above formula shows that the Bayes estimator *shrinks* the difference between the data and the prior mean $\hat{\beta} - \mu$ towards zero; the

matrix multiplying this difference is of the form $1 - 1/(1 + a)$ in the
scalar case. Estimating the decision rule above is, in a sense, equivalent
to estimating ν^2 as was done in the first type of empirical Bayes estimate. This is because the hyperparameters will typically appear in the
decision rule and hence estimating the decision rule will explicitly or
implicitly involve estimating the hyperparameters. The only difference
between the two approaches is that the second approach attempts to
find suitable estimators for the particular function of the hyperparameter which appears in the decision rule. The above representation is
suited to this goal in that the hyperparameter ν^2 appears in only one
place.

A suitable empirical Bayes rule of the second type can only be given
for the case of the g-prior, where the decision rule simplifies to

$$\hat{\beta}_B = \mu + \left\{ 1 - \frac{\sigma^2}{\sigma^2 + \nu^2} \right\} \left\{ \hat{\beta} - \mu \right\}.$$

Note that in the case of the g-prior, the marginal density of $\hat{\beta}$ is

$$\hat{\beta} \sim N(\mu, (\sigma^2 + \nu^2)(X'X)^{-1}).$$

It follows taht if $Q = (\hat{\beta} - \mu)'X'X(\hat{\beta} - \mu)$ then $Q/(\sigma^2 + \nu^2) \sim \chi^2_K$. If
$Z \sim \chi^2_K$ it is easily verified that for $K \geq 3$, $\mathbb{E}1/Z = 1/(K - 2)$. It
follows that

$$\mathbb{E}\frac{(K - 2)\sigma^2}{Q} = \mathbb{E}\frac{(K - 2)\sigma^2}{(\hat{\beta} - \mu)'X'X(\hat{\beta} - \mu)} = \frac{\sigma^2}{\sigma^2 + \nu^2}.$$

The quantity on the LHS is an unbiased estimate of the quantity on
the RHS, which appears in the formula for the Bayes estimate. By
substituting this estimate into the formula, we get an empirical Bayes
estimate

$$\hat{\beta}_{EB} = \mu + \left\{ 1 - \frac{(K - 2)\sigma^2}{(\hat{\beta} - \mu)'X'X(\hat{\beta} - \mu)} \right\} \left\{ \hat{\beta} - \mu \right\}. \qquad (3.3)$$

This estimator specialized to the right context is the James-Stein estimator and dominates the MLE under certain conditions discussed in
the next chapter.

To summarize, note that the second type of empirical Bayes estimate uses the quantity $(K - 2)\sigma^2/Q$ as an estimator of $\sigma^2/(\sigma^2 + \nu^2)$. If we wish, we could generate the implied estimate of ν^2 as follows. By taking reciprocals, $Q/(K - 2)\sigma^2$ estimates $1 + \nu^2/\sigma^2$ so subtracting one and multiplying by σ^2 yields $Q/(K - 2) - \sigma^2$ as the implicit estimate of ν^2. However, the rationale for the $K - 2$ comes from the particular form in which the ν^2 appears in the Bayes rule, and is valid only for $K \geq 3$. Thus the second type of empirical Bayes estimate requires paying much greater attention to the particular function of the hyperparameter which it is necessary to estimate to get the decision. In the present example, the second type of estimate was obtained under conditions substantially more restrictive than the first type (i.e. only with a g-prior, and only for $K \geq 3$).

The third type of empirical Bayes method is an attempt to estimate the posterior distribution rather than the prior. This is closely related to the Hierarchical Bayes procedure discussed in Section 3.6 and therefore will be discussed there. Before proceeding, we discuss briefly how the empirical Bayes estimators resolve the three difficulties with classical Bayes. As indicated above, the James-Stein estimator is a special kind of empirical Bayes estimator and has a risk function superior to that of the MLE in three or more dimensions. Typically, empirical Bayes estimators have bounded risk functions, although they may be substantially worse than ML estimates on certain portions of the parameter space. The second difficulty of choice of hyperparameters is also partially resolved, since we estimate the hyperparameter from the data. Note that it is typically not useful in small samples to estimate all features of the prior, and some parts of the prior must still be specified a priori. As to the third difficulty, conflicts between prior and data are minimized by estimating the prior from the data. Conflicts can still occur however. For example, the distributions may look quite different from the marginal normal density that should be observed if the prior is normal and the data are normal. In such cases of conflict, the empirical Bayes estimator will typically estimate ν^2 to be very high, and reduce to the ML estimator.

3.6   Hierarchical Bayes Estimates

Neither of the first two types of empirical Bayes (EB) estimates directly consider the uncertainty associated with estimating the hyperparameter. This problem is substantially reduced in large samples but can be serious in small samples. Another problem with the empirical Bayes procedures is that they are typically not admissible. While Bayes procedures are usually admissible, estimating the hyperparameter puts empirical Bayes outside the class of Bayes procedures and leads to inadmissibility in most cases. A third problem is that, at least for the crude type of EB estimates discussed above, the effect of the (estimated) prior does not go to zero asymptotically. This leads to undesirable asymptotic properties as discussed in Section 15.3.1. These problems can be fixed by using the hierarchical Bayes procedure, which is described below.

Instead of estimating the hyperparameter by ML or MOM or some other method, we can estimate it by a Bayesian procedure. This will require us to put a prior density on the hyperparameter, which is the reason for the name *hierarchical Bayes* given to this approach. To be explicit, consider $\hat{\beta}|\beta \sim N(\beta, \sigma^2(X'X)^{-1})$. The first stage prior is on β, where we assume that $\beta \sim N(\mu, \nu^2\Omega)$. Take μ and Ω as fixed and known for simplicity, so that ν^2 is the only hyperparameter. In the hierarchical Bayes approach we regard the prior density as being a conditional density given the hyperparameters and introduce a second stage prior to describe the marginal density of the hyperparameter. This second stage may also have deeper hyperparameters which may then again have priors.

In the present introduction we consider the simplest case, where we only have one hyperparameter ν^2, and its density has no further hyperparameters. In most applications, the hyperparameters are equipped with uninformative densities. In cases where there is past experience on the process, previous estimates of the hyperparameters may permit use of an informative prior (see for example Hoadley (1981)). Consider, for example, a diffuse prior density $f(\nu^2) = 1$. In order to make inferences about β, we need to calculate the posterior density of β given $\hat{\beta}$. We have already calculated this density under the assumption of fixed ν^2 around Eq. (3.1) in the previous section. In the present setup, we regard Eq. (3.1) as the conditional density $f(\beta|\hat{\beta}, \nu^2)$ of β given $\hat{\beta}$ and ν^2. To get the posterior density $f(\beta|\hat{\beta})$ we must integrate out ν^2. Let $f(\nu^2|\hat{\beta})$ be the posterior density of ν^2. Then $f(\beta|\hat{\beta})$ can be written as $f(\beta|\hat{\beta}) = \int f(\beta|\hat{\beta}, \nu^2) f(\nu^2|\hat{\beta}) \, d\nu^2$. Also the mean of this density, which will be the hierarchical Bayes estimator for β, can be calculated by integrating $m = m(\hat{\beta}, \nu^2) = \mathbb{E}(\beta|\hat{\beta}, \nu^2)$ with respect to this same posterior density:

$$\hat{\beta}_{HB} = \int \left\{ V \left(\frac{1}{\sigma^2} X'X\hat{\beta} + \frac{1}{\nu^2}\Omega^{-1}\mu \right) \right\} f(\nu^2|\hat{\beta}) \, d\nu^2. \qquad (3.4)$$

To compute this estimate, we need the posterior density of ν^2. Even though the prior density $f(\nu^2) = 1$ is *improper* (that is, it does not integrate to one), the method of computing the posterior density remains the same. Let us specialize to the simplest case, that of the g-prior. Define $Q = (\hat{\beta} - \mu)'X'X(\hat{\beta} - \mu)$. The posterior density is given by

$$
\begin{aligned}
f(\nu^2|\hat{\beta}) &= \frac{f(\hat{\beta}|\nu^2)f(\nu^2)}{\int f(\hat{\beta}|\nu^2)f(\nu^2) \, d\nu^2} \\
&= \frac{(\sigma^2 + \nu^2)^{-K/2} \exp\left(-Q/\{2(\sigma^2 + \nu^2)\}\right)}{\int (\sigma^2 + \nu^2)^{-K/2} \exp\left(-Q/\{2(\sigma^2 + \nu^2)\}\right) \, d\nu^2}.
\end{aligned}
$$

The integral in the denominator cannot be evaluated analytically and hence must be done numerically. The hierarchical Bayes estimator is the mean of the posterior, displayed explicitly in Eq. (3.4). Computing it requires further numerical integration.

From the above calculations it should be clear that the hierarchical Bayes estimator is quite difficult to obtain, even in the simplest cases. The complexity becomes even greater if we make the model somewhat more realistic by allowing for unknown variance σ^2 and also introducing unknown parameters in μ and Ω as is frequently necessary in applications. For this reason, applications of Hierarchical Bayes estimation have been very few. See the study by Nebebe and Stroud (1988) discussed in chapter 17 of part IV. There are two ways around this obstacle of numerical complexity. The third type of empirical Bayes estimation procedure is more or less a method of approximating the posterior resulting from the hierarchical Bayes analysis. There are a number of different ad-hoc methods for carrying out such approximations, each adapted to particular applications. Hoadley (1981) used a technique like this to approximate the posterior in his scheme for quality control at A. T. & T. In the past this was the only method of obtaining approximations to hierarchical Bayes estimators. However, the invention of the Gibbs sampler technique for computing hierarchical Bayes estimators has rendered this type of approximation considerably less useful than before. If the cruder empirical Bayes techniques of the previous section are not considered adequate, then it is probably preferable to calculate the hierarchical Bayes estimates by the Gibbs sampler technique rather than approximate it in some ad-hoc fashion (which is typically quite complex itself).

3.7 Gibbs Sampler

The Gibbs sampler is a technique which permits the computation of hierarchical Bayes estimates (in some important cases) by Monte-Carlo methods. This permits us to avoid the cumbersome numerical integration required for direct calculations as discussed in the previous section. The idea of the technique is very simple. Let $\theta = (\theta_1, \ldots, \theta_p)$ be the parameters we wish to estimate. In order to use the usual Bayesian technique, we need to compute the posterior density of θ given the data y. The Gibbs sampler bypasses numerical integrations required to compute the posterior, obtaining instead a *sample* or a random vector $\tilde{\theta} = (\tilde{\theta}_1, \tilde{\theta}_2, \ldots, \tilde{\theta}_p)$ from the posterior distribution of $\theta|y$. By obtaining a large sample from this posterior distribution, we can estimate the posterior density. Usually the full posterior density is not needed, the posterior mean and covariance suffice. These are also easily estimated on the basis of a random sample from the posterior density.

In order to use the Gibbs sampler, we must have certain conditional distributions available. Fortunately, in many situations of interest the conditional density of θ_i given y *and* the remaining parameters $(\theta_1, \ldots, \theta_{i-1}, \theta_{i+1}, \ldots, \theta_p) \equiv \theta_{(i)}$ is easily calculated. Denote this distribution by $f_i \equiv f_i(\theta_i|y, \theta_{(i)})$. In order to use the Gibbs sampler, we need to have these distributions available. Let $\theta^0 = (\theta_1^0, \ldots, \theta_p^0)$ be an arbitrary initial value for the parameter θ. Now generate θ_i^1 from the conditional density f_i of θ_i given y and $\theta_{(i)} = \theta^0_{(i)}$. If this procedure is iterated enough times, the distribution of θ^n becomes close to the posterior distribution of θ given y. Below we sketch the procedure in the hierarchical Bayes model of the previous section.

The original model has $\hat{\beta}|\beta, \nu^2 \sim N(\beta, \sigma^2(X'X)^{-1})$. The prior density of β is $\beta|\nu^2 \sim N(\mu, \nu^2\Omega)$. The prior density of ν^2 is improper with $f(\nu^2) = 1$. The parameters σ^2, μ, and Ω are taken to be known, for simplicity. Now note that $\beta|\hat{\beta}, \nu^2 \sim N(m, V)$ where m, V are given in Eq. (3.1). Also, the conditional distribution of $\nu^2|\beta, \hat{\beta}$ is simple (in contrast to the posterior density of ν^2): this is given by $f(\nu^2|\beta, \hat{\beta}) \propto \nu^{-K} \exp(-Q/2\nu^2)$, where $Q = (\beta - \mu)'\Omega^{-1}(\beta - \mu)$. We write $\nu^2|\beta, \hat{\beta} \sim IG(K, Q)$, an inverse Gamma density with parameters K and Q.

With these two conditional densities in hand, it is easy to apply the Gibbs sampler to get a sample from the posterior density of β and ν^2. Fix ν_0 arbitrarily and generate β_0 from the density $N(m, V)$ (where m and V depend on ν_0). Next generate ν_1 from the inverse Gamma density with parameter Q computed using the β_0 generated. Now iterate the procedure to get $\beta_1, \nu_2, \ldots$ Discard the values of β_i for the first 100 iterations, say, since the distribution will take some time to converge to the posterior. On subsequent iterations, keep track of the running average of the generated β_i. This will eventually converge to the mean of the posterior distribution and provide the hierarchical Bayes estimator for β.

The procedure sketched above is a minimal but complete description. The models for which the Gibbs sampler works can be substantially more complicated than the one above. Also the random sampling methods can be made much more sophisticated. Some of these complications will be treated in conjunction with applications in Part IV of the text.

3.8 Exercises

1. Prove the 'Law of Iterated Expectation':

$$EX = E^Y \left(E^{X|Y}(X|Y) \right).$$

2. Prove the 'Law of Total Variance':

$$\text{Var}(X) = E^Y\left(\text{Var}^{X|Y}(X|Y)\right) + \text{Var}^Y\left(E^{X|Y}(X|Y)\right).$$

(Hint: use the fact that $\text{Var}^Y Z = E^Y Z^2 - (E^Y Z)^2$.) Use this result in Theorem 1 to prove that the covariance matrix of the marginal density of y must be the covariance of y given θ plus the covariance matrix of θ.

3. *Proof of Theorem 1: The posterior density:* The several steps involved in this proof are given below.

   (a) Suppose the marginal density of θ is $\theta \sim N(\mu, \Sigma_\theta)$ and the conditional density of y given θ is $N(0, \Sigma_{y|\theta})$. Write out the joint density of θ and y. $f^{(y,\theta)}(y,\theta) = f^{(y|\theta)}(y|\theta)$.

   (b) To get the other factorization $f^{(y,\theta)} = f^{(\theta|y)}f^y$, note that all terms involving θ can only enter into the density $f^{(\theta|y)}$. Collect all terms not involving θ into one function $H(y)$ and show that with $P = \Sigma_\theta^{-1} + \Sigma_{(y|\theta)}^{-1}$,

$$f^{(\theta,y)} = H(y)\exp\left(-\frac{1}{2}(\theta'P\theta - 2\theta'PP^{-1}\left[\Sigma_{(y|\theta)}^{-1}y + \Sigma_\theta\mu\right]\right).$$

   (c) Define $m = P^{-1}\{\Sigma_{y|\theta}^{-1}y + \Sigma_\theta^{-1}\mu\}$. Let $H_1(y) = H(y)$ $\times \exp(m'Pm/2)$ and multiply the rest of the expression by $\exp(-m'Pm/2)$ to get the following equality:

$$f^{(\theta,y)} = H_1(y)\exp\left(-\frac{1}{2}(\theta - m)'P(\theta - m)\right).$$

   (d) Let $\phi(\theta, m, P)$ stand for the multivariate normal density of θ with mean vector m and covariance matrix P^{-1}. Let $H_2(y) = H_1(y) \times (2\pi)^{k/2}(\det(P))^{-1/2}$. Show that $f^{(\theta,y)} = H_2(y)\phi(\theta, m, P)$. Note that although m depends on y, for all values of y, $\int \phi(\theta, m, P)d\theta = 1$.

   (e) Show that for any pair of random variables X, Y with joint density $f^{(X,Y)}(x,y)$, if $f^{(X,Y)}(x,y) = H(y)G(x,y)$ where G

has the property that $\int G(x, y)dx = 1$ for all y, then $H(y)$ is the marginal density of y and $G(x, y)$ is the conditional density of X given Y. Conclude from the previous exercise that $\phi(\theta, m, P)$ is the posterior density of θ given y.

4. *Theorem 1: The marginal density.* The following exercises are needed to establish the marginal density of y is as stated in Theorem 1.

   (a) Show that the function $H_2(y)$ from the previous exercise is

   $$H_2(y) = (2\pi)^{-k/2} \left\{ \det \left(\Sigma_{y|\theta} \Sigma_\theta \left[\Sigma_{y|\theta}^{-1} + \Sigma_\theta^{-1} \right] \right) \right\}^{-1/2}$$
   $$\times \exp \left(-\frac{1}{2} (y' \Sigma_{y|\theta}^{-1} y + \mu' \Sigma_\theta^{-1} \mu - m' Pm) \right).$$

   (b) As shown in the previous exercise, $H_2(y)$ must be the marginal density of y. By Theorem 1, $H_2(y)$ should be a normal density with mean μ and covariance matrix $\Sigma_{y|\theta} + \Sigma_\theta$. Show that this follows from the two equalities

   $$\Sigma_{y|\theta} + \Sigma_\theta = \Sigma_{y|\theta} \Sigma_\theta \left\{ \Sigma_{y|\theta}^{-1} + \Sigma_\theta^{-1} \right\}$$
   $$y' \Sigma_{y|\theta}^{-1} y + \mu' \Sigma_\theta^{-1} \mu - m' Pm = (y - \mu)' \left\{ \Sigma_{y|\theta} + \Sigma_\theta \right\}^{-1} (y - \mu).$$

   (c) Use the matrix equalities from the next exercise to prove the two identities above.

5. Derive the matrix equalities given below. As a rule, these are easy to derive using the following procedure. First pretend all the matrices are scalars and do a careful step-by-step derivation of the desired result . Then attempt to imitate the same steps for matrices, keeping in mind that matrix multiplication is not commutative, and that division is like multiplying by an inverse. Show that

   (a) $\Sigma_Z^{-1} - \Sigma_Z^{-1} \left(\Sigma_Z^{-1} + \Sigma_W^{-1} \right)^{-1} \Sigma_Z^{-1} = (\Sigma_Z + \Sigma_W)^{-1}$.
   (b) $\Sigma_Z^{-1} \left(\Sigma_Z^{-1} + \Sigma_W^{-1} \right)^{-1} \Sigma_W^{-1} = (\Sigma_Z + \Sigma_W)^{-1}$.

(c) $[(X^tX + k\mathbf{I})^{-1}X^tX - \mathbf{I}]^2 = k^2(X^tX + k\mathbf{I})^{-2}.$

(d) $(P\Lambda P^t + k\mathbf{I})^{-1}P\Lambda P^t(P\Lambda P^t + k\mathbf{I})^{-1}$
$= P(\Lambda + k\mathbf{I})^{-1}\Lambda(\Lambda + k\mathbf{I})^{-1}P^t.$

6. Suppose $X|\theta \sim N(\theta, \Sigma_0)$ and let $\theta \sim N(\mu, \Sigma_1)$ be the prior density (denoted π) for θ. The Bayes estimator δ_π corresponding the prior can be computed from Theorem 1. The Bayes risk of this estimator is computed as follows. Show that for any estimator δ $B(\pi, \delta) = \mathbb{E}^X \mathbb{E}^{\theta|X}(\theta - \delta(X))^2$ Since $\delta_\pi = \mathbb{E}^{\theta|X}\theta$ is the mean of the posterior density, show that $B(\pi, \delta_\pi) = \mathbf{tr}\,\mathbb{C}ov(\theta|X)$. Hint: first show that $B(\pi, \delta) = \mathbb{E}^\theta \mathbb{E}^{X|\theta}(\theta - \delta)^2.$

7. Suppose $T\hat{\sigma}^2/\sigma^2 \sim \chi^2_{T-k}$, and the prior density of σ is the following:

$$f(\sigma^2) = \frac{\lambda^m}{\Gamma(m)}\sigma^{-2(m+1)}\exp(-\frac{\lambda}{\sigma^2}).$$

Show that this is a density; it is called the inverse gamma density. Calculate the posterior density for σ^2 and find the Bayes estimate of σ^2 based on this prior. Compare the mean square error of this estimate with that of the MVUE.

8. Calculate the risk of $\hat{\beta}_k$ with the general quadratic loss function : $L(\hat{\beta}, \beta) = (\hat{\beta} - \beta)^t Q(\hat{\beta} - \beta)$, where Q is a positive definite matrix.

9. In the special case that $X^tX = \mathbf{I}$, find the set of values of the parameter β for which the ridge estimator $\hat{\beta}_k$ has lower risk than the MLE $\hat{\beta}_0$. Is the ridge estimator admissible?

10. Suppose the conditional density of Y given β is $f(y|\beta)$ and that of β given μ is $f(\beta|\nu)$. If the marginal density of ν is $m(\nu)$ show that

(a) The joint density of y, β, ν is $f(y, \beta, \nu) = f(y|\beta)f(\beta|\nu)m(\nu).$

(b) If $m(y)$ is the marginal density of y, the joint density above can be rewritten as $f(y, \beta, \nu) = f(\beta|y, \nu)f(\nu|y)m(y).$

(c) The conditional density of β, ν given y is therefore $f(\beta|y, \nu)f(\nu|y)$ and the conditional density of β given y is obtained by integrating out ν:

$$f(\beta|y) = \int f(\beta, \nu|y)\, d\nu = \int f(\beta|y, \nu) f(\nu|y)\, d\nu.$$

(d) It follows that $\mathbb{E}\beta|y = \int \left(\int \beta f(\beta, \nu|y)d\beta \right) f(\nu|y)\, d\nu.$

11. *Reciprocal of a normal mean.* Suppose $Z \sim N(\theta, 1)$ and we are interested in estimating $1/\theta$. An estimate δ incurs the loss $L(\theta, \delta) = (1 - \theta\delta)^2$.

   (a) Show that the Bayes estimate of $1/\theta$ is the mean of the posterior divided by one plus the variance of the posterior. Hint: look at the first-order conditions for minimizing posterior expected loss with respect to δ; differentiate 'through' the expectation.

   (b) Given the prior $\theta \sim N(0, \nu^2)$, calculate the Bayes estimate.

   (c) Consider the rule $\delta^*(y) = (y^2 + 1)^{-1}$. Is this an extended Bayes rule? (Difficult:) Is it admissible ?

   (d) What is the MLE of $1/\theta$? Consider a particular problem (e.g., estimating a multiplier) where you have to estimate the reciprocal of some parameter. Which among the above estimators would you choose, and why?

12. Let (K_i, P_i) be a sequence of ordered pairs of Swedish Kroners K_i and Sudanese Piastres P_i. At the i-th stage, we must choose a number $w_i \in [0, 1]$ which is the proportion of Kroners we choose. Our payoff in Swedish Kroners at the end of $i = 1, 2, \ldots, K$ decisions is $\Pi = \sum_{i=1}^{K} w_i K_i + \theta(1 - w_i) P_i$, where θ is the number of Kroners per Piastre. A sequence of decisions $w_1^*, \ldots, w_K^*$ is *consistent* with an exchange rate θ^* if $w_i = 1$ whenever $\theta^* P_i < K_i$ and $w_i = 0$ whenever $\theta^* P_i > K_i$. Show that every sequence which is rational is consistent with some exchange rate.(Hint: prove that if a sequence is inconsistent, we can improve the payoff uniformly for all θ by modifying the sequence of decisions to reduce or eliminate the inconsistency.) Does this mean that every rational decision maker knows the exchange rate between Swedish Kroners and Sudanese Piastres?

Hamudi was sitting with friends when a man came running with the news that his ship had sunk. After a little reflection, Hamudi said, 'All praise is for God'. A little bit later, another man came with the news that the earlier report was wrong, and his ship, laden with goods, was due to arrive shortly. Again after reflection, Hamudi said 'All praise is for God.' His companions were puzzled, and asked how he could say the same thing on both occasions. Hamudi said that I reflected to see if my heart had been affected by either the good or the bad news. Finding it not so, I praised God who has freed me from the love of the transient and blessed me with the love of the permanent.

Chapter 4

Minimax Estimators

4.1 Introduction

Since the MLE had all imaginable nice properties, the problem of estimating a multinormal mean was long considered one of the 'solved' problems in statistics. Stein's discovery of the inadmissibility of the MLE in 1955 created considerable confusion. Since the vast majority of applications of statistics revolve around estimating a multinormal mean, the soundness of all such analysis was called into question. In over thirty years of intensive research on these issues, several questions about the validity and relevance of Stein's results were settled.

First, there was a suspicion that the Stein phenomenon is a mathematical artifact, a peculiar property of the normal distribution, or else related to the unrealistic assumption of quadratic loss. Brown (1966)

showed that the same phenomenon was true for a very wide range of
densities and loss functions. Second, there was the feeling that in practical situations gains from using the Stein-type estimators would be
negligible. A series of papers by Efron and Morris addressed the issues of adapting Stein estimation to practical settings and showed that
dramatic gains over conventional estimators were possible in real estimation problems. In these papers, they also showed how to adapt the
Stein estimator for use in practical situations, where many of the ideal
conditions assumed for the derivations did not hold.

The bottom line of this research, summarized by Berger (1982),
is that it is virtually impossible to make significant improvements on
the MLE without having prior information. Any of the estimators
which dominates the MLE has significantly superior risk only over a
small region of the parameter space. If it is possible to pick this small
region in such a way that it has a modest chance of containing the true
parameter, then significant improvements on the ML are possible. This
will only be possible in the case that there is prior information. Without
prior information, the chances of an arbitrarily chosen small subset of a
(high-dimensional) parameter space containing the true parameter are
negligible. Thus the extra effort of selecting a minimax procedure is
unjustified by the potential gain.

This outcome of the tons of research effort poured into the Stein
estimation may appear disappointing. The fact that prior information
permits improved estimates was never in doubt. However, the minimax
theory pointed the way towards estimates which permit exploitation
of prior information in a risk-free fashion. If the prior information is
of dubious validity (as is frequently the case in applications), Stein
estimators will work well if it is correct, but will not be disastrous
even if the prior information is completely wrong. This contrasts with
classical Bayesian estimators, which frequently have unbounded risk
and can be arbitrarily worse than ML depending on how wrong the prior
information is. Furthermore, the type of prior information required
by Stein-type procedures is substantially weaker than that required
by classical Bayes procedures. Situations where such information is
available are correspondingly more common.

Section 4.2 shows that the ML estimator of the regression coefficients is minimax for any quadratic loss function. That is, the maxi-

mum risk of all estimators must be at least as large as the maximum risk of the ML. This is proven using the fact the ML has constant risk and can be arbitrarily closely approximated by Bayes estimators. Section 4.3 shows that the MLE is admissible for estimating a single parameter; this also uses the approximation of ML by Bayes estimators. Section 4.4 shows that the MLE is not admissible for estimating three or more parameters. A large class of estimators, all of which dominate the MLE, are constructed. These estimators have peculiarities which are discussed and explained in Section 4.5 titled Stein's Paradox and Explanations. Given that there is a large class of Bayes and minimax estimators, the issue of how to choose among these estimators is discussed in Section 4.6. Main considerations are the loss function and the type of prior information available. There is a conflict between empirical Bayes-type estimates and minimax estimates. The attempt to be minimax sacrifices most of the potential gains available from weak prior information of the type exploited by empirical Bayes procedures. Choosing between the two, and some compromise estimators such as limited-translation and robust Bayes are also discussed.

4.2   Minimaxity of the MLE

Throughout this chapter, we will continue to consider the case of known variance, so that $\hat{\beta} \sim N(\beta, \sigma^2(X'X)^{-1})$ is the sufficient statistic for the regression model. All of the results to be obtained do generalize without difficulty to the case of unknown variance, where the estimate $\hat{\sigma}^2 = \|y - X\hat{\beta}\|^2/(T - K)$ is available. The object of this section is to show that the ML estimate $\hat{\beta}$ is minimax with respect to quadratic loss: $L_Q(\delta, \beta) = (\delta - \beta)'Q(\delta - \beta)$, where δ is any estimate, and Q is a fixed positive semidefinite matrix. It will be convenient to abbreviate $x'Qx = \|x\|_Q$.

It will be convenient to transform the data in the following way. Let P be a matrix satisfying $P(X'X)^{-1}P' = \mathbf{I}_K$ and define $\hat{\alpha} = P\hat{\beta}$ and $\alpha = P\beta$. Then $\hat{\alpha} \sim N(\alpha, \sigma^2\mathbf{I})$. For any estimator $\delta(\hat{\beta})$ of β, let $P\delta(\hat{\beta}) = P\delta(P^{-1}\hat{\alpha}) = \tilde{\delta}(\hat{\alpha})$ be the corresponding estimator of α. Define $\tilde{Q} = P'^{-1}QP^{-1}$. Using the fact that $\|x\|_Q = \|Px\|_{(P')^{-1}QP^{-1}} = \|Px\|_{\tilde{Q}}$, we can relate the risks of these two estimators:

$$R(\delta(\hat{\beta}), \beta) = \mathbb{E}\|\delta(\hat{\beta}-\beta\|_Q^2 = \mathbb{E}\left\{P\delta(P^{-1}\hat{\alpha}) - P\beta)\right\}'\|_{\tilde{Q}} = \mathbb{E}\|\tilde{\delta}(\hat{\alpha})-\alpha\|_{\tilde{Q}}^2.$$

Note that if $Q = X'X$ then $\tilde{Q} = \mathbf{I}_k$. The maximum likelihood estimator of α is just $\delta_{ML}(\hat{\alpha}) = \hat{\alpha}$. Using the properties of the trace that $x'Qx = \mathrm{tr}(x'Qx) = \mathrm{tr}(xx'Q)$, the risk of the MLE can be calculated as follows:

$$
\begin{aligned}
R(\delta_{ML}(\hat{\alpha}), \alpha) &= \mathbb{E}(\hat{\alpha} - \alpha)'\tilde{Q}(\hat{\alpha} - \alpha) = \mathbb{E}\,\mathbf{tr}(\hat{\alpha} - \alpha)'\tilde{Q}(\hat{\alpha} - \alpha) \\
&= \mathbb{E}\,\mathbf{tr}(\hat{\alpha} - \alpha)(\hat{\alpha} - \alpha)'\tilde{Q} = \mathbf{tr}\left(\mathbb{E}(\hat{\alpha} - \alpha)(\hat{\alpha} - \alpha)'\right)\tilde{Q} \\
&= \mathbf{tr}\left(\mathbb{C}ov(\hat{\alpha})\right)\tilde{Q} = \sigma^2\,\mathbf{tr}(\tilde{Q}) \equiv M.
\end{aligned}
$$

Note that the risk does not depend on the parameter β, and hence the maximum risk over all β of the ML is also M. This is the best possible maximum risk for any rule.

Theorem 4.1 *Suppose we observe $\hat{\alpha} \sim N(\alpha, \sigma^2\mathbf{I}_k)$. For any estimator $\delta(\hat{\alpha})$, we have*

$$\max_{\alpha\in\mathbb{R}^K} R(\delta(\hat{\alpha}), \alpha) \geq M.$$

It follows that the ML $\hat{\alpha}$ is a minimax estimator of α and also that $\hat{\beta}$ is a minimax estimator of β.

Remark: The use of the article *a* (minimax estimator) is significant. Very surprisingly, it turns out that in three or more dimensions (i.e., $K \geq 3$), there also exist estimators other than the ML which are minimax. Thus there exists a large class of estimators satisfying $\max \alpha \in \mathbb{R}^K R(\delta^*(\alpha), \alpha) = M$. Since the ML has constant risk, any other minimax estimator δ^* must actually dominate the ML. This will be discussed at greater length later in the chapter. We note in the passing that the chapter story illustrates the achievement of constant risk (and hence minimaxity) in a real-life situation.

Proof: The strategy of proof is as follows. We will show that the ML is approximately Bayes, and that any approximately Bayes rule with constant risk must necessarily be minimax.

To show that the ML is approximately Bayes, consider the prior $\pi_\nu(\alpha) \sim N(0, \nu^2 \mathbf{I}_K)$. Then the posterior is $\alpha | \hat{\alpha} \sim N(\nu^2 \hat{\alpha}/(\nu^2 + \sigma^2), \nu^2 \sigma^2/(\nu^2 + \sigma^2) \mathbf{I}_K)$. Thus the Bayes rule with respect to this prior is $\delta_\nu(\hat{\alpha}) = (\sigma^2 + \nu^2)\hat{\alpha}/\nu^2$. Recall that the Bayes risk of any rule is defined to be the risk function of the rule averaged with respect to the prior density:

$$B(\delta(\hat{\alpha}), \pi_\nu) = \mathbb{E}^\alpha \, \mathbb{E}^{\hat{\alpha}|\alpha} L(\delta(\hat{\alpha}), \alpha) = \mathbb{E}^{\hat{\alpha}} \, \mathbb{E}^{\alpha|\hat{\alpha}} L(\delta(\hat{\alpha}), \alpha).$$

These equalities show that the Bayes risk can be evaluated either by integrating the risk function with respect to the prior density, or by integrating the posterior expected loss[1] with respect to the marginal density of $\hat{\alpha}$.

We now show that the Bayes risk of the ML is close to the Bayes risk of δ_ν if ν is large, so that ML is approximately Bayes for large ν. The Bayes risk of δ_ν can be calculated from the posterior expected loss of δ_ν which is

$$
\begin{aligned}
\mathbb{E}^{\alpha|\hat{\alpha}} \|(\alpha - \delta_\nu(\hat{\alpha})\|_{\tilde{Q}}^2 &= \mathbb{E}^{\alpha|\hat{\alpha}} \, \mathbf{tr}(\alpha - \delta_\nu(\hat{\alpha})(\alpha - \delta_\nu(\hat{\alpha})' \tilde{Q} \\
&= \mathbf{tr} \, \mathbb{C}ov(\alpha|\hat{\alpha})\tilde{Q} = \mathbf{tr} \, \frac{\nu^2 \sigma^2}{\nu^2 + \sigma^2} \tilde{Q}.
\end{aligned}
$$

Note that this is approximately $M = \sigma^2 \, \mathbf{tr}(\tilde{Q})$ for large ν^2. Now the risk of δ_{ML} is constant at M and integrating this constant against the prior density will just give M, so that $B(\delta_{ML}, \pi_\nu) = M$.

Now that we have established that δ_{ML} is approximately Bayes we can proceed by contradiction. Assume some rule $\delta^*(\hat{\alpha})$ has maximum risk equal to $m < M$. Since for all values of the parameter α, $R(\alpha, \delta^*) \leq m$ while $R(\alpha, \hat{\alpha}) = M$, the difference in the risks of these rules is at least $M - m > 0$. This same inequality continues to hold when this risk is integrated with respect to any prior density π:

$$0 < M - m \leq B(\pi, \hat{\alpha}) - B(\pi, \delta^*) = \int \{R(\alpha, \hat{\alpha}) - R(\alpha, \delta^*)\} \, \pi(\alpha) \, d\alpha.$$

---

[1]That is, the loss function integrated with respect to the posterior density of α given $\hat{\alpha}$.

Next note that for any prior π_ν, the Bayes rule δ_ν has the property of minimizing the Bayes risk. Thus δ^* must satisfy

$$B(\pi, \delta_\pi) \le B(\pi, \delta^*) \le B(\pi, \hat\alpha) - (M - m).$$

This shows that there must be a gap of the size $M - m$ between the Bayes risk of the MLE and the Bayes risk of the Bayes rule for any prior π. Formally,

$$B(\pi_\nu, \hat\alpha) - B(\pi, \delta_\nu) \ge M - m > 0. \tag{4.1}$$

But we have already shown that this difference of risks can be made arbitrarily small. This contradiction completes the proof of minimaxity of the MLE. We note for future reference that any estimator δ for which the difference $B(\pi, \delta) - B(\pi, \delta_\pi)$ can be made to approach zero is called and *extended Bayes* rule. It can be shown that all admissible rules must be either Bayes rules or arbitrarily close to being Bayes (i.e., extended Bayes).

4.3 Admissibility of ML in One Dimension

In the previous section, we proved that the ML was a minimax estimator. It is also the case that it is the only minimax estimator in dimensions one and two. However, there are large classes of minimax estimators in dimensions three or more. This section proves the admissibility (which is equivalent to nonexistence of other minimax rules) for the ML in one dimension. The next section displays rules which are minimax in three or more dimensions.

In the case of linear regression with only one regressor $\hat\beta$ is distributed as a univariate normal with mean β and variance $\sigma^2 x' x$. Recall that $\delta_{ML}(\hat\beta) = \hat\beta$ of β is *admissible* if, given any other estimator $\delta'(\hat\beta)$, there exists a parameter β^* such that $R(\delta'(\hat\beta), \beta^*) > R(\delta_{ML}, \beta^*) = M$. Note that the admissibility of δ_{ML} guarantees that all other estimators have higher maximum risk.

Theorem 4.2 *The estimator $\delta_{ML}(\hat\beta) = \hat\beta$ is* admissible.

Proof: Note that $\hat{\beta} \sim N(\beta, \sigma^2/x'x)$. We will assume that $\sigma^2/x'x = 1$ to simplify writing. Also, in the one-dimensional case, we can take $Q = \hat{Q} = 1$ in the loss functions without any effect on the argument. Let π_ν denote a normal density on β with mean zero and variance ν^2. The Bayes estimator for π_ν is $\delta_\nu = \nu^2 \hat{\beta}/(\nu^2 + \sigma^2)$ in the one-dimensional case. The difference in the Bayes risk of δ_ν and δ_{ML} was calculated in the last section to be

$$B(\pi_a, \delta_{ML}) - B(\pi_a, \delta_\nu) = \sigma^2 - \frac{\nu^2 \sigma^2}{\nu^2 + \sigma^2} = \frac{\sigma^2}{\sigma^2 + \nu^2}\sigma^2.$$

Suppose towards contradiction that some estimator δ' dominates $\delta_{ML} = \hat{\beta}$. It follows that $B(\delta_{ML}, \pi_\nu) \geq B(\delta', \pi_\nu)$. Furthermore, the Bayes rule δ_ν minimizes the Bayes risk over all rules so that $B(\delta', \pi_\nu) \geq B(\delta_\nu, \pi_\nu)$. It follows that

$$B(\delta_{ML}, \pi_\nu) - B(\delta', \pi_\nu) \leq B(\delta_{ML}, \pi_\nu) - B(\delta_\nu, \pi_\nu) = \frac{\sigma^4}{\sigma^2 + \nu^2}.$$

Expressing the Bayes risk as an integral with respect to the prior density, and multiplying both sides by the constant $\sqrt{2\pi}\nu$, yields

$$\frac{\sqrt{2\pi}\nu\sigma^4}{\sigma^2 + a^2} \geq \int_{-\infty}^{\infty} [R(\beta, \hat{\beta}) - R(\beta, \delta')] \exp(-\frac{\beta^2}{2\nu^2})\, d\beta.$$

This last inequality is valid for all decision rules δ'. Now consider the effect of making a very large. The left hand side goes to zero. On the right hand side, the exponential becomes close to 1 for most values of β so that we simply have the integral of the difference in risks. Since δ' is supposed to dominate $\hat{\beta}$, the difference in risks must be strictly positive for some values of β, so the integral must converge to some strictly positive quantity.[2] This contradiction proves the admissibility of $\hat{\beta}$ in one dimension.

---

[2] To be more precise, it is necessary to argue that all risk functions must be continuous. This follows from the fact that the risk function can be expressed as a Laplace transform. Now if δ' dominates δ_{ML} then its risk function must be strictly smaller than that of the ML on some interval (a, b). From this we can conclude that the limit as ν goes to infinity must be strictly positive.

What makes the above argument work? We note two important elements. The first is the need to multiply the difference in Bayes risk of δ_{ML} and δ' by ν. If we don't do this, then the density $\pi_\nu(\beta)$ goes to zero for all β. Thus, even though $R(\beta, \hat{\beta}) - R(\beta, \delta')$ is positive, the integral of this difference with respect to the density π_a will converge to zero as a goes to infinity. The density π_ν has the constant $1/(\sqrt{2\pi}\nu)$ which drives it to zero. Multiplying by ν has the effect of preventing this, so that the density now approaches the limiting value of one as ν goes to infinity. This means that the integral of the risk difference remains strictly positive in the limit. The second aspect is that the excess Bayes risk of $\hat{\beta}$ goes to zero at the rate of $1/(1 + \nu^2)$. This rate is fast enough so that even when multiplied by ν, which goes to infinity, the excess risk still goes to zero. This is what creates the contradiction proving the admissibility of $\hat{\beta}$. This same argument fails to work to prove the admissibility of the maximum likelihood estimate of a bivariate normal mean. The bivariate normal prior density has the constant $(2\pi\nu)^{-2}$ so that we need to multiply by ν^2 to prevent it from going to zero. On the other hand, the excess Bayes risk can be calculated to be the same as before: $1/(1 + \nu^2)$. This no longer goes to zero when multiplied by a^2. Thus the proof of admissibility in two dimensions requires the use of a more complicated sequence of priors. Finally, as we now show, in three or more dimensions the estimator is inadmissible.

4.4   A Class of Minimax Estimators

We will show that the empirical Bayes estimator, given in Eq. (3.3) derived in the previous chapter, as well as a large class of other estimators are minimax in dimensions three or more. Note that if δ is minimax, it satisfies $\max_\beta R(\delta, \beta) = M = R(\delta_{ML}, \beta)$. That is, all minimax estimators must have the same risk as δ_{ML}, which has been proven to be minimax. It follows that if δ is minimax, $R(\delta, \beta) \leq R(\delta_{ML}, \beta) = M$ for all β, so that δ dominates the ML if and only if it is minimax.

An important tool in establishing the minimaxity of estimators is the following lemma. Recall that ∇_θ is a $k \times 1$ gradient vector: $\nabla_\theta = (\partial/\partial\theta_1, \partial/\partial\theta_2, \ldots, \partial/\partial\theta_k)'$. Given a vector of functions

$\mathbf{g}(\theta) = (g_1(\theta), \ldots, g_k(\theta))'$, we will write $\nabla'_\theta \mathbf{g}$ for $\sum_{i=1}^k \partial g_i(\theta)/\partial \theta_i$.

Lemma 4.1 *[Stein's Formula] Suppose $X \sim N_k(\theta, \Sigma)$ and the piecewise differentiable vector function $\mathbf{h}(x) = (h_1(x), h_2(x), \ldots, h_k(x))'$ satisfies the following hypotheses for $i = 1, 2, \ldots, k$:*

$$\lim_{|x|_i \to \infty} |\mathbf{h}(x)| \exp\left\{-\frac{1}{2}(x - \theta)'\Sigma^{-1}(x - \theta)\right\} = 0 \qquad (4.2)$$

$$\mathbb{E}|h_i(X)|^2 < \infty, \qquad (4.3)$$

$$\mathbb{E}\|\nabla_x h_i(x)\|^2 < \infty. \qquad (4.4)$$

Then we have $\mathbb{E}(X - \theta)'\mathbf{h}(X) = \mathbb{E}\nabla'_x \Sigma \mathbf{h}(x)$.

The proof is based on integration by parts; see exercise 2 for details. Use of this lemma permits easy development of a large class of minimax estimators all of which dominate the MLE $\hat{\beta} \sim N(\beta, \sigma^2(X'X)^{-1})$.

Theorem 4.3 *Given $\hat{\beta} \sim N(\beta, \sigma^2(X'X)^{-1})$ we wish to estimate β with quadratic loss $L(\delta, \beta) = (\delta - \beta)'Q(\delta - \beta)$. Consider the class of estimators $\delta(\hat{\beta})$ defined by*

$$\delta(\hat{\beta}) = \hat{\beta} - \frac{r(\|\hat{\beta}\|_M^2)\sigma^2}{\|\hat{\beta}\|_M^2}Q^{-1}X'X\hat{\beta},$$

where $\|x\|_M^2 = x'Mx$. If r is piecewise differentiable, monotone nondecreasing (so that $r' \geq 0$), satisfies $0 \leq r \leq 2k - 4$, and $M - (X'X)Q(X'X)$ is a positive definite matrix, then δ is minimax.

Proof: Define the vector function $\gamma(\hat{\beta}) = \{r(\|\hat{\beta}\|_M^2)\sigma^2/\|\hat{\beta}\|_M^2)\}Q^{-1}X'X\hat{\beta}$. The risk of the estimator $\delta(\hat{\beta}) = \hat{\beta} + \gamma(\hat{\beta})$ above can be calculated as

$$
\begin{aligned}
R(\delta, \beta) &= \mathbb{E}\|\hat{\beta} + \gamma(\hat{\beta}) - \beta\|_Q^2 \\
&= \mathbb{E}\|\hat{\beta} - \beta\|_Q^2 + 2\mathbb{E}(\hat{\beta} - \beta)'Q\gamma(X) + \mathbb{E}\gamma(\hat{\beta})'Q\gamma(\hat{\beta}) \\
&= \sigma^2 \, \mathbf{tr}((X'X)^{-1}Q) + 2\mathbb{E}\nabla_{\hat{\beta}}\left\{\sigma^2(X'X)^{-1}Q\gamma(\hat{\beta})\right\} + \mathbb{E}\gamma(\hat{\beta})'Q\gamma(\hat{\beta}) \\
&= E_1 + E_2 + E_3.
\end{aligned}
$$

We will now evaluate the three terms separately. Note that the first term E_1 is the risk of the MLE $\hat{\beta}$, which is $M = \sigma^2 \, \mathrm{tr}(X'X)^{-1}Q$, the minimax risk. We will show below that $E_2 + E_3 \leq 0$, so that δ has smaller risk than ML, and hence is also minimax. To evaluate E_2 note that $\nabla'_{\hat{\beta}}\hat{\beta} = k$, $\nabla_{\hat{\beta}}\|\hat{\beta}\|^2_M = 2M\hat{\beta}$, and $\nabla_{\hat{\beta}} r(\|\hat{\beta}\|^2_M) = 2r'(\|\hat{\beta}\|^2)M\hat{\beta}$. In the following, we will omit the argument of $r(\cdot)$, which is always $\|\hat{\beta}\|^2_M$. Thus we have

$$
\begin{aligned}
E_2 &= -2\sigma^4 \nabla_{\hat{\beta}} \frac{r}{\|\hat{\beta}\|^2_M} \hat{\beta} \\[2mm]
&= \frac{-2r\sigma^4)}{\|\hat{\beta}\|^2_M} \nabla'_{\hat{\beta}}\hat{\beta} - \left(\nabla_{\hat{\beta}} \frac{r}{\|\hat{\beta}\|^2_M} \right)' \hat{\beta} \\[2mm]
&= -4r'\sigma^4 + \frac{-2(k-2)r\sigma^4}{\|\hat{\beta}\|^2_M}.
\end{aligned}
$$

Note that since $r \geq 0$ and also $r' \geq 0$, both terms above are negative. Defining $P = (X'X)Q^{-1}(X'X)$, note that

$$
E_2 = \frac{r^2\sigma^4}{\|\hat{\beta}\|^4_M} \hat{\beta}'(X'X)Q^{-1}(X'X)\hat{\beta} \leq \frac{r^2}{\|\hat{\beta}\|^2_M} \frac{\|\hat{\beta}\|^2_P}{\sigma^4\|\hat{\beta}\|^2_M} \leq \frac{(2k-4)r}{\|\hat{\beta}\|^2_M}.
$$

To get the second inequality, use the fact that $r \leq 2k - 4$ implies $r^2 \leq (2k-4)r$ and that our assumption on M implies that $\hat{\beta}'X'XQ^{-1}X'X\hat{\beta} \leq \hat{\beta}'M\hat{\beta}$. It follows that $E_2 + E_3 \leq -4r' \leq 0$.

4.5 Stein's Paradox and Explanations

The cold mathematics of Theorem 4.3 gives no hint of paradox. Nonetheless, a special case of this result created considerable controversy when originally obtained by Stein. Consider the case where $\sigma^2(X'X)^{-1} = Q = M = \mathbf{I}$, and $r(z) = (k-2) < 2(k-2)$ for $k > 2$. According to Theorem 4.3,

$$
\delta_{JS}(\hat{\beta}) = \left(1 - \frac{k-2}{\|\hat{\beta}\|^2} \right) \hat{\beta}
$$

is easily seen to be minimax, and hence superior to $\hat{\beta}$, the MLE. We first discuss why this appears paradoxical. The basic puzzle about the Stein result is that when $\hat{\beta} \sim N(\beta, I_k)$, the components $\hat{\beta}_i$ and $\hat{\beta}_j$ are independent of each other, and the distribution of $\hat{\beta}_j$ provides no information about β_i whatsoever. Nonetheless, the James-Stein (J-S) estimator uses the value of $\hat{\beta}_j$ in estimating β_i! The J-S estimate of the i-th component of β is $\hat{\beta}_i^{JS} = (1 - (k-2)/\|\hat{\beta}\|^2)\hat{\beta}_i$, which involves all coordinates of $\hat{\beta}$, even though coordinates other than $\hat{\beta}_i$ are completely unrelated to β_i. In the debate over whether the Stein effect was a mathematical artifact or a genuine statistical phenomenon, critics pointed out early the following disturbing implication. Suppose we were interested in estimating wheat yields in Iowa, Japanese voter turnout in the next elections, and the number of sunspots to appear on a given day. We have independent data on each of the three quantities which provide us with estimates. The Stein phenomenon suggests that combining all three of these estimators will yield better estimates. However, it is contrary to common sense to use Japanese voter and sunspot data to adjust (and 'improve') estimates of wheat yield in Iowa.

To answer this objection, we must realize that it is not claimed that our estimators of each coordinate are superior. We only claim overall superiority of the estimator with respect to the loss function of sum the squared error in each component. In fact, for each coordinate, $\hat{\beta}_i$ is an admissible estimator of β_i, and thus it is impossible for the modified estimator of James-Stein to dominate. The risk of the James-Stein estimate for one coordinate is compared with the risk of $\hat{\beta}_i$ in Figure 4.1. The top curve in each graph, labeled 'Sum,' is the overall risk $\mathbb{E}\|\hat{\beta}^{JS} - \beta\|^2$ which is less than the ML risk of 3. The second curve, labeled '1,' is the risk of the first coordinate or $\mathbb{E}(\hat{\beta}_1^{JS} - \beta_2)^2$. Note that this crosses over 1, the ML risk of the first coordinate. The bottom curve, labeled '2 & 3,' is the risk of the second and third coordinates, or $\mathbb{E}(\hat{\beta}_2^{JS} - \beta_2)^2 = \mathbb{E}(\hat{\beta}_3^{JS} - \beta_3)^2$. These two coincide since $\beta_2 = \beta_3 = 0$ in all four graphs. The first coordinate β_1 is increased, making shrinkage a worse choice for the first coordinate, from $\beta_1 = 0$ to $\beta_1 = 3$ in the four graphs. When the first coordinate is poorly estimated (i.e., has risk larger than 1), the second two are well-estimated so that the sum of the risks always remains below 1.

As we would expect, the James-Stein estimator is worse for a large set of values than the ML $\hat{\beta}_1$; using Japanese voter and sunspot data does *not* improve our estimate of the wheat yield in Iowa. However, poor estimates of wheat yield are compensated for by better estimates of voters and sunspots, so that the sum of the losses comes out better for the James-Stein estimate. This is really the key to the Stein paradox: even though different coordinates are independent, everything is related via the loss function, $\sum_{i=1}^{k}(\delta_i - \beta_i)^2$. The loss function permits us to trade off losses in one component for gains in others. It is hard to imagine a situation where sum of squared errors would be a realistic loss function for the three problems (wheat, voters, sunspots) above. If three groups of people are each interested in their own subproblem, and do not feel that an improved sunspot estimator will compensate for a potentially worse forecast of wheat yields, we cannot use Stein estimates. There do exist other situations where we are only interested in the overall accuracy in a group of independent subproblems. These are the cases where Stein-type procedures are relevant.

A paper by L. D. Brown (1973) confirms the essentiality of the possibility of trading off losses in one coordinate for gains in another. Brown considers the estimation of a multinormal mean θ given an observation $Z \sim N(\theta, \mathbf{I}_k)$ with loss function $L(\theta, \delta) = \sum_{i=1}^{k} w_i(\theta_i - \delta_i)^2$. Instead of treating all coordinates equally, we assign them weights w_i. We can fix $w_1 = 1$ without loss of generality. The i-th weight then measures the relative importance of the i-th coordinate relative to the first, and also determines a rate at which we can trade off losses between coordinates. If the w_i are fixed and known, it is possible to construct an estimator which dominates the MLE in this problem. Brown considers the situation where the w_i cannot be fully specified. Suppose instead that we can only determine a range of possible values, say $\underline{w}_i < w_i < \overline{w}_i$. This creates uncertainty as to the rate at which gains in one coordinate compensate for losses in another. Brown shows that it is still possible to find an improvement over the MLE as long as $0 < \underline{w}_i$ and $\overline{w}_i < \infty$. If this condition is violated for any coordinate, we cannot determine either an upper or a lower bound to the rate at which we can trade losses in that coordinate for gains in others. For such coordinates, it does not pay to abandon the MLE, and the Stein phenomenon disappears.

The above arguments show that it is possible to have the Stein para-

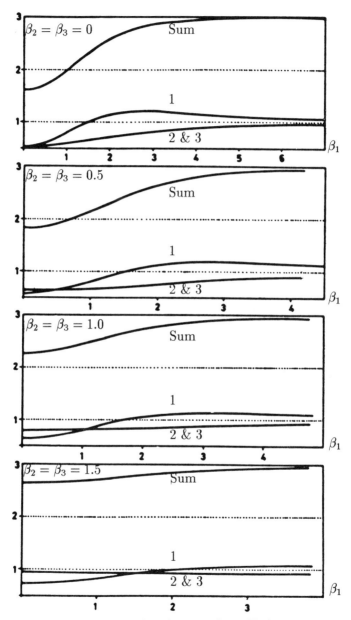

Figure 4.1: Coordinate Risks of James-Stein Estimate

dox, but do not make it plausible. It is still surprising that in three essentially independent problems, it is possible to arrange our estimation so that we achieve overall improvement over (more or less) the best estimators in the subproblems. One heuristic explanation for the good performance of the Stein estimator has already been given in the form of the empirical Bayes approach. The essence of the empirical Bayes explanation is as follows. Suppose that the three or more unknown means are exchangeable; it is then clear that each observation will provide information on the common distribution of the means and will be relevant to the estimation of each mean separately. The degree of exchangeability is actually estimated from the data, and when the means are related an improved estimate is used while when the means are unrelated the estimate is essentially equivalent to least squares. This argument will become clearer in the context of the applications to be presented in Part IV of the text.

A second heuristic argument shows the significance of dimensionality more clearly. If $Z \sim N(\theta, \mathbf{I}_k)$, the squared length of Z is, on the average,

$$\mathbb{E}\|Z\|^2 = \sum_{i=1}^{k} \mathbb{E}Z_i^2 = \sum_{i=1}^{k} (\theta_i^2 + 1) = \|\theta\|^2 + k.$$

Thus the vector Z is always longer than the vector θ, and the discrepancy between their lengths increases with the dimension k. This suggests that 'shrinking' the length of Z should result in an improved estimate, and that this should be more useful in higher dimensions. This heuristic can be taken further. Realizing that Z is 'too long', suppose we are interested in finding a constant $c < 1$ so that 'cZ' gives an improved estimate of θ. The best possible choice for c would minimize $\|\theta - cZ\|^2$; that is, cZ would be the projection of θ on the vector Z. We can easily calculate that $c = \theta^t z / z^t z$ is optimal. Of course cZ is not an estimator because c involves the unknown quantity θ. However, it is an interesting fact that

$$\mathbb{E}c = \mathbb{E}(\theta^t z / z^t z) = \mathbb{E}\{1 - (k-2)/z^t z\}.$$

Thus c is the same, on the average, as $1 - (k-2)/z^t z$; replacing c by this unbiased estimate yields precisely the James-Stein estimator of θ.

4.6   Choosing an Estimator

After Stein threw a monkey wrench into the established structure of theory for estimating the multinormal mean, over thirty years of heated research has been poured into exploring alternatives to the MLE in depth. By now considerable clarity has been achieved on the issues involved in choosing an estimate for the multinormal mean. The main factors which affect the choice of an appropriate estimator, whether the MLE or some alternative, are prior information and the loss function (or the goal of estimation).

4.6.1   Suitable Prior Information

To begin with, it appears virtually impossible to improve on MLE (despite its inadmissibility) without some sort of prior information. All estimators which dominate the MLE do so on a small subset of the typically high-dimensional parameter space. If there really is no prior information, the chances of hitting the true value by selecting a small subset of $\mathbb{R}^n$ are virtually zero, and so there is no gain from attempting to improve on the MLE. Witness to this is a small number of econometric studies which attempted to exploit the Stein estimator at a time when this phenomenon was imperfectly understood. The implicit prior for the original Stein estimator is $\beta \sim N(0, \nu^2 \mathbf{I})$. This prior makes no sense in most econometric applications.[3] First of all, the units of measurement for each regressor are typically different and are more or less arbitrary. This means that each coefficient can be arbitrarily rescaled, making the assumption of identical prior variance for each coefficient untenable. In addition, the choice of zero as the prior mean is not reasonable in most applications.

As we will show later, the Stein estimator has the following representation:

$$\delta_{JS} = \frac{1}{F}\hat{\beta} + \left(1 - \frac{1}{F}\right)\beta^+,$$

---

[3] In contrast, this and similar priors related to Stein estimates make a lot of sense in several statistical applications. This accounts for the success of Stein estimates in statistics and their relative failure in econometrics in early applications.

where β^+ is an estimator which imposes the prior constraints, and F is a test statistic for the validity of the prior constraints. Thus the Stein estimator is a weighted average of a constrained and unconstrained estimator, and the weight depends on a test statistic for the constraint. Large values of F suggest that the prior is not valid and the estimator reduces to OLS. Small values of F lead to greater weight on the constrained estimator. When the prior is not correct, the F test implicit in the Stein estimator puts very low weight on the RLS and thus reduces to the OLS. This was the finding in the early applications — that the Stein estimates were very close to the OLS estimates. This shows that in these applications the implicit prior constraints were rejected by the data. This problem can be remedied by using appropriate types of prior information, as will be discussed in much greater detail in Part IV of this text.

If prior information does exist (and we shall show in Part IV of the text that it exists in many problems in forms not commonly realized), appropriate utilization depends on many factors, principally on the objective of estimation, which can often be described by a loss function. We now discuss this relationship.

4.6.2   An Appropriate Loss Function

The second reason why the use of Stein estimators was discouraged by several authors was a misunderstanding regarding the appropriate loss function. As we have discussed in the previous section, the Stein estimator dominates the OLS when the loss function is $L_{X'X}(\hat{\beta}, \beta) = (\hat{\beta} - \beta)'(X'X)(\hat{\beta} - \beta)$, but not necessarily for usual quadratic loss which is $L_{\mathbf{I}}(\hat{\beta}, \beta) = \|\hat{\beta} - \beta\|^2$. Some importance was given to this phenomenon and conditions on $X'X$ were worked out for which the Stein estimator continues to dominate OLS with respect to usual quadratic loss. It turns out that this condition is: $2\lambda_{max}(X'X)^{-1} \leq \mathbf{tr}(X'X)^{-1}$. When $X'X$ satisfies this condition, the Stein estimator continues to dominate OLS with respect to usual quadratic loss, $L_{\mathbf{I}}$. See Greenberg and Webster (1983) for an extensive discussion of this and related issues regarding applications of Stein estimators in econometrics. Aigner and Judge (1971) evaluated several econometric models and discovered that this condition is frequently violated by matrices of regressors in typical applications. Thus, typically Stein estimators will not dominate OLS relative to usual quadratic loss in econometric application.

In fact, quadratic loss function is almost never a suitable loss function in applications, so that which of the estimators dominates with respect to quadratic loss is not relevant to the choice of estimators. Again this is due to the fact that the scale or units of measurement of the regressor are arbitrary in most applications. Suppose the loss is quadratic $(L(\hat{\beta}, \beta) = \sum_{i=1}^{K} (\hat{\beta}_i - \beta_i)^2)$. By changing the units of measurement of the regressors, we can transform the loss to any weighted loss function of the type $L(\hat{\beta}, \beta) = \sum_{i=1}^{K} w_i (\hat{\beta}_i - \beta_i)^2$. Because of this lack of invariance to units of measurement, it makes no sense to talk of quadratic loss without first specifying how the units of measurement for different regressors are to be chosen.

The central issue here is the trade off in losses among the different coordinates. As a simple example, consider a consumption function of the type $C_t = \alpha + \beta Y_t + \gamma i_t$, where the consumption C and the disposable income Y are measured in million of dollars and average around 1000, while the interest rate is measured in percentages and averages around 0.08. If the interest rate is to be of economic interest, it is necessary for γ to be at least around 1000 (in order for the interest rate to make measurable impact on consumption), while β is less than 1. In such a case it is obvious that a unit error in γ is not at all important, while an error of one unit in β is catastrophic. A quadratic loss function fails to capture the relative importance of errors in estimation for the different coefficients.

Different loss functions may be suitable for different purposes, but the implicit loss function for which the Stein estimator dominates OLS is quite reasonable for the following reason. Suppose we assess the relative importance of errors in different coefficients by the difference it makes for forecasting purposes. The optimal forecast for period $T + 1$ is $x'_{T+1}\beta$ where β is the true parameter, and x_{T+1} is the value of the regressors in period $T + 1$. We can measure the loss due to an estimate $\hat{\beta}$ by the squared difference from the optimal forecast:

$$L(\hat{\beta}, \beta) = (x'_{T+1}\beta - x'_{T+1}\hat{\beta})^2 = (\hat{\beta} - \beta)' \left\{ x_{T+1} x'_{T+1} \right\} (\hat{\beta} - \beta).$$

This loss function assigns weights to the different coordinates according to how important they are for forecasting and is automatically invariant to changes in units of measurement for different regressors. It is, however, specific to the particular value of regressors x_{T+1} (which may not be known). If x_t are i.i.d. then a good estimate for the average value of $x_{T+1}x'_{T+1}$ would be the average value of this quantity in the observed sample, namely $(1/T)\sum_{t=1}^{T} x_t x'_t = (X'X)/T$. If x_{T+1} is unknown, or if one wishes to avoid weights specific to one particular observation, then the appropriate loss function becomes $L(\hat{\beta}, \beta) = \frac{1}{T}(\hat{\beta} - \beta)'X'X(\hat{\beta} - \beta)$. This is sometimes called the 'predictive loss' function, since it assigns weights to the estimates $\hat{\beta}$ according to their ability to predict y. In decision theoretic analysis, the constant $(1/T)$ makes no difference, and for this loss function, the Stein estimator dominates OLS.

4.6.3   Conflict between EB and Minimaxity

In the special case of orthogonal regressors ($X'X = \mathbf{I}$), the Stein estimator is uniformly superior to the MLE in terms of mean square error. It is also an empirical Bayes estimator. This coincidence suggests that we may be able to incorporate prior information of certain kinds without loss; the estimator will be better if the prior information is valid, and will not be worse if it is wrong. Unfortunately, this possibility occurs only in some special cases. In typical applications, the two concepts of minimaxity and empirical Bayes lead to different estimators with different properties. This section explores the conflict between these concepts with a view to deciding which of the two is more relevant for applications.

To spell out the nature of the conflict, consider a simple minimax rule for the unequal variances case. Suppose $\hat{\alpha} \sim N(\alpha, \sigma^2\Lambda^{-1})$, where $\Lambda = \text{diag}(\lambda_1, \ldots, \lambda_k)$. Consider $\delta^* = (\delta_1^*, \ldots, \delta_k^*)$ where the coordinate estimate δ_i^* for α_i is defined by

$$\delta_i^*(\hat{\alpha}) = \left(1 - \frac{(k-2)\lambda_i\sigma^2}{\sum_{i=1}^{k}\lambda_i^2\hat{\alpha}_i^2}\right)\hat{\alpha}.$$

By Theorem 4.3, δ^* is minimax, and reduces to the James-Stein rule in the equal variances case. However, when the eigenvalues are unequal, the shrinkage factors $F_i = (k-2)\lambda_i/(\sum_{i=1}^{k}\lambda_i^2\hat{\alpha}_i^2)$ are ordered in the same way as the eigenvalues λ_i. Thus coordinates with small variances are shrunk less and coordinates with high variances are shrunk more. In contrast, Bayes and empirical Bayes rules shrink in the opposite fashion: high variance coordinates are shrunk more and low variance coordinates less. Intuitively, this conflict arises because the Stein effect depends on trading off losses in one coordinate for gains in another. In the unequal variance case, large shrinkages for high variance components can lead to potentially large losses which cannot be compensated for by smaller changes in the low variance components. This accounts for the tendency of minimax estimators to reverse the natural order of shrinkage.

More formally, Thisted (1976) has given several theorems illustrating the fact that rules which satisfy the empirical Bayes shrinkage property cannot be minimax for all variance structures. Some of his results are based on the following theorem of Morris, which establishes minimaxity for classes of estimators not covered by Theorem 4.3. For this theorem suppose $\hat{\alpha} \sim N(\alpha, \sigma^1 \Lambda^{-1})$ as before.

Theorem 4.4 *Consider the estimator $\delta_i(\hat{\alpha}) = (1 - B_i)\hat{\alpha}_i$ defined by the shrinkage factors $B_i = \{\tau_i \sigma^2 \lambda_i w_i\}/\{c + \sum_{j=1}^{k} w_j \hat{\alpha}_j^2\}$ where $c \geq 0$, $w_i \geq 0$, and $\tau_i \geq 0$ are constants. This estimator is minimax if and only if for all $i = 1, 2, \ldots, k$*

$$\frac{w_i \tau_i (4 + \tau_i)}{\lambda_i^2} \leq 2 \sum_{j=1}^{k} \frac{w_j \tau_j}{\lambda_j^2}.$$

In the case that the constant c is replaced by c_i, if the resulting estimator is minimax then the inequalities above must hold.

For a proof see Thisted (1976). We cite the theorem mainly as a tool for establishing conditions under which Stein-type estimators continue to be minimax in the case of unequal variances. It is easily established as a consequence of this result that the Stein estimator used unaltered in the unequal variance case is minimax if and only if

$$\max_{i=1,\ldots,k} \frac{1}{\lambda_i} \leq \frac{2}{k+2} \sum_{j=1}^{k} \frac{1}{\lambda_j^2}. \tag{4.5}$$

See also Berger (1985) for a more detailed study of this conflict between minimax and empirical Bayes estimation procedures. His conclusion is that the attempt to force minimaxity on an estimator will sacrifice a large part of the gains obtainable from implementing the prior information. Given the conflict between minimax and empirical Bayes in the case of unequal variances, how should we choose between the two in applications? This is discussed in the next section.

4.6.4   Compromises between EB and Minimax

An important reason why Bayesian techniques have been avoided in econometrics is that reliable prior information is frequently unavailable, and traditional Bayesian estimators perform very poorly if the prior is somewhat misspecified. One of the main virtues of Stein-type minimax estimators is that they permit the incorporation of prior information in risk-free fashion; they dominate the MLE. Empirical Bayes type techniques are typically not minimax in the case of unequal variances and can therefore be worse (sometimes substantially so) than the MLE. Sensible choice between minimax and empirical Bayes estimators requires assessment of the reliability of the prior information. For very reliable prior information, it is reasonable to ignore minimax considerations. To the extent that the prior information is unreliable, it is worthwhile to compromise between full utilization of the prior and minimaxity, or safeguarding against possible errors in the prior information. It is important to note that minimax estimators (unlike empirical Bayes) depend on the structure of the loss function. Thus, guarding against losses requires explicit knowledge of the loss structure, something which is frequently unavailable in applications. The fact that empirical Bayes (and robust Bayes, discussed below) approaches do not require specification of the loss function makes them more attractive for applications.

It is possible for empirical Bayes estimators to fail to be minimax by a large amount (i.e., to have risk much larger than OLS at certain points of the parameter space which are considered unlikely relative to the posterior distribution). If the prior information is uncertain, it is desirable to ensure that the risk of the estimator is not too high even when the prior is wrong. One of the simplest ways of achieving this goal is the 'limited translation' idea of Efron and Morris (1971, 1972).

The idea is to limit the amount by which any component is shrunk to a maximum of, for example, 2 standard deviations. This modification will in fact improve the EB procedure in terms of its minimaxity and also has strong intuitive appeal (given that the prior restrictions are of uncertain validity). Thus EB combined with limited translation has in fact been used in a number of applications. There exists no theory regarding the extent to which truncation can overcome the failure of minimaxity of the EB estimates, although the intuition for supposing this to be the case is quite strong.

There exist more sophisticated theoretical approaches to this problem. The question these approaches seek to answer is roughly the following. How can we utilize a given set of prior information while retaining minimaxity or approximate minimaxity? Berger (1985) has studied this question in considerable detail. He has shown (see Section 5.4.3 of Berger (1985)) that attempts to utilize empirical Bayes-type prior information subject to a minimaxity type constraint will usually sacrifice most of the gains available from the prior information. Nonetheless, such estimators are available and may be useful in cases where the prior information is suspect, so that it is not desirable to incur risks when it is wrong. An alternative approach to this situation has also been developed by Berger (1985). Labeled 'robust Bayes,' this approach develops estimators which incorporate the prior information while assuming that the prior may be wrong to some extent. It turns out that these estimators are frequently close to minimax, and hence cannot be substantially worse than the MLE. At the same time they permit exploitation of prior information. Another advantage of Berger's robust Bayes estimators over standard types of minimax estimators is that it is not necessary to specify a loss function. Because of the complexity of the estimators in question, they have not been widely applied. This seems a promising avenue for future research, especially since the Gibbs sampler discussed in Part IV of the text may permit implementation of otherwise complex formulas for the estimators.

4.7   Exercises

1. Verify that if $Z \sim \chi_k^2$ then $\mathbb{E}(1/Z) = 1/(k-2)$ for $k \geq 3$; what is the expected value for $k = 1, 2$?

2. (Stein's Formula: Multivariate) Show that if $Z \sim N(\theta, \sigma^2 \mathbf{I}_k)$, then
$$\mathbb{E}\sum_{i=1}^{k}(\theta_i - Z_i)f_i(Z) = -\sigma^2 \sum_{i=1}^{k} \mathbb{E}\frac{\partial}{\partial Z_i}f_i(Z).$$
   Hint: first prove Lemma 4.1.

3. Use Stein's Formula to prove that
$$\mathbb{E}\frac{\theta^t z}{z^t z} = \mathbb{E}(1 - \frac{k-2}{z^t z}).$$

4. Use Stein's Formula above to show that if $Z \sim N(\theta, \sigma^2 \mathbf{I}_k)$ and $ms^2/\sigma^2 \sim \chi_m^2$, then
$$\mathbb{E}\|\theta - \left(1 - \frac{(k-2)s^2}{Z^t Z}\right)Z\|^2 = k\sigma^2 + \mathbb{E}\frac{(k-2)^2 s^4 - 2(k-2)\sigma^2 s^2}{Z^t Z}.$$

5. (Continuation:) Calculate $\mathbb{E}[(k-2)^2 s^4 - 2(k-2)\sigma^2 s^2]$, and use to show that the MLE is and inadmissible estimator even when σ^2 is unknown.

6. Suppose $Z \sim N(\theta, \mathbf{I}_k)$, and $L(\theta, \delta) = \sum_{i=1}^{k} l_i(\theta_i - \delta_i)^2$. Consider the estimator of $\theta = (\theta_1, \ldots, \theta_k)$ given by $\delta(Z)$, where
$$\delta_i(Z) = (1 - \frac{k-2}{l_i Z^t Z})Z_i.$$
   Show that this dominates the MLE.

7. Show that if $Z \sim N(\theta, \mathbf{I}_k)$ the estimator $\delta(Z) = Z + (1 - (k - 2)/\|Z - \mu\|^2)(Z - \mu)$ dominates the MLE (where μ is a $k \times 1$ vector of constants). This estimator gives the best performance for θ near μ rather than 0.

8. Suppose different varieties of wheat seeds are planted in various plots of land. Each variety has a 'true' yield θ_i, and the observed yield is $y_i = \theta_i + \epsilon_i$. Here ϵ_i is a random error due to varying weather conditions, soil conditions etc. Assume the ϵ_i can be considered i.i.d. normal with mean zero and variance σ^2. Suppose eight different varieties are observed to yield (in 000 bushels/acre)

$$10, 14, 8, 19, 22, 7, 27, 15.$$

Give the MLEs for the θ_i and suggest improved estimators for the yields.

9. Read Efron and Morris (1975) Data analysis using Stein's estimator and its generalizations, *Journal of the American Statistical Association* **70** 311–390.

It is said that on Judgement Day, the men will be asked to form two lines. All will be in the line labeled 'Those who feared their wives' while there will be only one skinny man in the line for 'Those who did not fear their wives.' Amazed angels will ask the man how he managed this feat. He will respond 'You see, my wife said she would kill me if I joined the other line.'

Chapter 5

Robust Regression

5.1 Introduction

We have discussed maximum likelihood, Bayesian, and minimax estimators for regression models. All of these estimators have been derived under the hypothesis that the errors are multivariate normal. The *robustness* of an estimator refers to how well the estimator works under failures of this assumption. There are two major reasons why the normal model is central in statistical analysis. One is the central limit theorem (CLT), which suggests that many random variables may be well approximated by normals. The second is the more important fact that computational procedures associated with normal distributions are easy to carry out. For example, the Least Absolute Error estimator obtained by minimizing the sum of absolute residuals (instead of squared residuals) is intuitively appealing and was suggested early in the history of regression. However, minimizing the sum of squares is substantially easier both analytically and computationally, and therefore won out.

Theoretical and technical developments have changed the significance of both aspects of the centrality of normal distribution theory.

Rapid advance in computational power has made feasible exploration of large numbers of alternative estimation procedures which were impossible before. This has been important in the development of robust procedures, all of which require more computer power than least squares. An even more important spur to the development of robust procedures has been the discovery that classical procedures, which are optimal for normal distributions, can perform extremely poorly for distributions which are very close to normal. This makes the CLT justification for use of classical procedures invalid, as the CLT only produces approximate normality, while classical procedures require exact normality.

Developments in the field of robustness have been rapid and are continuing, making it impossible to give final answers. Therefore, this chapter has been patterned on the 'Neti Neti' technique of Indian philosophy, according to which truth may be reached by a sequence of 'Not this ... ' and 'Not that ... ' kind of observations. Robust techniques will be discussed in a sequence such that each technique addresses some shortcoming of earlier techniques. Due to the stylistic requirements of this technique, we will necessarily indulge in finding defects. We warn the reader that the defect may not be serious enough to warrant abandoning the technique. We also apologize in advance to the several authors who have spent substantial research efforts in one or the other of the areas which we will cover below by hit-and-run type criticisms. It should be clear that the method cannot reach conclusions unless we turn it on itself and deny the the validity of 'Neti Neti.'

Section 5.2 shows that the OLS can be heavily influenced by outliers. Section 5.3 introduces the breakdown point, a measure of how many outliers it takes to make an estimate completely unreliable. Regression estimators which are affine, scale, and regression equivariant can have breakdown points of up to $1/2$ the sample size. The OLS breaks down with just 1 out of T bad observation. Since the median has the highest possible breakdown point of $1/2$, it is natural to look to its generalization, the LAD estimator for high breakdown properties in regression models. Unfortunately, this does not turn out to be the case. Another popular approach to handling the low breakdown value of OLS is to do a sensitivity analysis of the estimate. We show that sensitivity analysis will pick up the effects of one bad observation, but may not detect any problems if there are more than one bad observations. This is known as the masking effect. Two estimators which can cope with large numbers (up to one half of the sample) of bad observations are presented in Section 5.4; these are the Least Median Squares (LMS) and the Least Trimmed Squares (LTS) estimators introduced by Rousseeuw (1984). Unfortunately, the LMS has 0% efficiency while the LTS has 50% efficiency at normal models. Since there are typically fewer then 50% outliers, an interesting suggestion is to just remove the outliers and then proceed with a standard analysis. This can lead to robust procedures in some special simple situations, but is unsuitable for regression. Apart from difficulties in identifying outliers, some outliers (the good leverage points) contribute substantially to efficiency in estimation. Their removal will substantially decrease efficiency. The techniques of bounded influence estimation work very well in simpler statistical problems to produce robust and highly efficient estimates. Unfortunately, the standard estimators used for this purpose all have very low breakdown values when adapted to regression models. Coakley and Hettmansperger (1993) recently achieved a breakthrough in robust estimation. They produced an estimator with a high breakdown value, bounded influence, and high efficiency at the normal distribution. It is likely that further developments in this area will lead to practical and useful estimation procedures. The methods of semiparametric and nonparametric estimation are also briefly discussed.

5.2 Why Not Least Squares?

This section is meant to illustrate the need for robust estimators. Cases where OLS results are thrown off by a single outlier occur with some frequency in practice. One example is given in exercise 4. With one regressor and a small data set, a plot of the data (done all too infrequently in practice) will usually reveal the problem. A case where a group of outliers distorts OLS and masks important structure in the data is discussed below.

Table 5.1 gives data, taken from Rousseeuw and Leroy (1987) on body weight and brain weight for 28 selected species. It seems plausible that a larger brain is required to control a larger body. OLS regression results for the dependent variable $y = \log(\text{brain weight})$ regressed on a

constant and $X = \log(\text{body weight})$ supporting this hypothesis are as
follows:

$$y = \begin{matrix} 1.11 \\ 0.18 \end{matrix} + \begin{matrix} 0.496 \\ 0.078 \end{matrix} X + \begin{matrix} \epsilon. \\ 0.665 \end{matrix}$$

$$R^2 = 0.61 \quad SSR = 11.50 \quad F(1,26) = 40.26$$

Table 5.1: Brains Data and Standardized Residuals

Index	Species	Body(Kg.)	Brain(g.)	OLS Res.	LMS Res.
1	Mountain Beaver	1.350	8.100	-.26	.02
2	Cow	465.000	423.000	.19	-.64
3	Gray Wolf	36.330	119.500	.19	.51
4	Goat	27.660	115.000	.23	.81
5	Guinea Pig	1.040	5.500	-.37	-.34
6	Diplodocus	11700.000	50.000	-1.42	-9.00
7	Asian Elephant	2547.000	4603.000	.86	1.46
8	Donkey	187.100	419.000	.38	.59
9	Horse	521.000	655.000	.35	.02
10	Potar monkey	10.000	115.000	.45	2.20
11	Cat	3.300	25.600	.04	.93
12	Giraffe	529.000	680.000	.37	.07
13	Gorilla	207.000	406.000	.35	.40
14	Human	62.000	1320.000	1.12	4.23
15	African Elephant	6654.000	5712.000	.75	.55
16	Triceratops	9400.000	70.000	-1.23	-8.08
17	Rhesus monkey	6.800	179.000	.73	3.55
18	Kangaroo	35.000	56.000	-.12	-.85
19	Hamster	0.120	1.000	-.65	-.54
20	Mouse	0.023	0.400	-.69	.02
21	Rabbit	2.500	12.100	-.22	-.08
22	Sheep	55.500	175.000	.26	.64
23	Jaguar	100.000	157.000	.09	-.37
24	Chimpanzee	52.160	440.000	.68	2.43
25	Brachiosaurus	87000.000	154.500	-1.37	-9.66
26	Rat	0.280	1.900	-.55	-.51
27	Mole	0.122	3.000	-.17	1.47
28	Pig	192.000	180.000	.01	-1.01

Note: Adapted from Rousseeuw and Leroy (1987). *Robust Regression and Outlier Detection.* Reprinted by permission of John Wiley & Sons, Inc.

Table 5.1 also lists the standardized OLS residuals — these are the residuals divided by the estimated standard error of ϵ, which is 0.665. All of the residuals are less than 1.5 standard deviations away from zero and one may well think that there are no outliers on the basis of the OLS residuals. However, a robust analysis done later in this chapter shows that there are three extreme outliers — these are three species of dinosaurs included in the data. A robust analysis of the data makes the residuals for these species 9 standard deviations below the regression line. Also, the human species has an extremely high positive residual (+4 standard deviations), showing an exceptionally large brain to body ratio. Could it be that the failure of OLS to identify dinosaurs as exceptional dooms the technique to share their fate?

This kind of phenomena occurs frequently, due to the extreme sensitivity of OLS to 'outliers,' or data points which deviate from the general pattern. See Rousseeuw and Leroy (1987) for several additional examples. In applied work it is frequently necessary to introduce 'special purpose' dummies, which in effect remove the deviant data from the observations. Unfortunately, it frequently takes some detective work to find the poorly fitting observations. A commonly used technique for this purpose is to look at the OLS residuals. However, the OLS residuals may not reveal the deviant observation/s. In the example above, the residuals from the OLS regression do not reveal any unusual pattern. Careful analysis of this data set and visual inspection will reveal the fact that the three dinosaur species are out of line with the others. However, this is quite difficult to do when we have high dimensional data. It is then not possible to plot the data, or otherwise somehow look at the data to find anomalies. More automatic techniques are needed. Some are discussed below.

5.3   The Breakdown Point

The concept of the *breakdown point* is a way of formalizing the difficulty with the OLS heuristically described in the previous section. This is defined in general as follows. Given observations $(y_1, \ldots, y_n)$, let $\hat{\theta}(y_1, \ldots, y_n)$ be any estimator (of any quantity) based on the data. Suppose that one of the observations, say y_1, is replaced by a corrupted value x_1. How much damage can a single bad observation do? There are numerous ways to assess this, but we take the following simple approach. If the value of $\hat{\theta}(x_1, y_2, \ldots, y_n) - \theta_0$ can be made arbitrarily large by choosing x_1 appropriately, then we say that $\hat{\theta}$ breaks down in the presence of 1 corrupt observation out of n data points. In short, we say that $\hat{\theta}$ has a breakdown point of $1/n$. On the other hand, if the value of $\hat{\theta}$ remains bounded regardless of how we choose x then we say that $\hat{\theta}$ does not break down at $1/n$. In this case, we put in one more corrupt observation, replacing y_2 by x_2. If the value of $\hat{\theta}(x_1, x_2, y_3, \ldots, y_n) - \theta_0$ can be made arbitrarily large by choosing suitable values for the corrupt observations x_1 and x_2, then $\hat{\theta}$ has the breakdown point of $2/n$. We continue replacing the genuine observations by corrupt ones until the estimator breaks down; the number of observations that need to be corrupted to make the estimator unbounded is called the breakdown value of the estimator. An example is useful to clarify this concept.

Example: Given observations $y_1, \ldots, y_n \overset{iid}{\sim} N(\theta_0, 1)$, consider the ML estimator $\hat{\theta} = (1/n) \sum_{i=1}^{n} y_i$. Obviously if we replace y_1 by a corrupt observation x and drive x to ∞, $\hat{\theta}$ will also go to infinity. This means that one observation is enough to break it down. Consider a slightly modified estimator, sometimes used in applied situations, where we omit the highest and smallest observation before taking the mean. In this case, one corrupt observation will not break down the 'trimmed' mean. As soon as the value of x becomes an extreme value, it will not be averaged in and therefore the trimmed mean will remain bounded. On the other hand, if we put in two corrupt observations the trimmed mean will break down. Note that if the two corrupt observations are on the opposite extremes then the trimmed mean will ignore both of them. However, the concept of the breakdown value requires us to assess the worst damage that can be done by the corrupt data. Obviously the worst case here occurs when both corrupt values are on the same extreme. In this case the trimmed mean will ignore the higher one of the two corrupt values but it will put the second one in the average. By making both of these corrupt values very large, we can make the trimmed mean arbitrarily large.

In contrast to the mean, the median $\tilde{\theta}(y_1, \ldots, y_n)$ has the high breakdown value of about $1/2$. Consider the case of odd n. Suppose we take $(n-1)/2$ observations and put them all at $+\infty$ (or close to it). The median will then be the largest point of the remaining $(n+1)/2$ observations. It follows that the median cannot be made arbitrarily large even if we corrupt slightly less than half of the data. Intuitively it should be clear that reasonable estimators cannot have breakdown values more than $1/2$. If more than half the data is corrupt, and the corrupt portion of the sample is made to imitate a perfect mock sample somewhere out in the neighborhood of infinity, any *reasonable* estimator should base itself on the mock sample and thereby be deceived. A formal version of this argument is given later.

Suppose we decide to arbitrarily truncate the mean so as to ensure that it lies between -1 and 1. If we don't know the value of the parameter is between these bounds than this estimator is ridiculous. It nonetheless has a very high breakdown value, namely 100 %. Certain natural restrictions can be placed on the class of estimators that rule out such unreasonable behavior. We now describe these restrictions.

5.3.1 Regression Equivariance

Given a regression model $y = X\beta + \epsilon$, consider the transformation $y' = y + X\beta_0$. It is easily seen that $y' = X(\beta + \beta_0) + \epsilon$. Thus the observation of y' is equivalent to the effect of changing the parameters from β to $\beta + \beta_0$. If an estimator based on the observations (y, X) estimates β to be $\hat{\beta}$, then based on y', X it should estimate β to be $\hat{\beta} + \beta_0$. An estimator $\delta(y, X)$ is called *regression equivariant* if it satisfies this condition:

$$\delta(y + X\beta_0, X) = \delta(y, X) + \beta_0.$$

These definitions rule out silly estimators with 100 % breakdown values of the kind considered above. It can be shown that if an estimator is regression equivariant, then its breakdown point must be less than $1/2$. A special case of regression is the location parameter estimation problem discussed earlier. If we let X be the column of ones, the regression model can be written as $y_t = \beta + \epsilon_t$, or equivalently $y_t \overset{iid}{\sim} N(\beta, \sigma^2)$. If we require an estimator of β to be regression equivariant (which is called translation equivariance in this special case), it can be shown that such estimators must have breakdown points less than $1/2$. It therefore follows that the median has the highest possible breakdown point in the class of location invariant estimators.

5.4   Why Not Least Absolute Error?

The median is an extremely robust estimate of the normal mean, while the mean is extremely nonrobust. The mean is obtained by finding the value of θ which minimizes the sum of squared residuals $\sum_{j=1}^{n}(y_j - \theta)^2$. On the other hand the median is obtained by minimizing the sum of absolute deviations: $\tilde{\theta}$ minimizes $\sum_{j=1}^{n}|y_j - \theta|$. See exercise 2 for a proof of this. This leads to a natural suggestion for a 'robust' estimator for regression models. The OLS $\hat{\beta}$ is obtained by minimizing $\sum_{t=1}^{t}(y_t - x_t'\beta)^2$. We can define an analog to the median to be the value $\tilde{\beta}$ which minimizes $\sum_{t=1}^{T}|y_t - x_t'\beta|$. We will call this the LAE (least absolute error) estimator.[1]

Surprisingly, this estimator *does not* have the same robustness property as the median does in one-dimensional situations. When the values of the regressors contain no outliers, the LAE does resist outliers. Exercise 4 gives data on annual fire claims made on insurance companies in Belgium from 1976 to 1980. Figure 5.1 (A) plots the LAE fit to the data for four different values of the 1976 data point. The four lines plotted are the total extent of the change — moving the 1976 outlier arbitrarily far up or down produces no further change. Evidently the LAE is robust to outliers, at least on this data set. This is because there are no x-outliers; the data are sequential from 1976 to 1980. If we move the datum at 1976 to 1975, the year becomes an x-outlier. Figure 5.1 (B) shows that the behavior of LAE changes dramatically. Far from resisting the outlier, LAE now follows it slavishly.

---

[1]This estimator somehow inspires affection; consequently, it has acquired a large number of nicknames, or acronyms. It is known by the colorful name MAD (for minimum absolute deviations), the youthful LAD (least absolute deviations), and the hot and smooth LAVA (least absolute value). Other acronyms are LAR (least absolute residual) and LAV. The acronym MAD is on its way to becoming a standard abbreviation for a type of robust scale estimate, and therefore should probably be avoided here.

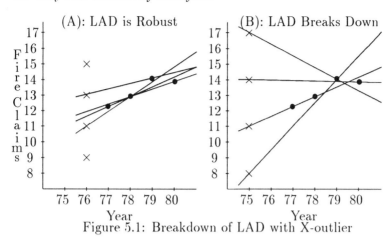

Figure 5.1: Breakdown of LAD with X-outlier

An important sense in which LAE is superior to the OLS is its behavior in the presence of error distributions without moments. It can be shown that as long as the median of the errors is zero, even if the mean and the variance does not exist, the LAE will be consistent, while the OLS fails to be consistent in this situation. Nonetheless, in the presence of x outliers (discussed in greater detail in Section 5.8), the LAE breaks down with just one corrupt value. Furthermore, as illustrated by Figure 5.1, only a slightly outlying value of x is needed to break down the LAE estimate.

5.5 Why Not Sensitivity Analysis?

Given that OLS is extremely sensitive to outliers and abberant data, a natural way to proceed is to assess this sensitivity. Several books and articles have been devoted to this topic; see, for example, Belsley *et al.* (1980). The main idea is to delete one or several observations and study the impact on various aspects of the regression. When an observation is deleted, all of the regression statistics change. Furthermore, there are T observations. Measuring the impact of deletion of each observation on each of the outputs necessarily generates a lot of output. Numerous measures have been proposed to assess the impact of dropping an observation on various regression statistics; many such measures are automatically available as outputs of regression packages. Unfortunately, the interpretation of these measures, and what to do with them if some of them turn out significant remains far from clear. Furthermore, as we will shortly illustrate by an example, due to the presence of the masking effect, these measures cannot achieve their intended objective.

The intended objective of sensitivity analysis is to assess whether the OLS regression results are seriously affected (or distorted) by the presence of a small group of observations; such observations may be termed *influential*. While the sensitivity measures taken in combination can, in hands of experts, achieve this objective, there now exist simple ways which can routinely identify the extent to which OLS is affected by outlying subgroups of observations. These include the class of high breakdown regression estimators, including LMS and LTS, both introduced by Rousseeuw (1984). Application of these techniques immediately reveals the presence of subgroups of observation which differ substantially from, or exert undue influence on the regression results. In contrast, in presence of masking, it requires some detective work to arrive at the same place using the sensitivity analysis techniques.

Consider the brains data presented in Table 5.1. Because there are three dinosaur species, dropping any one observation has little effect on the regression results. Dropping all three at the same time does have a big effect, but this is hard to discover using the usual sensitivity analysis tools. This data set passes all sensitivity tests in the sense that omitting any one of the observations has small effect on the results. Sensitivity analysis fails to reveal the large negative residuals of the dinosaurs, and also fails to reveal the large positive residuals for the three species of monkeys, as well as humans, in this data set. One is tempted to say that the brains data make a monkey out of sensitivity analysis of OLS.

5.6   Why Not Least Median Squares?

As we shall see, least median squares (and related high breakdown regression techniques) have several properties which make them very attractive for preliminary data analysis. In particular, if results from an LMS analysis are similar to OLS, we can safely conclude that no small subgroup of the data is causing undue distortion of the OLS results. Nonetheless, the relatively low efficiency of these techniques in the case of normal errors makes it difficult to recommend them for final data analysis.

The problem with OLS is that it fails if any one of the observations is bad. Sensitivity analysis can cope with one bad observation, but not with two bad observations in the same place. This is called the 'masking' effect; each bad observation masks the effects of dropping the other. We obviously need techniques which can cope with larger numbers of bad observations. The median works in this way in the location problem, but its obvious generalization, the LAE, has zero breakdown value. More sophisticated generalizations of the median, the LMS and LTS devised by Rousseeuw (1984), are the first regression equivariant estimators which achieve 50 % breakdown value. These estimators are described below.

For any β, let $e_t^2(\beta) = (y_t - x_t'\beta)^2$ be the squared residual at the t-th observation. Rearrange these squares residuals in increasing order by defining $r_{(1)}^2(\beta)$ to be the smallest residual, $r_{(2)}^2(\beta)$ to be the second smallest, and so on ($r_{(h)}^2(\beta)$ is called the h-th order statistic of the residuals). The LMS estimator is defined to be the value of β for which the median of the squared residuals is the smallest possible. The LTS estimator is defined to be the one minimizing $\sum_{t=1}^{T/2} r_{(t)}^2$; that is, it ignores all residuals above the median in forming the sum of squares to minimize. Proofs that both of these estimators have 50 % breakdown value and further details are available in Rousseeuw and Leroy (1987). Intuitively, this seems very plausible since if less than half the data is corrupted, the residuals corresponding to the uncorrupted data will remain fixed and will determine the median residual, as well as all residuals below it.

5.6.1 Computational Details

The finite sample breakdown value is slightly less than 50 % and converges to an asymptotic value of $1/2$. The finite sample breakdown value increases slightly if instead of minimizing the median of the squared residuals, we minimize the h-th order statistic, where

$h = [T/2] + [(K + 1)/2]$, where T is the sample size, K is the number of regressors, and $[x]$ is the largest integer smaller than x. Similarly, the finite sample breakdown value of the LTS is increased if we minimize $\sum_{i=1}^{h} r_{(h)}^2$, where $h = [T/2] + [(K + 1)/2]$.

Computation of LMS, LTS, as well as other robust estimators and quantities is difficult and the following simple algorithm, related to the bootstrap, is of some value. In a regression with K regressors and a sample of size T, any subsample of size K is called an *elemental subset*. Randomly select an elemental subset, and let $\hat{\beta}$ be the coefficients of the unique line which fits exactly (elemental subsets which are not in general position are discarded). Calculate the value of the criterion function (i.e., LMS, LTS, or others to be considered later) for this $\hat{\beta}$. Repeat the process with other random samples, keeping track of the minimum achieved. At the end of a fixed number of samples (determined by computational cost), stop and declare the best achieved minimum to be an approximation to the answer. A PC-based program implementing this algorithm for LMS is available from Rousseeuw.

We now describe Hawkins (1994) algorithm to compute the LTS exactly — similar algorithms are available for other robust quantities such as the LMS. Let S be the set of all subsamples of size h of original sample of size T. For $s \in S$ let $\beta(s)$ and $SSR(s)$ be the least squares estimate and sum of squared residuals for the subsample. The LTS minimizes $SSR(s)$ over $s \in S$ (by definition). The optimal subset s^* must satisfy the following condition: if one element inside the set s is swapped with an element outside the set, the criteria must increase. This is a necessary condition for a local maximum. Now start with a randomly selected subset $s \in S$ and test it by swapping each inside element with an outside element. Any time the criterion is reduced, make the swap and start over. Eventually, since the criterion is reduced at each step, we will reach a set where no further swap produces any reduction. This set is a candidate for the LTS fit. Hawkins (1994) recommends doing many random starts and following the swapping procedure to its minimum. He shows that with a sufficiently high number of starts, we will with high probability hit the global minimum of the criterion. A cheaper option is to use the elemental subsets algorithm to select a good starting point and follow it to minimum.

5.6.2 Applications

The proof of the pudding is in the eating. We study the performance of the LMS in the context of the brains data of Table 5.1. The LMS provides a high breakdown estimator $\hat{\beta}^{LMS}$ but a statistical analysis requires more, analogous to standard errors, t and F statistics, etc. Based on an empirical study, Rousseeuw recommends the following estimate for the scale[2] of the errors:

$$s^0 = 1.4826(1 + \frac{5}{T-K})\sqrt{\operatorname{med}_i r_i^2(\hat{\beta}^{LMS})}. \qquad (5.1)$$

An analog of the R^2 can also be defined. It is also possible, though computationally demanding, to assess the variability of coefficient estimates by bootstrap.

An alternative quick method of getting out the standard statistics is to standardize the residuals from a robust fit, remove observations corresponding to large residuals (2.5 standard deviations is the cutoff recommended by Rousseeuw), and then run OLS on the remaining data. This provides us with all the usual statistics, although the effect of the first round trimming step is not taken into account. Nonetheless, if the data are truly normal the effect should be small, and it should be beneficial if they are not.

Figure 5.2 shows the LMS and OLS fits in the regression of log brain weight on a constant and log body weight. The standardized LMS residuals from the fit (i.e. residuals divided by the scale estimate given above) are listed in Table 5.1. The results clearly show the superiority of human brains, especially those who use LMS. The residuals for the Dinosaurs stand out far below the norm, providing a clear explanation for the cause of their extinction.

An application of the LMS to economic data is given in Rousseeuw and Wagner (1994). They analyze earnings functions, relating earnings to human capital, on various international data sets. Not surprisingly, these data sets have some outliers in the form of Ph.D.'s who are unemployed, or earning little, as well as uneducated people making a great

---

[2]The variance of a random variable X is a non-robust measure of the 'scale' or the size of fluctuations of X. Very small changes in the density, which may have no observable implications, can drastically alter the variance. In robustness studies, it is essential to use less sensitive measures of scale.

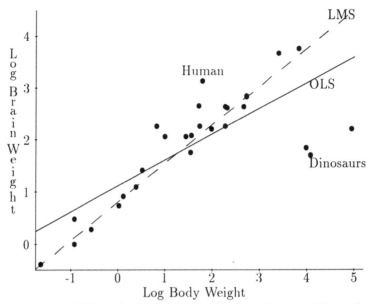

Figure 5.2:   LMS and OLS Fit to Brains Data: Adapted from
Rousseeuw and Leroy (1987). *Robust Regression and Outlier Detection*. Reprinted by permission of John Wiley & Sons, Inc.

deal. Even though these outliers are a small proportion of the sample,
the OLS results are distorted by their presence. Reanalysis by LMS
yields different conclusions. It seems clear that the LMS is a valuable tool for identifying the central patterns of the data. Because the
LMS finds a good fit for a majority of the data, rather than finding a
mediocre fit for all of the data, the contrasts between good and bad
fitting points are brought much more sharply into focus. This idea can
be clarified by considering the 'Exact Fit Property' of LMS.

Theorem 5.1 *Given data points (y_t, x_t) for $t = 1, 2, \ldots, T$, suppose
that a subset p of the points fits a linear relationship exactly: $y_t = x_t'\beta_0$
for some β_0 for these p points. If $P \geq (T+1)/2$, then the LMS estimate
is β_0.*

The proof is almost trivial since β_0 generates 0 residuals for more
than half the data points so that the median of the absolute residuals

is 0, which cannot be beat by any alternative. This has the implication that regularities in the majority of the data which may be hidden by large numbers of nonconforming points will nonetheless be picked up by the LMS. For an illuminating discussion of the exact fit property, see Yohai and Zamar (1988).

Since strong endorsement of the LMS would be contrary to the philosophy of this chapter, we discuss the problems with the LMS. Because the LMS bases the fit on a relatively small portion of the data and utilizes it relatively inefficiently, it may miss important clues to the fit provided by high leverage points; this issue is discussed in greater detail in Section 5.8. The main problem is that the LMS estimator β^* has the 'wrong' rate of convergence. While usual estimators have variances converging to zero at the rate of T^{-1}, the variance of the LMS converges to zero at rate $T^{-2/3}$. The implication is that the efficiency of LMS relative to typical estimators is zero. An alternative superior to the LMS in this respect is LTS. Here, instead of minimizing the median of the absolute residuals, we minimize the sum of squares of the $(T+1)/2$ smallest residuals. This also turns out to have a breakdown value of $1/2$. To consider the efficiency of this estimator, suppose that the data are perfectly normal and our concern with robustness is unjustified. Then the LTS unnecessarily discards half of the data and therefore has efficiency $1/2$ relative to the best estimator using all the data. This is a great improvement over the zero efficiency of LMS (corresponding to throwing away a percentage converging to 100 % of the data asymptotically). Nonetheless, it is still relatively poor compared to what is achievable. One can get higher efficiency from the LTS by trimming less of the data but this will be at the cost of reducing the breakdown value. Such a trade-off may be acceptable in some situations. We will later discuss more complex procedures which achieve high efficiency and also high breakdown values.

5.7 Why Not Just Remove Outliers?

The thrust of the previous section has been that the LMS and LTS are good methods for preliminary data analysis, mainly in identifying outliers. One suggestion for robustness consists of the simple idea that somehow we should identify the outliers and remove them and then proceed with standard analysis. This suggestion has the merit of simplicity in that once outlier detection is accomplished, there is no further need for robustness theory. Of course, as we have seen, the detection of outliers is a nontrivial problem in high dimensions (one example of an apparent outlier which turns out not to be one is given in the chapter story). Furthermore, we will typically have three types of cases: clear outliers, clear nonoutliers, and in-between. Handling of in-between cases is not clear in the "reject outliers" theory and this can make a big difference in any particular analysis.

There are more fundamental problems with the simple outlier rejection approach to robustness, especially in the context of regression. Outliers may contain very precious information. One point far away from the rest gives very accurate indications of the slope of the line. If such a point confirms the tendency of the rest of the data, then it will substantially increase the accuracy of the estimates. Such a point is called a *good leverage point* (details in following section). If, on the other hand, the point goes against the central tendency then it is essential to think through several possibilities. One possibility, namely that the data point is in error, is the one which suggests outlier rejection. A second possibility is that the model is in error. For example, it may be linear over the range of the major subgroup but quadratic over the wider range between the outlier and the remaining data. A third possibility is that both the model and the data are correct. The conflict between the estimates with and without the outlier arises due to the small amount of precision of the estimate based on the major group of data — in other words, the outlier is right and the majority of the data points the wrong way. One cannot decide between these without knowing much more about both the model and the data. Thus it is very useful to have techniques like LMS, LTS, and others to be discussed which highlight the outliers, but decisions on what is to be done cannot be made in such an automated fashion. The phenomena discussed here qualitatively can be quantified as discussed in the following section.

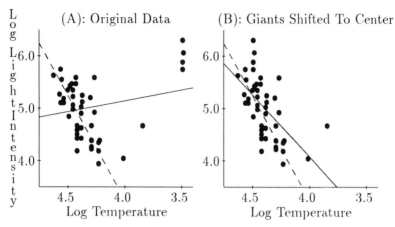

Figure 5.3: The Leverage of Giant Stars: Adapted from Rousseeuw and Yohai (1984) with permission of Springer-Verlag

5.8   Good and Bad Leverage Points

Intuitively, a leverage point can be defined as an observation which exerts disproportionally large influence on the OLS regression results. In general, values of regressors which are far from the crowd have this effect. A simple example is given in Figure 5.3. The log of light intensity is plotted against the temperature for a cluster of 47 stars near Cygnus. The four points at the top right hand corner of Figure 5.3 (A) are exceptionally cool[3] and bright — characteristics of giant stars. These giant stars form leverage points, since they correspond to extreme values of the regressor, which is the temperature. As expected, the OLS (solid line) gravitates towards the leverage points, forsaking the linear pattern of the majority of the stars. The dashed line of the LMS fit ignores the giants and fits the main sequence. To illustrate the effect of leverage, the four giant stars are shifted in (or heated up) by exactly one unit in Figure 5.3 (B). Since they are still the brightest, they can be distinguished at the top of the star cluster. After the

---

[3]This type of a diagram is called a Hertzsprung-Russell diagram in astronomy, and the temperature increases from right to left.

shift, the temperature of the giants is no longer an x-outlier, and their influence in determining the OLS fit is substantially reduced.

In this example we have a cluster of leverage points which many types of analyses will fail to detect. Just one leverage point can completely change the direction of the OLS fit. It is important to note that leverage measures a potential effect — values of y corresponding to an x-outlier exercise a significant effect on the regression coefficients. Leverage points can be classified as 'good' or 'bad' according to whether they confirm the general tendency of the remaining data or otherwise. Since a good leverage point is in conformity with the data, inclusion does not affect the estimates very much. However, the variance of the estimates can be dramatically reduced since a point some distance away on the same line as the central data provides fairly strong confirmation of the regression results. In contrast, inclusion of a 'bad' leverage point typically yields a large bias, since the estimates change dramatically. For a proper understanding of the effects of outliers, it is important to be able to classify the data points according to their leverage and also their conformity with the main tendencies of the data.

In the literature on sensitivity analysis of the OLS a measure (indeed the definition) of leverage of the t-th observation is taken to be the (t,t) diagonal entry of the projection matrix $X(X'X)^{-1}X'$. Roughly speaking, this measures the distance of the observation from a central value of all the x's, and hence seems to capture our intuitive idea (see exercise 5). Unfortunately, this measure is also subject to the masking effect so that the presence of a small group of outliers may cause the corresponding diagonal entries to be small. Both in the brains example and in the star cluster example, the outlying values, which are high leverage points, are not picked by this measure because of the masking effect. Furthermore, the mean of the regressors is not a good indicator of the central observations in the presence of outliers. Less important, it is a bit difficult to decide on what is large and small. For these reasons, we now describe an alternative proposal of Rousseeuw and van Zomeren (1990) for measuring leverage.

First, an appealing definition of center is the Minimum Volume Ellipsoid (MVE). Considering the data $x_1, x_2, \ldots, x_T$ as T points in k-dimensional space, the MVE is the smallest (in volume) ellipsoid containing half of the data points. Any positive semidefinite matrix

C can be used to define an ellipsoid centered at some $x_0 \in \mathbb{R}^k$ by the formula:

$$E(C, x_0) = \{x \in \mathbb{R}^k : (x - x_0)'C^{-1}(x - x_0) \leq 1\}.$$

The volume of $E(C, x_0, a)$ can be shown to be proportional to $\sqrt{\det(C)}$. Thus the MVE is defined by choosing C, x_0 to minimize $\sqrt{\det(C)}$ subject to the constraint that the set $E(C, x_0)$ should contain at least half of the observations. The vector x_0 is a robust estimate for a center of the observations and C is a robust estimate for the covariance matrix. With these in hand, the distance from the center for any observation x_j is naturally defined to be $d^2(x_j, x_0) = (x_j - x_0)'C^{-1}(x_j - x_0)$. If x is multivariate normal with mean x_0 and covariance matrix C, then it is easily seen that $d^2(x_j, x_0) \sim \chi_k^2$. Thus high leverage points can be determined by looking at distances which are large in comparison with the critical points of the χ_k^2 distribution.

Use of the MVE permits us to decide which of the regressor values are far from the central observations, and hence points of high leverage. In addition, we need to know which of these are good leverage points and which are bad. An effective way to decide on this is to use the residuals from a high breakdown regression estimator. Rousseeuw and van Zomeren (1990) suggest the use of residuals from the LMS but it seems preferable to use the LTS in this context; it has the same high breakdown point but is more efficient. Atkinson (1986) showed that the LMS may produce a high residual at a good leverage point. This is because small fluctuations in the regression coefficients have a large impact on residuals at high leverage points. Thus the Atkinson effect may be reduced by increased efficiency, leading to less fluctuations in the estimator.

The plot of robust leverage versus robust residuals introduced by Rousseeuw and van Zomeren (1990) appears to be a valuable tool for data analysis, but has not yet been sufficiently tested on real data. This plot permits us to identify good and bad leverage points readily. High leverage points with high robust residuals are the 'bad' leverage points, and must be eliminated for any sensible analysis of the data set. In the presence of such points, the OLS fit is heavily influenced by a very small amount of the data, and hence is unreliable. Robust methods automatically ignore such points. Some routes to robustness involve reducing the influence of leverage points. Since good leverage points carry a lot of information, indiscriminate downweighting of leverage points results in inefficiency. The twin goals of high efficiency and high robustness require a discriminating between good and bad leverage points, as is done by the Coakley-Hettmannsperger estimator discussed in Section 5.10.

5.9   Why Not Bounded Influence?

Bounded Influence estimation has been an important approach to robustness in statistics. Roughly speaking, the influence function measures the impact of changing the assumed distributions slightly. An estimator has a bounded influence function if an ϵ change in the assumed data density (typically normal) produces at most a $M\epsilon$ change in the estimator; M is the bound on the influence of a small change. For example the OLS in the regression model is the optimal estimator for normal errors. However it does not have a bounded influence function – this means that there exist densities which differ very slightly from the normal but cause large changes in the performance of the OLS. The goal of bounded influence estimation is to produce estimators which optimize various criterion subject to some bound on the influence function. Heuristically, such estimators behave well not only exactly at the assumed model (like OLS) but also at densities close to the assumed density. Details of the approach can be obtained from Hampel *et al.* (1986); their book provides an excellent and eminently readable account not only of the bounded influence approach but also of competitors. Unfortunately, as the authors indicate, the classes of estimators considered have very high breakdown values in the context of regression models. For example, Krasker and Welsch (1982) develop an optimal bounded influence estimator for regression models. However, on a real data set with several outliers, the Krasker-Welsch estimator is almost identical with OLS, but high breakdown estimators correctly identify the outliers and provide an accurate fit to the central data; see, for example, Figure 2 of Yohai (1987).

Apart from low breakdown values, Morgenthaler and Tukey (1991) criticize the bounded influence approach to robustness on other accounts. They suggest that in small samples we cannot determine with any accuracy the approximate distribution of the error terms. Thus achieving good behavior in a neighborhood of the normal (or any other assumed central model) is not a relevant goal. We should try for good behavior regardless of the underlying density. In large samples, we may be able to determine the error distribution fairly closely. However, it is likely to be the case that this data-determined distribution will not be close to any of the standard distributions. In either case, bounded influence estimation appears fundamentally wrong. Instead, Morgenthaler and Tukey (1991) offer a new approach labeled 'Configural Polysampling.' Although the details are intricate, the main idea is quite simple. Classical techniques give optimal estimators under the assumption that errors are normal. Bounded influence techniques allow the errors to be slightly nonnormal. The goal of polysampling is to optimize for a finite number of distinct error densities. Due to computational difficulties, the approach boils down to finding estimators which are *biefficient* – that is, simultaneously efficient at the normal and also at a distribution labeled the Slash (a normal divided by a Uniform (0,1)). The idea is that if estimators are good at these two very different densities then they will also be good everywhere in between. Some Monte-Carlo evidence is presented to support this claim.

Currently there is no theoretical basis to support any optimality claims for configural polysampling. The goal of achieving good performance independent of the exact error distribution is certainly laudable, but the extent to which it is achieved by the suggested procedures is unknown (though there are some promising Monte Carlo performance reports). Furthermore, the procedures are largely untested on applied data sets. A third strike against these procedures is that they are tremendously burdensome computationally, especially for regression models.

5.10   Combining Different Goals

One of the goals of estimation is to find estimators which perform well when the errors are normal. For regression models, OLS is the winner in this regard. Unfortunately, even one bad observation can drastically alter the situation. Thus a second goal is to find procedures which work well in the presence of large numbers of deviant observations. The high breakdown procedures LMS and LTS fulfill this goal by achieving the largest possible breakdown value of 50%. Unfortunately, these have very low efficiency when the errors are really normal. Yohai and Zamar (1988) succeeded in finding an estimator which had both high efficiency when the errors were normal and also had 50 % breakdown value. Unfortunately, their estimator does not have a bounded influence function. Since then, the race has been on to find an estimator which simultaneously achieves the three goals of (a) high breakdown point, (b) high efficiency at the normal distribution, and (c) bounded influence function. Estimators fulfilling any two of the three objectives were found, but the triple crown eluded the grasp of researchers until recently, when it was taken by Coakley and Hettmansperger (1993). We describe their estimator in the present section.

Consider the standard regression model $y = X\beta + \epsilon$. For later use let x_t denote the $1 \times k$ row vector which is the t-th row of x, and let $e_t(\beta) = y_t - x_t\beta$ be the t-th residual. Define $e(\beta^*) = y - X\beta^*$ to be the vector of residuals associated with any estimate β^* of β. Note that OLS is obtained by solving the equations $X'e(\hat{\beta}) = 0$. In order to obtain the desired estimates, we have to modify this estimating equation somewhat. The first step is to attach weights to the regressors. Let W be a diagonal matrix of weights, with $W_{tt} = w(x_t)$ being the weight attached to the t-th observation on the regressors. We will choose these weights so that bad leverage points get small weight. Furthermore, we need to replace residuals by a function of residuals which decreases the weights attached to the high residuals. Let $\psi : \mathbb{R} \to \mathbb{R}$ be a bounded, monotonic increasing function satisfying $\psi(x) = -\psi(-x)$. We would like to replace the residuals $e_t(\beta)$ by $\psi(e_t(\beta))$. Because ψ is bounded, the high residuals will automatically be reduced. However, it is necessary to measure the residuals on a common scale. This involves dividing the residuals by $\hat{\sigma}$, an estimate of the variance, and also by W_{tt}, the weight being attached to the t-th observation. Define $\Psi(\beta)$ to be the $T \times 1$ vector of adjusted residuals with $\Psi_t(\beta) = \psi(e_t(\beta)/(\sigma W_{tt}))$. Then the modified estimating equation can be written as

$$0 = X'W\Psi(\beta). \tag{5.2}$$

A standard strategy to solve nonlinear implicit equations like Eq. (5.2) is to linearize by making a Taylor expansion. Let $\psi'(x) = \partial\psi(x)/\partial x$ be the first derivative of ψ. Note that

$$\partial\psi(e_t(\beta)/(\sigma W_{tt}))/\partial\beta = \psi'(e_t(\beta)/(\sigma W_{tt}))x_t'/(\sigma W_{tt}).$$

Defining $\Psi_1(\beta)$ to be the $T \times T$ diagonal matrix with (t,t) entry $\psi'(e_t(\beta)/(\sigma W_{tt}))$, we can write the derivative in matrix form as follows:

$$\frac{\partial}{\partial\beta}\Psi(\beta) = \sigma^{-1}W^{-1}\Psi_1(\beta)X.$$

This is a $T \times k$ matrix with j-th column being the derivative of $\psi(e_j(\beta)/(\sigma W_{jj}))$ with respect to the vector β. Given an initial estimate $\hat{\beta}_0$ of β, we can expand Eq. (5.2) in a Taylor series around $\hat{\beta}_0$ as follows:

$$0 = X'W\left(\Psi(\hat{\beta}_0) + \sigma^{-1}W^{-1}\Psi_1(\hat{\beta}_0)X(\beta - \beta_0)\right). \tag{5.3}$$

The solution $\hat{\beta}_1$ of the above linearized equation should be an improvement over the initial guess $\hat{\beta}_0$. By iterating this procedure, it is possible to get an exact solution (under certain conditions). This is known as the Newton-Raphson procedure, and each step of the procedure is known as a Newton-Raphson iteration or step. The recommended procedure of Coakley and Hettmansperger involves starting with a high breakdown estimator $\hat{\beta}_0$ and taking just one Newton-Raphson step towards the solution of Eq. (5.2). Thus their estimator is given by the solution of Eq. (5.3), which is just

$$\hat{\beta}_1 = \hat{\beta}_0 + \sigma(X'\Psi_1(\hat{\beta}_0)X)^{-1}X'W\Psi(\hat{\beta}_0).$$

Note that the unknown σ must be replaced by a suitable estimator $\hat{\sigma}$.

The properties of the Coakley-Hettmansperger (C-H) estimator depend on the initial estimator $\hat{\beta}_0$, $\hat{\sigma}$, the chosen ψ function, and the

chosen weight function $w(x_t)$. In particular, if $\hat{\beta}_0$ is affine, scale, and
regression equivariant, then so is $\hat{\beta}_1$. If the weights $w(x_t)$ are chosen
appropriately, then $\hat{\beta}_1$ inherits the breakdown properties of the initial
estimates $\hat{\beta}_0$ and $\hat{\sigma}$. In particular if these estimates have breakdown
point close to $1/2$ then so does $\hat{\beta}_1$. In order for $\hat{\beta}_1$ to achieve $\sqrt{T}$ convergence rates, which is essential for efficiency, we need $\hat{\beta}_0$ to be a $\sqrt{T}$
consistent estimator. This rules out LMS, but the LTS is eminently
suitable for this purpose.

We also need a high breakdown estimator $\hat{\sigma}$ of σ. A scale estimate
associated with LTS residuals may be suitable but this case has not
been studied as much as the one of the LMS residuals. For this reason,
C-H propose to use the high breakdown scale estimate associate with
LMS define as s^0 in Eq. 5.1.

To specify the C-H estimator, we also need to choose the weight
function w and the function ψ. Any functions satisfying certain assumptions listed by Coakley and Hettmansperger will work. We give
here a particular choice satisfying the assumptions. Let x_0 and C be the
center and the covariance matrix for either the Minimum Volume ellipsoid, or the Minimum Covariance Determinant. Then weights defined
by

$$w(x_t) = \min \left\{ 1, b \left((x_t - x_0)' C^{-1} (x_t - x_0) \right)^{-a/2} \right\},$$

where $a, b > 0$ are arbitrary constants, will work. One simple choice
of $\psi(x)$ is $\psi_c(x) = x$ for $|x| \le c$, $\psi_c(x) = -c$ if $x < -c$ and $\psi_c(x) = c$
for $x > c$. Another choice which works is the CDF of the logistic
distribution. Suggested values for the constants a, b, and c are $a = b = 1$
and $c = 1.345$. These definitions completely specify the estimator.

Coakley and Hettmansperger show that the weights attached to the
regressors (which is $w(x_t)\psi(r_t/\sigma w(x_t))\sigma/r(x_t)$) is high for good leverage
points, but low for bad leverage points. Earlier robust estimators did
not achieve high efficiency because they put low weights on all leverage
points. The discrimination of C-H pays off in producing an efficient estimator. To show efficiency, they derive the covariance matrix of this estimator, which can be described as follows. Let V be a diagonal matrix
with $V_{tt} = w(x_t)\psi(r_t(\hat{\beta}_0)/\hat{\sigma}w(x_t))$. Let $C = \text{plim}_{T \to \infty}(1/T)(X'V^2X)$
and let $D = \text{plim}_{T \to \infty}(1/T)(X'BX)$. Then $\sqrt{T}(\hat{\beta}_T - \beta)$ is asymptotically $N(0, \sigma^2 D^{-1} C D^{-1})$. The relative efficiency of this covariance,

compared to OLS, depends on the choice of the constant c determining the cutoff for the ψ_c function. The higher the c the greater the efficiency at the normal. However, c also determines the bound on the influence function. This means that larger values of c will also lead to larger biases at approximately normal distributions. This is why a compromise value of c is suggested.

Numerous alternative estimators which also achieve the same goal can (and will) undoubtedly be developed. Among these, some may be superior to others. It seems very likely that practically useful estimators will emerge from this line of research. Further theoretical development and empirical experience are needed before firm recommendations can be made. One issue of substantial importance about which little is known is how to choose between high efficiency at normal versus efficiency at nearby or not-so-nearby distributions. The choice of c trades off efficiency at normal with bias at approximately normal distributions. There are also unknown and unquantified trade-offs between efficiencies at approximately normal and efficiency at distinctly abnormal densities. We next discuss semiparametric estimation which attempts to achieve full efficiency at all distributions.

5.11 Why Not Semiparametrics?

The Coakley-Hettmansperger estimator achieves high breakdown value, as well as high efficiency for normal or approximately normal errors. Could we ask for more? As discussed in connection with the criticism of bounded influence by Morgenthaler and Tukey (1991), one might ask for high efficiency not only at the normal but at all distributions. This fourth crown can actually be achieved by some procedures, but there are some costs. The breakdown properties and the influence functions of such procedures have not been investigated. Nearly all development in the field is asymptotic, and it appears that the high efficiency will be achieved only for very large sample sizes, making the procedures of little practical value.

The idea of semiparametric estimation is very simple (although the implementation is very complex). Consider the regression model $y_t = x_t'\beta + \epsilon_t$, where $\epsilon_t \overset{i.i.d.}{\sim} F$. The common distribution of the errors is some unknown F. The model is semiparametric since the regression coefficients are modeled parametrically, but the error distribution is nonparametric. In large samples it is possible to estimate exactly what the error distribution is. Thus we can estimate the error distribution and then use an estimate of the parameters which is optimal for the estimated error distribution. If done right (and this involves considerable care), it is possible to develop an estimator which is the most efficient possible regardless of what the true error distribution is. Because the estimation procedure adapts itself to the type of error density estimated, this is also called adaptive estimation.

Semiparametric estimation is theoretically challenging and the prospect of automatically adapting the estimation procedure to be the most suitable for the underlying density is exciting. From a practical point of view, however, it does not seem appealing. The gains of the procedure depend on the reliable estimation of the underlying common density of the error. Reliable estimates of density tend to require substantially larger sample sizes than are normally available in econometric applications. Even if suitable sample sizes are available, there is the following conceptual difficulty. If the errors do not have a common density, then there can be no gain from semiparametric estimates. It seems reckless in applications to pin one's hopes on the possibility that all errors have an identical distribution. It seems more likely that in applications we will have ϵ_t independent with distribution F_t which may well resemble each other but need not be *identical*. It is not known how semiparametric procedures would behave in this situation. It is known that the best procedure is very sensitive to the exact shape of the distribution, and can be quite different for distributions quite similar to each other. This suggests that slight deviations from the i.i.d. assumption crucial to semiparametric estimation may cause vast degradations in performance. Furthermore, we can never tell, even with infinite samples, the difference between an i.i.d. model and one which has slight deviations of the type described above. If theorems can be established to show that semiparametric procedures behave well for all i.n.i.d. errors observationally indistinguishable from the i.i.d. one, there may be some hope for this type of estimation. Otherwise its superiority may rest on metaphysical assumptions.

5.12 Nonparametric Regression

In nonparametric regression, we start with the idea that y is influenced by some set of regressors $x_1, \ldots, x_k$, but would like the data to tell us the form of the relationship. This makes the basic concept substantially different from the type of robustness we have been studying in this chapter. Our study of robustness is predicated on the assumption that we know the form of the relationship and distrust the data to some extent. In particular, we do not want our estimates to be affected by (i) data recorded with errors, (ii) data outside range of validity of the relationship, or possible (iii) data satisfying some relationship different from the target relationship being estimated. In each case, some data will be 'outlying' and will lead to distorted estimates of the target relationship. In contrast, in nonparametric regression, we rely completely on the data, and let it guide the shape of the relationship being estimated to a large extent. Thus, the topic does not really belong in the present chapter; nonetheless, we take the opportunity to make a few comments about this procedure here.

A survey of econometric applications and pointers to the literature are given in Härdle and Linton (1993). Instead of taking a linear form $y = \sum_i \beta_i x_i + \epsilon$, we estimate a relationship of the form $y = f(x_1, \ldots, x_k, \epsilon)$, where the function f itself is to be estimated. We would like to note an elementary but neglected fact about nonparametric estimation. Given a finite data set, it is a priori obvious that any estimation procedure will estimate only a finite dimensional model. To be more precise, consider the set of all possible estimated models as the data varies over all possible outcomes. Since the set of T data points is finite dimensional, the set of estimated models will be a map from the finite dimensional vector space of observations into some subset of an infinite dimensional space. Thus qualitatively the nonparametric estimation technique will in fact produce estimates confined to a finite dimensional class of models. A suitable finite dimensional (and hence parametric) estimation technique should be able to mimic the performance of the nonparametric techniques.

What this means is that there is no qualitative difference between nonparametric and parametric techniques. Rather, the difference is better understood as a difference in the attitude of the econometrician.

It is clear that parametric model users expand the number of regressors and therefore the dimension of model in response to increasing data. Nonparametricians are more aggressive in terms of quickly expanding the dimensionality of the models, and being more willing to consider more complex models with less data. Series estimators provide the clearest demonstration of the point we are making here — in these models (discussed in Härdle and Linton (1993)) the dimensionality is systematically expanded as the data increases so as to enable estimation of an arbitrarily complex function asymptotically. Other nonparametric techniques can also be interpreted in this fashion, where a class of finite dimensional parametric models is expanded so as to asymptotically become infinite dimensional. The crucial questions involve the issue of how many dimensions should be estimated with how much data, and how should the dimension be increased in response to increasing data. Nonparametric methods answer these questions in a systematic way, whereas conventional parametric methods address these issues in an indirect and ad-hoc way.

One can increase the number of parameters rapidly in parametric models to achieve essentially the same effect as that of nonparametric estimation. Nonetheless, a large variety of different nonparametric techniques have been developed and these permit an easier and automatic approach to the same problem. Our objective here is to point out that nonparametric techniques are not radically different from parametric techniques and cannot offer any panacea to estimation problems. There is merit to the suggestion of nonparametric econometricians that the difficulties of quickly expanding dimensionality of parametric models in response to increasing data mean that in practice significant non-linear structures may escape notice of parametric econometricians. On the other hand, rapidly increasing the dimensionality can also lead to meaningless 'curve-fitting' — it is clear that a very high dimensional model can exactly fit any finite dimensional data set, but the model will not provide any information.

Another way to phrase this problem is to note that fitting an infinite dimensional nonparametric model to a finite data set must involve supplying adding some information, usually from the class of models considered, to the finite amount of information in the data. The resulting curve mixes sample information with difficult to quantify prior information and it will be difficult to differentiate between what the data is saying and what has been added by our modelling technique. On the other hand, using parameter space of too low a dimension can similarly distort inference by not allowing the data to express itself.

5.13 Exercises

1. Prove that if H is positive semidefinite, then $h_{ij}^2 \leq h_{ii}h_{jj}$. Hint: show that $H = PP'$ for some matrix P. Let $X \sim N(0,\mathbf{I})$ and define $Y = PX$. Use the Cauchy-Schwartz inequality for $\mathbb{C}ov(Y_i, Y_j)$.

2. Show that the value m which minimizes $\sum_{t=1}^{T} |y_t - m|$ is the median of the y_t. Hint: this function can be differentiated except for $m = y_t$.

3. There are several algorithms for computing LAE estimates of regression parameters. This is the value of β which minimizes $\sum_{t=1}^{T} |y_t - x_t\beta|$, where x_t is $1 \times K$ and β is $K \times 1$. It is possible to show, as in the previous exercise, that the LAE estimate $\tilde{\beta}$ must pass exactly through K of the data points. A simple strategy which works for very small data sets is just to look through all subsets of size K, solve for the unique β fitting the particular subset, and find the one which minimizes the sum of absolute residuals. Devise a suitable algorithm for larger data sets.

4. Rousseeuw and Leroy (1984, page 50) give the following data regarding the number of reported claims of Belgian fire insurance companies:

Year	1976	1977	1978	1979	1980
Claims	16,694	12,271	12,904	14,036	13,874

   Is the number of claims increasing or decreasing? Compare OLS with a robust analysis.

5. Let X be a $T \times K$ matrix with first column $\mathbb{E}$, a vector of 1's. Let $P = X(X'X)^{-1}X$ be the projection matrix. Define the *centering* matrix $C = \mathbf{I} - (1/T)\mathbb{E}e'$. This exercise shows that diagonal

elements of the projection matrix reflect the distance of the regressors from their mean value.

(a) For any vector y, Cy centers y by subtracting the mean of y from each element.

(b) Define $\tilde{X}$ to be the $T \times (K-1)$ matrix obtained by dropping the first column from CX. The space spanned by the columns of $\tilde{X}$ is orthogonal to the space spanned by E.

(c) It follows that $\Pi_{\mathbf{X}}(y) = \Pi_{E}(y) + \Pi_{\tilde{\mathbf{X}}}(y)$.

(d) Let h_{tt} denote the (t, t) diagonal entry of $X(X'X)^{-1}X'$. Let $\tilde{h}_{tt}$ denote the (t, t) diagonal entry of $\tilde{X}(\tilde{X}'\tilde{X})^{-1}\tilde{X}'$. It follows from the previous question that

$$h_{tt} = \frac{1}{T} + \tilde{h}_{tt} = \frac{1}{T} + (x_t - \bar{x})(\tilde{X}'\tilde{X})^{-1}(x_t - \bar{x})',$$

where x_t is the t-th row of X, and $\bar{x}$ is the average of all the rows.

Part II

Hypothesis Tests for Regression Models

A seeker came to Mevlana Jalaluddin Rumi and requested to be allowed to participate in his mystical ceremonies which had acquired some fame. The Mevlana asked him to fast for three days and then come back. Upon his return, he found a table with a delicious and ample meal, and the Mevlana, waiting for him. The Mevlana said that if your desire is strong enough that you can forego the meal, you are welcome to join us in the ceremony now.

Chapter 6

Stringent Tests

6.1 Introduction

In the first part of the text, a large variety of estimators for linear regression models were introduced and evaluated. This second part introduces and evaluates a large variety of hypothesis tests. In normal distribution theory, optimal tests and confidence intervals can be based on the distributions of optimal estimators. Thus the three topics can be (and in typical econometric texts, are) presented in a unified way. However, this coincidence breaks down in general situations, and optimal tests need have no relation to optimal estimates. Thus it is essential to consider these questions separately, as we do here. In fact, we consider only hypothesis tests here; confidence intervals are closely related to, and can be derived from, hypothesis tests, as shown in Lehmann (1986).

Our first concern, dealt with at length in this and the next chapter, is to set up criteria to be used in comparing the performance of tests.

Subsequent chapters explore compromises between stringent ideals and the demands of practice. In this chapter we introduce and study a new criterion, 'stringency' as a performance measure for hypothesis tests. This provides a unified view of optimality properties of tests. The present chapter explores the stringency of a large class of tests in the context of a single example. Our object is to introduce the reader to a variety of principles used in developing test statistics. At the same time, these principles are evaluated in terms of stringency.

Three principles, the LM (Lagrange Multiplier), the LR (Likelihood Ratio), and the Wald test, are widely used for the construction of tests in econometrics. A popular criterion for evaluation of test performance, Pitman efficiency, judges these three tests to be equivalent. In the context of the testing example discussed in this chapter, the three are very different, with LR dominating by far the other two tests. Reasons for this difference are also discussed here and in Section 15.5.

6.2   Shortcomings of Tests

A general testing problem can be described as follows. We observe $y = (y_1, \ldots, y_n)$, distributed according to some family of densities $f(y, \theta)$. The null hypothesis is $H_0 : \theta \in \Theta_0$ while the alternative is $H_1 : \theta \in \Theta_1$. Any function $T(y)$ taking values $\{0, 1\}$ can be regarded as a test. The interpretation is that when y is observed, we decide that H_1 is true if $T(y) = 1$, and H_0 is true $T(y) = 0$. The set of values of y for which $T(y) = 1$ is called the *rejection region* of the test T. Instead of thinking of tests as functions, we can equivalently characterize tests via their rejection regions $R^*(T) = \{y : T(y) = 1\}$. For a fixed value of $\theta = \theta^*$, the probability that a test $T(y)$ rejects the null will be denoted $R(T, \theta^*)$, defined formally as $R(T, \theta^*) = \int_{R^*(T)} f(y, \theta^*) \, dy = \mathbb{P}\{y \in R^*(T)|\theta^*\}$.

Two characteristics, *size* and *power*, are of essential importance in assessing a test. The size or level L of a test is the maximum probability with which a test rejects the null when the null hypothesis is true (a type I error):

$$L(T) \equiv \max_{\theta \in \Theta_0} R(S, \theta).$$

It is a commonly accepted principle that only tests of equivalent levels should be compared. Let $\mathcal{T}_\alpha$ be the set of all tests of size α. The power (function) of a test is just $R(S, \theta)$ evaluated at points θ belonging to the alternative hypothesis $\theta \in \Theta_1$. This measures the probability of correctly rejecting the null. It is usual to compare tests T_1 and T_2 belonging to $\mathcal{T}_\alpha$ on the basis of their powers. Unfortunately, this rarely yields satisfaction, since typically T_1 is more powerful for some set of points while T_2 is more powerful for others.

In order to evaluate tests, we first set up a benchmark for test performance. A natural benchmark is the power envelope $\beta_\alpha^*(\theta_1)$ defined as the maximum possible power any test of size α can attain at $\theta_1 \in \Theta_1$:

$$\beta_\alpha^*(\theta_1) = \sup_{T \in \mathcal{T}_\alpha} R(T, \theta_1).$$

The *shortcoming* S of a test $T \in \mathcal{T}_\alpha$ at $\theta_1 \in \Theta_1$ is defined as the gap between the performance of T and the best possible performance at θ_1:

$$S(T, \theta_1) \equiv \beta_\alpha^*(\theta_1) - R(T, \theta_1).$$

The shortcoming of T, denoted $S(T)$, without reference to a particular alternative $\theta_1 \in \Theta_1$, will be used for the *maximum* shortcoming of T over all $\theta_1 \in \Theta_1$:

$$S(T) = \max_{\theta_1 \in \Theta_1} S(T, \theta_1)$$

We will compare tests on the basis of their shortcomings. If test T_1 has a smaller shortcoming than T_2 (and both are of the same level) then we will say that T_1 is *more stringent* than T_2. If a test T_1 has a smaller (or equal) shortcoming than all other tests of the same level, we will say that T_1 is the most stringent test of level α. This set of definitions appears a bit abstract at first encounter. This chapter is devoted to making it operational in the context of a very simple example. Some special cases are discussed first.

6.2.1 Tests with Zero Shortcoming

Consider first the simplest case where both the null and the alternative are simple hypotheses, so that $H_0 : \theta = \theta_0$ and $H_1 : \theta = \theta_1$. The Neyman-Pearson theory discussed in Appendix B characterizes the most powerful test of level α as follows. Define the rejection region

$NP(\theta_0, \theta_1, c)$ to be the set of all y such that $LR(y) = f(y, \theta_1)/f(y, \theta_0) >$ c. As c increases from zero to infinity, the rejection region shrinks from being the set of all values of y to the empty set. Find c_α such that the probability of the set $NP(\theta_0, \theta_1, c_\alpha)$ is exactly equal to α (this can always be done if y has a continuous density) under the null hypothesis. Then the test NP_α with rejection region $NP(\theta_0, \theta_1, c_\alpha)$ is the most powerful test of level α at θ_1. Since the test is most powerful, it has zero shortcoming at θ_1.

Next suppose $\theta \in \mathbb{R}$ and consider the slightly more complicated situation where the null is simple $H_0 : \theta = \theta_0$, but the alternative is $H_1 : \theta > \theta_0$. Suppose that by a fortunate coincidence the rejection region $NP(\theta_0, \theta_1, c_\alpha)$ is the same for all $\theta_1 > \theta_0$. Then it is obvious that the test with this rejection region is simultaneously most powerful for all $\theta_1 > \theta_0$. Such a test is a called a *Uniformly Most Powerful* (UMP) test. This test also has zero shortcoming at all $\theta_1 > \theta_0$. Conditions when this coincidence occurs are explored in detail in Appendix B. Briefly, UMP tests exist if and only if the the family of densities $f(y, \theta)$ has *monotone likelihood ratio (MLR)*: for some monotonic function $H_{\theta_1}(\cdot)$, $H_{\theta_1}(f(y, \theta_1)/f(y, \theta_0))$ is a function of y independent of θ_1.

6.2.2   Recommended Approach

Since UMP tests have zero shortcoming at all points in the alternative, their maximum shortcoming is also zero, and hence they are automatically most stringent. When such tests exist, there is universal agreement that (only) they should be used. It can be proven (see Lehmann (1986) or Pfanzagl (1968)) that this fortunate circumstance requires three conditions: (i) the densities must be from an exponential family, (ii) there must be only one parameter, and (iii) the alternative hypothesis must be one-sided; that is, $H_1 : \theta_1 > \theta_0$ and not $H_1 : \theta_1 \neq \theta_0$. Appendix B contains a detailed discussion of how to find UMP tests. The present chapter discusses the more common case of what to do when no test of zero shortcoming exists.

In some problems it is possible to find the test with the smallest shortcoming (or the most stringent test). This is the case when methods of invariance developed by Hunt and Stein (1946) can be applied. This approach will be discussed in the next chapter. There is a larger group of problems to which invariance theory cannot be applied. In this second group, it is typically not possible to find the most stringent test by theoretical methods. It is this second group which will be the focus of our discussion in the present chapter. There exist a large number of different test statistics which can be constructed by various ad-hoc methods to be discussed.

We will see below that the Likelihood Ratio (LR) test has the smallest shortcoming among all the tests considered in the example of this chapter. Section 15.5 shows that LR has several asymptotic optimality properties . Our study of UMP invariant tests which are most stringent will also frequently lead to LR tests. Thus it seems reasonable to take the LR test as the first choice of a test statistic. The arguments for LR have a heuristic flavor, and counterexamples where the LR performs poorly have been devised, so that its good performance cannot be taken for granted. If possible, we should evaluate the shortcoming of the LR (as is done in a special case in this chapter). If this is satisfactory (e.g. 5% or below), then no test can improve on the LR by much and we can confirm our choice. If the shortcoming is unsatisfactory there are several options. We can look for other tests in the hope of improving the shortcoming, or we could restrict the class of alternatives to the null — by focusing on a narrower set of alternatives one can frequently improve power.

Our proposal would create greater uniformity in the approach to testing since current practice in econometrics allows for an extremely large variety of tests, alternatives to the LR. This is because:

- It is widely believed that LR is asymptotically equivalent to LMP and Wald as well as several other tests.

- The LR test statistic is frequently difficult to compute, and its distribution is even more difficult to compute.

In fact, the asymptotic equivalence of LR and other tests is valid only up to the second term in the asymptotic expansion of the local power function of the three tests. Differences appears in the third term of the power function, and these favor the LR — see Section 15.5 for discussion and references. The computational difficulty was indeed a serious problem but has become much less so with the recent rapid decline in computing costs. While increased computation power has substantially changed the set of feasible choices, inertia has resulted in continuation of testing strategies which can be proven to be poor and should be abandoned. Several fast and cheap techniques for computation of the ML estimates required for the LR test statistic have emerged, including the powerful EM algorithm (see Dempster *et al.* (1977)). The Bootstrap-Bartlett correction discussed in Section 14.6 should provide adequate approximations to the distribution of the LR in most problems of interest. Thus the practice should move towards greater use of LR and less use of supposedly 'asymptotically equivalent' procedures.

6.3   Tests for Autocorrelation

The example we use throughout this chapter is a simplified model of autocorrelation. Suppose $y_0 \sim N(0,1)$, and for $t = 1, 2, \ldots, T$ we observe y_t satisfying $y_t = \rho y_{t-1} + \epsilon_t$, where $\epsilon_t \overset{i.i.d.}{\sim} N(0, \sigma^2)$ for $t = 1, 2, \ldots, T$. This time series is *stationary* if and only if $\sigma^2 = 1 - \rho^2$ and $-1 < \rho < 1$. Stationarity means that the joint distribution of a finite set of k observations $y_m, y_{m+1}, y_{m+2}, \ldots, y_{m+k}$ does not depend on the value of m. In other words, the distributions remain the same as time passes. We will assume this in what follows in order to have a simple case to deal with.

Note that if we are interested in testing the null $\rho = 0$ versus the alternative $\rho = \rho_1$ for some fixed $\rho_1 > 0$, the Neyman-Pearson theorem provides an optimal test. Forming the likelihood ratio, it is easily seen that the Neyman-Pearson statistic is

$$NP(y, \rho_1) = \frac{((1 - \rho_1^2))^{-T/2} \exp\left(-\frac{1}{2(1-\rho_1^2)} \sum_{t=1}^{T} (y_t - \rho_1 y_{t-1})^2\right)}{\exp\left(-\frac{1}{2} \sum_{t=1}^{T} y_t^2\right)}$$

For the Neyman Pearson test, ρ_1 is treated as a constant. Dropping a constant and taking logs, we see that the Neyman Pearson statistic is a monotone increasing function of

$$NP'(y, \rho_1) = -\frac{1}{2(1 - \rho_1^2)} \sum_{t=1}^{T} (y_t^2 - 2\rho_1 y_t y_{t-1} + \rho_1^2 y_{t-1}^2) + \frac{1}{2} \sum_{t=1}^{T} y_t^2$$

$$= \frac{\rho_1}{2(1-\rho_1^2)}\left(2\sum_{t=1}^{T}y_ty_{t-1} - 2\rho_1\sum_{t=1}^{T-1}y_t^2 - \rho_1(y_T^2 + y_0^2)\right)$$

Ignoring the constant factor $\rho_1/(2(1-\rho_1^2))$, we can rewrite the Neyman-Pearson test statistic as

$$NP''(y,\rho_1) = 2\sum_{t=1}^{T}y_ty_{t-1} - \rho_1\left(2\sum_{t=1}^{T-1}y_t^2 + y_0^2 + y_T^2\right). \qquad (6.1)$$

Note that the test statistic depends on the parameter ρ_1. Since the Neyman-Pearson test depends on the alternative, there is no UMP test. It is also not possible to find the most stringent test by invariance theory (to be discussed in the next chapter). We now discuss, in increasing order of complexity, several strategies for testing which have been proposed for this problem. Each strategy attempts to remedy a defect in the previous one.

6.4 The Locally Most Powerful Test

In problems where UMP tests do not exist, a popular choice is to use a Neyman-Pearson test with alternatives very close to the null, resulting in tests which are called *locally most powerful* (LMP). Using asymptotic methods, it is very easy to develop the theory of locally most powerful tests and derive the distribution of the test statistic. Prior to the computer age, an *essential* requirement for a test statistic was that it should be easy to compute and also have a readily computable distribution. Statistics which did not fulfill these requirements could not be used in practice. The locally most powerful test typically fulfills both of these requirements, accounting for its popularity. The LMP tests are equivalent to Lagrange multiplier tests; Section 1.9 gives a general development of the test statistic and its asymptotic distribution.

As a test for autocorrelation of the error terms in the regression model, the locally most powerful test is famous under the name of the Durbin-Watson statistic. The major difficulty in the regression model was that all such tests (including the Durbin-Watson) have distributions depending on the matrix of regressors X. This makes it impossible to tabulate critical values. In a high computing cost era, this makes

it impossible to use such a test.   Durbin and Watson (1950) showed
that bounds could be developed for their test which did not depend on
the X matrix. This made it possible to tabulate critical values, and
hence gave the first usable test. Nowadays tabulation is unnecessary
since the computer can easily compute significance levels on the spot
for any desired test statistic and any give X matrix. Furthermore, it
has been shown that the Durbin-Watson test is seriously deficient in
power for certain types of X matrices. Another important theoretical
difficulty with the use of Durbin-Watson, and an alternative test, are
discussed in Sections 8.3 and 9.7.

The Neyman-Pearson test statistic was calculated in Eq. (6.1). To
get the locally most powerful test, we take the limit of the NP tests as
ρ_1 approaches zero, getting

$$LMP(y) = \sum_{t=1}^{T} y_t y_{t-1}.$$

Figure 6.1 plots the power envelope, the power of the LMP test, and the
shortcoming, which is the difference between the two, for three sample
sizes, $T = 30, 60$, and 90. All the tests are level 5 % tests. The algorithm
used to compute the numbers in the figure is described in some detail in
the last section of this chapter. Here we discuss the conclusions we can
draw from Figure 6.1. Note that the power envelope shifts upward as T
increases. This reflects the fact that increased sample size makes better
discrimination possible. The power of the LMP test also shifts upward,
due to the same phenomenon. The gap between the two shrinks, so
that the shortcoming of the LMP test decreases as the sample size
increases. Tests of different levels are qualitatively similar — at level
10% the LMP is closer to the power envelope at all sample sizes, while
at level 1% the shortcomings are larger. These features are typical of
most tests.

There is one unusual feature of the LMP, namely that its power
curve dips down near $\rho = 1$. Furthermore, the dip is quite sharp and
the power of the LMP is affected only in a very small neighborhood of
$\rho = 1$. This makes the test have low stringency, for the gap between the
power of the LMP and the maximum power is very large and declines
very slowly.

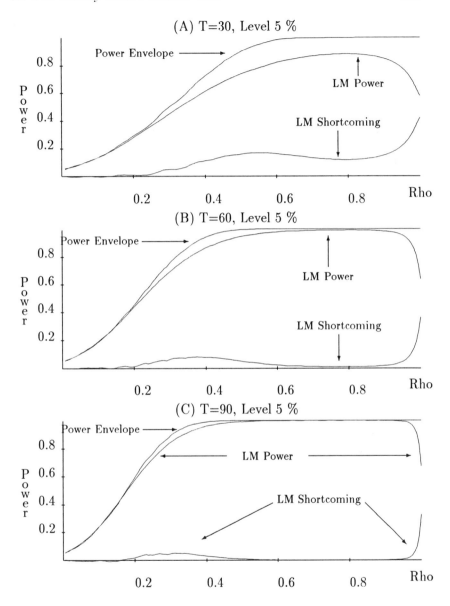

Figure 6.1: Shortcomings of LMP Tests

This feature makes the stringency of the LMP easy to calculate; all we have to do is evaluate the power of the LMP at $\rho = 1$. For all but the smallest sample sizes, the maximum shortcoming of the LMP is just 1 minus its power at $\rho = 1$. From the curves sketched in Figure 6.1, we can calculate the stringency of the LMP test for $T = 30, 60$, and 90. By sketching such curves for each T, we can plot the stringency of the LMP as a function of T. This is done (after a *long* computer run) in Figure 6.2 (C). Values of stringency were computed by direct simulation for $T = 5, 10, \ldots, 95$, while higher values were computed by asymptotic calculations (see exercise 2 for details).

The LMP test does extremely poorly in terms of stringency. The shortcoming of the level 5 % LMP test starts out at the extremely low values of 61 % for T=5. At the relatively large sample size of 100, the shortcoming is still around 32 %. This means that optimal tests achieve power more than 25% larger than the LMP even at samples sizes of $T = 100$. It can be proven that the stringency of the LMP increases asymptotically to zero. However, this happens extremely slowly. If we use the rule of thumb that 5% stringency is an acceptable value, the level 10% LMP test achieves 5% stringency for a sample size greater than 10,000. Higher levels require even larger sample sizes. We will see that other testing principles achieve satisfactory values of stringencies at very low sample sizes.

A heuristic explanation for the poor performance of the LMP tests is as follows. For alternatives very close to the null, *all* tests will have essentially equivalent and low power approximately the same as the level of the test. To choose a test with power of say .1001 for a nearby point, in preference to one which has power .10001 at the same point does not seem very reasonable. For practical purposes, essentially trivial differences in power near the null should not be the basis of choice among tests. Nonetheless, the LMP test makes these small differences the criterion of choice. In order to achieve good performance in a neighborhood of the null (where it is more or less impossible to achieve) the LMP test sacrifices good performance at distant points. However, for practical purposes it is more important to detect a violation of the null hypothesis when the violation is large (i.e., the alternative is distant) than when the violation is small. Tests which maximize power over suitably distant alternatives are also preferable for this reason.

6.5 Berenblutt-Webb-Type Tests

It is evident from the graphs of power that the locally most powerful (LMP) test has very poor power in the neighborhood of $\rho = 1$. Berenblutt and Webb (1973) argued that it was more important to detect higher values of ρ in applications. They suggested the use of the test based on taking the limit of the Neyman Pearson tests as ρ approaches 1. This is the opposite extreme to the LMP test, in the sense that the LMP maximizes power near the null while this test maximizes power at the farthest point from the null. If we evaluate the expression for NP'' given in Eq. (6.1) at the value $\rho_1 = 1$ we get the analog of the Berenblutt-Webb (BW) test in the context of our simplified model.

$$BW(y) = \sum_{t=1}^{T} y_t y_{t-1} - \sum_{t=1}^{T-1} y_t^2 - \frac{1}{2}(y_T^2 + y_0^2)$$

Figure 6.2(A) and (B) plot the power of a level 5 % BW test against the power envelope for $T = 30$ and $T = 60$. As we would expect, the BW test behaves opposite to the LMP test, with high shortcomings near $\rho = 0$ and low shortcomings near $\rho = 1$. It becomes clear from the plots of the power that the Berenblutt-Webb test is not very good for small and intermediate values of ρ. The reason for this is that nearly all reasonable tests achieve power very close to unity for large values of ρ. Thus, for example, Berenblutt and Webb end up choosing between two tests with powers 0.9990 and 0.9999 at $\rho = 1$. In this case, it does not seem reasonable to prefer the latter test because of its superior performance in the fourth decimal place. In practice one would consider this to be essentially equivalent power and choose among the tests on the basis of their performance elsewhere. The price paid for using the wrong criterion for choice becomes clear from the graph of the stringency of the BW test (Figure 6.2 (C)). For small sample sizes $\rho = 1$ is not such an extreme value, and the BW test is substantially superior to the LM test in terms of stringency. The value of $\rho = 1$ becomes more and more extreme as the sample size rises, and the BW shortcomings increase, until it equals the LM test shortcomings for a sample size of 95. Asymptotically the shortcomings of both should decline to zero, but this may require substantially larger sample sizes.

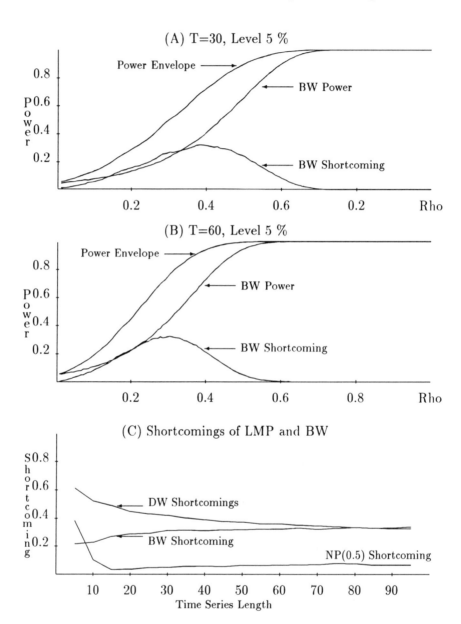

Figure 6.2: Berenblutt-Webb Test Shortcomings

Even though the basic idea of using a 'large' alternative is superior to that of the using an extremely small alternative, our analysis shows that it does not lead to a satisfactory test. Shortcomings of 30% and more for large sample sizes are unacceptable. An idea which improves on both close and far alternatives is discussed in the next section.

6.6 Fraser *et al.* and King's Test

Since both extreme choices for the alternative ρ_1 lead to poor performance of the Neyman-Pearson Test, it is natural to try an intermediate value. The natural intermediate value of $\rho_1 = 0.5$ was suggested by Fraser *et al.* (1976) and Evans and King (1985) after noting the poor performance of the Durbin-Watson test and the Berenblutt-Webb test. The stringency of this test, labeled NP(0.5), as a function of sample size is plotted in Figure 6.2 (C), together with the stringencies for the LMP and the BW. Clearly the NP(0.5) test is superior to both LMP and BW, except for the case of $T = 5$. For samples sizes ranging from 50 to 100, the stringency of the NP(0.5) test varies between 6% and 7%. These values could be considered borderline acceptable but for the fact that substantially superior tests are available, as described in the sections to follow. The basic idea, namely that intermediate values will perform better than either extreme, is correct. However, appropriate intermediate values depend on the sample size. For small sample sizes, the value of 0.5 is indeed intermediate, and the performance of the NP(0.5) test improves rapidly up to $T = 20$. For larger sample sizes, the Neyman-Pearson tests already achieve power essentially equal to unity at small values of ρ; at $\rho = 0.45$ for sample size $T = 75$ for a level 10 % test for example. This means that as sample size increases, the NP(0.5) test will share the characteristics of the Berenblutt-Webb test. That is, since large classes of tests will have power near one at 0.5, the NP(0.5) test will discriminate between tests on the basis of differences in power in the fourth decimal place.

The large sample performance of the NP(0.5) test, as displayed in Figure 6.2 (C), is governed by two factors. One is that tests improve as the sample size increases. The second is that the value of the alternative $\rho = 0.5$ becomes more and more an extreme value as the

sample size increases. This second factor degrades the performance of
the test in large samples. The combined effect of the factors is that
over the range examined, the stringency of the NP(0.5) test becomes
worse (after falling initially), but remains below 10%, so that the test
could be considered marginally acceptable. However, alternatives with
substantially superior performance are available, as we will soon see.

6.7   Efron's Test

Efron(1975) recommends using testing $\rho = 0$ versus an alternative that
is statistically reasonable. Details of this suggestion are as follows. In
Part III of the text, we show that if $\hat{\rho}_{ML}$ is the maximum likelihood
estimator of ρ based on a sample of size T, then it has asymptotic
variance the reciprocal of the information $\mathcal{I}_T^{-1}(\rho)$. If $f^y(y, \rho)$ is the joint
density of all the observations given the parameter ρ, the information
is defined as

$$\mathcal{I}(\rho) \equiv \mathbb{E}\left(\frac{\partial \log f^y(y, \rho)}{\partial \rho}\right)^2 = -\mathbb{E}\frac{\partial^2 \log f^y(y, \rho)}{\partial \rho^2}. \qquad (6.2)$$

The first term is the definition of the information, while the second is an
equality which frequently holds and sometimes facilitates computation
of the information.

   For $\rho = 0$ it can be calculated that the information in a sample of
size T is $\mathcal{I}_T(0) = T$ (see exercise 1). Thus under the null hypothesis
$\rho = 0$, the MLE has standard deviation $1/\sqrt{T}$ in large samples. It
follows that attempting to discriminate between $\rho = 0$ and $\rho = \epsilon$ where
ϵ is much smaller than $1/\sqrt{T}$ is not statistically feasible; there is not
enough information to permit such fine discrimination. Similarly, values
of ρ much bigger than $3/\sqrt{T}$ will be too easily distinguished. This latter
is the case with the Berenblutt-Webb test and also with the NP(0.5) test
for large values of T. This explains why the performance of NP(0.5)
(and also the BW) declines with increasing sample size — the value
0.5 is 'intermediate' in small samples but becomes extreme as sample
size increases. Efron suggests using an alternative that is 2 standard
deviations away from the null as leading to a statistically reasonable
problem. In the particular case of autoregressive model discussed above,

this amounts to testing $\rho = 0$ versus $\rho = 2/\sqrt{T}$. The stringency of this test was evaluated by numerical methods. At level 5 % the test has maximum shortcoming below 5% for all sample sizes $T \geq 15$. The maximum shortcoming is below 2.5% for sample size $T \geq 30$. Thus Efron's suggestion yields an acceptable test for all but small sample sizes, and improves on all three of the earlier proposals.

Since the constant $c = 2$ is somewhat arbitrary, we explored numerically the difference made by choosing c differently. Interestingly, the results were very much in accordance with the intuitive principle suggested by Efron. At significance level 10%, the testing problem is easier. It turns out that discriminating against $\rho_1 = 2/\sqrt{T}$ is too easy, and $\rho_1 = 1.5/\sqrt{T}$ gives generally better results in terms of stringency. At significance level 5% the testing problem is somewhat harder and it pays to shift the alternative to $\rho_1 = 1.7/\sqrt{T}$. At level 1% higher values of ρ_1 are needed (it is too difficult to discriminate against closer values). Values of $\rho_1 = 2/\sqrt{T}$, corresponding to Efron's original suggestion, worked well at the 1 % significance level.

The particular constants work well for the problem at hand for small samples. Different testing problems may well require different choices. It is worth noting that all reasonable choices for the constant give similar performance in large samples; it is only in small samples that there are noticeable differences in performance.

6.8   Davies Test

As we have seen, Efron's choice of an intermediate value of ρ is an improvement over $\rho_1 = 0.5$, since it takes into account the sample size and the information in the sample about the parameter. Nonetheless, the suitable intermediate value also depends on the level of the test. To put it heuristically, the suggested value $c = 2$ is 'too easy' to discriminate for a level 10% test, and test performance is improved by changing to $c = 1.5$. However, at level 1%, $c = 1.5$ performs poorly in terms of stringency; at this level it is too close to the null, and hence too difficult to discriminate against for optimal test performance. Obviously the optimal choice for c will vary from problem to problem, in addition to depending on the level of the test. The next proposal for the choice of an alternative in the Neyman-Pearson test automatically adjusts for the level of the test.

Davies (1969) develops a criterion called beta optimality for evaluating tests. He suggested that we fix some desired power, for example 80%, and attempt to develop a test which attains this power as close as possible to the null. In practical cases, this boils down to finding a value of $\rho_1 > 0$ for which the Neyman-Pearson test has 80% power. We can look at the Davies proposal from a different point of view, corresponding to the suggestion of Efron. Consider the problem of quantitatively assessing how difficult it is to discriminate between $H_0 : \rho = 0$ and $H_1 : \rho = \rho_1$. An obvious answer is to measure the power of the most powerful test for discriminating between the two hypotheses. If the power of the most powerful test is near unity, the testing problem is *too easy*, since nearly perfect discrimination is possible – this is what happens when we use the Berenblutt-Webb test, or the NP(0.5) test for larger sample sizes. If the power of the test is near the level of the test, the testing problem is *too difficult* — this corresponds to the case of the locally most powerful test. By using an alternative at which the maximum power attainable is some intermediate number — 80% is suggested by Davies — we get a problem of intermediate difficulty.

Following the concept of Efron and Davies, Figure 6.3 plots the stringency of an $NP(\rho_1)$ test versus the level of difficulty of testing $\rho = 0$ versus $\rho = \rho_1$, as measured by the maximum power attainable at ρ_1. Figures 6.3 (A), (B), and (C) plot this curve for sample sizes $T = 20$, $T = 40$, and $T = 60$, respectively. All plots are for tests of 5% significance level. These figures provide a lot of information about the performance of the Neyman-Pearson tests in this problem and will repay careful study, which we now undertake.

Ragged Curves: The curves are not smooth because they were drawn using Monte-Carlo methods, which give random errors in the final result. The true curves should be smooth, and some smoothing of the results was done in an ad-hoc way. Since oversmoothing is dangerous and can distort results, we preferred to leave some jags in the curves. In this connection, it should be noted that each point in the curve is the result of a *substantial* amount of computations and summarizes a lot of information. On the bottom horizontal axis, we plot the maximum power corresponding to a value of ρ which is displayed on the top horizontal axis. This involves computing the power envelope. Separately, for each rho, we need to compute the power curve for that rho, and calculate the largest difference between the maximum power and the power curve to get the shortcoming. Each point therefore requires calculation of a power curve and a maximum difference. Each power curve requires several Monte Carlo for each point on the power curve. To make the task possible we had to compromise on the Monte Carlo sample size.

U Shape: The general U shape of the stringency curves is the most important feature, showing the validity of Efron's idea that we should use tests of intermediate difficulty. As the power goes towards the extremes of 0 and 1 the shortcoming increases sharply. These extremes correspond to ρ values which are too easy and too difficult respectively. The shortcomings are fairly small for intermediate values of difficulty, and below the level of 5 % for a range of such values.

Correspondence between ρ and Difficulty: We have made the x-axis measure the difficulty of the test (via the maximum power). The maximum power increases nonlinearly with ρ. Values of ρ corresponding to the difficulty levels of $20, 40, 60,$ and 80% are given on the top x-axis. It should be understood that the top scale is not linear in ρ. Also note that as the sample size increases the difficulty level of a particular ρ decreases. For example, the maximum power attainable at $\rho = 0.5$ is only 75% for $T = 20$. Thus the test is of intermediate difficulty. However, the maximum power increases to 96% for $T = 20$, suggesting that this is too easy a problem at this sample size.

The Effect of Sample Size: Increasing sample size has the effect of decreasing the shortcomings, and the curve is substantially lower at $T = 60$ than at $T = 20$. Also, the set of difficulty levels which produce acceptable tests (i.e., with shortcomings less than 5%) becomes larger. This means that it becomes less and less critical to specify the constant c in Efron's suggestion and also to specify the difficulty level in Davies' suggestion (80% is the original suggestion). It has already been mentioned that the range of ρ values corresponding to intermediate difficulty levels shrinks towards 0, with increasing sample size.

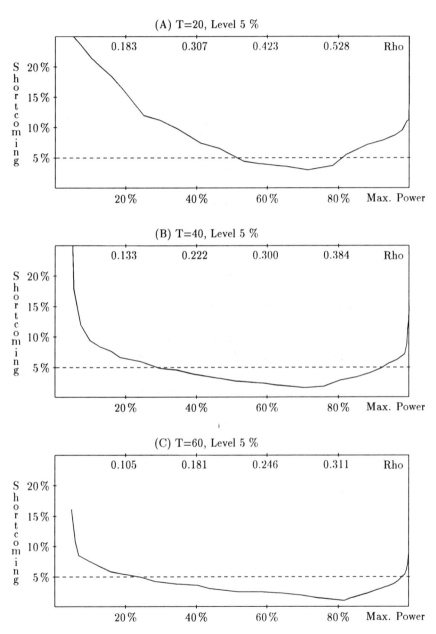

Figure 6.3: Shortcomings versus Maximum Power

The Best Point Optimal Test: Once we have a performance measure, such as shortcoming, for tests, we can immediately ask for the 'best' test. This is the value of ρ corresponding to the bottom of the shortcoming curve. In typical examples, it will be extremely time-consuming to compute the best. Fortunately, at least in the present example, the bottom of the curve is fairly flat. This means that it is not so critical to get the minimum exactly — being in a rough neighborhood is good enough. Both Davies' suggestion and Efron's suggestion are adequate to get us in the rough neighborhood of the minimum. Alternatively, one may be able to calculate the best point optimal test using asymptotic approximations. If this is not too tedious, it may be worthwhile.

The Effect of Level: All of the tests plotted are level 5%, so as to avoid crowding the graphs. Similar calculations were also made for tests of level 10% and 1% and led to qualitatively similar and intuitively plausible results. At level 10 %, difficulties of tests are reduced and correspondingly the shortcomings are also reduced. The shortcoming curve is U-shaped as usual, but the bottom of the curve is closer to $\rho = 0$, reflecting the smaller difficulties. Similarly, at level 1%, difficulties and shortcomings increase. The bottom of the U-shaped shortcoming curve is farther from 0. This shows that the constant c in Efron's suggestion of $c/\sqrt{2}$ should be adjusted for level.

6.9 Point Optimal Tests

So far we have confined attention to Neyman-Pearson tests for different values of the alternative. Such tests have been named 'Point Optimal' tests by King (1987-1988), who argued for their inclusion in the econometricians toolkit. We discuss some factors for and against the use of point optimal tests in general.

In the next section we will study two tests, the Wald test and the likelihood ratio (LR) test, which are outside this class. The LR test dominates, in this particular example, all the point optimal tests, with respect to stringency. However, the differences are not large, especially for $T \geq 20$. Thus we may well prefer the point optimal tests in this example over the more complex LR test. This highlights the most important virtue of point optimal tests, which is simplicity. These tests are generally easy to compute and their distributions are frequently manageable. Also, we are assured of their optimality in a neighborhood of the selected ρ. It must be recognized that this class of tests is of importance only in the case of a one-dimensional parameter of interest with a one-sided alternative hypothesis. In higher dimensions, or with two-sided alternatives, it is easily seen that a point optimal test cannot perform reasonably.

An important extension of the point optimal tests is a test which may be thought of as point optimal against a finite collection of points. Suppose $y \sim f(y, \theta)$. Consider testing $H_0 : \theta = \theta_0$ against a finite collection of alternatives $H_1 : \theta \in \{\theta_1, \ldots, \theta_p\}$. It can be shown that optimal tests all take the following form. Define the statistic

$$NP^*(y) = \left(\sum_{i=1}^{p} \pi_i f(y, \theta_i) \right) / f(y, \theta_0),$$

and reject H_0 whenever $NP^*(y) \geq c$. This statistic can be thought of in two different ways, each of which offers additional insight.

One way is to think that the several alternatives are replaced by a 'combined alternative' which is a mixture of the alternative densities. Define $f(y, \theta^*) = \sum_{i=1}^{p} \pi_i f(y, \theta_i)$ and consider the problem of testing $\theta = \theta_0$ versus $\theta = \theta^*$. This is now a simple-vs-simple problem. It can be shown that by varying the weights π_i one can trace out all of the optimal solutions for the original problem, exactly as we can trace out all optimal solutions for the simple-versus-simple problem by varying the significance level.

An alternative, Bayesian, point of view is to think that π_i are the prior probabilities for the θ_i. As a side benefit, we get $\mathbb{P}(\theta_i|y) = \pi_i f(y, \theta_i) / \sum_{j=1}^{p} \pi_j f(y, \theta_j)$ which are the *posterior probabilities* of the θ_i. These provide important information on the relative likelihood of the various alternative probabilities. See also Section 10.3 for an illustration of a Bayesian hypothesis test.

Unlike the point optimal tests which are of limited applicability as discussed earlier, it can be proven that we can approximate the most stringent test to an arbitrarily high degree of accuracy by using a test of this type. In general the most stringent test can be represented by using a prior density $\pi^*(\theta)$ over the space of alternatives and testing $\theta = \theta_0$ versus the mixture density $\int_{\theta \in \Theta_1} f(y, \theta) \pi^*(\theta) \, d\theta$. This integral can be approximated by a weighted sum over a finite subset of the space of alternatives Θ_1. This class of tests, where we artificially restrict a high-dimensional alternative to a finite subset, is only slightly more complex in structure than the point optimal tests, but is capable of providing arbitrarily good approximations to the most stringent test. The author is not aware of any applications in the literature of this principle so this promise is only potential so far.

6.10   Likelihood Ratio and Wald Test

All of the previous approaches fix a $\rho_1 > 0$ independently of the data by some method and then use the Neyman-Pearson test for $\rho = 0$ versus $\rho = \rho_1$. The likelihood ratio approach fixes ρ_1 to be the most likely of the values in the alternative. To be more precise, let $\hat{\rho}_{ML}$ be the ML estimator of ρ. Then the LR test is the Neyman-Pearson test for $\hat{\rho}_{1ML}$ against the null. In the case of the autoregressive process, its performance is plotted in Figure 6.4. As can be seen, it has very small stringencies and comes quite close to having the maximum power attainable. Of all the alternatives studied, LR appears the best in this particular case.

The Wald test uses the asymptotic normal distribution of $\hat{\rho}_{ML}$ to form a test statistic for the null. In this investigation, we gave Wald his best chance by using the exact finite sample distribution of the ML (determined by simulations) to get the critical values of the test. Despite this, the test did not perform very well. In Figure 6.4 the stringencies of the Wald, LR, and LMP tests are compared. These three tests have acquired the status of a classical trio in econometrics. Since these have equivalent local asymptotic power (up to the second term in the asymptotic expansion), it has been suggested that the choice among them should be according to computational convenience. However, our plots of stringency show that these tests differ substantially. The LMP is extremely poor, the Wald test comes in second, and the LR is a clear winner. The stringencies of the Best Point Optimal test are also plotted on this graph. These represent the best one can do by judicious choice of the alternative using a Neyman-Pearson point optimal test. It is clear that we lose something by restricting ourselves to the point optimal tests, since even the best of them cannot compete with the LR. However, the difference is not of practical importance for $T \geq 20$.

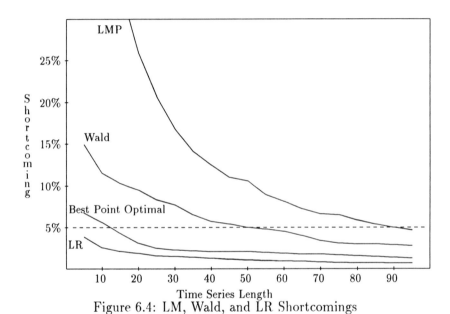

Figure 6.4: LM, Wald, and LR Shortcomings

6.11   Algorithms

We need to be able to generate random time series of the type described in Section 6.3. This is easily done. Assume we have a program GEN(rho,T) which does the following. It generates an initial value y_0 and for each $t = 1, 2, \ldots, T$ it recursively generates $y_t = \rho y_{t-1} + \epsilon_t$, where ϵ_t is a random number with density $N(0, 1 - \rho^2)$. It is useful to note that it is not necessary to keep the whole time series; the program can work with the following sufficient statistics:

$$S = \sum_{t=1}^{T} y_t y_{t-1} \quad S0 = \textstyle\sum_{t=0}^{T-1} y_t^2 \quad S1 = \sum_{t=1}^{T} y_t^2$$

The program GEN should either set these global variables or return these three numbers as output. Every time it is called a new random sample should be generated.

It is sufficient to describe how to calculate a single point on the power envelope; the entire envelope can then be traced out by an iterative procedure. Suppose we wish to find the maximum power at some value ρ. The NP statistic at ρ can be written as

$$NP(\rho, S, S0, S1) = 2S - \rho(S0 + S1).$$

In order to calculate the power of the, for example, level 10% test, we need to first calculate the critical value. Write a program $NP0(\rho,T)$ which calls GEN(0,T) and then computes $2S - \rho(S0 + S1)$. Thus we get a sample value of the NP statistic under the null hypothesis. Repeat the process 1000 times (for example) and keep the observations in order (with 1 being the smallest and 1000 the largest). Then the 900th value is an estimate of the desired critical value. Call this estimate CV90. Then use a second program $NP1(\rho,T)$ which calls $GEN(\rho,T)$ and computes $2S - \rho(S0 + S1)$. This is now a sample under the alternative hypothesis. For a sample of size 1000, count the number of times the value of NP1 exceeds CV90. This gives an estimate of the maximum power.

Obviously this procedure can be repeated with any statistic. For example, suppose we have a program ML(S,S0,S1) which computes the maximum likelihood estimate of ρ on the basis of these sufficient statistics. Then we can calculate the critical value of the Wald test (based on the ML) by generating a sample of 1000 under the null, and getting the 900th (or the 950th, or the 990th, for 5% or 1% probability of type I error) value as CVML. To trace out the power, we generate samples for each rho, and estimate the probability the ML(S,S0,S1) > CVML.

The above is a rough outline of what needs to be done. To implement it in detail, several refinements are needed. First we must assess how large a sample size is needed. Let T be any statistic and $CV(p)$ be a constant such that $\mathbf{P}(T > CV(p)) = p$. Let $X_1, \ldots, X_n$ be a sample of i.i.d. binomials where $X_i = 1$ if $T > CV(p)$ and $X_i = 0$ if $T \leq CV(p)$. We will estimate p by $S_n = \sum_{i=1}^{n} X_i/n$. It is easily checked that $\mathrm{Var}(S_n) = p(1-p)/n$. If $p = 0.1$, it is easily seen that the standard error of an estimate of p based on a sample size of $n = 10,000$ will be $\sqrt{(0.1 \times 0.9)/(100 \times 100)} = 0.003$. Normally this would be adequate, but in repeated tabulation, we can expect errors around 2 to 3 standard deviations every once in a while; with $n = 10,000$ we would expect estimates e to have accuracy $e \pm 0.009$. An error of 0.009 is about 10% of the number 0.10 which is a rather large error. To reduce the standard error by a factor of 3, we need 9 times the sample size, or $n = 90,000$. This will give us a standard error of 0.001 which should certainly be adequate for practical purposes. If $n = 90,000$ is too burdensome, we can keep the sample sizes low but do smoothing after making the grid size suitably fine. This requires more careful programming. Things become worse if the level of the test is 1%. With a sample size of $n = 10,000$, the standard error is only 0.001, which is an improvement. However, 3 standard errors or 0.003 is now about 30% of 0.01, which is too large to be acceptable. Thus the standard error must be reduced to get acceptable estimates of these critical values.

To write a program which computes the powers, we just need two counters, e.g. m and n. Given a critical value CV and a ρ at which we need to compute the power of some statistic T, we generate a new value of T and update the total counter $n = n + 1$. The success counter m is updated to $m = m + 1$ only if $T > CV$. At the end we calculate m/n to be the probability of rejection. Putting in a loop over the ρ will generate the power curve of the statistic T. The only point worthy of attention in programming this task is that it may be useful to stop the loop in n if the standard error becomes very small. The standard error can be calculated as $p(1-p)/\sqrt{n} \approx (m/n)(1 - m/n)/\sqrt{n}$. If this is implemented, it is important to ensure that a minimum sample size of, e.g., $n = 500$ is always obtained, since in small sample sizes purely random fluctuation may lead to an excessively small estimate of standard error. This refinement prevents excessive computations for those values of rho for which the statistic has achieved power close to one; the range of such values is large, especially for large sample sizes.

A second refinement is that the size of the grid should grow finer as sample size increases; roughly the tabulation should be in units of, e.g., $0.05/\sqrt{T}$. This is because the powers increase faster in larger sample sizes.

Somewhat more difficult programming is required for the computation of the critical values. Due to limited memory, unnecessary sorting for bottom of the sample, and lack of elegance, the whole sample cannot be retained if we are getting a sample of size 100,000 for example. There are two different methods which can be used to get the 90,000th element, which will give us the 90% critical value. One need merely keep the top 10% of the sample and throw away the rest. This still requires a memory capacity of about 10,000 which may be too large for some machines. The second alternative is to take an initial sample of 1000, for example, and get preliminary estimates of $cv85, cv86, \ldots, cv90, cv91, \ldots, cv99$. These 15 to 18 numbers are kept in an array, e.g., $cv[i]$. Now for the subsequent sample of 99000, for each of the $cv[i]$ we calculate the probability of $T < cv[i]$ using the technique for calculating power already discussed. This does not use much memory; we merely need a counter for each $cv[i]$. Finally, interpolate in this table to get the desired values.

6.12 Exercises

1. Given observations $y_0, \ldots, y_T$ with density $f^y(y_0, \ldots, y_T, \rho)$ depending on the parameter ρ, the score function for ρ is defined to be:

$$S(y, \rho) = \frac{\partial \log f^y(y_0, \ldots, y_T, \rho)}{\partial \rho}.$$

(a) Show that the score function for the autoregressive process is given by

$$S(y, \rho) = \frac{T\rho}{1 - \rho^2} - \frac{\rho}{(1 - \rho^2)^2} \sum_{t=1}^{T} (y_t - \rho y_{t-1})^2$$

$$+ \frac{1}{1 - \rho^2} \sum_{t=1}^{T} (y_t - \rho y_{t-1}) y_{t-1}.$$

(b) Show that the score function evaluated at $\rho = 0$ yields the locally most powerful test. Explain why this must be so.

(c) Show that the expected value of the score function $S(y, t)$ is zero. Hint: to evaluate the expectation of a function of y_t and y_{t-1}, first condition on y_{t-1}. That is, $\mathbb{E}H(y_t, y_{t-1}) = \mathbb{E}^{y_{t-1}} \, \mathbb{E}^{y_t|y_{t-1}} \, (H(y_t, y_{t-1})|y_{t-1})$.

(d) Show that the information about ρ in observations $y_0, \ldots, y_T$ is T. Hint: use the second expression in Eq. (6.2). The derivative of S is messy but not difficult to evaluate at $\rho = 0$.

2. The main idea of the central limit theorem is that the sum of a large number of random variables is well approximated by a normal distribution. In this exercise, we apply this to study the performance of the Durbin-Watson test.

(a) Let $\mathbf{y} = (y_0, \ldots, y_t)'$ be the column vector of observations on the autoregressive process. Show that $\mathbf{y} \sim N(0, \Omega)$, where $\Omega_{i,j} = \rho^{|i-j|}$. Hint: to show that y is multivariate normal, show that $y = M\epsilon$ for some matrix M. It is easy to show that $\mathbb{E}y_t = 0$ for all t. To calculate $\Omega_{i,j} = \mathrm{Cov}(y_i, y_j)$, first show that $\mathrm{Cov}(y_i, y_i) = 1$ for all i. Next, assume $i > j$, and write

$$y_i = \epsilon_i + \rho\epsilon_{i-1} + \cdots + \rho^{|i-j|-1}\epsilon_{j+1} + \rho^{|i-j|}y_j.$$

From this the result follows immediately.

(b) Show that $\mathbb{E}DW = \mathbb{E}\sum_{t=1}^{T} y_t y_{t-1} = T\rho$.

(c) Calculate $\mathrm{Var}(DW)$. This part is a bit messy so we break it up into steps.

   i. Show that

$$\mathrm{Var}(DW) = \left(\sum_{t=1}^{T}\sum_{s=1}^{T} \mathbb{E}y_t y_{t-1} y_s y_{s-1}\right) - T^2\rho^2.$$

   ii. For any indices $\alpha, \beta, \gamma, \delta$ show that

$$\mathbb{E}y_\alpha y_\beta y_\gamma y_\delta = \Omega_{\alpha,\beta}\Omega_{\gamma,\delta} + \Omega_{\alpha,\gamma}\Omega_{\beta,\delta} + \Omega_{\alpha,\delta}\Omega_{\beta,\gamma}.$$

Hint: let $m(\theta) = \mathbb{E}\exp(\theta'\mathbf{y}) = \exp[(1/2)(\theta'\Omega\theta)]$ be the m.g.f. of $\mathbf{y}$. Note that $\partial m(\theta)/\partial\theta_\alpha = m(\theta)(w_\alpha\theta)$, where w_α is the α-th row of the matrix Ω. Differentiate four times in this manner and evaluate at $\theta = 0$ to get the desired formula.

iii. Show that $\Omega_{t,t-1}\Omega_{s,s-1} = \rho^2$, $\Omega_{t,s}\Omega_{t-1,s-1} = \rho^{2|t-s|}$ and finally

$$\Omega_{t-1,s}\Omega_{t,s-1} = \rho^{|(t-s)-1|+|(t-s)+1|} = \rho^{2\min(|t-s|,1)}.$$

iv. Let $S = \sum_{t=1}^{T}\sum_{s=1}^{T}\rho^{|t-s|}$. Show that

$$S = 2\left(\sum_{t=2}^{T}\sum_{s=1}^{t-1}\rho^{t-s}\right) - \sum_{t=1}^{T}\sum_{s=t}^{T}\rho^{|t-s|}$$

The second term is clearly T. Use the fact that $1 + a + \cdots + a^k = (1 + a^{k+1})/(1-a)$ to prove that

$$S = 2\left(\frac{T+1}{1-\rho^2} - \frac{1-\rho^{2(T+1)}}{(1-\rho^2)^2}\right) - T.$$

v. Define $R = \sum_{t=1}^{T}\sum_{s=1}^{T}\rho^{2\min(|t-s|,1)}$. Show that

$$R = T\rho^2 + 2\left(\frac{T+1}{1-\rho^2} - \frac{1-\rho^{2(T+1)}}{(1-\rho^2)^2}\right) - 2T.$$

vi. Show that $\mathrm{Var}(DW) = T\rho^2 + S + R$.

(d) With formulas for $\mathbb{E}(DW)$ and $\mathrm{Var}(DW)$ in hand, we can approximate $DW \sim N(\mathbb{E}(DW), \mathrm{Var}(DW))$ Since DW is a sum of many variables. This approximation should be accurate for large T and small values of ρ. For large ρ, the dependence between successive terms makes convergence to normality quite slow. Use computer simulations to find out how large a T is needed for the normal approximation to be reasonable accurate for (i) $\rho = 0$ and (ii) $\rho = 0.1$.

> *Lost in the desert, the companions were anxiously search-*
> *ing for water. In desperation they decided to dig. Most*
> *dug a hole here, and another one there, abandoning their*
> *holes quickly to look for a better place in their anxiety.*
> *One man however put his trust in God and calmly keep*
> *digging deeper and deeper at the same spot. He found the*
> *water and also gave the others to drink.*

Chapter 7

UMP Invariant Hypothesis Tests

7.1 Introduction

We discussed in the previous chapter various ad-hoc approaches which can be used to find tests of high stringency. This chapter is devoted to an important situation where one can find the most stringent test by theoretical means. This is where the theory of invariance, developed by Hunt and Stein (1946) for this purpose, can be applied.

Invariant tests are based on the generalization of a very simple idea. Suppose we have observations on heights of people, and wish to test some hypothesis. Our procedures should not depend on the units of measurement — we should get the same result whether the height is measured in centimeters, inches, or some other unit. More generally, suppose we observe $X \sim f^X(x, \theta)$ and wish to test $H_0 : \theta \in \Theta_0$ versus $H_1 : \theta \in \Theta_1$. Make a change of variables $Z = g(X)$. For each density $f^X(x, \theta)$ we can calculate (via change of variables formula) the density for Z to be $f'^Z(z, \psi)$, where f' is some new parametric family, and the parameter ψ is some transformation of the parameter θ. For each $\theta \in \Theta_0$ we can find an equivalent $\psi \in \Psi_0$ and similarly find a transformation Ψ_1 of Θ_1. The hypothesis testing problem posed in terms of Z is called a *transformation* of the original problem. The invariance principle says that one should take the same decision in both the original and the transformed problem: if we reject $H_0 : \theta \in \Theta_0$ after observing X, we should also reject $H_0' : \psi \in \Psi_0$ after observing Z. If the transformation $g(X) = \lambda X$ then this is exactly a change in the units of measurement, but invariance allows for more general changes.

Apart from the direct intuitive appeal of invariance, the Hunt-Stein theorem establishes an important optimality property of invariant tests. The theorem shows that the best invariant test is, under certain conditions, the most stringent test. Several examples of such tests will be given in this and subsequent chapters.

7.2   A Simple Example

By changing units of measurement or other similar transformations, invariance permits us to transform one problem to another form, but this is rarely helpful, since the new problem is of equal difficulty with the old. However, there is one special case in which the invariance principle acquires added bite. This is when the new problem is effectively the same as the old one. An example will be helpful in illustrating this.

Suppose $X \sim N(\theta, 1)$ and we wish to test $H_0 : \theta = 0$ versus the alternative $H_1 : \theta \neq 0$. As discussed earlier, there is no UMP test for this problem. Consider the transformation $Z = -X$. Then $Z \sim N(\psi, 1)$ where $\psi = -\theta$. The transformed hypotheses are $H_0' : \psi = 0$ and $H_1' : \psi \neq 0$. Except for name changes (i.e. Z for X and ψ for θ) the new problem is *identical* to the old. This is called an *invariant transformation*. Let $\delta(X)$ be a hypothesis test: $\delta(X)$ takes values '0' or '1' accordingly as we reject or accept the null for the observed value X. According to the invariance principle, we must do the same thing for X as for Z, so that $\delta(X) = \delta(Z)$ must hold for all invariant hypothesis tests. Since $Z = -X$, all invariant tests satisfy $\delta(X) = \delta(-X)$ and hence are symmetric. If hypothesis tests are restricted to be invariant, then there is a unique best test, which is called the UMP invariant test. The key to proving the existence of a UMP invariant test is to note that all invariant tests are functions of $|X|$, or, equivalently, functions of $|X|^2$. In Lemma B.1 we show that if $Z = |X|^2$, we can find a Poisson random variable $J \sim Po(\psi^2/2)$ such that the conditional distribution of Z given J is central chi-square with $2J + 1$ degrees of freedom. Thus the density of Z is

$$f^Z(z, \theta) = \sum_{j=0}^{\infty} f^Y(y|J = j)\mathbf{P}(J = j)$$

$$= \sum_{j=0}^{\infty} \left(\frac{(\theta^2/j)^j}{j!} \exp(-\theta^2/2) \right) \frac{2^{-(j+1/2)}}{\Gamma(j + 1/2)} y^{j-1/2} e^{-y/2}. \quad (7.1)$$

For the case $\psi = 0$, the density of Z is χ_1^2, or just the first term of the series above. Thus the likelihood ratio is

$$LR(Z, \psi, 0) = \frac{f^Z(Z, \psi)}{f^Z(z, 0)}$$

$$= \sum_{j=0}^{\infty} \left(\frac{(\psi^2/j)^j}{j!} \exp(-\psi^2/2) \right) \frac{2^{-j}\Gamma(1/2)}{\Gamma(j + 1/2)} Z^j.$$

This is obviously monotone increasing in Z. By the monotone likelihood ratio criterion, a UMP test exists and rejects $H_0 : \psi = 0$ for large values of Z. The constant is chosen according to the null distribution of Z which is just χ_1^2. We note for future reference that the alternative distribution of Z is noncentral chi-square with one degree of freedom and noncentrality parameter $\theta^2/2$; this is denoted $\chi_1^2(\theta^2/2)$.

7.3 Data Reduction by Invariance

A better understanding of invariance is possible by comparing it with sufficiency. Suppose that we observe random variables (X, Y) with joint distribution $F^{(X,Y)}(x, y, \theta)$, and the distribution of $Y|X$ does not depend on the parameter of interest θ. In this case we say that X is sufficient for θ and to make inference about θ we can safely ignore Y.

Thus the principle of sufficiency permits us to throw away redundant information and simplify the problem. The principle of invariance is very similar in that it also attempts to throw away irrelevant information and simplify the problem.

As an example consider a linear regression model with $y = X\beta + \epsilon$, where $\epsilon \sim N(0, \sigma^2 I_T)$. By sufficiency, the data can be reduced to $\hat{\beta} = (X'X)^{-1}X'y$ and $s^2 = \|y - X\hat{\beta}\|^2$. These statistics are independent and $\hat{\beta} \sim N(\beta, \sigma^2(X'X)^{-1})$ and $s^2/\sigma^2 \sim \chi^2_{T-K}$. Suppose we wish to test a hypothesis regarding σ^2; for example $H_0 : \sigma^2 = 1$ versus $H_1 : \sigma^2 > 1$. In such situations, the unknown parameter β is called a *nuisance parameter*. It is not of interest to the experimenter, but he must deal with it. If β was known, then $S(\beta) = \|y - X\beta\|^2$ would be a sufficient statistic for σ^2 and there would be a UMP test. When β is unknown, no UMP test exists. To check this, note that the optimal UMP test is based on the value of $S(\beta)$ and is different for different values of β. Therefore, there is no test simultaneously most powerful for different values of β.

Since the distribution of $\hat{\beta}$ depends on σ^2, the parameter of interest, sufficiency does not permit us to ignore the value of $\hat{\beta}$. However, we will soon see that invariance allows us to ignore the value of $\hat{\beta}$ and reduce the data to just s^2. Of course, based only on s^2 there is a UMP test, and this test is UMP among all invariant tests (UMP invariant for short). Consider the transformation which maps $\hat{\beta}$ to $\hat{\psi} = \hat{\beta} + c$ and β to $\psi = \beta + c$. Then s^2 is unchanged and $\hat{\psi} \sim N(\psi, \sigma^2)$ has the same distribution as that of $\hat{\beta}$ except that β has been transformed to $\psi = \beta + c$. A transformation which only affects the nuisance parameter is automatically an invariant transformation, since it does not affect the null and the alternative hypotheses. The problem based on observing $s^2, \hat{\psi}$ is the same as the original one except for relabelling β as ψ, and hence by invariance, we should take the same decision in both. An invariant rule must satisfy for all c,

$$\delta(s^2, \hat{\beta}) = \delta(s^2, \hat{\beta} + c).$$

In this case, it is clear (by setting $c = -\hat{\beta}$, for example) that an invariant rule can only depend on the value of s^2. Thus, as stated earlier, invariance tells us to ignore the value of $\hat{\beta}$ and simplify the problem by considering only s^2.

Intuitively speaking, invariance tells us that the information in $\hat{\beta}$ is used up in estimating the nuisance parameter β, and there is nothing left over to help in estimating σ^2. Thus even though the distribution of $\hat{\beta}$ depends on σ^2 we can still ignore the value of $\hat{\beta}$ for hypotheses relating to σ^2. The example studied in the previous section can also be analyzed in these terms. If $X \sim N(\theta, 1)$, we can consider dividing the parameter and the observation into the magnitude and the sign: $\theta = (|\theta|, \text{sgn}(\theta))$ and $X = (|X|, \text{sgn}(X))$. Then, since the null hypothesis only involves $|\theta|$, effectively the sign of θ is a nuisance parameter, and the value of $\text{sgn}(X)$ only gives information pertaining to the nuisance parameter, and hence can safely be ignored from the invariance point of view. Further examples of data reduction by invariance will be presented in the sections to follow.

Sufficiency is a 'bulletproof' method of data reduction. All theories of inference, classical, Bayesian, and others, respect data reduction by sufficiency. Invariance is not so bulletproof; there exist examples where the best invariant procedures are not as good as noninvariant procedures. The reason for this is as follows. For the example of this section, suppose that some small amount of prior information about β is available. In this case it is clear that $\hat{\beta}$ will now provide information about σ^2 and hence it cannot be ignored, contrary to invariance. Similarly there are examples where even though no prior information about the nuisance parameter is available, one can still improve on the invariant procedure by making up some such information. The noninvariant procedure, which depends on the information about this nuisance parameter, works better than the invariant procedure when the made up information happens to be correct, and the same as the invariant when this information is incorrect. This is why data reduction by invariance must be applied more cautiously than sufficiency.

7.4 The Maximal Invariant

Applying invariance theory requires finding that part of the data which remains unaltered by the invariant transformations. Roughly speaking, all information about data which is preserved by the invariant transformations is called the maximal invariant. For example, the transfor-

mation X to $-X$ of our first example loses information about the sign
of X. However, $|X|$ or X^2 is unchanged by this transformation and
is therefore a maximal invariant. In the example of the linear regression model of the previous section, the transformations mapping $\hat{\beta}, s^2$
to $\hat{\beta} + c, s^2$ are invariant for testing hypotheses about σ^2. Clearly no
information about $\hat{\beta}$ is preserved by this transformation so that s^2 is
a maximal invariant. We will now discuss this concept in a somewhat
more formal way.

Given $X \sim f^X(x, \theta)$ with $H_0 : \theta \in \Theta_0$ and $H_1 : \theta \in \Theta_1$, a
transformation g is called *invariant* if it satisfies (i) for all $\theta \in \Theta_0$,
$Z = g(X) \sim f^X(x, \psi)$ for some $\psi \in \Theta_0$, and (ii) for all $\theta \in \Theta_1$,
$Z \sim f^X(x, \psi)$ for some $\psi \in \Theta_1$. In words, whenever the density of X
satisfies the null hypothesis, the density of $Z = g(X)$ must also satisfy
the null, and similarly for the alternative hypothesis. Given invariant
transformations g for a hypothesis testing problem, we are able to make
statements of the following type:

(A) A decision rule δ is invariant if and only if for all functions g
belonging to a given set (of invariant transformations) G, $\delta(X) = \delta(g(X))$.

Such characterizations of invariant rules are somewhat abstract and
difficult to use. Using the idea that invariance performs a reduction of
the data, it is possible to get a more concrete description of the set of
invariant rules which is of the following type:

(B) A decision rule $\delta(X)$ is invariant if and only if it can be written
as $\delta(X) = \delta'(m(X))$ for some fixed function $m(X)$.

The quantity $m(X)$ which occurs in (B) is called a *maximal invariant*.
More precisely, $m(X)$ is a maximal invariant for the transformations
$g \in G$ when a decision rule δ is invariant if and only if it can be written
as a function of $m(X)$. In our first example, the statement of type
(A) was that $\delta(X) = \delta(-X)$ for all X. To proceed, we argued that
this is true if and only if δ is a function of $|X|$ or X^2. Note that
both $m_1(X) = |X|$ and $m_2(X) = X^2$ can serve as maximal invariants.
Nonuniqueness is a general feature of maximal invariants. In fact, it
is easily verified that every one-to-one function of a maximal invariant

must also be a maximal invariant. One can prove that maximal invariants always exist in these kinds of problems. Hence one can always go from abstract statements of type (A) to more concrete statements of type (B). While there are no elementary methods for finding maximal invariants, applications in econometrics require only a small set of examples where the maximal invariant is easily guessed or derived. We will now use these techniques to derive best invariant tests for a variety of regression models, starting from simple cases and progression to more complex ones.

7.5 Nuisance Location Parameters

In a previous section, we studied an example where invariance was used to eliminate a location parameter which did not affect the null and the alternative hypothesis. Many examples of this kind are of interest. In this section we study one more example of this process.

If vector $X = (X_1, X_2, \ldots, X_n)$ has known densities $f_0^X(x)$ under the null and $f_1^X(x)$ under the alternative, an optimal test can be based on the likelihood ratio $LR(X) = f_1^X(X)/f_0^X(X)$. Sometimes the density can be specified only up to an unknown location parameter. That is, for some unknown $\mu \in \mathbb{R}$, $X - \mu\mathbb{1} = (X_1 - \mu, X_2 - \mu, \ldots, X_n - \mu)$ has the density f_0 and f_1 under the null and alternative, respectively (where $\mathbb{1} = (1, 1, \ldots, 1)$ is a vector of ones). The Neyman-Pearson Lemma is no longer applicable to this problem because the likelihood ratio for the observations X depends on the unknown parameter μ. Nonetheless, invariance can be used to eliminate this parameter and obtain a UMP invariant test.

Lemma 7.1 (Nuisance Location Parameter) *For $i = 0, 1$ suppose that under hypothesis H_i $X = (X_1, \ldots, X_n)$ has density $f_i^X(x_1 + \mu, \ldots, x_n + \mu) = f_i^X(x + \mu\mathbb{1})$. A UMP invariant test for H_0 versus H_1 rejects for large values of the statistic:*

$$T = \frac{\int f_1^X(X_1 + \mu, \ldots, X_n + \mu)\, d\mu}{\int f_0^X(X_1 + \mu, \ldots, X_n + \mu)\, d\mu}.$$

Proof: The proof of the lemma is given in a sequence of steps which are typical in the use of invariance to solve testing problems.

Step 1: Find Invariant Transformations Consider the transformation

$$g(X) = X + c\mathbb{1} = (X_1 + c, X_2 + c, \ldots, X_n + c).$$

Let $\mu' = \mu - c$ and $X' = g(X)$. Then $X' \sim f_i^X(X' + \mu'\mathbb{1})$ This shows that g is an invariant transformation for this problem. All invariant hypothesis tests must satisfy

$$\delta(X_1, X_2, \ldots, X_n) = \delta(X_1 + c, X_2 + c, \ldots, X_n + c). \qquad (7.2)$$

Step 2: Find Maximal Invariant This gives us a description of type A for the invariant hypothesis tests. To get a description of type B, we must find a maximal invariant. Finding the maximal invariant involves finding characteristics of the data which are not changed by the transformation. A little reflection shows that the differences $X_i - X_j$ are not affected by adding a common constant to each of the X_i. Thus a maximal invariant under the transformation is

$$Z_1 = X_1 - X_n, Z_2 = X_2 - X_n, \ldots, Z_{n-1} = X_{n-1} - X_n.$$

We can prove that $Z_1, \ldots, Z_n$ are maximal invariants by setting $c = -X_n$ in (7.2); then we see that all invariant rules must satisfy

$$\delta(X_1, X_2, \ldots, X_n) = \delta(Z_1, Z_2, \ldots, Z_{n-1}, 0).$$

This proves that all invariant rules can be written as functions of $Z_1, Z_2, \ldots, Z_{n-1}$.

Step 3: Density of Maximal Invariant To get the density of the maximal invariant $(Z_1, Z_2, \ldots, Z_{n-1})$, we must first introduce an auxiliary variable, e.g., $Z_n = X_n + \mu$. We can now use the change of variables formula to calculate the density of $Z = (Z_1, Z_2, \ldots, Z_n)$. Let $X_i' = X_i + \mu$ so that $X' = (X_1', \ldots, X_n')$ has density f_j^X with $j = 0$ under the null and $j = 1$ under the alternative. Since $Z_i = X_i' - X_n'$ for $i = 1, 2, \ldots, n-1$ while $Z_n = X_n'$, we have for $j = 0, 1$:

$$f_i^Z(z_1, \ldots, z_n) = f_j^X(z_1 + z_n, \ldots, z_{n-1} + z_n, z_n).$$

In order to get the density of the maximal invariant, we must integrate out the Z_n, getting

$$f_j^{(Z_1,\dots,Z_{n-1})}(z_1,\dots,z_{n-1}) = \int_{-\infty}^{\infty} f_j^X(Z_1 + Z_n,\dots,Z_{n-1} + Z_n, Z_n)\,dZ_n.$$

Step 4: Evaluate Likelihood Ratio The test can now be based on the likelihood ratio f_1/f_0. If we substitute $X_j - X_n$ for Z_j for $j = 1, 2,\dots, n - 1$, and make the change of variables $X_n + \alpha = Z_n$, $d\alpha = dZ_n$, we can write the likelihood ratio as:

$$LR(X) = \frac{\int_{-\infty}^{\infty} f_1(X_1 + \alpha, X_2 + \alpha,\dots, X_n + \alpha)\,d\alpha}{\int_{-\infty}^{\infty} f_0(X_1 + \alpha, X_2 + \alpha,\dots, X_n + \alpha)\,d\alpha}.$$

This proves the lemma.

7.6 Nuisance Scale Parameter

Similar to the case of a nuisance location parameter is that of a nuisance scale parameter, handled in this section. For $i = 0, 1$ suppose $X_1,\dots, X_n$ have density $\sigma^{-n} f_i(X_1/\sigma, X_2/\sigma,\dots, X_n/\sigma)$ under H_i. Note that if the scale parameter σ was known, the Neyman-Pearson lemma would be applicable. This testing problem remains invariant under changes in the units of measurement of the X_i as we shall show. This invariance can be used to eliminate the unknown parameter σ, and reduce both the null and alternative to a simple hypothesis.

Lemma 7.2 *The best invariant test of H_0 versus H_1 rejects for large values of the statistic T defined by*

$$T = \frac{\int_0^{\infty} \sigma^{-(n+1)} f_1(X_1/\sigma,\dots, X_n/\sigma)\,d\sigma}{\int_0^{\infty} \sigma^{-(n+1)} f_0(X_1/\sigma,\dots, X_n/\sigma)\,d\sigma}.$$

Proof: As before, four steps are needed.

Step 1: Invariant Transformations By calculating the transformed densities using the change of variables formula, it can be checked that the problem is invariant under scale changes $X'_i = \lambda X_i$ and $\sigma' = \lambda \sigma$. It follows that invariant transformations must satisfy $\delta(X_1, \ldots, X_n) = \delta(\lambda X_1, \ldots, \lambda X_n)$. Thus δ is homogeneous of degree 0 in the X_i, so it is a function of the ratios $R_1 = X_1/X_n$, $R_2 = X_2/X_n, \ldots$, and $R_{n-1} = X_{n-1}/X_n$.

Step 2: Maximal Invariant Thus $R = (R_1, \ldots, R_n)$ is a maximal invariant, as is easily proven by setting $\lambda = 1/X_n$. Of course X_n is arbitrary and any of the other variables would serve just as well. Note that the ratios effectively eliminate the unit of measurement of the X_i from the problem. As we shall now calculate, the density of the ratios R_i does not depend on the common scale parameter σ. Thus the null and alternative reduce to simple hypotheses, and the Neyman-Pearson lemma yields a most powerful test.

Step 3: Density of Maximal Invariant To find the density of the maximal invariant R introduce the auxiliary variable $R_n = X_n$. The inverse transformation is $X_i = R_i R_n$ for $i = 2, \ldots, n$ and $X_n = R_n$. The Jacobian of the transformation is $|R_n|^{n-1}$. For $i = 0, 2$ the joint density of R, R_n under H_i is, by the change of variables formula,

$$f_i^{(R, R_n)}(r_1, r_2, \ldots, r_{n-1}, r_n) = |r_n|^{n-1} \sigma^{-n} f_i^X(r_1 r_n/\sigma, \ldots, r_{n-1} r_n/\sigma, r_n/\sigma).$$

The density of the maximal invariant is obtained by integrating out r_n.

Step 4: Likelihood Ratio Thus, the Neyman-Pearson test rejects for large values of the following statistic:

$$
\begin{aligned}
NP(R_1, \ldots, R_{n-1}) &= \frac{\int f_i^{(R,R_n)}(r_1, r_2, \ldots, r_{n-1}, r_n)\, dr_n}{\int f_i^{(R,R_n)}(r_1, r_2, \ldots, r_{n-1}, r_n)\, dr_n} \\
&= \frac{\int \sigma^{-n} |r_n|^{n-1} f_1^X(r_1 r_n/\sigma, \ldots, r_{n-1} r_n/\sigma, r_n/\sigma)\, dr_n}{\int \sigma^{-n} |r_n|^{n-1} f_0^X(r_1 r_n/\sigma, \ldots, r_{n-1} r_n/\sigma, r_n/\sigma)\, dr_n}. \quad (7.3)
\end{aligned}
$$

This actually completes the derivation of the UMP invariant test, but a few change of variables will put the statistic into a nicer form. The formula of the lemma has the advantage over Eq. (7.3) in that it is expressed in terms of the original variables X_i, symmetric, and is clearly independent of σ. It can be obtained from Eq. (7.3) by a sequence of change of variables in the integral. Let $A_i = \int f_i^{(R,R_n)}(r_1, r_2, \ldots, r_{n-1}, r_n) dr_n$, so that the NP statistic is just A_1/A_0. Introduce $\lambda = r_n/\sigma$. Then $d\lambda = dr_n/\sigma$ and the integrals in the NP statistic can be written as

$$A_i = \int \lambda^{n-1} f_i^X(r_1\lambda, \ldots, r_{n-1}\lambda, \lambda) \, d\lambda.$$

This shows that the densities of the maximal invariant given in (7.3) do not in fact depend on the value of the nuisance parameter σ^2. In this integral, we can replace r_i by the original variables x_i/x_n. To make the expression symmetric in the x_i it is convenient to make a second change of variable $\sigma = x_n/\lambda$. Then $d\lambda = (-x_n/\sigma^2) \, d\sigma$ and the integral above transforms to

$$
\begin{aligned}
A_i &= \int \lambda^{n-1} f_i^X(x_1\lambda/x_n, \ldots, x_{n-1}\lambda/x_n, \lambda) \, d\lambda \\
&= \int (x_n/\sigma)^{n-1} f_i^X(x_1/\sigma, \ldots, x_{n-1}/\sigma, x_n/\sigma)(x_n d\sigma/\sigma) \\
&= |x_n|^n \int \sigma^{-(n+1)} f^X(x_1/\sigma, \ldots, x_n/\sigma) \, d\sigma
\end{aligned}
$$

Substituting this last expression into the numerator and denominator of Eq. (7.3) results in the formula of the lemma.

It is also possible to eliminate both location and scale nuisance parameters. Such a case will be considered in a later chapter.

7.7 Normal Mean: Known Variance

In this section we consider an important example of invariant testing which will also prove useful to us later. We observe $X \sim N(\beta, \sigma^2 I_K)$ and wish to test $H_0 : \beta = 0$ versus the alternative $H_1 : \beta \neq 0$. The best invariant test can be described as follows.

Theorem 7.1 *The UMP invariant test rejects for large values of the statistic $Z = \|X\|^2/\sigma^2$. Under the null hypothesis, $Z \sim \chi_K^2$ so to get a* *level α UMP invariant test we reject H_0 for $Z > c_\alpha$ where c_α is the α upper tail probability for a χ_K^2 density.*

Proof: As usual, the proof follows a sequence of standard steps.

Step 1: Find Invariant Transformations Let P be any orthogonal matrix and consider the transformation $Y = g(X) = PX$. Let $\theta = P\beta$, so $Y \sim N(\theta, \sigma^2 \mathbf{I})$, since $PP' = \mathbf{I}$. Now $\theta = 0$ if and only if $\beta = 0$, in terms of Y we have $H_0 : \theta = 0$ and $H_1 : \theta \neq 0$. Except for a change of names (Y for X and θ for β) the problem is the same. For all orthogonal matrices P, invariant rules must satisfy

$$\delta(X) = \delta(PX).$$

Step 2: Find Maximal Invariant To find the maximal invariant, we must ask ourselves which characteristics of a vector remain unchanged under premultiplication by an arbitrary orthogonal matrix. If we think of representing vectors by their (generalized) spherical coordinates, the answer becomes obvious. All of the angles can be changed by the orthogonal transformations, but the length of the vector remains invariant. We will now prove that $Z = \|X\|^2$ is the maximal invariant under the orthogonal transformations described above. Let P be an orthogonal matrix with the first row equaling $(1/\|X\|^2)X$. Then $PX = (\|X\|, 0, 0, \ldots, 0)$, so that for invariant rules δ, we have

$$\delta(X) = \delta(PX) = \delta((\|X\|, 0, 0, \ldots, 0)).$$

This shows how to write invariant rules as functions of $\|X\|$ alone. It follows from this that an invariant rule can only depend on the length $\|X\|$ or equivalently, $\|X\|^2$.

Step 3: Find Density of Maximal Invariant To show that the density of $Z = \|X\|^2$ has monotone likelihood ratio, we first derive the density.

Lemma 7.3 *Let J be Poisson with parameter $\|\beta\|^2/2\sigma^2$ and let Y given J have conditional density χ_{2J+K}^2. Then the density of $Z = \|X\|^2/\sigma^2$ is the same as the marginal density of Y.*

Proof: Let Q be orthogonal with the first row proportional to β, and define $W = QX/\sigma$, so that $W \sim N(Q\beta/\sigma, I)$. It is easily checked that $Z = W'W = W_1^2 + W_2^2 + \cdots + W_K^2$. Now W_i are independent and $W_1 \sim N(\|\beta\|/\sigma, 1)$, while $W_i \sim N(0, 1)$ for $i > 1$ since $Q\beta = (\|\beta\|, 0, 0, \ldots, 0)'$. We know from Lemma B.1 that if J is Poisson with parameter $\|\beta\|^2/2\sigma^2$ and Y' given J has conditional density χ^2_{2J+1}, the marginal density of Y' is the same as that of W_1^2. Now $Y'' \equiv W_2^2 + W_3^2 + \cdots + W_K^2$ is obviously χ^2_{K-1}, and hence, conditional on J, $Z = Y' + Y''$ is the sum of independent χ^2_{2J+1} and χ^2_{K-1} variables. It follows that conditional on J, Z has χ^2_{2J+K} distribution, proving the lemma.

Step 4: Check Density for MLR Having found the distribution of the maximal invariant, we check to see if it has monotone likelihood ratio. The density can be written explicitly as

$$
\begin{aligned}
f^Z(z, \beta) &= \sum_{j=0}^{\infty} f^Y(y|J = j) \mathbf{P}(J = j) \\
&= \sum_{j=0}^{\infty} \left\{ \frac{(\|\beta\|^{2j}}{j!\, (2\sigma^2)^j} \exp\left(-\frac{\|\beta\|^2}{2\sigma^2}\right) \right\} \frac{(y/2)^{j+(k/2)-1}}{2\Gamma(j + k/2)} e^{-y/2}.
\end{aligned}
$$

Note that the density only depends on $\|\beta\|^2/2\sigma^2$. With $\beta = 0$ the density reduces to the first term of the series, a central chi-square. Fixing $\beta^* \neq 0$, the likelihood ratio is

$$
\begin{aligned}
LR(Z, \beta^*, 0) &= \frac{f^Z(Z, \beta^*)}{f^Z(Z, 0)} \\
&= \sum_{j=0}^{\infty} \left\{ \frac{(\|\beta\|^{2j}}{j!\, (2\sigma^2)^j} \exp\left(-\frac{\|\beta\|^2}{2\sigma^2}\right) \right\} \frac{2^{-j}\Gamma(k/2)}{\Gamma(j + k/2)} Z^j.
\end{aligned}
$$

This is monotone increasing in Z. It follows that a UMP invariant test exists and rejects for large values of Z. Finally, it is easily checked that Z has a χ^2_K density under the null, completing the proof of the theorem.

7.8 Normal Mean: Unknown Variance

We consider exactly the same hypothesis as in the previous section, but drop the assumption that the variance is known. Suppose now that $X \sim N(\beta, \sigma^2 \mathbf{I}_k)$, where σ^2 is unknown. However, we have an estimate $\hat{\sigma}^2$ independent of X with the property that $m\hat{\sigma}^2/\sigma^2 \sim \chi_m^2$. The best invariant test is now characterized as follows.

Theorem 7.2 *The best invariant test for $H_0 : \beta = 0$ versus the alternative $H_1 : \beta \neq 0$ rejects for large values of the statistic $V = \|X\|^2/K\hat{\sigma}^2$. Under the null this has an $F(K, m)$ distribution (a central F distribution with K and m degrees of freedom).*

Proof: The usual four steps are needed.

Step 1: Invariant Transformations   In addition to the orthogonal group of invariant transformations mapping X to PX, we also have change in units of measurement, also called scale change. If $\hat{\psi} = \lambda X$, $\psi = \lambda \beta$, $t^2 = \lambda^2 s^2$ and $\nu = \lambda \sigma$, then the transformed problem can be stated as follows. We observe $\hat{\psi} \sim N(\psi, \nu^2 \mathbf{I}_K)$ and t^2 independent of $\hat{\psi}$ with $t^2/\nu^2 \sim \chi_{T-K}^2$. Obviously this is identical to the original problem except for relabelling and hence the scale change is an invariant transformation. Thus invariant rules must satisfy two conditions. For all orthogonal matrices P and all positive scalars λ,

$$\delta(X, s^2) = \delta(PX, s^2)$$
$$\delta(X, s^2) = \delta(\lambda X, \lambda^2 s^2)$$

Step 2: Maximal Invariant   Invariance relative to the orthogonal transformations requires δ to be a function of $\|X\|^2$: $\delta(X, s^2) = \phi(\|X\|^2, s^2)$. When two groups of invariant transformation operate on the same problem, as in the present case, some care is required in reductions by invariance. It is necessary to do the sequence of operations in the 'right order'. In the present case, we must check that the scale invariance continues to operate relative to the distributions of the invariants $\|X\|^2$ and s^2. Now $\|X\|^2/\sigma^2 \sim \chi_K^2(\|\beta\|^2/2)$ and $s^2/\sigma^2 \sim \chi_{T-K}^2$. Making the scale change described in the previous paragraph, we find that $\|\hat{\psi}\|^2/\nu^2 \sim \chi_K^2(\|\psi\|^2/2)$ and $t^2/\nu^2 \sim \chi_{T-K}^2$, so that the scale change continues to be an invariant transformation. Invariance now requires tests ϕ to satisfy $\phi(\|X\|^2, s^2) = \phi(\lambda^2 \|X\|^2, \lambda^2 s^2)$. It should be obvious that the only information which will survive an arbitrary scale change intact is the ratio of the two quantities, $M = \|X\|^2/s^2$. This is the maximal invariant, as is easily proven by setting $\lambda = 1/s^2$ in the previous equation to show that all invariant rules can be written as functions of this ratio.

Step 3: Density of Maximal Invariant It is convenient to define the *unadjusted F(m,n) distribution* denoted $F^*(m,n)$ hereafter to be the distribution of the ratio X/Y where X and Y are independent chi square variables with m and n degrees of freedom respectively. The usual F distribution is of course 'adjusted for degrees of freedom' and defined as $F = (X/m)/(Y/n) = (n/m)F^*$. Thus the unadjusted F or F^* is just a scalar multiple of the usual F distribution. Using this relationship (or directly from the definition) it is easily shown that if $R \sim F^*(m,n)$ the density of R is

$$f^R(r) = C_{m,n}\left(\frac{r^{m-(1/2)}}{(1+r)^{m+(n/2)}}\right), \tag{7.4}$$

where the constant $C_{m,n} = \Gamma(m+(n/2))/\{\Gamma(m+(1/2))\Gamma((n-1)/2)\}$. We can now give a formula for the density of the maximal invariant.

Lemma 7.4 *Let J be Poisson with parameter $\|\beta\|^2/2\sigma^2$. Suppose Y conditional on J has an unadjusted F distribution $Y|J \sim F^*(2J + K, T - K)$. Then the marginal distribution of Y is the same as that of the maximal invariant $M = \|X\|^2/s^2$.*

Proof: From Lemma 7.3, we can find J Poisson with parameter $\|\beta\|^2/2\sigma^2$ and Y' with conditional density $Y|J \sim \chi^2_{2J+K}$ such that Y has the same marginal density as $Z = \|X\|^2/\sigma^2$. Now $M = Z/(s^2/\sigma^2)$, and hence, conditional on J, M is the ratio of two independent chi-square variables with $2J + K$ and $T - K$ degrees of freedom. It follows that conditionally on J, M has an unadjusted F distribution with degrees of freedom as claimed in the lemma.

Step 4: Monotone Likelihood Ratio Finally we check the density of M for monotone likelihood ratio. Define $\theta = \|\beta\|^2/2\sigma^2$. Using

Lemma 7.4 and Eq. (7.4), we can write the density of M as

$$f^M(m,\theta) = \sum_{j=0}^{\infty} \left(\frac{\theta^j}{j!} e^{-\theta} \right) C_{2j+K,T-K} \left(\frac{m^{2j+k-1/2}}{(1+m)^{2j+k+(T-K)/2}} \right).$$

Under the null hypothesis the density of M is $F^*(K, T-K)$, or just the first term of the series. Thus the likelihood ratio can be written as

$$LR(M,\theta,) = \frac{f^M(M,\theta)}{f^M(M,0)} = \sum_{j=0}^{\infty} \left(\frac{\theta^j}{j!} e^{-\theta} \right) \frac{C_{2j+K,T-K}}{C_{K,T-K}} \left(\frac{m}{1+m} \right)^{2j}.$$

This is clearly monotone increasing in M. Thus based on the maximal invariant M there is a UMP test which rejects for large values of M. It is obviously equivalent to reject for large values of $V = (T - K)M/K$ which has a standard F distribution and is also equivalent to the likelihood ratio test.

7.9   The Hunt-Stein Theorem

We have now derived several tests which have the property of being best among the invariant tests. It is natural to ask whether such tests have any optimality property relative to class of all tests. This question is answered by the Hunt-Stein theorem which shows that such tests are most stringent. This is an optimality property relative to the class of all tests.

An invariant test satisfies $\delta(X) = \delta(gX)$ for all invariant transformations g belonging to some set G. Given a noninvariant test, we can find the average value of $\delta(gX)$ over all $g \in G$ and fix δ^* to take this averaged action. Then δ^* is an invariant version of δ. By the general properties of averaging, δ^* will have a smaller maximum risk (and a larger minimum risk) than the noninvariant δ. A few additional steps are required to show that this implies that δ^* will be more stringent than δ. This implies the desired result; if for every noninvariant δ we can find an invariant δ^* which is more stringent, than the best invariant test will automatically be the most stringent test. Formalizing this result using the Hunt-Stein theorem requires a deeper study of the collection of invariant transformations, which we now undertake.[1]

We observe $X \sim f^X(x, \theta)$ and wish to test $H_0 : \theta \in \Theta_0$ versus the alternative $H_1 : \theta \in \Theta_1$. An invariant map $g(X)$ has the property that if $Y = g(X)$ then $f^Y(y) = f^X(y, \theta')$ for some $\theta' = g'(\theta)$ satisfying the same hypothesis as θ. Thus an invariant map g on the sample space induces a map g' on the parameter space. Now note that if g is an invariant transformation, the inverse transformation g^{-1} is also invariant. Furthermore, given invariant transformations g and f, the compositions $f \circ g$ and $g \circ f$ are also invariant. Of course, the identity map is always an invariant transformation. This shows that the collection of invariant transformations G forms a group under function composition.

The process of averaging tests necessitates the consideration of randomized tests. Any function $\delta(X)$ taking values in the closed interval $[0, 1]$ is a randomized test with the interpretation that if $\delta(X) = p$ we use some randomization mechanism independent of the statistic X to reject the null with probability p and accept with probability $1 - p$. With this interpretation, $R(\delta(X), \theta)$, the probability that δ rejects the null when θ is the true parameter is given by

$$R(\delta(X), \theta) = \int \delta(X) f^X(x, \theta)\, dx.$$

An important implication of the invariance property is that for any test $\delta(X)$, the rejection probability can be written as (using change of variables formula in the integral)

$$
\begin{aligned}
R(\delta(g(X)), \theta) &= \int \delta(g(X)) f^X(x, \theta)\, dx \\
&= \int \delta(Y) f^X(y, g'(\theta))\, dy = R(\delta(X), g'(\theta)).
\end{aligned}
$$

With this preparation, we first present an elementary version of the Hunt-Stein theorem.

Theorem 7.3 (Hunt-Stein:Elementary Version) *If the group of transformations is finite or compact, then the best invariant test is most stringent.*

---

[1] Readers willing to take this optimality property of best invariant procedures on faith can skip this section without loss of continuity.

Proof: First suppose G is a finite group[2] with $G = \{g_1, g_2, \ldots, g_m\}$. For any decision rule $\delta(X)$ define the average $\delta^*(X) = \frac{1}{m} \sum_{i=1}^{m} \delta(g_i(X))$. It is easily verified that δ^* is invariant: $\delta^*(X) = \delta^*(g_i(X))$ for any $i = 1, 2, \ldots, m$. This follow from the fact that the map $g : G \to G$ defined by $g(g_i) = g \circ g_i$ is a permutation of the group G, for every $g \in G$. We now show that δ^* is more stringent than δ.

We will first show that δ^* has the same or smaller probability of type I error than δ. It is easily seen that

$$
\begin{aligned}
\sup_{\theta \in \Theta_0} R(\delta^*(X), \theta) &= \sup_{\theta \in \Theta_0} \frac{1}{m} \sum_{i=1}^{m} R(\delta(g_i(X)), \theta) \\
&= \sup_{\theta \in \Theta_0} \frac{1}{m} \sum_{i=1}^{m} R(\delta(X), g_i'(\theta)) \\
&\leq \sup_{\theta \in \Theta_0} R(\delta(X), \theta)
\end{aligned}
\tag{7.5}
$$

Next, for any $\theta \in \Theta_1$ define the points $\theta_i = g_i'(\theta)$ and consider the stringency of δ over the collection of alternative points $\theta_1, \theta_2, \ldots, \theta_m$. We claim that the power envelope is invariant: $\beta_\alpha^*(\theta) = \beta_\alpha^*(g'(\theta))$. Let $\delta(X)$ be a level α rule achieving maximum power at $g'(\theta)$. Then the rule $\delta(g(X))$ has exactly the same power at θ. This shows that the maximum power at $g'(\theta)$ cannot be higher than that at θ. By symmetrical reasoning, it cannot be lower either. Since the power envelope is constant on $\theta_1, \ldots, \theta_m$, the stringency of any rule is determined by its minimum power over this set. By reasoning exactly analogous to that in Eq. (7.5), the minimum power of δ^* must be higher than the minimum power of δ establishing that invariant rules are more stringent than noninvariant ones.

For the case in which G is compact, the same reasoning as in the finite case works after defining the invariant version of $\delta(X)$ as $\delta^*(X) = \int \delta(g(X))\mu(dg)$ where μ is the normalized Haar measure on the compact group G. In the more general case of noncompact groups G, we have the following result.

---

[2] In our first example, where $X \sim N(\mu, 1)$, the invariant transformations are $X' = -X$ and the identity, $X' = X$. This is isomorphic to the finite group of integers $\{-1, 1\}$ under the group operation of multiplication.

Theorem 7.4 (Hunt-Stein:Advanced Version) *If the group G of invariant transformations is* amenable *then the best invariant hypothesis test is most stringent.*

The difficulty in this version of the theorem arises because Haar measure on noncompact groups is infinite and hence cannot be normalized to be a probability measure. This difficulty is handled in a straightforward way by approximating the Haar measure by a sequence of finite measures. The hypothesis of amenability ensures that a suitable sequence of approximate finite measures exists. The proof of this theorem will not be presented here. For the definition of amenability, references to the literature, and proofs, the reader is referred to Bondar and Milnes (1981). For our applications, it is sufficient to note that:

1. All compact groups are amenable.

2. All Abelian (i.e., commutative) groups are amenable.

3. Direct products of amenable groups are amenable.

In all of the examples of this chapter, as well as all best invariant tests presented later, the groups of transformations satisfy the conditions above. It follows that in all cases considered, the best invariant test is also the most stringent test of the hypothesis in question.

7.10 Exercises

1. Suppose X has the Cauchy density with parameter μ:

$$X \sim \frac{1}{\pi} \frac{1}{1 + (X - \mu)^2}.$$

(a) If $H_0 : \mu = 0$ and the alternative is $H_1 : \mu = 4$ or $\mu = -4$, show that there is no UMP test.(Hint: show that the MP 10% test for $H_1 : \mu = 4$ is different from the MP 10% test for $H_1 : \mu = -4$). Show that there is a UMP invariant test, and calculate it for the 10% level.

(b) Suppose we have a prior density for μ of the following form: with probability α, $\mu = 0$ while with probability $1 - \alpha$, μ has the prior density $\pi(\mu)$. Show that the Bayes test for $H_0 : \mu = 0$ versus $H_1 : \mu \neq 0$ rejects H_0 whenever

$$\int_{-\infty}^{\infty} \frac{1}{1 + (X - \mu)^2} \pi(\mu) d\mu > \frac{1}{1 + X^2}.$$

(c) Show that the test in (b) is invariant if π is a symmetric density: $\pi(\mu) = \pi(-\mu)$.

2. $X_1, X_2, \ldots, X_n \overset{iid}{\sim} N(\mu, \sigma^2)$ $H_0 : \sigma^2 = \sigma_0^2$ $H_1 : \sigma^2 = \sigma_1^2$. This is like a simple hypothesis test, except for the fact that there is an unknown translation parameter μ. Since the value of μ is not relevant in either hypothesis, μ is often called a nuisance parameter.

(a) Show that this problem is invariant under the transformation

$$g(X_1, X_2, \ldots, X_n) = (X_1 + c, X_2 + c, \ldots, X_n + c).$$

(b) Reduce by sufficiency first, then find a maximal invariant for this hypothesis testing problem.

(c) Calculate the density of the maximal invariant, and show that it has MLR. What is the level 10 % UMP invariant test?

3. If a density has the form $f^X(x) = (1/\lambda)f((x - \mu)/\lambda)$ then this is called a location scale family, and μ is the location parameter, while λ is the scale parameter.

(a) Suppose $(X_1, X_2, \ldots, X_n)$ is an i.i.d. sample from a location scale family $f^X(x) = (1/\lambda)f((x - \mu)/\lambda)$. Show that the family of distributions is invariant under the transformations $g(X_1, \ldots, X_n) = (aX_1 + b, aX_2 + b, \ldots, aX_n + b)$.

(b) Show that a maximal invariant under the transformations in (c) is: $(X_1 - X_{n-1}/X_n), (X_2 - X_{n-1})/X_n, \ldots, (X_{n-2} - X_{n-1})/X_n$.

(c) Calculate the density of the maximal invariant.

4. Suppose we wish to test $H_0 : \mu = 0$ versus $H_1 : \mu \neq 0$, given an i.i.d. sample $X_1, \ldots, X_n \overset{i.i.d.}{\sim} N(\mu, 1)$. Show that $S = |\sum_{i=1}^{n} X_i|$ is a maximal invariant (after reduction by sufficiency). Calculate the density of S (not S^2), and show that it has MLR. Construct the level 90% test when $n = 10$.

5. $X_1, X_2, \ldots, X_n \overset{iid}{\sim} N(\beta, I_k)$. $H_0 : \beta = 0$ $H_1 : \beta \neq 0$. Find the UMP invariant tests. Assume $k = 10$ and $n = 25$, find the level 95% test among these.

6. $X_1, X_2, \ldots, X_n \overset{iid}{\sim} N(\mu_1, \sigma^2), Y_1, Y_2, \ldots, Y_m \overset{iid}{\sim} N(\mu_2, \sigma^2)$. Suppose all X's and Y's are independent. $H_0 : \mu_1 = \mu_2, H_1 : \mu_1 \neq \mu_2$. Define $\overline{X} = \frac{1}{n} \sum X_i, \overline{Y} = \frac{1}{m} \sum Y_i$, and $S^2 = \sum_{i=1}^{n}(X_i - \overline{X})^2 + \sum_{j=1}^{m}(Y_j - \overline{Y})^2$. Show that $(\overline{X}, \overline{Y}, S^2)$ are sufficient statistics for μ_1, μ_2, and σ^2. Show that the transformations $g(\overline{X}, \overline{Y}, S^2) = (\overline{X} + C, \overline{Y} + C, S^2)$ and $g(\overline{X}, \overline{Y}, S^2) = (\lambda \overline{X}, \lambda \overline{Y}, \lambda^2 S^2)$ leave the problem invariant. Find the maximal invariant and the best invariant test.

When Hoja Nasruddin arrived in a small village, every-
one was talking of a recent theft. The Hoja informed the
villagers that his donkey had the power to detect thieves.
Following his instructions, a secluded and enclosed tent
was set up with the donkey inside, and all the villagers
were lined up. Everyone was to go into the tent one by
one and stroke the donkey's tail. Supposedly the don-
key would bray loudly on being stroked by the thief. All
villagers went in and out, but the donkey did not bray.
'Your donkey has failed,' several people said. 'Not so
fast,' said the Hoja, and asked all the villagers to hold
out their hands. All hands but one were shiny with oil,
and the Hoja immediately caught the dry-handed person,
who confessed to the theft. 'I had oiled the tail of the
donkey,' the Hoja explained.

Chapter 8

Some Tests for Regression Models

8.1 Introduction

In the previous chapters, we have studied finite sample optimality properties of tests. Cases where UMP invariant tests can be found are rare
in practice, and situations where numerical evaluation of stringency is
possible are also few and far between. Even though direct applicability
of the theory is limited, it is useful to have some clarity on the qualities

we are looking for in tests. Furthermore, asymptotic theory reduces very general classes of problems to a sufficiently simple form where the tools developed earlier are applicable. The present and subsequent chapters on hypothesis testing are devoted to the interface between the rarefied realms of theory and the demands of practice.

An issue of essential importance in practice, which does not emerge naturally from theory, is that the goal of testing strongly influences the choice of the test statistic. Hypothesis tests are done for diverse, multiple, and occasionally conflicting purposes. For the sake of clarity in the following discussion, we will somewhat arbitrarily categorize goals of testing as follows: note that these are rather broad and general categories, and we shall see later that tests can be conducted for other purposes as well.

1. *Tests of Auxiliary Assumptions:* In formulating a statistical or a regression model, we inevitably make several assumptions solely for convenience of analysis. Assumptions not required by our theory nor strictly necessary for the statistical analysis are called *auxiliary.* Since our aims and interests lie elsewhere, tests of auxiliary assumptions are not carried out as frequently as they should be — a point stressed by Hendry (1991) in his discussion of methodology. Tests of auxiliary assumptions differ from other types of tests since we are not directly interested in the truth or falsity of the assumption. Rather, we would like to be reassured by such tests that there is no gross conflict between our auxiliary assumptions and that data, and be able to get on with our job. Occasionally a test will reveal a gross conflict and force us to rethink and reformulate more suitable auxiliary assumptions.

2. *Tests of (Pure) Significance:* On many accounts, the essence of scientific method consists of confronting a theory with data and seeing if the data conform to the predictions of the theory. Repeated confirmations add weight to, but can never conclusively prove a theory. A single decisive refutation may be enough to force reformulation and development of alternative theories. A test of pure significance is supposed to tell us of the degree of compatibility between a given theory and the data. A strong

conflict, corresponding to a rejection of the null at a high significance level, signals the need to reformulate the theory, or to find other suitable explanation.

3. *Comparison of Alternative Theories:* Here we would like to decide which of two (or several) competing theories is correct. This differs somewhat from the traditional setup in that acceptance or rejection are not the only alternatives. We may find that both theories are more or less compatible with the data (if one theory is more and the other is less compatible, this may or may not be of significance depending on the context). Thus we may end up accepting both the null and the alternative. It is also possible that both theories are grossly in conflict with the data, and thus we reject both the null and the alternative. Finally, we have the more conventional possibility that one theory is strongly in conflict with the data and the other is more or less compatible with the data, so that we reject one hypothesis and accept the other. These issues are dealt with in the theory of nonnested hypothesis tests, a topic which we will only refer to tangentially in this text. For an evaluation of three types of tests for nonnested hypotheses, see Zabel (1993). For an expository discussion with extensive references, see Judge *et al.* (1985).

How should we proceed if no hypothesis test with finite sample optimal properties is available? Numerous testing principles have been developed in response to this question. We suggest that whenever possible, the likelihood ratio principle should be used to construct tests. Other methods, such as Lagrange multiplier and Wald, are asymptotically equivalent to LR when evaluated by Pitman's criterion for asymptotic efficiency of tests. However, several other criteria reveal that the LR is preferred to these tests in general. These are discussed in greater detail in Chapter 15. Our study of stringencies in Chapter 6 also revealed the superiority of the LR in a particular example. As a general principle, alternatives to the LR should be only considered or used when we can find a good reason to justify not using the LR.

Chapter Summary: Three important assumptions (typically of an auxiliary type) in the regression model are the normality, homoskedasticity, and independence of the errors. Tests for each of these assumptions are discussed. We derive the likelihood ratio test for linear restrictions on regression coefficients in a linear model. The F test so obtained is very flexible and can be used in many different ways. The next chapter is devoted to a variety of applications of the F test. For each test derived, we also discuss any known optimality properties associated with the test.

8.2 Normality of Errors

When we assume $\epsilon \sim N(0, \sigma^2 \mathbf{I}_T)$, there are three separate assumptions involved: (i) normality, (ii) equal variances, or homoskedasticity, and (iii) independence. These are important auxiliary assumptions in a regression model, and we will discuss testing all three, starting with a test for normality.

   This problem of testing whether or not the error terms in a regression are normal does not fit nicely into the theoretical frameworks we have developed. See Madansky (1988) for a complete description and details of how to perform a variety of tests for normality, together with a discussion of their relative powers. Tests based on an idea due to Kolmogorov compare the empirical distribution function (see Section 12.5 for a definition) of the errors with the normal distribution function. One can obtain an asymptotic distribution for the maximum difference and base a test on this. In regression, matters are complicated by the fact that errors are not directly observed, and the parameters are estimated. While the Kolmogorov test is asymptotically effective against *all* alternative distributions, it is somewhat cumbersome to compute. Furthermore, it may be possible to improve the power of the test by focusing on some smaller class of suitable alternatives.

   What class of alternatives to normality deserves special attention? This requires consideration of the purpose of testing for normality. Typically our concern will be with the distribution of estimators and test statistics derived under the assumption of normality. The theory of Edgeworth expansions, studied in Chapter 14, shows that the first term in the difference between smooth functions of sums of normal and non-normal random variables arises due to differences in the third moment.

Higher order moments have successively smaller effects. Thus a good practical possibility is to check the third and fourth moments of the errors and see if they match the moments expected for the normal distribution.

Given a random variable X, let $Z = (X - \mathbb{E}X)/\sqrt{\mathbb{V}ar(X)}$ be the standardized version of X. Then

$$\kappa_3(X) = \mathbb{E}Z^3 = \mathbb{E}\frac{(X - \mathbb{E}X)^3}{\mathbb{V}ar(X)^{3/2}}$$

is defined to be the *skewness* of X. It is clear that skewness is zero for variables symmetric around their mean (including normals). If X is normal, then it is easily calculated that $\mathbb{E}Z^4 = 3$. For general random variables, we define the fourth *cumulant*

$$\kappa_4(X) = \mathbb{E}Z^4 - 3 = \frac{(X - \mathbb{E}X)^4}{\mathbb{V}ar(X)^2}$$

to be a measure of the *kurtosis* of X. Roughly speaking, this is a measure of the heaviness of the tail of the distribution compared to the normal. See Stuart and Ord (1987, §3.32 and exercises 3.20-21, 3.25) for further detail on the relation between the shape of the distribution and the kurtosis measure. The concept of cumulants and their relation to normality is discussed further in Section 1.2. As indicated earlier, it makes sense to focus on these two measures of difference from normality, as the practical effect of lack of normality makes itself felt most often through these measures.

Many tests devised for normality do in fact focus on these measures. For example, the Jarque-Bera (1987) test is based on a weighted average of the skewness and kurtosis measure. Let $\hat{\kappa}_3$ and $\hat{\kappa}_4$ be the estimates of the third and fourth cumulants. These are obtained by setting $\hat{\mu} = \sum_{i=1}^{n} X_i/n$ and $\hat{\sigma}^2 = \sum_{i=1}^{n}(X_i - \hat{\mu})^2/(n-1)$ and defining the standardized variables $Z_i = (X_i - \hat{\mu})/\hat{\sigma}$. We can then estimate $\hat{\kappa}_3 = \sum_{i=1}^{n} Z_i^3/n$ and $\hat{\kappa}_4 = (\sum_{i=1}^{n} Z_i^4/n) - 3$. Then the Jarque-Bera test rejects the null hypothesis of normality for large values of

$$JB = n\left(\frac{\hat{\kappa}_4^2}{24} + \frac{|\hat{\kappa}_3|}{6}\right)$$

The critical value at which we reject the null must be determined by Monte Carlo methods, since the exact distribution is too complex. Jarque and Bera provide tables of critical values based on Monte Carlo sample sizes of 250. Because of the small sample size, their tabulated values are relatively inaccurate, and more precise values are given by Deb and Sefton (1994), who also discuss other tests and make power comparisons.

Jarque and Bera (1987) show that the test is a Lagrange multiplier test for normality against a class of alternatives known as the Pearson family of distributions. As already discussed in Chapter 6, and further detailed in Section 11.4.2 and Section 15.5, this is not a reliable optimality property. Nonetheless, the test can tentatively be recommended since it is based on skewness and kurtosis, performs reasonably in empirical power studies, and good alternatives are not known. Applying this test to regression models is easy, since we simply use the skewness and kurtosis estimates of the regression residuals based on a least squares fit.

One issue which appears not to have been explored is the effect of using least squares estimates (as opposed to more robust estimates) on testing residuals for normality. It seems clear from our study of robust estimators that the use of OLS can easily cause masking of outliers. If a robust regression reveals outliers, this gives a relatively clearer indication of lack of normality of residuals, and so should provide sharper tests.

8.3   Independence of Errors

A child struggling with the concept of π in grade school appealed to 'daddy' for help. The father, who experienced similar difficulties, was very sympathetic. He decided to exploit his position in the Indiana legislature to put an end to this difficulty once and for all. He proposed a bill setting the value of π equal to 3. The bill almost passed. Vigorous appeals by some professors from Purdue prevailed in the end, and the bill was defeated by a narrow margin.

Lack of independence of the errors, a standard auxiliary assumption in regression, causes serious damage to all results associated with regression. To rule out this possibility, it is traditional to test for first order (and higher order if necessary) autocorrelation of the error terms. The standard test for this, implemented in all computerized regression packages, is the Durbin-Watson. We would like to propose that the use of autocorrelated errors be henceforth banned. Once this is done, there will be no further need to test for autocorrelated errors. At first glance, this proposal appears to resemble the proposal to set π equal to 3. However, the use of errors which form an *innovations process*, which amounts to the same thing, has been recommended by many authors. Perhaps Hendry (1991) has presented and defended the case most vigorously. The case is based on the following proposition:

Ban on Autocorrelated Errors: For any regression model with autocorrelated errors, there exists an observationally equivalent model with martingale difference (and hence uncorrelated) errors.

We illustrate this by a simple example. Suppose $y_t = \beta' x_t + \epsilon_t$, where $\epsilon_t = \rho \epsilon_{t-1} + \nu_t$ and ν_t is i.i.d. By substituting $y_{t-1} - \beta' x_{t-1}$ for ϵ_{t-1} we can rewrite the model as $y_t = \beta' x_t + \rho(y_{t-1} - \beta' x_{t-1}) + \nu_t$. This model no longer has autocorrelated errors. A similar procedure can *always* be carried out, and we can ensure the errors are uncorrelated with previous errors. We first give an intuitive explanation of this. Any observable correlation among the errors shows that it is possible to predict the current period error using the past observations on the errors. This shows that the current model is deficient since it does not give the best possible prediction for the current observations. It is always possible to rectify this deficiency by incorporating any systematic relationship among the errors into the model (as was done in the example above). To be more technical, let $\mathcal{I}_t$ be the information set at time t. This includes all current variables available to predict y_t, as well as past values of all variables. Our object is to forecast y_t optimally given the currently available information. A good forecast must have the property that $\hat{y}_t - y_t$ should be orthogonal (or uncorrelated) with the information set $\mathcal{I}_t$; otherwise, it would be possible to improve.

A commitment to the use of models in which the errors form an innovations process involves rethinking tests for autocorrelation. The object of such a test is to discover if there are unmodeled effects coming from last period which influence variables in the current period. One piece of information from the last period is the error $y_{t-1} - \beta'x_{t-1}$, but that is not the complete information set for the last period. Unless we have strong theoretical reasons to believe the y_{t-1} and x_{t-1} can only influence the current outcome y_t through the residual, the logical approach is to include both. Thus, instead of testing for autocorrelation, we should carry out a test of the null hypothesis that $H_0 : \alpha = 0, \beta_1 = 0$ in the following augmented model:

$$y_t = \beta'x_t + \alpha y_{t-1} + \beta_1'x_{t-1} + \epsilon_t. \qquad (8.1)$$

If we cannot reject the null that $\alpha = 0$ and $\beta_1 = 0$, it is automatic that the residuals will not display noticeable correlation. This is because the first order error autocorrelation model $\epsilon_t = \rho\epsilon_{t-1} + \nu_t$ is a special case of the model above with parameters $\alpha = \rho$ and $\beta_1 = -\rho\beta$. Whereas an autocorrelation test picks up only a very special kind of dynamic misspecification, the suggested test picks up a very general set of misspecifications. See Hendry (1991) for further details on the generality of Eq. (8.1), which he terms the ADL model. An important problem with the conventional methodology of autocorrelation testing is that any type of dynamic misspecification can be reflected by a significant Durbin-Watson. Thus the Cochrane-Orcutt procedure, which corrects for autocorrelation, can easily lead to faulty deductions. An illustration of the errors which are possible is detailed by Hendry (1991). To test the null hypothesis $H_0 : \alpha = 0, \beta_1 = 0$ in Eq. (8.1), we can use an F test. Details of the procedure are presented in Section 9.7.

8.4   Types of Heteroskedasticity Tests

Tests of equality of variances for the regression errors can be carried out for different reasons. Three main reasons are listed below:

1. We carry out a standard OLS analysis. We would like to know if failures of the auxiliary assumptions invalidate or cast doubts on our results.

2. We are directly interested in the question of whether or not the error terms in a given equation display heteroskedasticity. For example, the question is of theoretical interest in studies of stock price variations.

3. We have some regressors in mind which may account for the fluctuations in variance of the error terms. Or, we may wish to develop such a model, in order to improve efficiency of estimation by taking greater advantage of low variance errors, and discounting high variance errors in a suitable way.

As we will see, three different tests are suitable for the three different goals. For the first goal, we will see that fluctuations in error variances which are uncorrelated with the regressors and the squares and cross products of the regressors do *not* affect the asymptotic distribution of OLS. Thus in order to validate an OLS analysis, it is necessary to test for a special type of heteroskedasticity. A test for this, called the Eicker-White test, is developed in White (1980) (It is a special case of the Breusch-Pagan test to be discussed later). See any recent econometrics text, for example Greene (1990), for a more digestible presentation of the test statistic. There is a common misunderstanding regarding this test, propagated in certain texts, that it is effective against all types of heteroskedasticity. It is also an interesting fact that if the regression coefficients fluctuate randomly (instead of being constant), the errors in a model which assumes them to be constant will display precisely the type of heteroskedasticity which the Eicker-White test tests for. Thus the Eicker-White test is also a test for random parameter variation. See Chesher (1984) for a proof of this fact. See also Section 1.4 and exercise 7 on page 388 for further details.

If we are interested in detecting arbitrary fluctuations in variances, not just those related to the regressors in the regression equation, a different test is more suitable. This test falls out as a special case of a general technique due to Pitman (1938), and is discussed in Section 8.5. This test will detect arbitrary fluctuations in variances provided that the errors are normal. Dutta and Zaman (1990) have shown that one cannot detect arbitrary fluctuations in variance if the density of the errors is not specified. More precisely, in finite samples, there always exist patterns of heteroskedasticity which will escape detection by any test (in the sense that the power of the test will be less than or equal to the size of the test) if the common error density is unknown. If one can specify the common density as normal, it is possible to detect arbitrary patterns of heteroskedasticity using the Pitman principle. Asymptotically, this problem disappears and one can detect arbitrary patterns of heteroskedasticity without knowing the common error density.

The Eicker-White test is useful for validating OLS results. Pitman's test is a general test for detecting any kind of heteroskedasticity, when we have no special model for heteroskedasticity in mind. If we have some explanation for heteroskedasticity then a third type of test is suitable. If we suspect that fluctuations in variance are related to some particular explanatory variable or variables, the Breusch-Pagan test is useful in testing this type of hypothesis. This test is discussed later. Some other tests for heteroskedasticity are also discussed below.

8.5 Pitman's Unbiased Test

We digress to develop a general testing principle which sheds light on properties of several different types of tests to be discussed. First consider a situation where the null hypothesis specifies that random variable X has density $f_0(x)$. We have no particular alternative in mind, but would like to assess the extent to which the null hypothesis is compatible with the data. Such a test is called a test of pure significance, and the most common way of carrying it out is via a *tail test* (different in type from the tail test of the chapter story). We accept the null if the observed value of X is in the center of the density and reject if it is in the tail. Formally, reject the null whenever $f_0(X) < C$, where C is a constant chosen to give the desired significance level under the null. There has been considerable philosophical debate over whether it makes sense to test a hypothesis when no alternatives are contemplated. We will see that the tail test falls out naturally from an approach due to Pitman (1938). This also clarifies the implicit alternative being considered in a tail test (see also exercise 3). We begin by defining the concept of an unbiased test.

Definition 8.1 *Let T be a test for $H_0 : \theta \in \Theta_0$ versus the alternative $H_1 : \theta \in \Theta_1$. Let $R(T, \theta)$ be the probability of rejecting the null when*

θ *is the true parameter. T is called an unbiased test if for all $\theta_1 \in \Theta_1$ and $\theta_0 \in \Theta_0$, $R(T, \theta_1) > R(T, \theta_0)$.*

In other word, a test is unbiased if it has a higher probability of rejecting the null when the null is false than when it is true. We will now develop the result of Pitman in stages, starting with the simplest situation.

Lemma 8.1 *Suppose random variables $X_1, \ldots, X_n$ have joint density $f^X(x_1 - \theta_1, \ldots, x_n - \theta_n)$. We wish to test $H_0 : \theta_1 = \theta_2 = \cdots = \theta_n = 0$ against the alternative that $H_1 : \theta_i \neq 0$ for some i. The test which rejects the null for small values of $P_0 = f^X(X_1, \ldots, X_n)$ is unbiased.*

Remark: Note that the test of the lemma is a tail test. It has the intuitively plausible property that it rejects the null when likelihood of the null is smaller than some specified constant — equivalently, it rejects the null when the observation is in the tail of the density. The lemma shows that such a test is unbiased against translates of the density. This clarifies the implicit alternatives of this classical test of pure significance.

Proof: To prove the lemma, define the set $A_c = \{(x_1, x_2, \ldots, x_n) : f^X(x_1, \ldots, x_n) \geq c\}$. This is the acceptance region of the test. For any subset A of $\mathbb{R}^n$ and $\theta = (\theta_1, \ldots, \theta_n) \in \mathbb{R}^n$, define $A + \theta = \{y \in \mathbb{R}^n : y = x + \theta \text{ for some } x \in \mathbb{R}^n\}$. The set $A + \theta$ is called a translate of A. It is intuitively obvious that A_c is more probable than any of its translates under the null hypothesis. See exercise 1 for details. Under the null hypothesis X has density $f^X(x)$. Under the alternative θ, $X + \theta$ has the same density $f^X(x)$. Therefore, probability of accepting the null when θ holds is $\mathbf{P}(X \in A_c | \theta) = \mathbf{P}(X + \theta \in A_c | H_0) = \mathbf{P}(X \in A_c - \theta | H_0) < \mathbf{P}(X \in A_c | H_0)$. Thus probability of accepting the null is highest when the null is true. This is equivalent to the lemma.

Next we show how the tail test can be used to derive an unbiased test for equality of the translation parameters θ.

Theorem 8.1 (Pitman) *Suppose random variables $X_1, \ldots, X_n$ have joint density $f^X(x_1 - \theta_1, \ldots, x_n - \theta_n)$. We wish to test $H_0 : \theta_1 = \theta_2 =$*

$\cdots = \theta_n$ *against the alternative that* $H_1 : \theta_i \neq \theta_j$ *for some pair* i, j. *The test which rejects the null for small values of*

$$P_1 = \int_{\mathbb{R}} f^X(x_1 + \theta, \ldots, x_n + \theta)\, d\theta$$

is unbiased.

We will prove this theorem by reducing it to the previous case using invariance. Note that if we define $Y_i = X_i + \delta$ and $\psi_i = \theta_i + \delta$, the hypothesis testing problem remains invariant. Thus invariant hypothesis tests must satisfy $T(X_1, \ldots, X_n) = T(X_1 + \delta, \ldots, X_n + \delta)$. As already discussed in Section 7.5, a maximal invariant is $Z_1 = X_1 - X_n, Z_2 = X_2 - X_n, \ldots, Z_{n-1} = X_{n-1} - X_n$. Let $Z = (Z_1, \ldots, Z_{n-1})$ be the maximal invariant. Using the same techniques as in Chapter 7, the density of the maximal invariant can be written as

$$f^Z(Z) = \int_{-\infty}^{\infty} f_j^X(Z_1 + Z_n - \theta_1, \ldots, Z_{n-1} + Z_n - \theta_{n-1}, Z_n - \theta_n)\, dZ_n.$$

Now define $\psi_j = \theta_j - \theta_n$ for $j = 1, 2, \ldots, n-1$, and make the change of variable $\theta = Z_n - \theta_n - X_n$ in the above integral to get

$$f^Z(Z) = \int_{-\infty}^{\infty} f_j^X(Z_1 + X_n + \theta - \psi_1, \ldots, Z_{n-1} + X_n + \theta - \psi_{n-1}, X_n + \theta)\, d\theta.$$
(8.2)

Note that under the null hypothesis $\psi_j = 0$ for all j while $\psi_j \neq 0$ for some j under the alternative. Since the form of this problem is exactly that of the previous lemma, we conclude that an unbiased test can be obtained by rejecting for small values of the density evaluated at $\psi_j = 0$. Substituting $\psi_j = 0$ and $Z_j = X_j - X_n$ into Eq. (8.2) yields the desired result. It is easily checked that a test which is unbiased for the transformed hypothesis regarding ψ_j is also unbiased for the original problem. For some other optimality properties of the Pitman test, see exercise 2.

Since scale parameters can be converted to translation parameters by taking logs, we can devise a test for equality of scale parameters using Pitman's procedure. This test also corresponds to the well-known Bartlett test for equality of variances in the special case when the underlying distribution is normal. Details are given in the following section.

8.6 Bartlett's Test for Equality of Variances

We now consider the problem of testing the equality of variances in a regression model, when we have no information about the reasons for fluctuations in the variances. That is, we have no model of the type $\sigma_t^2 = f(z_t)$, where z_t are regressors which explain fluctuations in the variances. Before carrying out such a test, it is essential to be clear on the purpose of the test. If we are testing heteroskedasticity as an auxiliary assumption, with the idea of ensuring that the OLS results we have obtained are sound, then the appropriate condition to test for is that of *generalized homoskedasticity relative to the regressors X*, as discussed in detail in Section 1.4. Only a special type of heteroskedasticity, namely fluctuations in variance correlated with cross-products of regressors, has an adverse effect on OLS results. To carry out a test of the type under consideration here, we must be interested in heteroskedasticity for its own sake. The author can think of two possible situations where this might hold; there may of course be others. Suppose that OLS estimates have variances which are too high relative to our goals — that is, we wish to know the value of some parameters with greater accuracy than is possible using OLS. If we have reason to believe that the errors are heteroskedastic, we may be able to improve efficiency of OLS by placing greater weight on the errors with low variances. As a first step, we will need to learn whether or not the errors are indeed heteroskedastic. A second situation is where heteroskedasticity is of intrinsic interest. For example, the average return to stocks may be fixed by rational expectations, and the variance may respond to various types of market conditions. In that case identifying whether the variances fluctuate and the variables which explain such fluctuations would be of direct economic interest. A general test for heteroskedasticity may be suitable in such circumstances.

Consider first a situation simpler than regression. Suppose that for $j = 1, 2, \ldots, J$, we observe s_j^2 which are independent, each having distribution $\sigma_j^2 \chi_{t_j}^2$, and wish to test the null hypothesis **EV** of equality of variances, that $H_0 : \sigma_1^2 = \cdots = \sigma_J^2$. The likelihood ratio test is easily derived.

Theorem 8.2 *The likelihood ratio statistic for the test of* **EV** *based on* $s_1^2, \ldots, s_s^2$ *rejects the null for large values of*

$$P = \frac{\left(\prod (s_j^2)^{t_j}\right)^{1/t}}{(1/t)\sum_{j=1}^J s_j^2}.$$

The proof, based on straightforward maximization of the likelihood under the null and alternative, is given in exercise 5. We now discuss the properties of this test. The relation to regression models is discussed separately in Section 8.6.1.

We now show that the test described above in Theorem 8.2 is a special case of the Pitman test and hence must be unbiased. The Pitman test as described above applies to testing equality of translation parameters. Suppose we observe $Z_i = X_i/\sigma_i$ with some joint density $F^Z(Z_1, \ldots, Z_n)$ and wish to test equality of the scale parameters σ_i. The transform $Y_i = \log Z_i = \log X_i - \log \sigma_i$ converts the scale parameters σ_i to the location parameters θ_i. It is a straightforward though somewhat tedious computation to show that making this transformation and applying the formula of Theorem 8.1 leads to the likelihood ratio statistic already derived in Theorem 8.2 earlier. See exercise 4 for details. This shows that the likelihood ratio statistic is unbiased. It is interesting to note that if we apply the likelihood ratio test directly to the original problem (instead of reducing by invariance first, as done in the previous section), this leads to a biased test statistic. See exercise 6.

Ostensibly, this test is a test for heteroskedasticity (or failure of the assumption **EV** equality of variances). However, there are several reasons why it may be more appropriate to regard this as a test for normality of the errors. First, the test is very sensitive to normality, in the sense that (unlike the F test for linear restrictions to be discussed later) the rejection probability that $P > c$ can change substantially depending on the density of the errors. In particular, even for densities fairly 'close' to normal, the rejection probability can be quite different from the nominal size of the test.[1] This problem has been much discussed in the statistics literature, and several alternative tests for

---

[1] The term *nominal size* is used to refer to the type I probability that a test is supposed to have. The *actual size* is the real type I error probability. The two can

homogeneity of variances have been constructed which are less sensitive to the assumption of normality. One possibility which turns out to have good performance is to use the bootstrap distribution of the test statistic (rather than the theoretical asymptotic distribution) to construct the critical value. See Boos and Brownie (1989).

A more fundamental issue is the following. Suppose for example that σ_j^2 are themselves i.i.d. according to some density g. If the error term ϵ_j is $N(0, \sigma_j^2 \mathbf{I}_{T_j})$ conditionally on σ_j^2 and σ_j^2 has the marginal density g, then the (marginal) density of the error term can be written as

$$f(\epsilon_j) = \int_0^\infty \left(2\pi\sigma_j^2\right)^{-T_j/2} \exp\left(\|\epsilon_j\|^2/2\sigma_j^2\right) g(\sigma_j^2) d\sigma_j^2.$$

This is some nonnormal density (known as a mixture of normals), and the errors are actually i.i.d. according to this density! Thus a problem which appears to be heteroskedasticity for normal errors is caused by i.i.d. nonnormal errors. In general, whenever the variances fluctuate for reasons *uncorrelated* with the regressors and their cross products it may be possible to reparametrize the model in terms of different error densities and eliminate heteroskedasticity. Even if the error terms have different densities, the result will be a sequence ϵ_t with independent but not identical densities f_t. These densities may or may not differ in their variances; even if they have identical variances, they may vary with respect to other features. The issue of testing homoskedasticity versus heteroskedasticity as currently formulated is closely linked to the assumption that the errors are normal. An issue which has not been explored is how matters are affected if we first test for normality of errors, and then use Bartlett's test. If normality is not to be taken as fundamental, then it may be better to reformulate the question in the following way. Suppose that ϵ_t are i.n.i.d. with distributions F_t; do these distributions have a common scale factor? In this reformulation, a measure of scale more robust than variances may be suitable. This type of variant has not been explored, and it is unclear whether or not it would be superior to existing approaches.

---

frequently differ since we construct a test on the basis of several assumptions and approximations which may fail to hold.

8.6.1   Extensions to Regression Models

The Bartlett test as derived above is not directly applicable to regression models. Several variants have been proposed which are applicable to regression models. There is no clear evidence on the relative merits of the various proposals. Again this is an area where research is needed.

A simple idea is to split up a regression into several regimes, and estimate the variance separately in each regime. The separate estimates of variance are independent chi squares and hence the Bartlett test is directly applicable. The problem with this idea is that in testing for heteroskedasticity, a maintained hypothesis is that the regression parameters are constant across regimes. If regression parameters are estimated separately in each regime, there is considerable loss of efficiency — see Section 8.7 for a more detailed discussion. On the other hand, if the parameters are constrained to be equal then the variance estimates from separate regimes are no longer independent. Ad-hoc tests include extensions of a statistic of Dutta and Zaman (1990), and also Ramsey's BAMSET (see Judge *et al.* (1983) for references and discussion). It appears likely that the straightforward likelihood ratio test will be the best. It is computationally feasible, and the critical values can be computed by Monte Carlo. Some evidence on the power of this test in the case of two regimes is provided in Tomak (1994) and Tomak and Zaman (1995). See also Section 8.7 for further discussion of the case of two regimes.

8.6.2   Distributional Issues

Since Bartlett's test is a likelihood ratio test, its asymptotic distribution is χ^2 using standard arguments (see Section 1.8). Bartlett modified the statistic so as to make its finite sample distribution closer to its asymptotic distribution. See Stuart and Ord (1991, Section 23.9) for the derivation. This modification, known as Bartlett's correction, has become the basis for a large literature. See Section 14.6 for further discussion.

Instead of asymptotic approximations, the exact distribution of the Bartlett statistic can also be derived with some ingenuity and effort. Dyer and Keating (1980) give the distribution for the special case of equal sample sizes in each regimes, and provide approximations for the case of unequal sample sizes. Nagarsenker (1984) gives formulas for the general case of unequal sample sizes.

A major problem with the Bartlett test is that it is highly sensitive to normality. See Rivest (1986) for simulations of the test statistic with nonnormal distributions. There is a vast literature which devises robust alternatives to the Bartlett test. Boos and Brownie (1989) show that the problem of sensitivity to normality is substantially reduced if the critical values of the test are computed by bootstrapping. Furthermore, they show that the power of the bootstrapped test is comparable to that of the best robust alternatives.

8.7 Breusch-Pagan and Goldfeld-Quandt Tests

The Goldfeld-Quandt (GQ) is a widely used and popular test for heteroskedasticity. Given a regression model $y_t = x_t\beta + \epsilon_t$ for $t = 1, 2, \ldots, T$, Let $\sigma_t^2 = \text{Var}(\epsilon_t)$. Consider testing the null hypothesis $H_0 : \sigma_1^2 = \cdots = \sigma_T^2$ of homoskedasticity versus the alternative that the variances are increasing order: $\sigma_1^2 \leq \sigma_2^2 \leq \cdots \leq \sigma_T^2$. Goldfeld and Quandt (1965) suggested the following test. Form two regression models $Y_1 = X_1\beta_1 + e_1$ and $Y_2 = X_2\beta_2 + e_2$, where the first model is based on observations $1, \ldots, T_1$ and the second on $T_1 + K, \ldots, T$. K central observations are omitted to heighten the contrast between the low and high variances in the two regimes. The GQ test rejects for large values of SSR_1/SSR_2, where SSR_1 and SSR_2 are the sum of squared residuals from the first and second groups of observations respectively. This statistic has an F distribution under the null hypothesis.

This test does not use three important pieces of information about the setup, and hence cannot be efficient. The first and perhaps least important is that some information is thrown away by omitting the middle K observations. More important, we show in Section 10.5 that this is a UMP invariant test for a different setup — it is an optimal test for the equality of variances in the two subsets of data, when the regression coefficients are *not* assumed to be the same. Since the test

is optimal for the situation where common regression coefficients are
not assumed, it is clear that the information about this equality across
regimes is not being used by the test. Utilizing this information should
improve the test. In fact, when the data are split into two parts and the
regression coefficients are estimated separately on each half of the data
set, in effect we utilize only *half* of the data for each estimate. This is
a substantial amount of information which is neglected, and gains from
incorporating this information are potentially significant. Tomak and
Zaman (1995) study power loss from neglecting this information, and
show that it can be substantial. Finally, since the GQ test is optimal for
the situation where the variance is constant in each of the two regimes,
it is clear that the information that the variance is increasing from one
observation to the next is not being utilized. In fact, the ability to order
the observations according to increasing variance means that we know
some variable or variables which cause the variance to increase (or are
correlated with such increases). Taking these sources which determine
the variance into account explicitly must improve test performance.
When such information about variances is available, the Breusch-Pagan
test is frequently used, which we now discuss.

The Breusch-Pagan test can be derived as a Lagrange multiplier
test, but this form is very sensitive to normality of errors. An alternative form of the statistic is easily understood and motivated intu-
itively. Let $e_t^2 = (y_t - x_t\hat{\beta})^2$ be the OLS residuals. Asymptotically $\hat{\beta}$
converges to β, so that e_t^2 should be close to ϵ_t^2 which has expected
value σ_t^2. To discover whether σ_t^2 is adequately explained by regressors
$Z_1, Z_2, \ldots, Z_m$, we just regress ϵ_t^2 on the constant term plus the regressors $Z_1, \ldots, Z_m$ — our regression equation is $e_t^2 = \beta_0 + \sum_{i=1}^{m} \beta_i Z_{it} + \nu_t$.
An F test can be used to test the null hypothesis $H_0 : 0 = \beta_1 = \cdots =
\beta_k$. In fact, this is just the overall F test for the regression discussed
in the following chapter. Optimality properties of this test are not
clear; see also Section 10.2.4, which suggests that the test may be poor.
Derivation of the test as a Lagrange multiplier test makes it appear
that the test is effective against alternatives which are heteroskedastic
sequences based on arbitrary functions of the regressors Z_t. However,
this added generality is illusory, as discussed in Section 11.4.2. A second caveat is that the asymptotic distribution of the test is typically a
poor fit in finite samples. It is therefore preferable to use Monte Carlo

or bootstrap methods to derive the critical values.

The Eicker-White test for *generalized homoskedasticity* with respect to the OLS regressors is a special case of the Breusch-Pagan test. It can be obtained by taking the regressors Z_j to be all the products $X_i \times X_j$, for $i = 1, \ldots, K$ and $j = 1, \ldots, K$ of the OLS regressors. Typically the regressors include a constant term, and $X_1 = 1$ so that regressors themselves are also included in the products. In some cases, the products will have linear dependencies, in which the linearly dependent vectors should be taken out of the set of pairwise products. Optimality properties of these tests have not been evaluated and it appears likely that these will be inferior to suitable likelihood ratio tests.

8.8   The LR Test for Linear Restriction

We have seen earlier that the likelihood ratio test has high stringency in the simple autocorrelation model. The likelihood ratio test has several optimality properties which make it a good candidate for an initial choice of a test statistic. We will show in a later section that the likelihood ratio test for linear restrictions is most stringent.

In order to calculate the likelihood ratio test, we need the maximum likelihood estimates of β under the null and the alternative hypotheses. Let $\hat{\beta} = (X'X)^{-1}X'y$ be the unrestricted ML estimate of β (as calculated earlier). The probability of $\hat{\beta}$ satisfying the constraint $R\hat{\beta} = r$ is zero, since the set of points which satisfies this constraint is a lower dimensional subspace of the vector space $\mathbb{R}^K$; all lower dimensional subspaces have volume zero, and hence integrating the density of $\hat{\beta}$ over a set of volume zero will yield 0. This establishes that $R\hat{\beta} \neq r$ with probability one, and hence $\hat{\beta}$ is the ML estimate under H_1 with probability one. The following lemma gives the ML estimates for the regression model under the null hypothesis.

Lemma 8.2 *Suppose β satisfies the constraint $R\beta = r$ in the linear regression model $y = X\beta + \epsilon$, where $\epsilon \sim N(0, \sigma^2 I_T)$. Assume that R is an $m \times k$ matrix of rank m, while r is an $m \times 1$ column vector. Then the maximum likelihood estimates of the parameters $\tilde{\beta}, \tilde{\sigma}^2$ are given by*

$$\tilde{\beta} = \hat{\beta} - (X'X)^{-1}R' \left\{ R(X'X)^{-1}R' \right\}^{-1} \left(R\hat{\beta} - r \right) \quad (8.3)$$

$$\tilde{\sigma}^2 = \frac{1}{T}\|y - X\tilde{\beta}\|^2. \tag{8.1}$$

Proof: The likelihood function for the observations y is

$$l(y,\beta,\sigma^2) = \left(2\pi\sigma^2\right)^{-T/2}\exp\left(-\frac{1}{2\sigma^2}\|y - X\beta\|^2\right). \tag{8.5}$$

We will maximize $\log l$ subject to the constraint $R\beta = r$. It is convenient to introduce the Lagrange multipliers $\lambda = (\lambda_1,\ldots,\lambda_m)$ and write the m constraints as $(1/\sigma^2)(R\beta - r) = 0$. Then the Lagrangean can be written as

$$\mathcal{L}(\beta,\sigma^2,\lambda) = -\frac{T}{2}\log(2\pi\sigma^2) - \frac{1}{2\sigma^2}\|y - X\beta\|^2 + \frac{1}{\sigma^2}\lambda'(R\beta - r).$$

To calculate the constrained ML estimates, we must solve the first order conditions on Lagrangean $\mathcal{L}$. Taking the gradient with respect to β yields

$$0 = \nabla_\beta\mathcal{L} = -\frac{1}{2\tilde{\sigma}^2}\left(-2X'y + 2(X'X)\tilde{\beta}\right) + \frac{1}{\tilde{\sigma}^2}R'\tilde{\lambda}.$$

Solving for $\tilde{\beta}$, we get

$$\tilde{\beta} = (X'X)^{-1}\left(X'y - R'\tilde{\lambda}\right). \tag{8.6}$$

Taking the gradient with respect to λ just yields the constraint $R\tilde{\beta} = r$, which can be solved for λ as follows.

$$r = R\tilde{\beta} = R(X'X)^{-1}\left(X'y - R'\tilde{\lambda}\right) = R\hat{\beta} - R(X'X)^{-1}R'\tilde{\lambda}.$$

From this it follows that

$$\tilde{\lambda} = \left\{R(X'X)^{-1}R'\right\}^{-1}(R\hat{\beta} - r).$$

Substituting this value into Eq. (8.6) yields the expression in Eq. (8.3) for $\hat{\beta}$ in the lemma. To get Eq. (8.4), we differentiate $\mathcal{L}$ with respect to σ^2:

$$0 = \frac{\partial\mathcal{L}}{\partial\sigma^2} = -\frac{T}{2\tilde{\sigma}^2} + \frac{1}{2\tilde{\sigma}^4}\|y - X\tilde{\beta}\|^2 - \frac{1}{\sigma^4}\lambda'(R\tilde{\beta} - r).$$

The last term is zero since $R\tilde{\beta} = r$. Solving yields Eq. (8.4), proving the lemma.

Once we have the constrained and unconstrained MLEs, it is an easy matter to obtain the likelihood ratio test.

Theorem 8.3 *In the linear regression model $y = X\beta + \epsilon$, a likelihood ratio test for the hypothesis $H_0 : R\beta = r$ can be based on rejecting the null for large values of the following statistic:*

$$S = \frac{T - K}{m} \frac{\|y - X\tilde{\beta}\|^2 - \|y - X\hat{\beta}\|^2}{\|y - X\hat{\beta}\|^2}.$$

Under the null hypothesis, this statistic has the F distribution with m and $T - K$ degrees of freedom.

Proof: To see that the LR test can be based on S, we simply write out the likelihood ratio statistic. With l as in (8.5), this is:

$$
\begin{aligned}
LR(y) &= \frac{\sup_{\beta,\sigma^2} l(y, \beta, \sigma^2)}{\sup_{\beta,\sigma^2 : R\beta = r} l(y, \beta, \sigma^2)} = \frac{l(y, \hat{\beta}, \hat{\sigma}^2)}{l(y, \tilde{\beta}, \tilde{\sigma}^2)} \\
&= \frac{\left(2\pi\hat{\sigma}^2\right)^{-T/2} \exp\left(-\|y - X\hat{\beta}\|^2/2\hat{\sigma}^2\right)}{\left(2\pi\tilde{\sigma}^2\right)^{-T/2} \exp\left(-\|y - X\tilde{\beta}\|^2/2\tilde{\sigma}^2\right)} \\
&= \frac{\hat{\sigma}^{-T}}{\tilde{\sigma}^{-T}} = \left(\frac{\|y - X\tilde{\beta}\|^2}{\|y - X\hat{\beta}\|^2}\right)^{T/2}.
\end{aligned}
$$

From this it follows that $S = ((T - K)/m)((LR)^{2/T} - 1)$, so that S is a monotone function of LR, showing that the LR test can be based on S. To prove that this has the F distribution claimed, we will need the following lemma.

Lemma 8.3 *Let $\hat{u} = y - X\hat{\beta}$ and $\tilde{u} = y - X\tilde{\beta}$. Then the following equalities hold:*

$$\tilde{u}'\tilde{u} - \hat{u}'\hat{u} = (\hat{\beta} - \tilde{\beta})'(X'X)(\hat{\beta} - \tilde{\beta}) = (R\hat{\beta} - r)' \left\{R(X'X)^{-1}R'\right\}^{-1} (R\hat{\beta} - r).$$

Remark: Before proceeding note that Lemma 8.3 furnishes us with three different algebraically equivalent expression for the F statistic, each of which are useful in different contexts.   Yet another form is given in exercise 4.   In applications, the expression $F = \{(T - K)(\tilde{u}'\tilde{u} - \hat{u}'\hat{u})\} / \{m\hat{u}'\hat{u}\}$ is very useful, since $\hat{u}'\hat{u}$ is the sum of squared residuals from an unconstrained regression, and $\tilde{u}'\tilde{u}$ is the sum of squared residuals from a constrained regression.   Many computer packages for regression do not implement the F test directly, but the required statistic can be computed by running the two separate regressions and using this form.

Proof: $X\hat{\beta}$ is the projection of y on the space spanned by the columns of X. $X\tilde{\beta}$ is a further projection of $X\hat{\beta}$ on a subspace defined implicitly by the linear restriction.   The basic projection geometry yields the following relationship:

$$\|y - X\tilde{\beta}\|^2 = \|y - X\hat{\beta}\|^2 + \|X\tilde{\beta} - X\hat{\beta}\|^2. \qquad (8.7)$$

This is illustrated in Figure 1.2. It can also be proven directly by using Eq. (8.3) to substitute for $\hat{\beta}$ and noting that $X(y - X\hat{\beta}) = 0$. Then Eq. (8.3) for $\tilde{\beta}$ yields

$$X\tilde{\beta} - X\hat{\beta} = X(X'X)^{-1}R' \left\{ R(X'X)^{-1}R' \right\}^{-1} (R\hat{\beta} - r)$$

From this it is easily calculated that Eq. (8.7) is equivalent to the lemma.

Using the lemma, we can establish the distribution of the F statistic as follows. Let $Z = R\hat{\beta} - r$. Then $Z \sim N(R\beta - r, \sigma^2 R(X'X)^{-1}R')$. From the lemma, the LR statistic S can be written as

$$F = \frac{Z' \left\{ R(X'X)^{-1}R' \right\} Z/m\sigma^2}{\|\hat{u}\|^2/(T - K)\sigma^2}.$$

We have already established that $\|\hat{u}\|^2/\sigma^2$ is χ^2_{T-K}. The denominator is a function of $\hat{\beta}$ and hence independent of $\hat{u}$. The numerator has the form $Z'(\text{Cov}(Z))^{-1} Z$, where Z has mean 0 under the null hypothesis. This is easily seen to be χ^2_m. Thus S is the ratio of independent chi-squares divided by their degrees of freedom, and hence has the F distribution claimed.

8.9   The F Test Is UMP Invariant

Note that for the linear regression model $y = X\beta + \epsilon$ the sufficient statistics $\hat{\beta}, \hat{\sigma}^2 = \|y - X\hat{\beta}\|^2/(T - K)$ are independent with densities $\hat{\beta} \sim N(\beta, \sigma^2 \mathbf{X}'\mathbf{X}_K)$ and $(T - k)\hat{\sigma}^2/\sigma^2 \sim \chi^2_{T-K}$. We will prove that the likelihood ratio test derived in the previous section is UMP invariant and most stringent.

Theorem 8.4 *For the model above, rejecting for large values of the test statistic V given by*

$$V = \frac{1}{m\hat{\sigma}^2}(R\hat{\beta} - r)' \left\{ R(X'X)^{-1}R' \right\}^{-1} (R\hat{\beta} - r),$$

provides a best invariant test of the null hypothesis $H_0 : R\beta = r$ versus the alternative $H_1 : R\beta \neq r$. This test is also the most stringent test of this hypothesis.

Proof: First consider the special case $R = \mathbf{I}_K$ and $r = 0$. Then the null hypothesis is simply $H_0 : \beta = 0$. With sufficient statistics $\hat{\beta} \sim N(\beta, \sigma^2(X'X)^{-1})$ and $\hat{\sigma}^2$, this is very similar to the problem considered in the previous chapter. A simple change of variables transforms the covariance matrix which is $(X'X)^{-1}$ to the identity, reducing this problem to the solved case. Let Q be any square nonsingular matrix such that $Q(X'X)^{-1}Q' = \mathbf{I}_K$. Define $\hat{\psi} = Q\hat{\beta}$ and $\psi = Q\beta$. Then $\hat{\psi} \sim N(\psi, \sigma^2\mathbf{I}_k)$, and the null hypothesis is $H_0 : \psi = 0$ versus the alternative $H_1 : \psi \neq 0$. Now it follows from the result of the previous chapter that the best invariant test is based on $\|\hat{\psi}\|^2/K\hat{\sigma}^2 = \hat{\beta}'(X'X)^{-1}\hat{\beta}/K\hat{\sigma}^2$, and this has $F(K, T - K)$ distribution under the null.

Next consider the general linear hypothesis $H_0 : R\beta = r$ versus the alternative $H_1 : R\beta \neq r$, where R is $m \times k$ and r is $m \times 1$. First we do an intuitive derivation of the F test. Consider $\hat{\theta} = R\hat{\beta} - r \sim N(0, \sigma^2 R(X'X)^{-1}R')$, where $\theta = R\beta - r$. In terms of θ, the null hypothesis is simply $H_0 : \theta = 0$, so that the previous paragraph becomes applicable. The test can be based on

$$
\begin{aligned}
F_{m,T-k} &= \frac{T - k}{ms^2} \hat{\theta}' \left[R(X'X)^{-1}R' \right]^{-1} \hat{\theta} \\
&= \frac{T - k}{ms^2} (R\hat{\beta} - r)' \left[R(X'X)^{-1}R' \right]^{-1} (R\hat{\beta} - r) \quad (8.8)
\end{aligned}
$$

The only difficulty with this derivation is that we have based a test on $R\hat{\beta}$ which is an $m \times 1$ vector, and ignored a $k - m$ dimensional portion of the data which is orthogonal to R. However, we can justify reducing the data by an appeal to invariance. For any γ such that $R\gamma = 0$ consider the transformation $Z = g_\gamma(\hat{\beta}) = \hat{\beta} + \gamma$. Then $Z \sim N(\theta, \sigma^2(X'X)^{-1})$, where $\theta = \beta + \gamma$. Since $R\gamma = 0$, it follows that $R\theta = 0$ so the null hypothesis remains invariant under this transformation. What is the maximal invariant under this transformation? To answer this question decompose $\hat{\beta}$ as $y_1 + y_2$, where $y_1 = R'(RR')^{-1}R\hat{\beta}$ is the projection of $\hat{\beta}$ on the vector space spanned by the columns of R', and $y_2 = (\mathbf{I} - R'(RR')^{-1}R)\hat{\beta}$ is the projection on the orthogonal space. Adding γ to $\hat{\beta}$ can change y_2 but cannot affect y_1. It follows that the maximal invariant under this transformation is y_1, the projection of y on the rowspace of R. Since $R\hat{\beta}$ is a one-to-one transform of y_1, it too is maximal invariant. But we have already derived the UMP invariant test based on $R\hat{\beta}$.

8.10   The F Test Is Robust

In addition to possessing the optimality properties discussed, the F test is also robust to lack of normality of the errors. This can be seen from the form of the F statistic, which is $F = \epsilon'P\epsilon/\epsilon'Q\epsilon$, where both P and Q are projection matrices. Suppose that ϵ is replaced by $\alpha\epsilon$, where α is any random variable. It is clear that the F statistic will not be changed since the α will cancel from the numerator and denominator. Thus if the errors are i.i.d. with common density f, where f is the density of an arbitrary random variable multiplied by a normal, the F statistic remains exactly the same. The class of densities covered by this form is quite large, and includes all unimodal densities. Furthermore, lack of symmetry of the error term also has no effect on the F statistic, which reduces to $\{\sum_{i=1}^{k} \lambda_i(P)\epsilon_i^2\}/\{\sum_{j=1}^{m} \lambda_j(Q)\epsilon_j^2\}$ after diagonalization of the two matrices. Since only squares of the errors enter the statistic, asymmetry has no effect. Experiments by the author show that the critical values of F remain approximately the same for a vast class of error densities even outside the cases discussed above. Aman Ullah and Zinde-Walsh (1984, 1985) and Zinde-Walsh and Aman Ullah (1987) have explored the robustness of the F and related test in a number of different contexts.

The above discussion applies to situations where the regression errors ϵ_t are i.i.d. according to some common density F which may not be normal. If the errors are i.n.i.d. and we have heteroskedasticity, then the F statistic does not have the right distribution. In fact, in order for the F statistic to fail to have the right distribution, the changes in the variances must be correlated with the products of regressors — that is, the assumption of *generalized homoskedasticity* with respect to the regressors must fail. See Section 1.4 for a discussion of this assumption. The Eicker-White test discussed earlier is useful in assessing the validity of this assumption required for the F test.

A second situation where the F test does not perform well is when the errors are autocorrelated. As we have already discussed, all models can be transformed to dynamic models where the errors form an innovations process. In such models, it is known that the F test is valid asymptotically. In small samples the exact critical values of F deviate from the asymptotic values given by the standard F tables. Results of Tire (1995) suggest that bootstrapping is an effective way to get the right critical values. Van Giersbergen and Kiviet (1994) discuss alternative procedures of bootstrapping and show that only one of four possibilities work correctly.

8.11 Exercises

1. Define the subset $A_c = \{(x_1, \ldots, x_n) : f^X(x_1, \ldots, x_n) \geq c\}$. Show that if $\int_{x \in A_c} f^X(x) \, dx$ is finite than for any $\theta = (\theta_1, \ldots, \theta_n) \neq 0$,

$$\int_{x \in A_c} f^X(x) \, dx > \int_{x \in A_c + \theta} f^X(x) \, dx.$$

Hint: eliminate the intersection of A_c and $A_c + \theta$ from both sides. Over the remaining part, $f^X \geq c$ on A_c and $f^X(x) < c$ on $A_c + \theta$.

2. Show that the Pitman test can be derived as Bayes test based on an invariant prior. Hence show that it maximizes average power, where average is defined in a suitable sense. Hence show that if

there exists a UMP invariant test for the problem, then Pitman's test must be UMP invariant.

3. Suppose that X has density $f_0(x)$ under the null. Explain why it would not be advisable to use a tail test when the class of plausible alternative densities has the form $f_1(x) = f_0(x/\sigma)/\sigma$ for some $\sigma \neq 1$. These are alternatives based on scale changes to the null. Define $Z = \log(X)$ and show that a tail test based on the density of Z would be suitable in this circumstance.

4. Suppose $s_i^2/\sigma_i^2 \sim \chi_{t_i}^2$. Define $Y_i = \log s_i^2$. Assume that $\sigma_i^2 = 1$ for all i and calculate the density $f^{Y_i}(y_i)$ of Y_i. Then calculate

$$T(Y_1, \ldots, Y_n) = \int \left(\prod_{i=1}^{n} f^{Y_i}(y_i + \theta) \right) d\theta.$$

Show that a test based on T is equivalent to the LR test.

5. [Proof of Theorem 8.2] If $s_j^2 \sim \sigma_j^2 \chi_{t_j}^2$, the joint density of $s_1^2, \ldots, s_J^2$ can be written as

$$f(s_1^2, \ldots, s_J^2) = \prod_{j=1}^{J} \frac{(s_{t_j}^2)^{(t_j/2)-1} \exp(-s_j^2/2\sigma_j^2)}{(2\sigma_j^2)^{t_j/2} \Gamma(t_j/2)}.$$

(a) Maximize the likelihood $\ell(\sigma_1^2, \ldots, \sigma_J^2) = f(s_1^2, \ldots, s_J^2)$ (or log likelihood) with respect to $\sigma_1^2, \ldots, \sigma_J^2$ and show the the unconstrained ML estimates for these parameters are $\hat{\sigma}_j^2 = s_j^2/t_j$.

(b) Maximize the likelihood subject to the constraint $\mathbf{EV}$ that $\sigma_*^2 = \sigma_1^2 = \cdots = \sigma_J^2$. Show that $\hat{\sigma}_*^2 = (\sum_{j=1}^{J} s_j^2)/(\sum_{j=1}^{J} t_j)$. How can you derive this directly from the previous exercise without calculations? Hint : the sum of independent χ^2 random variables is also χ^2.

(c) The likelihood ratio test statistic for $\mathbf{EV}$ is defined as the ratio of the unconstrained maximum of the likelihood $\ell_1 = \ell(\hat{\sigma}_1^2, \ldots, \hat{\sigma}_J^2)$ to the constrained maximum $\ell_2 =$

$\ell(\hat{\sigma}_*^2, \ldots, \hat{\sigma}_*^2)$. Show that except for constants the log likelihood ratio can be written as

$$\log(\ell_1) - \log(\ell_0) = \sum_{j=1}^{J} \log \left(\sum_{j=1}^{J} s_j^2 \right) - \sum_{j=1}^{J} t_j \log(s_j^2).$$

6. Calculate the likelihood ratio test statistic for testing **EV** without assuming **EM**. Show that this is the same as the statistic of Theorem 8.2 *except* that s_i^2 are divided by T_i (leading to biased estimates of σ_i^2) rather than $T_i - K$ as in the theorem. It can be shown that this leads to a biased test statistic as well.

7. Show that if $Z_1 \sim \chi_{t_1}^2$ and $Z_2 \sim \chi_{t_2}^2$ are independent then $G = (Z_1/t_1)/(Z_2/t_2)$ has the following density function:

$$f(G) = \frac{\Gamma(t_1/2)\Gamma(t_2/2)(t_1)^{t_1/2}(t_2)^{t_2/2}G^{(t_1/2)-1}}{\Gamma((t_1 + t_2)/2)(t_1 G + t_2)^{(t_1+t_2)/2}}.$$

This is called the F density with t_1 and t_2 degrees of freedom.

8. Suppose that X_1, X_2 are independent random variables with Gamma distributions, $X_i \sim \Gamma(p_i, \alpha)$. Show that the ratio $X_1/(X_1 + X_2)$ has a Beta density independent of $S = X_1 + X_2$.

Hoja Nasruddin saw a traveler searching for something under a lamppost.
"What are you looking for?" the Hoja asked.
"A bag of gold which I have lost," replied the traveler.
"Where did you lose it?"
"I dropped it in a jungle a few miles back."
"Why are you looking for it here?" the Hoja was puzzled.
"Because there is light and it is safe here," the traveler responded, with some scorn at the simple-minded question.

Chapter 9

Applications: F and Related Tests

9.1 Introduction

Several different types of hypothesis testing problems which arise in the linear regression model can be treated by means of the F and related tests. This chapter gives many examples of application.

An important use of the F and the related t test is to decide whether or not a given regressor is a significant determinant in a given relationship. We can also use the F test to assess the significance of a subset of the regressors. If used for *all* the regressors, the F test is closely related to the R^2 statistic for the regression. By adding nonlinear terms and assessing their joint significance, the F test can be used to test for linearity. We can split the sample in two and test for equality of coefficients; this is known as Chow's test for structural stability. Several tests of this type will be studied in the following chapter. We show how to use the F test to fit splines of various types to the data. A procedure for fitting distributed lags due to Almon is also amenable to analysis via the F test. Finally, we discuss various issues related generally to the use and abuse of the F test.

9.2   Significance of a Single Regressor

Suppose $y_t = \beta_1 x_{1t} + \beta_2 x_{2t} + \cdots + \beta_K x_{Kt} + \epsilon_t$ for $t = 1, \ldots, T$. This is the standard model written out in coordinate form, rather than in vector form. It is frequently of interest to know whether or not some regressor, e.g. x_i, is significant in determining y. This question can be paraphrased as the null hypothesis that H_0: $\beta_i = 0$. If this can be rejected, we can safely conclude that the i-th variable plays a significant role in determining y_t. If it cannot be rejected, we can conclude that the data do not provide evidence for significant influence of x_i on y.

The F statistic for H_0 is easily computed. Set $R = e_i$, where e_i is $1 \times k$ vector with one in the i-th position and zeros elsewhere. Then $R\beta = 0$ is exactly the hypothesis that $\beta_i = 0$. Using Lemma 8.3 the F statistic can be written as

$$F_{(1,T-K)} = \frac{\hat{\beta}' R \left\{ R(X'X)^{-1} R' \right\}^{-1} R' \hat{\beta}}{\|y - X\hat{\beta}\|^2 / (T - K)}.$$

Let a_{ii} be the i-th diagonal entry of the matrix $(X'X)^{-1}$. Let $\hat{\sigma}^2 = \|y - X\hat{\beta}\|^2 / (T - K)$ be the standard unbiased estimate of the variance. Using the fact the $R = e_i$, we can easily reduce the above statistic to

$$F_{(1,T-K)} = \frac{\hat{\beta}_i^2}{a_{ii}\hat{\sigma}^2}.$$

In fact it is more convenient to use t statistic, which is a square root of this F:

$$t_{T-K} = \frac{\hat{\beta}_i}{\sqrt{a_{ii}\hat{\sigma}^2}}.$$

For testing the two-sided hypothesis H_0: $\beta_i = 0$ versus H_1: $\beta_i \neq 0$, these two statistics are equivalent, since $t_{T-K}^2 = F_{(1,T-K)}$. However, it is frequently of interest to test H_0 against a one-sided alternative like H_1: $\beta_i > 0$. If theory tells us the sign of a parameter (and this is frequently the case in economic applications) incorporating this information can substantially improve the power of the test. In such situations, a t test is superior since one can reject for $t > c$ rather than $t^2 = F > c^2$ which gives a two-sided test. It can be proven that this t test (which is a likelihood ratio test) is also UMP invariant and hence most stringent. For details see Lehmann (1986, Section 6.4).

Note also that the marginal density of $\hat{\beta}_i$ is $N(\beta_i, \sigma^2 a_{ii})$ and hence $\sqrt{\hat{\sigma}^2 a_{ii}}$ is an estimate of the standard error of $\hat{\beta}_i$. Regression outputs frequently provide both of these numbers, the standard error for each coefficient and also the t statistic. Since the t is the ratio of the coefficient to the standard error, one can easily get one from the other. Also of importance, and frequently available from packages, is the probability that a t_{T-k} density takes values bigger than the observed t value. If this probability is very low, we say that the estimate $\hat{\beta}_i$ is *significant*.

Table 9.1: Winning Speed at Indy 500

Year	Winning Speed	Year	Winning Speed	Year	Winning Speed	Year	Winning Speed
11	74.59	27	97.54	41	115.11	59	138.85
12	78.72	28	99.48	46	114.82	60	138.76
13	75.93	29	97.58	47	116.33	61	139.13
14	82.47	30	100.44	48	119.81	62	140.29
15	89.84	31	96.62	49	121.32	63	143.13
16	84.00	32	104.11	50	124.00	64	147.35
19	88.05	33	104.16	51	126.24	65	151.38
20	88.62	34	104.86	52	128.92	66	144.31
21	89.62	35	106.24	53	128.74	67	151.20
22	94.48	36	109.06	54	130.84	68	152.88
23	90.95	37	113.58	55	128.20	69	156.86
24	98.23	38	117.20	56	128.49	70	155.74
25	101.13	39	115.03	57	135.60	71	157.73
26	95.90	40	114.27	58	133.79		

If the probability is high (for example 12 %) then the hypothesis that X_i does not really belong in the regression cannot be rejected. Such failure to reject the null can occur for two different reasons. One is that X_i really does not belong in the regression and $\beta_i = 0$. A second reason is that even though the true β_i is different from 0 that data is not very informative about the role of X_i. Because of this lack of information, the estimate of β_i is relatively imprecise, making it hard to say for sure whether it is different from zero or not. These two cases have substantially different implications, but are sometimes confused in practice.

As an example of the t test and its uses, consider the time series of observations on the winning speed at the Indy 500 race (Table 9.1). This has been considered as a proxy for measuring technological progress. It seems plausible that the winning speed W_t could be a function of time t, say $W_t = a_0 + a_1 t + a_2 t^2 + a_3 t^3 + a_4 t^4$. Whenever possible, it is advantageous to orthogonalize the regressors. Define T_1 to be time differenced from the mean, and T_2, T_3, T_4 so that each T_j is orthogonal to the previous regressors (see exercise 1). A regression of W_t on the vector of ones, and these four regressors yields the following result:

$$W_t = 63.12 + 1.27T_1 + 0.00445T_2 + 0.000328T_3 - 0.000002T_4 + \quad \epsilon$$
$$\quad\quad 0.96 \quad 0.02 \quad\quad 0.00138 \quad\quad 0.000085 \quad\quad 0.000006 \quad\quad 2.816$$
$$T = 55 \quad R^2 = 0.986 \quad SSR = 396.419 \quad F(4, 50) = 909$$

The estimated standard deviation of the error $\hat{\sigma} = \sqrt{SSR/(T - K)} = 2.816$ is written under ϵ. The numbers under the estimated coefficients are the standard errors $\hat{\sigma} \times a_{jj}$ where a_{jj} is the (j, j) diagonal entry of the matrix $(X'X)^{-1}$. Since the regressors have been orthogonalized, $X'X$ is diagonal and hence $a_{jj} = 1/\|T_j\|^2$. The t statistics can be computed by dividing the coefficient by its standard error. All the t statistics are highly significant except for the coefficient of T_4 which has a t statistic of $0.000002/0.000006 = 0.333$ which is insignificant. Clearly the quartic term has no effect on the dependent variable. An advantage of orthogonal regressors is that the effects of removing a regressor can be read off directly from the numbers. The estimated coefficients of the other variables remain exactly the same, and none of the other statistics changes much if T_4 is removed.

9.3   The F Test and R^2

As a generalization of the problem of the previous section, consider
the problem of testing a set of regressors (instead of just one) for significance. To be precise partition the regressor into two subsets, e.g.
$X = [X_1|X_2]$, where X_1 is $T \times K_1$ and X_2 is $T \times K_2$ with $K_1 + K_2 = K$.
Partition β conformably as $\beta' = [\beta'(1), \beta'(2)]$, where $\beta(j)$ is $K_j \times 1$, so
that

$$y = X_1\beta(1) + X_2\beta(2) + \epsilon.$$

Now suppose it is of interest to test the null hypothesis H_0: $\beta(2) = 0$
versus the alternative $\beta(2) \neq 0$. Let $\hat{\beta}(1) = (X_1'X_1)^{-1}X_1'y$ and $\hat{\beta} =
(X'X)^{-1}X'y$ be the constrained and unconstrained estimates for β.
The F statistic for testing H_0 can be written as

$$F_{K_2, T-K_2} = \frac{\left(\|y - X_1\hat{\beta}(1)\|^2 - \|y - X\hat{\beta}\|^2 \right)/K_2}{\left(\|y - X\hat{\beta}\|^2 \right)/(T - K)}.$$

We will follow the convention that in regression models with constant terms the first regressor will always be the constant, so that the
first column of the X matrix is a column of ones. An important special case of this test arises when $K_1 = 1$, so that $\beta(1) = \beta_1$ is just
the constant term of the regression. In this case the F statistic for
H_0: $\beta(2) = 0$ measures the significance of *all* the nonconstant regressors in the model. Since X_1 is a column of ones, it is easily checked that
$\hat{\beta}(1) = (1/T)\sum_{t=1}^{T} y_t = \bar{y}$. The quantity $\|y - X_1\hat{\beta}(1)\|^2 = \sum_{t=1}^{T}(y_t - \bar{y})^2$
is the sample variance of y, and is called the Total Sum of Squares
(SST). Since $X_1\hat{\beta}(1)$ is the projection of $X\hat{\beta}$ onto a subspace, the projection geometry yields (the equivalent of Eq. (8.7), illustrated in Fig-
ure 1.2 of Chapter 1)

$$\|y - X_1\hat{\beta}(1)\|^2 = \|y - X\hat{\beta}\|^2 + \|X\hat{\beta} - X_1\hat{\beta}(1)\|^2.$$

Now $X\hat{\beta} = \hat{y}$ is the vector of *fitted values*, while $X_1\hat{\beta}(1) = \bar{y}X_1$ is
the vector of ones multiplied by the average value of y_t. Since the
residuals sum to zero (in regressions with constant terms), it must
be the case that the average of the observed values y_t is the same

as the average of the fitted values $\hat{y}_t$. The quantity $\|X\hat{\beta} - X_1\hat{\beta}(1)\|^2$
thus equals $\sum_{t=1}^{T}(\hat{y}_t - \bar{\hat{y}}_t)^2$.   This is called the Explained Sum of
Squares (SSE). As discussed earlier in Chapter 1, this yields the identity
$SST = SSR + SSE$. The F statistic can be written as

$$F_{K-1,T-(K-1)} = \frac{T-K+1}{K-1}\frac{SST-SSR}{SSR} = \frac{T-K+1}{K-1}\frac{SSE}{SSR}.$$

This F statistic is critical for the overall healthiness of the regression. If it is insignificant, there is no empirical evidence for any relation between y and any of the explanatory variables chosen as regressors. In addition to this F, a closely related statistic $R^2 \equiv SSE/SST$ is also of importance in assessing the overall performance of the regression. The *coefficient of determination* R^2 measures the proportion of the variation of y 'explained' by the regressors X. In earlier times, it was considered essential to have R^2 close to 90 % for a reasonable regression. With time series data where the regressors and the dependent variables share a common increasing trend, such high values of R^2 are easily achieved. With the emergence of larger and more unruly data sets, especially in cross sections, lower values of R^2 become much more common, all the way down to $R^2 = 0.10$ or even $R^2 = 0.01$. Attention shifted from R^2 to the F statistic as the appropriate measure of significance of a regression. Theoretically, this shift is justified. As an analogy consider the relation between winning and skill for some game:$W = F(S) + \epsilon$. In some games the room for random variation will be very small; for example chess should fall into this category. While Bridge is also a game of skill, one would expect a lower value of R^2 in an analogous regression. There are other games where skill plays only a minor role, and one would expect a very small R^2. Theoretically a significant relation between y and X can easily have an error term with a very high variance; one says that such a relationship is very *noisy*.

Practically speaking one note of warning is in order. If one can be sure (from theoretical considerations for example) that one has specified all relevant regressors, then there is no harm in disregarding the R^2 and focusing exclusively on the F statistic. If this is not the case then low values of R^2 may indicate that we have missed some significant explanatory variable. If so, this is an important specification error and would cast doubt on our results. Thus low R^2 should lead to consideration of alternative specifications and a search for omitted significant explanatory factors.

9.4   A Test for Nonlinearity

It is frequently of interest to detect the presence of nonlinearities in a regression relationship. Given a linear regression with y regressed on the constant X_1 and regressors $X_2, \ldots, X_K$ define variables $Z_1, \ldots, Z_{(K-1)K/2}$ to be the variables $X_i X_j$ for i, j with $1 < i \leq j$. It is often the case that there are some linear dependencies in this set. For example if $X_i = t^i$ as in polynomial regression, then $X_2 \times X_2$ will be the same as $X_1 \times X_3$. Another common case is when X_j is a dummy variable and X_1 is a constant term. Then $X_1 \times X_j$ is the same as X_j^2. After removing such redundant regressors, define $Z^*(X_1, \ldots, X_k)$ to be the matrix of linearly independent columns of products $X_i \times X_j$. Next, let $Z^{**}(X_1, \ldots, X_k)$ be the matrix formed from $Z^*(X_1, \ldots, X_k)$ after deleting all columns which are linearly dependent on the original regressors $X_1, \ldots, X_k$. If X_1 is the constant term, as is usual, then $X_j \times X_1 = X_1$ and all products of this type will be excluded from Z^{**}. Typically but not always $Z^{**}(X_1, \ldots, X_k) = Z^{**}(X_2, \ldots, X_k)$.

Consider the model $y = X\beta(1) + Z^{**}(X_1, X_2, \ldots, X_k)\beta(2) + \epsilon$, with all second order terms included, and test the hypothesis H_0: $\beta(2) = 0$. This will test whether or not there are significant second order effects. Linearity of the regression is frequently an auxiliary hypothesis, on par with normality, homoskedasticity, and independence. It should also be routinely tested. If this test fails to reject the null, the data do not provide evidence for any nonlinearities, and we may proceed with analysis of a linear model. If the test does reject, we must take nonlinearities into account. While the importance of this test has been recognized in the literature, it is not routinely implemented because costs of inverting high order matrices are high. For example, in a model with 10 regressors, one would need to invert a 66×66 matrix to perform this test. This is a difficult job. Furthermore, there may not be enough data to estimate a model with 66 regressors. In the presence of computational constraints, or insufficient data for all the quadratic terms, it would be useful to select, in addition to the quadratic terms, those interaction terms which are a priori likely to have importance. This should cut down substantially on the number of regressors and make it feasible to implement the test.

In cases where the number of additional regressors formed by all second order products is large compared to the number of observations, another option is to use the RESET test, suggested by Ramsey (1974). This test takes the square of the fitted values $\hat{y}^2$ (sometimes also higher powers) and includes it as an extra regressor. The coefficient of this regressor is then tested for significance. The theoretical basis of this test is unclear. It is not known for which types of alternative hypotheses it performs well and for which it does not. Optimality properties are also not known.

9.5   The Chow Test

It is frequently of interest to test whether the parameters of a regression model remain stable over time. After all, this is a fundamental hypothesis in our formulation of the linear regression model. There are numerous tests which have been devised for this purpose. One of the most popular is the Chow test, which is described in detail below. Roughly speaking, the Chow test breaks up the data into two subsets and tests for equality of regression coefficients across the two sets. A major weakness of the Chow test is the assumption that the regression variances are the same in both subsets. This is aesthetically displeasing since it is logical to allow for instability both in regression coefficients and in error variances simultaneously. More important, it turns out that the performance of the Chow test is extremely sensitive to this assumption and deteriorates substantially when it is violated. In addition, both theoretical and empirical results show that it is easy to account for the possibility of unequal variances — the 'cost' is low, in the sense that good tests allowing for unequal variances work nearly as well as the Chow test in the situation where the variances are equal, and the benefit is high, since substantial gains over the Chow test are achieved when the realistic possibility of unequal variances occurs. For the theoretical argument see Pfanzagl (1974), who shows that tests accounting for unequal variances perform, to a high order of approximation, equiv-

alently with tests based on known variances. For the empirical results
see a large-scale Monte Carlo study by Thursby (1992), which confirms
empirically this theoretical result. One could summarize this research
by saying that the Chow test should be used with caution and only
in special situations. In general, any of several alternatives capable of
coping with unequal variances should be preferred. Detailed discussion
of how to test with unequal variances (and this can no longer be done
via an F test) is given in Section 10.4.

The Chow test can be conducted as follows. Split the data y, X into
two halves (approximately), e.g., $y(1) = X(1)\beta(1) + \epsilon(1)$ and $y(2) = X(2)\beta(2)+\epsilon(2)$, where for $i = 1, 2$, $y(i), X(i), \epsilon(i)$ are $T_i \times 1, T_i \times K$, and
$T_i \times 1$ respectively, with $T_1 + T_2 = T$. Define Z to be the $(T_1 + T_2) \times 2K$
matrix:

$$Z = \begin{pmatrix} X(1) & 0 \\ 0 & X(2) \end{pmatrix},$$

and let $\gamma' = (\beta(1)', \beta(2)')$ be a $2K \times 1$ vector of parameters. Then
the two equations $y(1) = X(1)\beta(1) + \epsilon(1)$ and $y(2) = X(2)\beta(2) + \epsilon(2)$
can be combined as $y = Z\gamma + \epsilon$. The null hypothesis to be tested is
H_0: $\beta(1) = \beta(2)$. Let SSR_1 be the sum of squared residuals from the
(unconstrained) regression[1] of y on Z. Let SSR_0 be the sum of squared
residuals from the constrained regression; this can easily be obtained by
regressing y on X. Then the F statistic is $F(K, T - 2K) = ((SSR_0 - SSR_1)/K)/(SSR_1/(T - 2K))$. See Lemma 8.3 and also exercise 2 for
alternative forms of the statistic.

As an example consider the Indy 500 data discussed earlier. A
hypothesis of interest is whether structural change took place as a result
of the two world wars. We first examine the effect of World War I. For
this purpose we include data up to 1941, just prior to World War II.
The years 1911 to 1916 provide 6 years of pre-war data, while 1919 to
1941 provide 23 years of post-war data. Form the regressor matrix Z
as follows:

$$Z = \begin{pmatrix} e_1 & t_1 & 0 & 0 \\ 0 & 0 & e_2 & t_2 \end{pmatrix}.$$

---

[1]Note that this regression has no constant term, and adding one will make the
matrix of regressors singular. Regression packages frequently automatically add a
constant term and this must be avoided here.

Here e_1 is 6×1 vector of 1's, t_1 is 6×1 vector of chronological year minus 1900. e_2 is 23×1 vector of 1's and t_2 are the post war years 1919 to 1941, again with 1900 subtracted. Regression results for $y = Z\gamma + \epsilon$ are

$$W_t = 150.9\,Z_1 + 2.48\,Z_2 + 116.6\,Z_3 + 1.26\,Z_4 + \epsilon$$
$$ 19.9 \qquad 0.70 \qquad\quad 1.2 \qquad\quad 0.09 \qquad 2.95$$

$$T = 29 \quad R^2 = 0.946 \quad SSR = 217.0 \quad F(4,25) = 109.5$$

It is important to note that this regression is not exactly the same as splitting the data into two parts and estimating each part separately. The reason is that the variance parameter σ^2 has been assumed to be the same in both parts of the regression. Thus the estimated $\hat{\sigma}^2$ relies on all of the data. If the data is split into two parts and the regressions are run separately, the OLS estimates will be identical to the ones above, but the standard errors and t statistics will be different since in this case we will have separate estimates of variance for each portion of the data.

This model was tested for nonlinearity by including the quadratic term, the square of the year, for both pre- and post-war periods. Adding the two quadratics lowers the SSR from 217 to 204. To test the joint significance of the two quadratic terms, we check whether this drop is significant by forming the F statistic $F(2,23) = ((217 - 204)/2)/(204/(29 - 6)) = 0.79$. This is insignificant at any level, indicating that there is no serious nonlinearity. Then a Chow test was run to test if the null hypothesis that the pre-war coefficients are the same as the post-war ones can be rejected. This can be done by running a simple linear regression on the first 29 data points from 1911 to 1941, which results in

$$W_t = 116.8 + 1.27\,t + \epsilon_t$$
$$ 1.1 \quad 0.06 \qquad 3.00$$

$$T = 29 \quad R^2 = 0.940 \quad SSR = 242.8 \quad F(1,27) = 420$$

The Chow test compares SSR (=243) for the one line model to the SSR (=217) for the separate lines in each regime. This yields an $F(2, 29 - 4) = [(243 - 217)/2]/[217/25] = 1.5$ which is not significant. Thus the data do not provide strong evidence for structural change as a result of World War I. In contrast, following essentially the same procedure for World War II, the Chow test strongly rejects the null hypothesis of identical coefficients for pre- and post-war periods. This is left to the exercises.

9.6 Splines

It has been suggested by Poirier (1973), that structural changes may not be represented by sudden shifts in coefficients; smoother forms of change are also possible. For example, in the case of the Indy 500, instead of estimating two separate lines for pre- and post war periods, we could insist that the underlying function is continuous. This is called a *spline*. It is easy to implement spline estimation using restricted least squares. So far, our best model for the Indy 500 has the form:

$$\text{speed}_t = a_1 + b_1 \text{year}_t \quad t < 1942$$
$$\text{speed}_t = a_2 + b_2 \text{year}_t \quad t > 1944$$

The Chow test suggests that the pre-World War I era can be combined with the post-World War I, but that the coefficients for pre-World War II and post-World War II are definitely different. To check whether a spline is appropriate, we could require the two lines to join at 43.5, the midpoint of World War II. This imposes the following linear restriction:

$$H_0 : a_1 + b_1 43.5 = a_2 + b_2 43.5.$$

SSR without the constraint is 354, but imposing the constraints increases the SSR to 495. The total number of data points is 55 from 1911 to 1971 (since there are no data for war years), and the number of coefficients estimated (without constraint) is 4. This yields $F(1, 51) = [(495 - 354)/1]/[354/51] = 20$. This high a value is significant at the 0.999% level. Thus the null hypothesis of a join is rejected at a high level significance. One can also see that there is no significant nonlinearity by adding a quadratic term to both pre- and post-war periods. Doing so yields an SSR of 352, an insignificant drop; also, the t statistics for the quadratic terms are not significant.

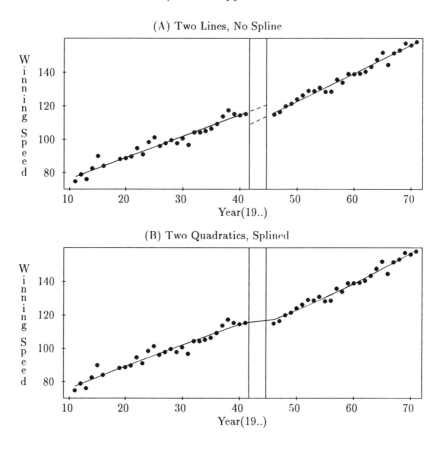

Figure 9.1: Indy 500 and Splines

We reach different, and erroneous, conclusions, if we nonetheless insist on splining. It turns out that using quadratic models on both pre- and post-war data improves the fit, and both quadratic terms are significant. The unsplined two lines model, and the splined quadratic model are pictured in Figure 9.1. Based on the spline estimate, Sampson (1979) reaches the incorrect conclusion that the rate of technological progress (i.e., the derivative of the rate of progress) was declining slowly prior to World War II, since the quadratic term has a significant negative coefficient. In fact we know that the quadratic term is not significant on unconstrained pre-war data. The reason for Sampson's result is that he forces the data to pass through a low point not consistent with the trend of the pre-war line, due to the spline constraint. Even though the data is basically linear, in order to pass through this low point to continuously join to the post-war data, we must fit a quadratic, and the quadratic term will be significant. Furthermore, this same constraint creates the illusion that there was a declining rate of technical progress since the curve is distorted to match this endpoint constraint. The SSR for the model with quadratics splined at 1913.5 is 356, slightly higher than the SSR for two lines. Using two quadratics adds two coefficients and the spline constraint effectively removes one. Thus we have added one parameter but the fit has become (slightly) worse, showing clearly that the splined quadratic model is not as good as the two lines without spline.

Thus the best fit to this set of data is obtained by two lines, one fit to pre-war and the second to post-war data. There is no evidence of significant nonlinearity in either part. Furthermore, the rate of progress post-war is slightly larger than pre-war. There is an average increase of 1.27 mph per year in winning speeds in the pre-war (i.e., pre-World War II) period. Post-war this rises to 1.66 mph per year. Interestingly, if we chop out the 4 year gap by subtracting 4 years from every year after 1946, these two line segments match up at the join point, even though they do have significantly different slopes. This suggests that Poirier's idea of continuity is valid, but that improvements in racing car technology were put in stasis for the duration of World War II.

9.7 Tests for Misspecified Dynamics

A *static* model is one in which all causality is contemporaneous. That is, the dependent variable y_t is a function of regressors x_t all of which are quantities measured in period t. If there are factors in period $t-1$ which influence events at time t then we have a dynamic model. Given a model $y_t = \beta x_t + \epsilon_t$, how should we assess whether the static specification is adequate? A more or less obvious way to proceed is to include the lagged variables y_{t-1} and x_{t-1} and see if they have a significant effect on the regression relationship.[2] Formally, we should test the null hypothesis that $\alpha = 0$ and $\beta_1 = 0$ in the following augmented model:

$$y_t = \beta' x_t + \alpha y_{t-1} + \beta_1' x_{t-1} + \epsilon_t \qquad (9.1)$$

If we cannot reject the null that $\alpha = 0$ and $\beta_1 = 0$, then we can conclude that there are no significant dynamic effects which we have failed to model. Since significant autocorrelation of the error term is a special type of unmodeled dynamics,[3] it is automatically tested for by this method, making it unnecessary to use Durbin-Watson or other such test procedures.

To illustrate some aspects of testing for dynamics and related issues, consider the following estimates of a Keynesian consumption function. The data are taken from International Financial Statistics, on nominal consumption and GNP in billions of US Dollars. We have quarterly data from 1957 to 1993 for a total of 148 observations. The regression below is run from 1st quarter 1960 to 4th quarter 1989. The first 3 years are omitted to allow for lagged equations to be run later with the identical dependent variable. The last 4 years are omitted to allow out of sample forecast performance tests. This part of the data, left out of the estimation process, is called a *holdout sample*:

$$C_t = -64.97 + 0.761\, Y_t + \epsilon_t$$
$$ 5.08 \quad\; 0.002 \qquad\;\; 31.59$$
$$T = 120 \quad R^2 = 0.999 \quad SSR = 117,764 \quad F(1,118) = 114,840$$

An important contribution of econometrics to economic theory is the discovery that, over different countries, time periods, and types of data sets, the simple Keynesian model does not adequately describe a consumption function. In the present instance, this can be checked by

---

[2]One reason this is not done routinely is that observations on time series are frequently highly autocorrelated. The author recalls being taught in an econometrics course that the use of the lagged value of the dependent variable can 'mask' economic relationships because of the high autocorrelation present in many time series. What this means is that variables predicted to be significant by economic theory may turn out insignificant when y_{t-1} is included, and this is an embarrassment for theory. Nonetheless, if the data tell us that the most significant determinant of y_t is y_{t-1} we cannot put aside this finding because theory tells us otherwise. Rather we must modify the theory to fit whatever information is given to us by the data.

[3]The first order error autocorrelation model $\epsilon_t = \rho \epsilon_{t-1} + \nu_t$ is a special case of the model above with parameters $\alpha = \rho$ and $\beta_1 = -\rho\beta$.

using the model for forecasting on the 16 time periods left out for this purpose. We find that the forecasts are systematically biased, and the standard error of forecasts is 102, four times the estimated standard deviation of the regression error term! A significant DW, not reported above, also signals misspecified dynamics. Numerous theoretical models, such as the Relative Income Hypothesis, Permanent Income, the Life Cycle model, and others, have been devised to explain the econometric finding of dynamics in the consumption function.

Since there is no agreement on the theoretical basis for the dynamics, we proceed in a general way. To establish the dynamic misspecification, we recommend (instead of DW and similar tests) running a second regression after including C_{t-1} and Y_{t-1} as additional regressors. This yields

$$C_t = -4.71 + 0.274\,Y_t + 0.917\,C_{t-1} - 0.202\,Y_{t-1} + \epsilon_t$$
$$ 2.05 \quad\ 0.033 \qquad\ 0.024 \qquad\quad 0.038 \qquad\quad 8.33$$

$$T = 120 \quad R^2 = 1.000 \quad SSR = 8051 \quad F(3,116) = 550{,}962$$

Both lagged regressors are highly significant, showing that the original equation was dynamically misspecified. The estimate of error standard deviation has gone down to 8.3 from 31.6, a substantial decrease. While it is unnecessary to carry out a test with such clear evidence, for illustrative purposes we indicate how to do an F test for the joint significance of C_{t-1} and Y_{t-1}. Note that in both regression, observations on the dependent variable C_t are taken from 1st quarter 1960. This makes the dependent variable the same in both regressions. The first regression is exactly the same as the second one subject to the constraint that the coefficients of C_{t-1} and Y_{t-1} are 0. The constrained SSR (CSSR) is 117,764 while the unconstrained SSR (USSR) is 8051. There are two constraints and hence the relevant F statistic is (see Lemma 8.3)

$$F(2,120-4) = \frac{(CSSR - USSR)/2}{USSR/(T-K)} = \frac{(117{,}764 - 11{,}897)/2}{11{,}897/116} = 776.74.$$

This is highly significant at any level we wish.

It is clearly possible that our second equation is also dynamically misspecified. Parenthetically we note that the DW test cannot be applied to the second equation — the presence of the lagged dependent variable C_{t-1} invalidates this test. Alternatives have been developed, but our methodology makes them unnecessary. We continue on our merry way, adding another pair of lagged variables, to get the following regression:

$$C_t = -5.84 + 0.320\,Y_t + 0.666\,C_{t-1} - 0.251\,Y_{t-1} + 0.232\,C_{t-2}$$
$$\quad\;\; 2.03 \quad\; 0.035 \qquad 0.095 \qquad\quad 0.055 \qquad\quad 0.091$$
$$+\, 0.021\,Y_{t-2} + \;\; \epsilon_t$$
$$\quad\; 0.041 \qquad\quad 8.04$$

$$T = 120 \quad R^2 = 1.000 \quad SSR = 7373 \quad F(5, 114) = 550,962$$

The SSR has decreased from 11,897 to 7373. The F statistic associated with the decline is $F(2, 114) = 5.24$ which is significant at the 1% level. At the next step, when we add C_{t-3} and Y_{t-3} to the previous equation, we find that these two are jointly insignificant. The sum of squared residuals goes to 7163, only a slight decline from 7373. The associated F statistic is $F(2, 112) = 1.67$ which has a p-value of 19 % and hence is not significant at all. Following a 'forward' stepwise regression strategy, we might call it quits, and say that the equation with three lags is not dynamically misspecified. Such a step would be unwise. Experience with quarterly data shows that there is frequently a significant correlation at 4 and occasionally 8 and 12 quarters. Going on to the 4th lag and adding C_{t-4} and Y_{t-4} to the regression results in a SSR of 6361, which is a significant decline from 7163:

$$C_t = -8.78 + 0.352\,Y_t + 0.552\,C_{t-1} - 0.230\,Y_{t-1} + 0.145\,C_{t-2}$$
$$\quad\;\; 2.19 \quad\; 0.037 \qquad 0.102 \qquad\quad 0.055 \qquad\quad 0.113$$
$$-\,0.008\,Y_{t-2} + 0.258\,C_{t-3} - 0.127\,Y_{t-3} - 0.108\,C_{t-4} + 0.146\,Y_{t-4} + \;\; \epsilon_t$$
$$0.058 \qquad\quad 0.113 \qquad\quad 0.058 \qquad\quad 0.096 \qquad\quad 0.039 \qquad\quad 7.61$$

$$T = 120 \quad R^2 = 1.000 \quad SSR = 6361 \quad F(9, 110) = 220,432$$

Further additions of pairs of lags are never significant, so that one can conclude that the above equation is not dynamically misspecified. We can again evaluate the models by their forecast performance on the holdout samples. Roughly speaking, the forecast performance for models including at least one lag is about two times worse than what one would expect from looking at the standard errors. It is well known that predictions of forecast error based on looking at standard error of regression are optimistic in typical applications. Part of the problem is due to model selection, but more generally regression fits are based on minimizing the sum of squared residuals and hence the true parameters *must* fit worse than the least squares fit — implying that there is a systematic bias when we estimate the forecast error from the regression fit. Ways to get better estimates of how models perform when used for forecasting are discussed in Efron (1983).

Table 9.2: Forecast Performance of Consumption Models

Lags	0	1	2	3	4	5	6
Std. Err.	31.6	8.3	8.0	8.0	7.6	7.6	7.7
For. Err.	102.4	18.1	16.8	17.2	18.6	18.6	19.0
Lags	7	8	9	10	11	12	13
Std. Err.	7.7	7.7	7.7	7.7	7.6	7.7	7.7
For. Err.	18.7	18.5	19.2	20.8	21.3	21.4	21.5

9.7.1 Refinements

A detailed and analytical approach to the bootstrap is presented in Part III of the text. However, the essential simplicity of the bootstrap is illustrated via two examples related to the problem discussed above.

One problem is that the F-statistic does not have an exact F distribution in finite samples in models with lagged regressors (or dynamic models) of the type discussed above. It is still asymptotically valid (as discussed in Section 1.5), and the sample size of 120 is large enough to expect that there should be no difficulty in applying the asymptotic result. For smaller sample sizes, one could assess the accuracy of the asymptotics via bootstrap as follows.

Take the estimated model as a proxy for the true model. Thus we assume the true regression coefficients equal the estimated regression coefficients and the true error variance equals the estimated error variance. With the true model in hand, we generate a thousand samples (say) of data. On each of these (bootstrap) samples we calculate the F statistic of interest and sort them in ascending order as $F_1, F_2, \ldots, F_{1000}$.

Then F_{900}, F_{950} and F_{990} are our bootstrap estimates of the level 90 %, 95 % and 99 % critical values of the F statistic. In the present context, the only difference between Monte Carlo and bootstrap is that Monte-Carlo is based on the true unknown parameters while bootstrap replaces them by estimates. Experiments revealed only minor differences between the critical values based on the F distribution and the bootstrap estimates for the sequence of inferences made on the above data set. Nonetheless, this technique may be useful on smaller data sets.

A second issue of some importance is that in this method of testing for misspecified dynamics, we add a lot of new parameters at each stage. Several methods for reducing parameters have been suggested. An interesting and popular general idea suggests that there is an equilibrium relationship $y^* = x^*\beta$ between the dependent variable and the regressors. If equilibrium is attained, it persists, and there are no spillover effects, or dynamics. Dynamics arise from disequilibrium, in the sense that only the differences $y_t - x_t\beta^*$ from previous periods can influence future periods. This idea, known as the *error correction model*, reduces the coefficients we need to estimate drastically, since it suggests the following dynamics:

$$y_t = x_t\beta^* + \alpha_1(y_{t-1} - x_{t-1}\beta^*) + \cdots + \alpha_k(y_{t-k} - x_{t-k}\beta^*) + \epsilon_t$$

Thus each lag adds only one coefficient. It is interesting that this model is formally equivalent to a model with autocorrelated error[4] terms. For a given static model, we first do a test for omitted dynamics in the manner described. If this test rejects the null, one can then attempt to see if the coefficient restrictions implied by the error correction mechanism are valid in the dynamic model.

Since the error correction restriction are nonlinear[5] there is a question of how to test them. For computational convenience Sargan (1980)

---

[4]Note that it is not possible to rehabilitate tests for autocorrelation as being tests for dynamics of the error correction type. The DW and similar tests will reject the null for most types of dynamics and do not particularly discriminate between the error correction type dynamics and other forms. This is the main reason why the Cochrane-Orcutt procedure can lead to misleading results — it assumes incorrectly, on the basis of a significant DW, that the dynamics must be of the error correction type.

[5]The restrictions are of the type $\beta\gamma_1 + \beta_1 = 0$ in the model $y_t = x_t\beta + \gamma_1 y_{t-1} +$

proposed the Wald test, which he labeled COMFAC (for common factors test). However, later authors have discovered that the Wald test has some unpleasant properties — see, for example, the LaFontaine and White (1986) article entitled 'Getting Any Wald Statistic You Want.' It would appear preferable to use a likelihood ratio test, although this makes heavier computational demands. However, Monte Carlo results of Tire (1995) suggest that a particular form of the Wald test has power greater than that of LR in this instance. This issue needs to be explored further. For a thorough discussion of this new and important class of dynamic economic models, see Bannerjee *et al.* (1992).

9.8 Almon Lags

Almon introduced a technique for estimating a type of dynamic model which has formed the basis for a huge literature; see Judge *et al.* (1985) for a review and references. In Almon's original model, we seek to explain values of the dependent variable y_t (current expenditures) by some regressor x_t together with its lagged values (current and past capital appropriations). The model of interest has the form:

$$y_t = \beta_0 x_t + \beta_1 x_{t-1} + \cdots + \beta_k x_{t-k} + \epsilon_t. \tag{9.2}$$

The problem addressed by the Almon technique is that in time series data, there is fairly strong multicollinearity between x and its lagged values, so that precise estimates of the coefficients are hard to get. The solution Almon proposed is to introduce a prior assumption that β_j change smoothly as a function of j. More specifically, β_j is specified to be a polynomial in j. Since Almon's original work, a large variety of alternative smoothness assumptions have been introduced and explored. Since nearly all such assumptions take the form of a linear restriction on the coefficients, the F test can be used to test such prior assumptions.

Unfortunately, there is a fundamental weakness in this class of models which appears to have been overlooked by workers in the field. If

---

$x_{t-1}\beta_1 + \epsilon$. These appear perfectly linear, but the product of two unknown coefficients appears in the restriction. Following through on the mathematics, we find that this cannot be treated as a linear restriction.

we believe that the right specification for y_t is a dynamic model, then we must give some theoretical justification for excluding the obviously relevant variables $y_{t-1}, \ldots, y_{t-k}$ from the model. It seems plausible in Almon's original example that a relevant variable is the stock of remaining capital, which is formed by summing the past capital appropriations and subtracting past expenditures. Just using this as the sole regressor in Almon's original regression improves on Almon's polynomial distributed lag procedure. Note that this variable involves both y_{t-1} and x_{t-1}. Below we report some regressions attempted as alternatives to the original distributed lag formulation of Almon. These alternatives reduce out-of-sample forecast error by 75 to 50% !.

An updated version of Almon's original data set is given in Judge *et al.* (1988). On this data set we tried three alternative estimation procedures. The last eight observations were set aside for forecast evaluation of models. On the first 80 observations, we first estimated an Almon type model as follows. To determine the value of k, the lag length, in Eq. (9.2), we ran a regression with $k = 11$. The coefficient β_{11} was not significant, suggesting a simpler model could be used. Insignificant coefficients were successively eliminated (a procedure known as 'backward selection,' discussed in the following section), until at $k = 7$ the estimated coefficient $\hat{\beta}_7$ was highly significant. Having fixed the lag length to be seven, we imposed the polynomial restrictions on the coefficients and tested the resulting linear constraint via the F test. We found that the constraint that the coefficients β_j lie on a cubic polynomial cannot be rejected. Thus we took this as our final Almon-type estimated model. The root mean squared error of forecast over the eight observations left out for this model was about 700.

As a second method, we tried the second order dynamic model:

$$y_t = \beta_0 x_t + \beta_1 x_{t-1} + \beta_2 x_{t-2} + \gamma_1 y_{t-1} + \gamma_2 y_{t-2} + \epsilon_t.$$

The order was selected to be 2 because that gives use five coefficients to estimate, which is roughly the same as the Almon procedure (which estimates 4 coefficients for a cubic polynomial). The root mean square error of forecasts for this model was about 140, more than a fourfold improvement. The fit, as judged by R^2 and standard error of regression, was also superior to the Almon lag. As a third method, we tried using the stock variable described earlier, together with a constant term. This gave a good fit, and a root mean square forecast error of about 350. It appears likely that the stock variable is the right idea, since the two coefficient model improves substantially on Almon lags. Introducing additional dynamics to take into account anticipated expenditures and appropriations into the stock model should improve its performance.

Another instance of this phenomena, where including a few lags of the dependent variable leads to substantial improvements over Almon's distribute lags, is discussed in Kelezoglu (1995). Kelezoglu re-estimates various investment functions formulated and estimated by Kopcke in a seris of papers discussed in Berndt (1991). These papers fit a distributed lag of capital stocks and other variables suggested by various theories. Typically the lag length is very high, and a relatively short polynomial lag is used. Reestimation after inclusion of lagged values of investment yields dramatic improvements in fit and forecast performance in each of the four models estimated by Kopcke. Furthermore, the needed lag length is quite short and the estimated dynamic models have substantially fewer parameters than the original models of Kopcke. We conclude that omission of y_{t-1} is a serious specification error in distributed lag models. Once this, and a few higher lags if necessary, are included, the dynamic structure becomes much shorter, obviating the need for polynomial-type specifications. If y_{t-1} is omitted, the lag appears larger than it is, since each of the lagged values x_{t-j} provides some information, by proxy, about the omitted y_{t-1}.

9.9 Use and Abuse of the F Test

In earlier sections we have illustrated the flexibility of the F test. It is useful in a wide variety of contexts, and frequently helpful in selecting an appropriate regression model. In this section we discuss several precautions necessary in applications of the F test.

9.9.1 The Data Mining Problem

Given a large collection of possible models, it is frequently tempting to search within the collection for models with the best statistics, for example the highest R^2 or the largest F statistic. Systematic procedures have been devised for adding and dropping variables on the basis of the F statistic. There also exist a variety of *model selection criteria* for regression models. These criteria are different functions of the F statistic and provide a single number for assessing a model. Among a collections of models, one can select a model which has the highest value of one's favorite selection criterion. Unfortunately, all such selection procedures invalidate the classical test statistics for the model finally selected. Issues and potential resolutions of this problem are discussed in the following section, but it can be illustrated briefly as follows. Suppose y is independent of a list of regressors $X_1, \ldots, X_{20}$. If we run 20 regressions of the type $y = a + bX_j$, one of these is very likely to be significant at the 95 % level. Thus the F statistic and R^2 of a model selected to maximize some criterion are artificially inflated.

This is an extremely serious problem with econometric analysis, considered by many authors (for example, Darnell and Evans (1990)) to be the fundamental problem of classical methods of analysis. 'Data-mining', as the process of choosing among models depending on their fit to data at hand is called, invalidates all classical testing procedures; worse, there is no known fix or solution to the problem. Because of this, numerous texts recommend a pristine approach — one should select a model on theoretical grounds, and never, or hardly ever, alter it. Unfortunately, experience with real data sets shows that such a strategy is neither feasible nor desirable. It is not feasible because it is a rare economic theory which leads to a unique model, and it is a rarer data set which will fit any given prespecified model. It is not desirable because a crucial aspect of learning from the data is learning what types of model are and are not supported by the data. Even if, by rare luck, the initial model shows a good fit, it is frequently important to explore and to learn the types of the models the data does and does not agree with. Since data mining must be done, pragmatic texts like Draper and Smith (1981) call the resulting difficulty of invalid statistics for the final model 'A Drawback to Understand but not be Unduly Concerned About' (title of Section 6.7).

Our attitude is close to that of Alan Blinder, as quoted in Berndt (1990, page 186), who says that the most harmful thing he was taught in econometrics at MIT was "You don't get the hypothesis from the data." He recommends "Learn to know your data." Recommendations not to do data mining are as futile as the efforts of the traveler in the chapter story; what we need to do is learn how to do it properly. Unfortunately, this is an area where the research is in its beginning stages. We describe some of the main strands below. New approaches may emerge with more experience. Since some creative effort is always involved in coming up with a good model, a part of effective data mining will always remain beyond the reach of theory. Below we discuss some aspects which can be treated more systematically.

Given that data mining is more or less necessary, how can we minimize the damage? One point to note is that the less experimentation we do, the less the statistics are affected. Thus it is useful to be disciplined in exploring the data, being guided more by theory and less by chance correlations. While starting with one model and sticking to it is not feasible, the *encompassing principle* is the next best thing. This approach shines in situations where a lot is known about suitable models in advance of looking at the data. For example, estimating a consumption function on US data has been done many many times and numerous variant models have been explored. Armed with knowledge of the types of models which do and don't work, a researcher could specify a sufficiently general family of models to include all possible outcomes and have half a chance of success. If he does not encounter unexpected difficulties, he could simplify the model down to a suitable one in a sequence of reduction steps. This process, recommended by Hendry (1991), does not eliminate the model selection problem, but makes the damage controllable. It is possible to evaluate the final significance level of a sequence of nested tests. Parenthetically, the technique of orthogonalization of regressors, used in the Indy 500 example, is also useful in disentangling the influences of different regressors without going through the woods of model selection; unfortunately, it has limited applicability.

In the more wild and wooly situations where the data sets are new and unexplored, it is not possible and in fact undesirable to specify a very general model in advance, as required by encompassing. In this case, any of a number of strategies are available for model selection, and the eventual outcome may be evaluated by one of the methods suggested below. To evaluate model selection procedures, all of which are based on functions of the F statistic, it is essential to be clear about the goal of model selection.

Somewhat paradoxically, one can get good forecasts without doing any model selection. More generally, it can be proven that model selection should be avoided for any convex loss function; see Zaman (1984). The right strategy in such situations is to use an optimal mix of all available models, a method known as 'forecast combination.' If the goal of model selection is to find the true model, this has a nonconvex loss function — the loss is 0 if we hit the truth and 1 otherwise, or some similar loss. In such cases, common in practice, forecast combination must be avoided, and genuine selection procedures used.[6] See also Diebold (1989) for insightful comments on the use and abuse of forecasting from combined models and its contrast with the encompassing methodology.

In cases where primary interest centers on the model itself, model selection is unavoidable. For example, in the case of the Indy 500, we were interested in finding out whether spline-type models are preferred to dummy variable-type models. The data seem to indicate quite clearly that there is a real discontinuity across World War II and hence spline-type models have a distinctly worse fit than models which allow for a jump discontinuity. This supports the learning-by-doing hypothesis in this case, since there was no progress in those years in which the race did not take place. In this and similar cases (which are frequent in practice) interest centers on which types of models fit best, and hence model selection is more or less unavoidable.

A general principle which is useful in evaluation of models is the use of forecasting on holdout samples. Since the holdout samples are not involved in the data mining phase, good forecasting performance provides genuine evidence regarding the relative worth of different models. What are efficient ways to use holdout samples, and how much of the sample should be reserved for this purpose? Some systematic research to answer these questions has been initiated. Recently, Ashley (1992, 1994) has developed a promising bootstrap methodology for evaluating

---

[6]Some model selection criteria explicitly set out to find a model which yields best prediction. Such criteria should not be used since their objective, good forecasts, can be achieved without doing any model selection.

models, using a holdout sample. Another bootstrap approach along somewhat different, though related lines, is given by Kipnis (1989a,b). Using techniques developed in the engineering and control systems literature, Penm *et al.* (1992) have developed a quite different type of approach to model selection using bootstrapping. Yet another approach is taken by Brownstone (1990). Since we are routinely deceived by good fits which disappear or deteriorate substantially out-of-sample, resolution of this issue appears critical — the Draper and Smith recommendation not to be unduly concerned does not appear reasonable. On the other hand, research on these problems is in its infancy, and only more experience with real data sets will tell which of a number of emerging techniques will prove of value in resolving the issues of inference posterior to model selection.

9.9.2 'Insufficient Evidence' Is Not 'Innocent'

Here we would like to repeat our earlier warning that failure to reject the null hypothesis does not mean that the null hypothesis holds. For example, it seems quite plausible that the rate and structure of technical progress in racing prior to World War I was different from this structure after World War I. However, the difference was not large enough to be detectable with only the six pre-war data points available to us. So on the basis of evidence available to us, we cannot definitively say (at conventional significance levels) that there was any difference in the rate of technical progress pre-World War I and post-World War I. Note that there is some evidence of difference since the coefficients estimated separately on the two parts of the data are quite different. However, the standard errors of estimate are large enough that the possibility that the two sets of estimates are the same cannot be confidently rejected. In contrast, there is enough data pre- and post-World War II that even though the difference between the coefficients is relatively small for OLS regressions run separately on pre- and post-war periods, the hypothesis that the two sets are the same can still be rejected.

9.10   Exercises

1. Given any sequence of regressors $X_0, X_1, \ldots, X_k$, we can make them orthogonal using the Gram-Schmidt procedure described in Chapter 1. Let $X_j = t^{j-1}$ be the t-th regressor in the INDY 500 data set. Write a program to orthogonalize the regressors using the following logic. Define $\tilde{X}_0 = X_0$. Loop through the following steps starting from $j = 1$ and ending with $j = k$.

   (a) Form the $T \times j$ matrix M_j with columns $\tilde{X}_0, \ldots, \tilde{X}_{j-1}$.

   (b) Let $\hat{\beta}_j = (M_j' M_j)^{-1} M_j' X_j$, and define $\tilde{X}_j = X_j - M_j \hat{\beta}_j$.

2. Divide data y, X into two regimes $y(1), X(1)$ and $y(2), X(2)$ as in the discussion of the Chow test. Define $H = \{X(1)'X(1)\}^{-1} + \{X(2)'X(2)\}^{-1}$. Let $\hat{\beta}_j$ for $j = 1, 2$ be the OLS estimates for each regime. Show that the Chow test described in the text is equivalent to the following statistic:

$$F(K, T - 2K) = \frac{T - 2K}{K} \frac{(\hat{\beta}_1 - \hat{\beta}_2)' H^{-1}(\hat{\beta}_1 - \hat{\beta}_2)}{\|y(1) - X(1)\hat{\beta}_1\|^2 + \|y(2) - X(2)\hat{\beta}_2\|^2}.$$

3. Let W_t be the winning speed in year t in the Indy 500. Suppose, as suggested by Barzel, that the correct model is a semilogarithmic one, with $\log(W_t) = f(t) + \epsilon$, where f may be a polynomial.

   (a) Ignore the world wars, and determine what order of polynomial fits best, using the F test.

   (b) Use the Chow test to determine whether the regression parameters remained constant before and after World War I. Do the same for World War II.

   (c) Check to see if a spline constraint is valid at either world war. To be more precise, if we estimate separate lines on both periods, is it possible to constrain the lines to join?

   (d) Test to see if the regression error variances are the same in both periods, a necessary condition for the validity of the Chow test.

   (e) Determine which of your models is the best. Compare the semilog and direct model by using forecasting techniques.

4. Consider the linear constraint $R\beta = r$, where R is a $m \times K$ matrix of full rank m and r is $m \times 1$.

   (a) Show there exists a $K \times 1$ vector β_0 and a $K \times (K - m)$ matrix of H of full rank $K - m$ such that $\beta = H\alpha + \beta_0$.

   (b) Show that the condition $\beta = H\alpha + \beta_0$ is equivalent to the original constraint $R\beta = r$.

   (c) If $\beta = H\alpha + \beta_0$, the formula for the constrained estimator $\tilde{\beta}$ is easily derived by substitution to be:

   $$\tilde{\beta} = (H'X'XH)^{-1}H'X'(y - X\beta_0)$$

   (d) It follows that if a constraint is given in the form $\beta = H\alpha + \beta_0$ where H is a $p \times K$ matrix of full rank, we can test it using the statistic of Theorem 8.3 by setting

   $$S = \frac{T - K}{K - p} \frac{\|y - X\tilde{\beta}\|^2 - \|y - X\hat{\beta}\|^2}{\|y - X\hat{\beta}\|^2}.$$

   This has an F distribution with $K - p$ and $T - K$ degrees of freedom under the null hypothesis.

5. In the linear model $y = X\beta + \epsilon$, we assume that the first regressor is the vector of ones. We will take $K = 2$ and test the hypothesis that $\beta_2 = 0$ using the F test. Thus $X = (e|x)$, where e is a $T \times 1$ vector of ones, and x is a $T \times 1$ vector of the one regressor in the model. We will assume that $\beta_1 = \beta_2 = 0$ so that $y = \epsilon$.

   (a) Generate the vector x as an i.i.d. standard normal for $T = 10, 20, 40, 80$. Keep this vector fixed for this exercise. Generate $\epsilon \sim N(0, \mathbf{I})$ and $y = \epsilon$ and calculate the F statistic for $\beta_2 = 0$. Look up the correct theoretical critical value for a test with type I error probability of 10%, 5%, and 1%. Compare the number of times your test rejects the null in the Monte Carlo with the theoretical values.

   (b) *Monte Carlo Standard Deviation or MCSD.* Let C be a critical constant for the F and let $F_1, F_2, \ldots, F_n$ be an i.i.d.

sample of F statistics. Let $p = \mathbf{P}(F > C)$ be the true theoretical probability of the event that $F > C$. Let $X_i = 1$ if
$F_i > C$ and $X_i = 0$ if $F_i \leq C$. Then

$$\hat{p} = \frac{1}{n} \sum_{i=1}^{n} X_i,$$

is an estimate of p. This estimate is exactly of the type that
you computed in the previous exercise. Calculate the theoretical standard deviation of $\hat{p}$. Note that this can be made
arbitrarily small by taking n large. Check that the difference between your estimates of the F rejection probability
and the theoretical values is within 2 MCSD's. If this is not
the case, you have probably done something wrong, since
the F test should work fine in the above example.

(c) Let $X_0, X_1, X_2, \ldots, X_k$ be i.i.d. $N(0,1)$ and define

$$T_k = \frac{X_0}{(1/k)\sqrt{\sum_{i=1}^{k} X_i^2}}.$$

Then T_k has a t density with k degrees of freedom. As k
goes to infinity, the denominator converges to 1 by the Law
of Large Numbers (prove), and hence T_n is approximately
$N(0,1)$ for large n. T_n is symmetric around 0 and hence has
zero mean (when the mean exists). Repeat the first exercise using t distribution errors instead of normals. Calculate
$\hat{p}$ the probability of falsely rejecting the null and also the
MCSD, which is an estimate of the error with which this
probability is estimated. Do this for $n = 5, 20, 50$.

(d) *Spurious Regressions* Generate both y_t and x_t as integrated
processes: let e_i, f_i be i.i.d. $N(0,1)$ and define $y_t = \sum_{i=1}^{t} e_i$
and $x_t = \sum_{i=1}^{t} f_i$. Run a regression of y on x as in exercise
1. Use Monte Carlo to compute the probability of finding a
significant relation between the two (unrelated by construction) series. Your results should show that an investigator
relying on F or t will be deceived in this case. Show that
if he tests the model for dynamic misspecification, he will
quickly be 'undeceived.'

Once when King Akbar was being particurly difficult, his minstrel Beerbul told him that three types of tantrums are hard to appease – the King's, the woman's, and the child's. Akbar immediately understood about the King, and had enough experience with his wives to understand the second, but did not believe that, at least with the kingly resources at his disposal, it would be difficult to satisfy a child. Beerbul proceeded to demonstrate the difficulty by starting to cry like a child, and asked for an elephant. Akbar had no difficulty fulfilling this wish. Next he asked for a pot, which was immediately brought. Lastly he started crying insistently that he wanted the elephant put inside the pot. At last Akbar understood.

Chapter 10

Similar Regression Models

10.1   Introduction

In this chapter we consider various tests of hypotheses regarding the equality of parameters for separate regression models. The basic format is as follows. Suppose $y_j = X_j\beta_j + \epsilon_j$ for $j = 1, \ldots, J$, where y_j is $T_j \times 1$, X_j is $T_j \times K$, β_j is $K \times 1$ and $\epsilon_j \sim N(0, \sigma_j^2 \mathbf{I}_{T_j})$. Thus for each $j = 1, 2, \ldots, J$ we have a separate regression model but all have the same number of regressors. The theory of this chapter is meant to apply to cases where these regressions are similar in some sense. For example, if a single regression model is broken up into two or more subsamples, by taking observations from periods $1, \ldots, T_1, T_1, T_1 + 1, \ldots, T_2$, up to $T_{J-1}, \ldots, T$. As another example, the production function for different firms in a given industry may be expected to be similar. We may have a macroeconomic relationship (such as an investment or consumption function) estimated in different but similar countries. interest. Let **EM** stand for the hypothesis of **E**quality of **M**eans and **EV** stand for the hypothesis of **E**quality of **V**ariances. In such situations, several combinations of the hypotheses that **EM**: $\beta_1 = \beta_2 = \cdots = \beta_J$ and that **EV**: $\sigma_1^2 = \cdots = \sigma_J^2$ are of practical interest.

Suppose we wish to test **EM**, the equality of means across regimes. This breaks down into two separate cases. The easier one occurs if we assume that **EV** or equality of variances holds as a maintained hypothesis. This leads to generalizations of the Chow test. We can find UMP F tests in this case. Unfortunately, as we will see, the elegant theory for this case is dangerously misleading. The more difficult case is if we do not assume **EV** and permit the variances to be different across regimes. This is a generalization of the notorious Behrens-Fisher problem, for which it has been proven that no exact solutions of a certain desirable type exist. However, excellent approximate solutions exist. Also, by a slight generalization of the concept of 'solution,' Weerahandi recently developed an exact solution; see Weerahandi (1994), for a nice exposition and references. These solutions to the Behrens-Fisher problem have the property that they work very well both when the variances are equal and also when they are not. This is in sharp contrast to the UMP F tests which generalize Chow's solution. These tests work very poorly if the variances are unequal. Both theory and simulations suggest that allowing for different variances across regimes does not cost much power in the case of equal variances and is very effective when the variances are unequal. Thus UMP F tests should be used with extreme caution, if at all.

Next consider testing the hypothesis **EV** of equal variances. When the regression coefficients differ across regimes, the LR test is Bartlett's test, discussed in the previous chapter. More frequent in econometric work is the problem of testing **EV** under the assumption of **EM**, which is a heteroskedasticity problem. Several of the tests suggested for this case are poor, for reasons to be discussed. Alternatives are suggested. The final case discussed in the chapter is when we are interested in testing both **EM** and **EV** simultaneously. In this case we can develop an exact test, but it suffers from nonrobustness like the Bartlett test.

Some remedies are explored. A model which shares the structure introduced is the Seemingly Unrelated Regressions (SUR) model introduced by Zellner. See Judge *et al.* (1985, Chapter 12) for an extensive discussion and references. This differs from the case we consider here in that we assume from the start that the error terms of the separate regimes are independent. Since the analysis given below is not valid when this assumption fails, it would be worthwhile doing a preliminary test of this assumption before using the techniques of this chapter, at least in contexts where doubt exists. The likelihood ratio test statistic for independence is developed in Section 12.1.2b of Judge *et al.* (1985). Because the statistic is highly nonlinear and the models are high dimensional, it is likely that the asymptotic distribution of the LR will be a poor fit in finite samples. Bootstrap-Bartlett corrections, discussed in Section 14.6, should give good results. The techniques discussed below can be modified to handle the case of error terms which are correlated across regimes, but considerable additional complexity is introduced.

10.2 General Chow-Type Tests

If we assume **EV** so that the variances are equal across the regressions, an optimal test for the hypothesis **EM** is the usual F test. We will refer to each of the separate regression equations as a *regime* in the sequel. Below we present the test in a form which allows for the possibility that some of the regimes may have insufficient observations to allow an estimate $\hat{\beta}_j$ of β_j. For this case, let $\hat{y}_j$ be the projection of y_j onto the column space of the matrix X_j. This is always well defined and unique, even though the estimates $\hat{\beta}_j = (X_j'X_j)^{-1}X_j'y_j$ are not well-defined in the case that $T_j < K$. Let $\hat{\beta}_*$ be the estimate of β under the constraint **EM** that all regimes have identical coefficients. When both **EM** and **EV** hold, all of the data can be aggregated into one regression as follows. Define

$$
y_* = \begin{pmatrix} y_1 \\ y_2 \\ \vdots \\ y_J \end{pmatrix} \quad X_* = \begin{pmatrix} X_1 \\ X_2 \\ \vdots \\ X_J \end{pmatrix} \quad \epsilon_* = \begin{pmatrix} \epsilon_1 \\ \epsilon_2 \\ \vdots \\ \epsilon_J \end{pmatrix}.
$$

Let β_* be the common value of β_j under the hypothesis **EM**. Then the regimes can be aggregated in the form $y_* = X_*\beta_* + \epsilon_*$, leading to the estimate for β_* under **EM** of

$$
\begin{aligned}
\hat{\beta}_* &= (X'_*X_*)^{-1}X'_*y_* \\
&= \left(\sum_{\alpha=1}^{s}(X'_\alpha X_\alpha)\right)^{-1}\left(\sum_{\alpha=1}^{s}(X'_\alpha X_\alpha)\hat{\beta}_\alpha\right).
\end{aligned}
$$

The second equation shows that the aggregated estimate is a matrix weighted average of the estimates for each separate regime, where the weights are proportional to the precision matrices (i.e., inverses of the covariance matrix).

We now present a general form of Chow's theorem which allows for any number of regimes as well as rank deficiencies in the matrices X_i. An equivalent formulation was first developed by Dufour (1982).

Theorem 10.1 (Dufour) *Assume that the variances are equal across the regimes so that **EV** is held as a maintained hypothesis. Let r_j be the rank of the matrix X_j, and assume that for at least one j, $T_j > K$ and $r_j = K$. Suppose we reject the null hypothesis of equality of regression coefficients H_0: **EM** for large values of the statistic*

$$
S = \frac{(T - \sum_{j=1}^{J} r_j)\sum_{j=1}^{J}\|\hat{y}_j - X_j\hat{\beta}_*\|^2}{\left(\sum_{j=1}^{J} r_j - K\right)\sum_{j=1}^{J}\|y_j - \hat{y}_j\|^2}.
$$

Under the null $S \sim F(\sum_{j=1}^{J} r_j - K, T - \sum_{j=1}^{J} r_j)$. This gives a UMP invariant and most stringent test of the null hypothesis.

Proof: We can obtain this hypothesis test as a special case of the F test of the previous chapter. To do this define

$$
X_0 = \begin{pmatrix} X_1 & 0 & \cdots & 0 \\ 0 & X_2 & \cdots & 0 \\ \vdots & \vdots & & \vdots \\ 0 & 0 & \cdots & X_s \end{pmatrix} \quad \beta_0 = \begin{pmatrix} \beta_1 \\ \beta_2 \\ \vdots \\ \beta_s \end{pmatrix}. \tag{10.1}
$$

With these definitions, we can write the model as $y_* = X_0\beta_0 + \epsilon_*$. Since **EV** is the maintained hypothesis, $\epsilon_* \sim N(0, \sigma_*^2 I_T)$, where σ_* is the common value of the variances. The null hypothesis **EM** can be written as a linear restriction on β_0. Let $\hat{y}_0$ be the projection of y onto the column space of X_0. Because of the diagonal form of X_0, $\hat{y}_0$ is obtained by stacking the $\hat{y}_j$ (the projections of y_j onto the column space of X_j) to form one vector. Let $\hat{y}_*$ be the projection of y_* onto the column space of X_*. Then $\hat{y}_*$ is obtained by stacking the vectors $X_j\hat{\beta}_*$ to form one vector. Note that the column space of X_* is a vector subspace of the column space of X_0. Let d_0 and d_* be the dimension of the column spaces of X_0 and X_*. Using Lemma 8.3 the standard F statistic for testing the constraint can be written as

$$F = \frac{\|\hat{y}_0 - \hat{y}_*\|^2/(d_0 - d_*)}{\|y - \hat{y}_0\|^2/(T - d_0)}.$$

Under the null hypothesis, this has an F distribution with $d_0 - d_*$ and $T - d_0$ degrees of freedom. It is easily checked that this is the statistic of the theorem.

The above formulation allows for the possibility that the matrices X_j fail to have full rank, and also that $T_j < K$ for some of the regimes. More explicit forms of the statistic can be given under the assumption that each X_j has full rank (see, however, exercise 2). There are a number of important special cases, which we discuss in the following sections.

10.2.1 All Regimes Have $T_j > K$

The most common special case arises when all regimes have sufficient observations so that each matrix X_j has rank $r_j = K$ and also $T_j > K$. In this case $\hat{\beta}_j$ is well defined for each j and $\hat{y}_j = X_j\hat{\beta}_j$. Furthermore, the column space of X_0 has dimension JK while that of X_* is K. It follows that in this case the F statistic for testing **M** with the maintained hypothesis **V** is

$$F = \frac{(T - JK)\sum_{j=1}^{J}(\hat{\beta}_j - \hat{\beta}_*)'(X_j'X_j)(\hat{\beta}_j - \hat{\beta}_*)}{K(J - 1)\sum_{j=1}^{J}\|y_j - X_j\hat{\beta}_j\|^2}.$$

This will have an F distribution with $K(J-1)$ and $T - JK$ degrees of freedom under the null hypothesis. In the case of $J = 2$, so there are only two regimes, it can be shown that the statistic reduces to the Chow test discussed in the previous Chapter.

10.2.2   Some Regimes Have $T_j \leq K$

It is sometimes desired to carry out a test of **EM** when some of the regimes have observations fewer than K. To estimate the variance, it is essential that at least one of the regimes should have $T_j > K$. We will show shortly that without this condition a valid test cannot be obtained. Assume that we order the regimes so that $T_j > K$ for $j = 1, \ldots, m$, while $T_j \leq K$ for $\alpha = m + 1, \ldots, J$. For $\alpha \geq m + 1$, since the number of observations is less than or equal to the number of parameters for these regimes, we can find a $\hat{\beta}_j$ which yields a perfect fit, so that $\hat{y}_j = y_j$. More formally, under the assumption that X_j is of full rank $T_j \leq K$, $y_j \in \mathbb{R}^{T_j}$ must lie in the column space of X_j. This means that the projection of y_j onto the column space of X_j is just y_j itself. Let $f_1 = K(m-1) + \sum_{j=m+1}^{J} T_j$ and $f_2 = T - (mK + \sum_{j=m+1}^{J} T_j)$. Then the test statistic can be written as

$$ S^* = \frac{f_2 \sum_{j=1}^{m} \|X_j \hat{\beta}_j - X_j \hat{\beta}\|^2 + \sum_{j=m+1}^{J} \|y_j - X_j \hat{\beta}\|^2}{f_1 \sum_{j=1}^{m} \|y_j - X_j \hat{\beta}_j\|^2}. \qquad (10.2) $$

Under the null hypothesis, this has an F distribution with f_1 and f_2 degrees of freedom. This is just a special case of Theorem 10.1 above. The dimension of the column space of X_0 is mK from the initial m regimes and $\sum_{j=m+1}^{J} T_j$ from the final $T - m$. For the (rare) case where the rank of X_j differs from both T_j and K, the above statistic must be adjusted, though the one of Theorem 10.1 remains valid. In the special case that there are only two regimes, and for one of them $T_j \leq K$ this is known as the *predictive Chow test*. Dufour (1980) showed that it is possible to carry out the test in standard regression packages using dummy variables. Briefly, let D_t be a dummy variable which takes value 1 in period t and 0 in all other periods. For the regime for which $T_j < K$ and X_j is of full rank, introduce a dummy D_t for every index t belonging to the regime. Testing the coefficients of the dummies jointly for significance is equivalent to the test described above. Roughly speaking, this is because the dummy makes the residual sum of squares zero for that regime, and that is exactly the form of the F test above, since $\hat{y} = y$ for such regimes. For details, see exercise 1. For the reason we need to require X_j to have full rank, see exercise 2.

10.2.3 All Regimes Have $T_j \leq K$

In the case that for all j, the rank of X_j satisfies $r_j \leq K$, it is not possible to test the hypothesis **EM**. If $T_j = K$ in each regime then it is possible to estimate β_j separately in each regime. However, a test for equality requires some estimate of the variances of the β_j and that is impossible when $T_j \leq K$ for every regime. The object of this section is to explain the impossibility of carrying out this test in a formal way. We formulate the principle in a simpler but essentially equivalent problem.

Suppose we observe $Z \sim N(\theta, \sigma^2 \mathbf{I}_T)$, and wish to test $H_0: \theta = 0$ versus the unrestricted alternative $H_1: \theta \neq 0$. Because σ^2 cannot be estimated under the alternative hypothesis, there does not exist a satisfactory level α test. To see this, let A and R be the acceptance and rejection regions of a level α test so that (i) $\mathbf{P}(Z \in R|H_0) \leq \alpha$ and (ii) $\mathbf{P}(Z \in A|H_0) \geq 1 - \alpha$. It is easily seen that A must be unbounded; for any bounded region large values of σ^2 will make the probability go to zero, violating (ii). But unbounded acceptance regions cause problems because then there exist $\theta^* \neq 0$ of arbitrarily large norm for which $\mathbf{P}(Z \in A|\theta^*)$ can be made arbitrarily close to unity. This can be done by choosing θ^* in the interior of A and letting σ^2 go to zero so as to make a small neighborhood of θ^* inside A have probability arbitrarily close to unity. Thus for any test, there exist arbitrarily large values of θ^* such that the probability of accepting the null when θ^* is true is arbitrarily close to 1.

This problem does not arise if σ^2 can be estimated under the alternative hypothesis. If the alternative hypothesis puts a restriction on θ (for example, that it lies in some vector subspace of $\mathbb{R}^T$) then it does become possible to estimate σ^2 and hence construct reasonable tests. This is why we require that at least one of the regimes should have X_α be of full rank and $T_\alpha > K$. When this condition is not fulfilled, it becomes impossible to estimate σ^2 under the null. This in turn makes it impossible to construct a reasonable test. The equivalence of the problem discussed above to the test of the previous section is discussed in the exercises.

10.2.4 Random Coefficients: $T_j = 1$ for all j

Now we consider the extreme case where every regime has exactly one observation. The model can be described as $y_j = x_j \beta_j + \epsilon_j$ for $j = 1, 2, \ldots, s$, where y_j is a scalar, x_j is $1 \times K$, β_j is $K \times 1$, and ϵ_j are i.i.d. $N(0, \sigma^2)$. As discussed in the previous section, it is impossible to test for the equality of all the β_j. However, if we assume that the β_j follow some pattern, then it may become possible to carry out a test. A simple assumption is that β_j are themselves i.i.d according to a normal distribution with common mean $\overline{\beta}$ and covariance matrix Ω. This model was introduced by Hildreth and Houck (1968) and is called the Hildreth-Houck random coefficients model. This model is closely related to the empirical Bayes models to be discussed in Part IV, although estimation techniques and motivation for the models are different. It is usual to assume that Ω is a diagonal matrix, though we will not do so here. The model can then be written as

$$y_j = x_j \overline{\beta} + x_j(\beta_j - \overline{\beta}) + \epsilon_j.$$

If, as usual, the first regressor is the constant term, then the error ϵ_j can, without loss of generality, be absorbed into the first random coefficient. If we write $\nu_j = x_j(\beta_j - \overline{\beta})$, it is easily seen that $\text{Var}(\nu_j) = x_j \Omega x_j' = \sum_{k=1}^{K} \sum_{l=1}^{K} x_{jk} x_{jl} \Omega_{kl} \equiv \sigma_j^2$. The variance simplifies to $\sum_{k=1}^{K} x_{jk}^2 \Omega_{kk}$ when the covariance matrix is diagonal.

In this formulation, the hypothesis that all the β's (except for the constant) are equal is equivalent to the hypothesis that $\Omega_{kl} = 0$ whenever (k, l) is different from $(1, 1)$. That is, only the constant term (which takes the place of the error term) has nonzero variance; all other coefficients have zero variance. If we knew the values of σ_j^2, it would be possible to calculate the values of Ω_{kl} from the equations

$$\sigma_j^2 = \sum_{k=1}^{L} \sum_{l=1}^{K} \Omega_{kl} x_{jk} x_{jl}.$$

Now $\mathbb{E}\nu^2 = \sigma^2$, so $\nu^2 = \sigma^2 + e_j$, where e_j has mean zero. If we observed ν^2, we could estimate Ω_{kl} from the regression equation

$$\nu_j^2 = \sum_{k=1}^{K}\sum_{l=k}^{K} \Omega_{kl}x_{jk}x_{jl} + e_j.$$

The regressors in this equation are all possible pairwise products of the regressors in the original equation The sum over l is from k to K so as to exclude duplicate regressors. Since $\Omega_{kl} = \Omega_{lk}$, all coefficients can still be estimated. It can be calculated that the variances of e_j will not be constant in this setup, so that the regression will display heteroskedasticity. Unfortunately, we do not have the luxury of observing ν_j, the true errors, so we must approximate them by $\hat{\nu}_j = y_j - x_j\hat{\beta}$ the residuals from the original regression equation. It is still true that $\mathbb{E}\hat{\nu}_j^2 \approx \sigma_j^2$ asymptotically, so that we could try the following regression:

$$\hat{\nu}_j^2 = \sum_{k=1}^{K}\sum_{l=k}^{K} \Omega_{kl}x_{jk}x_{jl} + e_j'.$$

In this equation, not only are the error terms heteroskedastic, but they are also correlated with each other. If we ignore this, we could use the F test for significance of the nonconstant regressors to check whether all Ω_{kl} other than Ω_{11} are 0. Recall that the F test is a monotonic transform of the R^2 for the regression. It has been shown that TR^2 for this regression is asymptotically distributed as chi-squared with degrees of freedom equal to $K(K+1)/2$. This test is known as the Eicker-White test of heteroskedasticity, and will also be discussed later.

We note a few among many reasons to believe this test is rather inefficient and inoptimal. First, the F statistic (or the monotonic transform TR^2) is not a suitable statistic for the situation where there is correlation and heteroskedasticity. Second, the Ω matrix is a positive definite matrix, but this restriction is not imposed in this method of estimating and testing. Third, this test has been shown to be a locally most powerful test. It is generally true that one can improve on a locally most powerful test by replacing it by a Neyman-Pearson test for a suitable distant alternative. Some tests based on this principle have been suggested by Brooks and King (1994), who also provide a review of nearly all available tests and studies for this model. Nonetheless, there is no alternative which takes care of all the difficulties (in particular, imposing the positive definiteness constraint on Ω which should be important for good power). At this time there is no alternative which is demonstrably superior from either a theoretical or a numerical point of view, even though it is clear that superior alternatives must exist. Thus we leave it to the ambitious reader to actually find a superior test.

10.3 Changepoint Problems

So far we have assumed that both J the number of regimes, and $T_1, \ldots, T_j$, the number of observations in each regime are known. Suppose that we observe a series (y_t, x_t) and we suspect that the parameters may change from time to time. In this case we do not know the value of T_j, the periods at which the regression parameters change. Nor do we know the number of changepoints. This makes the problem substantially more complex. The problem is similar in spirit to nonparametric regression, discussed briefly in Section 5.12. See Friedman (1991) for an algorithm adapted to this type of situation. In this section we discuss the substantially simpler case when it is known it advance that there is at most one change of regime, but the time of change (if there is one) is not known.

For a fixed t, let F_t be the F statistic for testing that there are two regimes with the data partitioned at the observation t. The test which rejects for large values of F_t is, as shown before, optimal if the changepoint is t. An intuitively plausible strategy to use in the case of unknown t is simply to look at the largest F_t value over all t. This is sometimes called the 'Rolling Chow' test, as the value of t is rolled over the time period. Two issues of interest arise in using this test. Even though for fixed t, the test is optimal, it is unclear whether any optimality properties persist for the case where t is unknown. The second issue is that the distribution of the maxima of the F_t needs to be known to calculate critical values for this test.

Andrews (1993) addresses these issues. He shows that the power of this test is compromised if we allow t to range over all possible values, and recommends that we should use $\max_{T_1 \leq t \leq T_2} F_T$ where $T_1 = 0.15T$

and $T_2 = 0.85T$. Intuitively it seems plausible that if T_1 or T_2 are allowed to get too close to the ends, the estimates of $\hat{\beta}$ and $\hat{\sigma}^2$ will have large variance in the smaller regime and this can cause difficulties in testing. Andrews also establishes a weak optimality property for the test. More important, he shows that the power of the rolling Chow test is superior to that of several other proposals in the literature. Finally, he obtains an asymptotic distribution for the test statistic. Diebold and Chen (1995) show that bootstrap critical values for this test are more accurate in finite samples.

As discussed in Chapter 2, conditioning on an ancillary statistic can substantially improve the quality of inference. Hinkley and Schechtman (1987) show that there exists an approximate ancillary statistic in the simplest form of the changepoint problem, and that inferences can be sharpened considerably by conditioning on it. In the generalization to regression model, following this route appears difficult. An alternative is to use a Bayes-type procedure; since the posterior density used for Bayesian inference is conditional on all the data, it automatically conditions on any ancillary information. In typical cases, use of a Bayesian technique results in improved inference only when the Bayesian technique incorporates additional valid prior information. In the present case, this is not required since the Bayesian technique automatically uses the ancillary information.

Following this line of argument, Başçı(1995) develops a Bayesian test for an unknown changepoint, and shows that it is more powerful than the maximum F test suggested by Andrews — which in turn is more powerful than popular and widely used CUSUM and related procedures for testing for model change. Further issues related to this Bayesian test and power comparisons are given in Başçı*et al.* (1995). The Bayesian test is given below, both to illustrate the procedure of deriving a Bayesian test and because it is interesting in its own right as the current champion in terms of power. This last status may change, since we don't have any optimality proofs. In particular, it is possible that Bayesian tests for different priors may be better, and also that equivalent frequentist test based on conditioning may be obtainable.

10.3.1   A Bayesian Test

For $t = 1, 2, \ldots, T$ we observe $y_t = x_t\beta_j + \epsilon_t$, where $\epsilon_t \overset{iid}{\sim} N(0, \sigma^2)$ and x_t is a $1 \times k$ vector of observations on the k regressions in period t. Under the null hypothesis, $\beta_j = \beta_0$ and under the alternative hypothesis, $\beta_j = \beta_1$ for $t \leq t^*$, the changepoint, and $\beta_j = \beta_2$ for $t > t^*$. It is convenient to write Y_a to denote the $a \times 1$ vector of observations $(y_1, \ldots, y_a)'$ and X_a the $a \times K$ matrix with rows $x_1, \ldots, x_a$. Also Y^a and X^a denote the $T - a \times 1$ vector and $(T - a) \times K$ matrix with rows $y_{a+1}, \ldots, y_T$ and $x_{a+1}, \ldots, x_T$ respectively. We will also need to define $SSR_a = \|y_a - X_a\hat{\beta}_a\|^2$ and $SSR^a = \|y^a - X^a\hat{\beta}^a\|^2$, where $\hat{\beta}_a = (X_a'X_a)^{-1}X_a'y_a$ and $\hat{\beta}^a = (X'^aX^a)^{-1}X'^ay^a$. Let SSR, Y, and X stand for SSR_T, Y_T, and X_T, respectively.

Let H_0 stand for the null hypothesis that there is no changepoint. Let H_a be the alternative hypothesis that change occurs at $t^* = a$. We wish to test H_0 against $H_1 = \cup_{a=K+1}^{T-(K+1)} H_a$. It is technically convenient to restrict the changepoint to be at least $K + 1$ units away from the extremes, so that all parameters can be estimated for both regimes. Under H_0 and H_a the likelihood function for the data can be written as $l_0(Y)$ and $l_a(Y)$ where

$$l_0(Y) = (2\pi)^{-T/2} \exp\left(-\frac{1}{2\sigma^2}\|Y - X\beta_0\|^2\right),$$

$$l_a(Y) = (2\pi)^{-T/2} \exp\left(-\frac{1}{2\sigma^2}\left\{\|Y_a - X_a\beta_1\|^2 + \|Y^a + X^a\beta_1\|^2\right\}\right).$$

In order to devise a Bayesian test, we need to specify the prior density for each parameter and H_0 and H_a. It is convenient to take a diffuse or uninformative prior density for each parameter. In the case of the regression coefficient β_i, this is just $f(\beta_i) = 1$. For the variance σ^2, the usual uninformative prior is $f(\sigma^2) = d\sigma^2/\sigma^2$. See Berger (1985) for further details regarding uninformative priors. Once these (improper) densities are specified, we need to compute the probability of observing Y under H_0 and under H_a.

Lemma 10.1 *The probability of observing Y under H_0 is*

$$\mathbb{P}(Y|H_0) = \pi^{-(T-K)/2}\Gamma((T-K)/2)\,\boldsymbol{det}(X'X)^{-1/2}\|Y - X\hat{\beta}\|^{-(T-K)}.$$

The probability of observing Y under H_a is

$$\mathbb{P}(Y|H_a) = \pi^{-(T-2K)/2}\Gamma((T-2K)/2)\,\mathbf{det}(X_a'X_a)^{-1/2}\,\mathbf{det}(X'^aX^a)^{-1/2}$$
$$\times \left(\|Y_a - X_a\hat{\beta}_a\|^2 + \|Y^a - X^a\hat{\beta}^a\|^2\right)^{-(T-2K)/2}.$$

The lemma is easily proven by integrating out the parameters β and σ^2 from the likelihood function from Y; see exercise 4. Now suppose π_0 is the prior probability of H_0 and π_a is the prior probability of H_a. We can get posterior probabilities of the hypotheses by a straightforward application of Bayes formula:

$$\mathbb{P}(H_i|Y) = \frac{\mathbb{P}(Y|H_i)\pi_i}{\mathbb{P}(Y|H_0)\pi_0 + \sum_{a=K+1}^{T-(K+1)}\mathbb{P}(Y|H_a)\pi_a}.$$

Let $t_{\min} \geq K + 1$ be the smallest possible changepoint, and let $t_{\max} \leq T - (K+1)$ be the largest possible changepoint. Assume that all of the H_a have equal probabilities. We should reject the null for small values of the posterior probability $\mathbb{P}(H_0|Y)$. This is equivalent to rejecting the null for large values of

$$B = \sum_{t^*=t_{\min}}^{t^*=t_{\max}} \frac{\mathbf{det}(X_{t^*}'X_{t^*})^{-1/2}\,\mathbf{det}(X'^{t^*}X^{t^*})^{-1/2}(SSR)^{(T-K)/2}}{\mathbf{det}(X'X)^{-1/2}(SSR_{t^*})^{(t^*-K)/2}(SSR^{t^*})^{(t^*-K)/2}}.$$

Başçı(1995) evaluates the power of this test by Monte-Carlo methods and shows that it is more powerful than the maximum F test suggested by Andrews(1990). The Bayesian method also provides estimates of $\mathbb{P}(H_a|Y)$ or the posterior probability of a change at time $t^* = a$. Plotting these probabilities frequently provides valuable information not easily available by any other approach. As an illustration, applying the above formulae, we computed the probability that a change occurs at period t for the Indy 500 data discussed in the previous chapter. The results are striking in that the posterior probability of a changepoint at 1941 is about 61 %, while at all other periods the probability of a changepoint is under 3.3%. See Başçi *et al.* (1995) for further illustrations and extensions.

10.4 The Behrens-Fisher Problem

The problem of testing the equality of means when the variances are not known to be equal enjoys some notoriety as the *Behrens-Fisher* problem. One reason for its fame is that it can be *proven* that there is no solution satisfying classical criteria for good tests. Somewhat more precisely, every invariant rejection region of fixed size for the problem must have some unpleasant properties. See Pfanzagl (1974) or Cox and Hinkley (1974) for an exposition and references. Briefly, first best solutions (that is, uniformly most powerful invariant tests) either do not exist or have strange properties. Thus we must look for second best solutions, of which there is an abundance. There are three basic types of solutions. One type abandons the requirement of invariance. This involves dropping some observations, or treating the regressors asymmetrically in an arbitrary fashion. Tests of this type adapted to the econometrics context were developed by Jayatissa (1977) and others. A virtue of these tests is that the probability of incorrect rejection of the null does not depend on the value of the nuisance parameters $\sigma_1^2, \ldots, \sigma_J^2$.[2] Such tests are called *similar*.[2] However, the lack of invariance is aesthetically unappealing; such tests involve making arbitrary choices which make a difference in the outcome. Even worse, simulation studies show that such tests have substantially less power than good tests of other types. For example, Thursby (1992) is a major simulation study and provides references to different types of solutions not discussed here.

The second type of test abandons the desirable property of similarity. This means that the exact probability of incorrectly rejecting the null depends on the value of the nuisance parameters. Since the nuisance parameters are unknown, this means that the exact size of the test will be unknown. Nonetheless, it turns out that we can get tests with good properties which are approximately similar — that is, the exact size fluctuates very little as a function of the nuisance parameters. We discuss one such solution below.

---

[2]Similar is short for similar to the sample space, which has the same property – that is, probability of an observation belonging to the the sample is 1, independent of any parameters.

The third type of test is built on the basis of a new idea. Conventional theory requires us to either accept or reject the null, and make the decisions so that we have a fixed probability of false rejection of the null. An alternative is to state the 'p-value' or the probability of the null hypothesis; if this is too low, we reject the null hypothesis. On the surface the two approaches appear similar (and lead to essentially the same solution) in many problems. In this particular case, however, the second approach yields a useful and clear solution, whereas the first approach does not. This approach was initiated by Weerahandi, and is discussed in Section 10.4.2.

10.4.1   A Good Approximate Test

A practical approximate solution can be obtained by the following line of reasoning. Suppose that σ_j^2 are not equal, but known. In this case, the sufficient statistics are just $\hat{\beta}_j \sim N(\beta_j, \sigma_j^2(X_j'X_j)^{-1})$. In this case of known variances, an optimal test for the hypothesis **EM** that $\beta_j = \beta_*$ for all j rejects for large values of the statistic:

$$WA = \sum_{j=1}^{J} \frac{1}{\sigma_j^2}(\hat{\beta}_j - \hat{\beta}_*)' (X_j'X_j)^{-1} (\hat{\beta}_j - \hat{\beta}_*). \tag{10.3}$$

The statistic WA has a chi-square density with K degrees of freedom under the null hypothesis. Since the theory of this test has already been worked out earlier, the derivation is left to exercise 5. In the original problems σ_j^2 are not known. To make this test statistic operational, one can replace the σ_j^2 by their estimates $\hat{\sigma}_j^2 = \|y_j - X_j\hat{\beta}_j\|^2/(T_j - K)$. Now the distribution of this statistic is no longer chi-squared under the null, though it can be shown that this distribution is asymptotically valid. As already discussed, the statistic WA is not similar and therefore the rejection probability $\mathbf{P}(WA > c|H_0)$ depends on the nuisance parameters $\sigma_1^2, \ldots, \sigma_J^2$. Welch and Aspin independently developed a higher order approximation to the distribution of W, which takes the nuisance parameters into account; see Section 14.4 for more details, and Cox and Hinkley (1977) for an exposition and references. Numerical work shows that these approximations substantially reduce the problem and stabilize the size of the test except for very small samples sizes.

Rothenberg (1984) has generalized these approximations to the more complex framework required in econometrics. Work of Pfanzagl (1971) shows that this type of approximate solution has asymptotic optimality properties. Even though Rothenberg's solution shares these optimality properties, and performs quite well in the first round of competitions studied by Monte Carlo in Thursby's (1992) simulation study, Thursby eliminates it from consideration on the grounds of its computational complexity. The bootstrap-Bartlett technique of Rayner (1990) is computationally simpler and provides equivalent accuracy. This technique is discussed in some detail in Section 14.6.

10.4.2   A Test Based on p-Value

Weerahandi initiated a new approach to testing which avoids some of the difficulties in a classical approach. See Weerahandi (1994) for an exposition, further developments, and references for the material discussed below. In particular, this approach yields a type of exact solution to the Behrens-Fisher problem. Simulations by Thursby (1992) and others show that it has good sampling properties. We give a heuristic development here.

Standard tests require finding an event E which has a fixed low probability (such as 5% or 1%) under the null hypothesis. If the event E occurs, we conclude that the null hypothesis is unlikely to be true. Since the probability of E is fixed, the probability of a false rejection of the null is also fixed. In many problems with nuisance parameters, suitable events of low probability under the null cannot be found because the nuisance parameters affect the probability of all events but are not known. In the present case, it can be shown that events E with fixed low probabilities under null suitable for testing cannot be found. In testing with the p-value, we relax classical restrictions on E by allowing it to depend on the observations and also on the nuisance parameters. The event E is 'unlikely' to occur under the null, and more likely to occur under the alternative. Furthermore we can calculate exactly the probability of E under the null, but, unlike classical tests, this probability is not fixed in advance. In the context of significance testing with p-values, such a region E is called an extreme region of the sample space; see Weerahandi (1994) for an extensive discussion and several examples. Weerahandi develops tests with the property that $\mathbf{P}(E|H_0) < \mathbf{P}(E|H_1)$, and we can calculate $p = \mathbf{P}(E|H_0)$. Thus we reject H_0 if p is too low, and not otherwise. To make testing by p-value comparable to fixed level testing, we could decide to reject the null whenever the p value is less than 5% (for example). In this case whenever we reject, the p value will be less than or equal to 5% so that the actual level of the test will be smaller than or equal to 5% but the exact difference is typically hard to calculate.

A region suitable for testing by p-value can be developed as follows. First consider for simplicity the case of two regimes. Let $\delta = \beta_1 - \beta_2$ and $\hat{\delta} = \hat{\beta}_1 - \hat{\beta}_2 \sim N(\delta, \sigma_1^2 \Sigma_1 + \sigma_2^2 \Sigma_2)$, where $\Sigma_1 = (X_1' X_1)^{-1}$ and $\Sigma_2 = X_2' X_2)^{-1}$. If we know σ_1^2 and σ_2^2, an optimal test would reject for large values of T defined by

$$T = \hat{\delta}' \left(\sigma_1^2 \Sigma_1 + \sigma_2^2 \Sigma_2 \right)^{-1} \hat{\delta}.$$

Furthermore, T has a χ_K^2 distribution under the null hypothesis. When σ's are unknown but equal, we divide by an estimate of σ to eliminate their influence, getting the ratio of independent chi-squares or an F distribution for out test statistics. When they are unequal, something similar is still possible. Let $s_1^2 = \|y_1 - X_1 \hat{\beta}_1\|^2$ and $s_2^2 = \|y_2 - X_2 \hat{\beta}_2\|^2$ be the sum of squared residuals for each of the regimes. Then s_1^2/σ_1^2 and s_2^2/σ_2^2 are independent chi-square random variables with $T_1 - K$ and $T_2 - K$ degrees of freedom. Their sum $S = s_1^2/\sigma_1^2 + s_2^2/\sigma_2^2$ is chi-squared with $T_1 + T_2 - 2K$ degrees of freedom. The ratio T/S is the ratio of independent chi-squares and therefore has an F distribution with K and $T_1 + T_2 - 2K$ degrees of freedom. We can rewrite the ratio in the following form:

$$\frac{T}{S} = \hat{\delta}' \left(\frac{S}{s_1^2/\sigma_1^2} s_1^2 \Sigma_1 + \frac{S}{s_2^2/\sigma_2^2} s_2^2 \Sigma_2 \right)^{-1} \hat{\delta}.$$

Defining $R = (s_1^2/\sigma_1^2)/S$, we see that

$$\frac{T}{S} = \hat{\delta}' \left(\frac{s_1^2}{R} \Sigma_1 + \frac{s_2^2}{1-R} \Sigma_2 \right)^{-1} \delta.$$

This statistic has a known distribution, and would provide an optimal test if we knew the value R. Let $\mathrm{CDFF}_{m,n}(x)$ be the cumulative distribution function of an F distribution with m and n degrees of freedom. We would then reject the null hypothesis if $p = 1 - CDFF_{K,T_1+T_2-2K}(T/S)$ was too low. Unfortunately, we cannot compute this p-value since we do not observe the variable R. However, R has a known distribution independent of the parameters σ_1^2, σ_1^2, and is also statistically independent of the sum S and hence also of the statistic T/S. It can be shown that R has a Beta distribution with parameters $(T_1 - K)/2, (T_2 - K)/2$; see exercise 6. Thus, instead of the p-value conditional on R, which is not available, we can calculate the expected value of p averaged over all possible values of the observation R, according to its Beta density. To summarize, the recommendation of Weerahandi is to reject the null of equality if the p-value given by the following formula is too low:

$$p \;=\; 1 - \mathbb{E}^R \left[\mathrm{CDFF}_{K,T_1+T_2-2K} \left(\frac{T_1 + T_2 - 2K}{K} \times \right. \right.$$
$$\left. \left. (\hat{\beta}_1 - \hat{\beta}_2)' \left\{ \frac{SSR_1}{R}(X_1'X_1)^{-1} + \frac{SSR_2}{1-R}(X_2'X_2)^{-1} \right\} (\hat{\beta}_1 - \hat{\beta}_2) \right) \right]$$

Here the expectation is taken with respect to the random variable R which has a Beta density with parameters $(T_1 - K)/2$ and $(T_2 - K)/2$.

In Koschat and Weerahandi (1992) (KW hereafter in this section), it is shown that this procedure can be extended to a test of the general linear restriction in models of the type under study in this chapter. Using the definitions in and around Eq. (10.1), we can write the equation for all J regimes as $y_* = X_0\beta_0 + \epsilon_*$. The covariance matrix of the combined error term is of the following form:

$$\mathrm{Cov}(\epsilon_*) = \Sigma_0 = \begin{pmatrix} \sigma_1^2 \mathbf{I}_{T_1} & 0 & \cdots & 0 \\ 0 & \sigma_2^2 \mathbf{I}_{T_2} & \cdots & 0 \\ \vdots & \vdots & & \vdots \\ 0 & 0 & \cdots & \sigma_J^2 \mathbf{I}_{T_J} \end{pmatrix}.$$

Suppose we wish to test the linear hypothesis $R_0\beta_0 = r_0$, where R_0 is $m \times JK$ with rank $m \le JK$. Note that this includes as a special case the hypothesis **EM**. The basic result of Weerahandi which gives a p-value test for this hypothesis is given as a theorem below. We introduce some necessary notation first. First, let $y_\dagger = \Sigma_0^{-1/2} y_*$, $X_\dagger = \Sigma_0^{-1/2} X_0$, and $\epsilon_\dagger = \Sigma_0^{-1/2} \epsilon_*$. Then $y_\dagger = X_\dagger \beta_0 + \epsilon_\dagger$, where $\epsilon_\dagger$ now has the identity as a covariance matrix. Let $\hat{\beta}_0 = (X_1' X_1)^{-1} X_1' y_1$ be the unconstrained OLS estimate in this model, and let $\tilde{\beta}_0$ be the maximum likelihood estimate subject to the constraint $R_0 \beta_0 = r_0$. An explicit formula for $\tilde{\beta}_0$ is given in Eq. (8.3). Let $s_j^2 = \| y_j - X_j \hat{\beta}_j \|^2$ be the sum of squared residuals in the j-th regime before the transformation. Let $s_0^2 = \| y_\dagger - X_\dagger \hat{\beta}_0 \|^2$ be the unconstrained sum of squared residuals and $s_\dagger^2 = \| y_\dagger - X_\dagger \tilde{\beta}_0 \|^2$ be the constrained sum of squared residuals. Note that s_0^2 and $s_\dagger^2$ both depend on the values of σ_i^2 and hence are not observable when the σ_i^2 are unknown. We will write $s^2 = s^2(\sigma_1^2, \ldots, \sigma_J^2)$ to make explicit this dependence on the unknown variances.

Theorem 10.2 (Koschat and Weerahandi) *Define random variables $R_0, \ldots, R_J$ as follows: $R_0 = 0$ with probability 1. If $1 \leq j \leq J-1$, R_j is Beta with parameters $\sum_{i=1}^{j}(T_i - K)/2$ and $(T_{j+1} - K)/2$, while R_J is Beta with parameters $\sum_{i=1}^{J}(T_i - K) = T_0 - JK$ and $(J-1)K/2$. Define $\tilde{\sigma}_i^2 = R_i s_i^2 / (1 - R_i) \prod_{j=i-1}^{j=J-1} R_j$. Consider the random variable $V(\sigma_1^2, \ldots, \sigma_J^2) = s_\dagger^2(\sigma_1^2, \ldots, \sigma_J^2) - s_0^2(\sigma_1^2, \ldots, \sigma_J^2)$. Let $\mathbf{F}_{m,n}$ be the cumulative distribution function of an F distribution with m and n degrees of freedom. Define $p(\sigma_1^2, \ldots, \sigma_J^2) = 1 - \mathbf{F}_{m, T_0 - KJ}((T_0 - KJ) V(\sigma_1^2, \ldots, \sigma_J^2)/m)$. A test based on p-values rejects for small values of $\mathbb{E} p(\tilde{\sigma}_1^2, \ldots, \tilde{\sigma}_J^2)$, where the expectation is taken with respect to (the known distributions of) the random variables R_j.*

For a proof, the reader is referred to KW. We note that for small values of J, e.g. 2 or 3, it is possible to calculate the expected value by numerical integration. For higher numbers of regimes, it will be easier to use Monte Carlo methods. That is, we generate at random $R_1, \ldots, R_J$ according the distributions described, and then compute the F statistic and its p-value. After repeating this process for say, one hundred times, we average over the p-values obtained. This will give a reasonably accurate approximation to the desired p-value. The form of the theorem above implicitly assumes that the regressor matrix in each regime is of full rank. The result of KW also covers cases where this full rank assumption fails; only a minor modification is required. The reader is referred to KW for details. We note that there is a misprint in KW: in expression (1.8b) the fraction $n - q$ over $p - q$ should be $(n - p)/(p - q)$.

10.5   Testing Equality of Variances

In the case where nothing is assumed about equality of the regression coefficients, Bartlett's test for equality of variances across regimes was discussed in Section 8.6. In the special case of two regimes, this test reduces to the Goldfeld-Quandt test, and has certain optimality properties discussed below. Frequently more useful, in applications, is the situation where regression coefficients are assumed equal across regimes. Testing for equality of variances in this case is discussed below. Finally, we also discuss a test for aggregation, which tests for simultaneous equality of means and variances across the regimes.

10.5.1   Goldfeld-Quandt Test

Consider the case of two regimes: $y_j = X_j\beta_j + \epsilon_j$ for $j = 1, 2$. We are interested in testing **EV**, the hypothesis of equality of variances $H_0: \sigma_1^2 = \sigma_2^2$. If the one-sided alternative, $H_1: \sigma_1^2 < \sigma_2^2$, is of interest, we can get a UMP invariant test in this case.

Theorem 10.3 *The test which rejects the null for large values of s_1^2/s_2^2 is UMP invariant and also most stringent for testing H_0 versus H_1.*

Proof: Since **EM** is not assumed, β_j are not known and not equal, and we can reduce the problem to the statistic $s_j^2 = \|y_j - X_j\hat{\beta}_j\|^2$ for $j = 1, 2$ by the same argument given for multiple regimes. Now note that the testing problem is scale invariant (this same idea is used in the Pitman test; after taking logs, scale invariance becomes translation invariance). That is, if $Z_1 = \lambda s_1^2$ and $Z_2 = \lambda s_2^2$ and $\theta_1 = \lambda\sigma_1^2$ and $\theta_2 = \lambda\sigma_2^2$, the testing problem framed in terms of observations Z_1, Z_2 and parameters θ_1, θ_2 is identical to the original problem. Invariant tests must satisfy $\delta(s_1^2, s_2^2) = \delta(\lambda s_1^2, \lambda s_2^2)$. The maximal invariant is just the ratio s_1^2/s_2^2, or equivalently $F = (s_1^2/t_1)/(s_2^2/t_2)$. Dividing by the degrees of freedom gives F the standard F distribution with degrees of freedom t_1 and t_2 under the null hypothesis. To show that the test based on the F statistic is UMP invariant, we must verify that the family of densities has a monotone likelihood ratio for the one-sided test.

We now derive the distribution of the F statistic under the null and alternative hypothesis. First note that if $Z_1 \sim \chi^2_{t_1}$ and $Z_2 \sim \chi^2_{t_2}$ are independent then $G = (Z_1/t_1)/(Z_2/t_2)$ has an F density with t_1 and t_2 degrees of freedom. This density is

$$f(G) = \frac{\Gamma(t_1/2)\Gamma(t_2/2)(t_1)^{t_1/2}(t_2)^{t_2/2}G^{(t_1/2)-1}}{\Gamma((t_1+t_2)/2)(t_1 G + t_2)^{(t_1+t_2)/2}} \qquad (10.4)$$

Let $\lambda = \sigma_2^2/\sigma_1^2$. Note that $\lambda F = (s_1^2/t_1\sigma_1^2)/(s_2^2/t_2\sigma_2^2)$; this has an F density with t_1 and t_2 degrees of freedom. In general if λX has density $f^X(x)$ then X has density $(1/\lambda)f^X(x/\lambda)$. Thus the density of F for $\lambda \geq 1$ can be expressed as $(1/\lambda)f(F/\lambda)$, where f is given by (10.4). We show that this density has monotone likelihood ratio as follows:

$$\begin{aligned} LR &= \frac{(1/\lambda)f(F/\lambda)}{f(F)} \\ &= \frac{\lambda^{-t_1/2}F^{(t_1/2)-1}(t_1 F/\lambda + t_2)^{-(t_1+t_2)/2}}{F^{(t_1/2)-1}(t_1 F + t_2)^{(t_1+t_2)/2}} \\ &= \lambda^{-t_1/2}\left(\frac{t_1 F + t_2}{t_1 F/\lambda + t_2}\right)^{(t_1+t_2)/2}. \end{aligned}$$

This shows that the LR test can be based on rejecting for large values of $(t_1 F + t_2)/(t_1 F/\lambda + t_2)$. By differentiation, it is easily established that this is monotonic increasing in F for all $\lambda > 1$. It follows that the one-sided test is UMP invariant.

To show that it is the most stringent test, we must verify that the transformation groups used are either Abelian or compact. We first eliminated the estimators $\hat{\beta}_i$ from consideration using the translation group, which is Abelian. Then we used scale invariance; the group of transformations involved is the multiplicative group of positive reals. This is also Abelian. Since all groups are Abelian, the hypotheses of

the Hunt-Stein theorem are fulfilled and the resulting best invariant
test is most stringent.

It is clear that a similar analysis applies when the alternative is
H_1': $\sigma^2 < \sigma_1^2$. However, the situation is different when we are interested
in a two-sided alternative. This is discussed next.

10.5.2   GQ: Two-Sided Test

Suppose we wish to test H_0: $\sigma_1^2 = \sigma_2^2$ versus the two-sided alternative
H_1: $\sigma_1^2 \neq \sigma_2^2$. When $t_1 = t_2$ then the distribution of F is the same
as that of $1/F$. This is an invariant transformation mapping the ratio
$\lambda = \sigma_2^2/\sigma_1^2$ to $1/\lambda$. It follows that invariant tests must satisfy $\delta(F) =
\delta(1/F)$. It follows easily that tests which reject the null hypothesis for
$F > c$ or $F < 1/c$ are UMP invariant and most stringent.

When $t_1 \neq t_2$ then F and $1/F$ do not have the same distribution
and the above argument does not apply. It is easy to establish that all
good tests reject the null for $F \leq c_1$ and $F \geq c_2$. If we fix the level
of the test at, e.g., 5%, we can choose the constants c_1 and c_2 so that
$\mathbf{P}(F > c_2|H_0) + \mathbf{P}(F < c_1|H_0) = 0.05$. One obvious way to choose the
constants is so that $\mathbf{P}(F > c_2|H_0) = \mathbf{P}(F < c_1)$. Such a test is called
an equal tails test. Equal tails tests have the undesirable property that
they are biased. Use of Pitman's procedure discussed earlier leads to
an unbiased test. Pitman's statistic for the case of two regimes can be
written as (where $t = t_1 + t_2$)

$$P = \frac{(s_1^2)^{t_1/t}(s_1^2)^{t_2/t}}{s_1^2 + s_2^2} = \frac{(s_1^2/s_2^2)^{t_1/t}}{(s_1^2/s_2^2) + 1} = \frac{(t_2 F/t_1)^{t_1/t}}{F + 1}.$$

P is a unimodal function of F and increases from 0 to $t_1/(t_2 + 2t_1)$ and
decreases thereafter. Thus rejecting for $P \geq c$ is equivalent to rejecting
for $F \geq c_1$ and $F \leq c_2$. The constants determined by using P have the
property of leading to an unbiased test.

How can we decide whether to use equal tails or the unbiased test?
A reasonable procedure is to look at the stringency of the tests. In this
problem the unbiased test is (by numerical methods) always more stringent then the equal tails test, but the most stringent test lies between
the two. However, the differences in powers are typically small so that

use of the unbiased test can be recommended without serious reserva-
tions; computing the constants for the most stringent test is somewhat
more complex.

10.5.3 Testing *EV* When *EM* Holds

In some applications, we wish to test equality of variances across
regimes when the regression coefficients are assumed equal. In fact,
the Goldfeld-Quandt test is designed for this type of situation, but it
does not use the information about equality of coefficients. A virtue of
the GQ test is that it is easy to compute and has a known exact distribution. As we shall see, utilizing the information of equality requires
substantially greater computational power, and hence procedures of the
type proposed have only recently become feasible.

The simplest way to use the information of equality of means is via
a likelihood ratio test. This test will also have asymptotic optimality
properties. Tests with finite sample optimality properties are not known
for this case. Some theoretical and Monte Carlo analysis of this test for
the special case of two regimes in given in Tomak (1994) and Tomak
and Zaman (1995). We show how to carry out the test in the general
case of J regimes below.

It is convenient to begin with some notation. Note that the likelihood function for the J regimes is

$$\ell(\beta_1, \ldots, \beta_J, \sigma_1, \ldots, \sigma_J) = \prod_{j=1}^{J} (2\pi\sigma_j^2)^{-T_j/2} \exp\left(-\frac{\|y_j - X_j\beta_j\|^2}{2\sigma_j^2}\right).$$

Let β_0 be the common value of β_j under the hypothesis **EM** that all
regression coefficients are the same across regimes. Let σ_0 be the common value of the σ_j under the hypothesis **EV** that all the variances are
the same across regimes. Our null hypothesis is that both **EV** and **EM**
hold. It is easy to estimate the parameters β_0 and σ_0 under this null,
simply by aggregating all the data and running an ordinary regression
on the full data set. The resulting ML estimates are

$$\hat{\beta}_0 = \left(\sum_{j=1}^{J} X_j'X_j\right)^{-1} \left\{\sum_{j=1}^{J} X_j'y_j\right\}$$

$$\hat{\sigma}_0^2 = \frac{\sum_{j=1}^K \|y_j - X_j\hat{\beta}_0\|^2}{\sum_{j=1}^J T_j}.$$

For the LR, we need to get the ML estimates $\tilde{\beta}_0$ and $\tilde{\sigma}_j^2$ under the alternative hypothesis that **EM** holds but **EV** does not. The first order conditions for maximum are easily seen to be

$$\tilde{\beta}_0 = \left(\sum_{j=1}^J \frac{1}{\tilde{\sigma}_j^2} X_j' X_j\right)^{-1} \left\{\sum_{j=1}^J \frac{1}{\sigma_j^2} X_j' y_j\right\}$$

$$\tilde{\sigma}_j^2 = \frac{1}{T_j}\|y_j - X_j\tilde{\beta}_0\|^2.$$

These equations cannot be solved analytically for the ML estimates. Nonetheless, we can find the ML estimates fulfilling these first order conditions by an iterative process. Fix σ_j^2 at arbitrary initial values and solve for $\hat{\beta}_0$ from the first equation. Solve for new values of σ_j^2 from the second equation. Iterate back and forth until convergence. This procedure is guaranteed to converge, and typically does so very fast.

Once we have the ML estimates under the null and the alternative, an LR test statistic is easily written down as

$$LTZ = -2(\log \ell(\tilde{\beta}_0, \ldots, \tilde{\beta}_0, \tilde{\sigma}_1^2, \ldots, \tilde{\sigma}_J^2) - \log \ell(\hat{\beta}_0, \ldots, \hat{\beta}_0, \hat{\sigma}_0^2, \ldots, \hat{\sigma}_0^2).$$

The issue of how to compute the distribution of the likelihood ratio, so as to determine a critical value, remains. The simplest method is to use asymptotic theory, according to which LTZ has a chi-square distribution with J degrees of freedom, the number of constraints imposed by the null hypothesis. This is typically unreliable except in very large samples. Instead, two alternatives based on bootstrap methods are recommended. Repeat the following sequence of steps 100 times:

1. For $j = 1, 2, \ldots, J$, fix $\beta_j = \hat{\beta}_0$ and $\sigma_j^2 = \hat{\sigma}_0^2$ to be the constrained ML estimates.

2. For $j = 1, 2, \ldots, J$, generate a new set of errors ϵ_j^* from a $N(0, \sigma_j^2 \mathbf{I}_{T_j})$ density.

3. For $j = 1, 2, \ldots, J$, generate new values of y_j as $y_j^* = X_j \beta_j + \epsilon_j^*$.

4. Compute a new value for the LTZ statistic based on the fresh data set and record it as LTZ^*.

After finishing this computation, we will have a collection of statistics $LTZ_1^*, \ldots, LTZ_{100}^*$, which have been generated under conditions which approximate the null hypothesis. Suppose that these are sorted in ascending order. Then a direct bootstrap estimate of the level 90 % critical value is simple LTZ_{90}^* — we would reject the null at the 90 % significance level if the observed LTZ statistic for the original sample is above this value. An indirect method is to let m be the mean value of the 100 bootstrap statistic LTZ_i^*, and modify the original statistic to $LTZ' = J \times LTZ/m$. Then LTZ' should have expected value J, which matches the expected value for the asymptotic χ_J^2 distribution of LTZ. Now we look up the critical value in the χ_J^2 table to see if the observed value is significant or not. This last method is known as a Bootstrap Bartlett correction, and is discussed in greater detail in Section 14.6. Typically it is more efficient, in the sense that a smaller bootstrap sample size is required for a given desired level of accuracy in computation of the critical value.

10.5.4   A Test for Aggregation

It may be desired to make a joint test of the null hypotheses that both **EM** and **EV** hold versus the alternative that at least one of the two fails. This can be called a test for aggregation since if we fail to reject the null, all of the data may be aggregated into one regression model. Some theoretical details and empirical study of the performance of this test are given in Maasoumi and Zaman (1995). Below we show how the test can be carried out and some of its properties.

Let $\hat{\beta}_j = (X_j' X_j)^{-1} X_j' y_j$ and $\hat{\sigma}_j^2 = \|y_j - X_j \hat{\beta}_j\|^2 / T_j$ be the unconstrained ML estimates for each regime separately. In the notation of the previous Section, the likelihood ratio statistic for jointly testing **EM** and **EV** versus the unconstrained alternative can readily be written down as

$$LMZ = -2(\log \ell(\hat{\beta}_0, \ldots, \hat{\beta}_0, \hat{\sigma}_0^2, \ldots, \hat{\sigma}_0^2) - \log \ell(\hat{\beta}_1, \ldots, \hat{\beta}_J, \hat{\sigma}_1^2, \ldots, \hat{\sigma}_J^2)).$$

The number of constraints imposed by the null hypothesis is $J + J(K - 1)$, where K is the number of regression coefficients and J is the number of regimes. Hence the LMZ has an asymptotic χ^2 distribution with JK degrees of freedom. As discussed before, the asymptotic distribution is likely to be unreliable. It would be preferable to make a bootstrap Bartlett correction, or else compute the critical value directly by bootstrapping.

It is also of some interest to note that $LMZ = LCH + LEV$, where LCH is a form of the Chow test for the equality of means under the condition that the variances are the same across regimes, and LEV is similar to a Bartlett's test for equality of variances (with no assumptions about means). Intuitively, we could consider testing for aggregation in a two-step way. First, use Bartlett's test to determine whether the variances are equal across regimes or not (note that Bartlett's test does not assume equal means). If Bartlett's test rejects the null, the regimes cannot be aggregated. If Bartlett's test fails to reject the null, we can proceed under the assumption that the variances are equal. In this case, a generalized Chow test can be carried out to test equality of means in all regimes. If this rejects the null, we conclude that aggregation fails, while if it accepts, we conclude the regimes can be aggregated. The aggregation statistic LMZ takes the sum of these two statistics rather than doing the tests in a sequential form, and avoids some awkward decisions regarding the level of the two tests.

10.6   Exercises

1. [Predictive Chow] Consider the case where $J = 2$ so we have two regimes. Assume the second regime has only *one* observation: $T_1 = T - 1$, $T_2 = 1$. We wish to test **EM** under the assumption that **EV**. In other words, we wish to find out if the last observation is suitably similar to the first $T - 1$. An alternative way of thinking about this test is to introduce a dummy variable D which takes value 1 in the last period and zero in all other periods. Next consider the regression $y_* = X_* \beta_* + D\gamma + \epsilon_*$. The * subscript indicates that the observations from all regimes have been stacked, as discussed in the text. Obviously if γ is signif-

icantly different from zero, the last observation is different from
the rest. Show that the F statistic for testing $\gamma = 0$ is equivalent to the statistic in Eq. (10.2) of the text. Hint: let $\tilde{\beta}$ be the
OLS estimate of β based on first $T-1$ observations. Let $\hat{\beta}, \hat{\gamma}$ be
OLS estimates of β and γ based on all observations. Show that
$\hat{\beta} = \tilde{\beta}$ and $\hat{\gamma}$ satisfies $y_T = X_t\tilde{\beta} + D\hat{\gamma}$; in other words, $\hat{\gamma}$ makes
the residual zero for the T-th observation.

2. Suppose that two of the rows of X_j are identical. If $T_j > K$ this
will not matter, since the rank of X_j will remain K. Whenever the
rank of X_j is K the projection $\hat{y}_j$ of y_j on the space $\mathbf{X}_j$ spanned
by the columns of X_j is $X_j\hat{\beta}_j$ However, if $T_j \leq K$, then the rank
of X_j will be less than K and the matrix $(X_j'X_J)^{-1}$ will not be
well defined. The case considered in Section 10.2.2 assumes that
each row of X_j is independent. In this case the column space
has maximum possible dimension T_j, and y_j is $T_j \times 1$. Since
we are projecting onto the entire space, the projection of y_j is
just $\hat{y}_j = y_j$. This is the basis for the formula in Section 10.2.2.
This formula is not valid if two rows of X_j are the same; more
generally if the rank of X_k is less than T_j and $T_j \leq K$. Show how
to compute $\hat{y}_j$ in this case, and derive the appropriate F statistic.

3. This computer exercise is designed to teach you how to compute
the critical values of the rolling Chow test using bootstrapping.
Pick a data set (y_t, x_t), and a range T_1, T_2. We wish to test
whether or not there is a changepoint T between the points T_1
and T_2. Let $MC = \max_{T_1 \leq T \leq T_2} F_T$ where F_T is the F statistic for testing whether there is a change point at T. We wish
to determine the distribution of MC under the null hypothesis
that there is no changepoint. Let $\hat{\beta} = (X'X)^{-1}X'y$ be the estimate of regression coefficients based on all the data (this is a
valid estimate under the null hypothesis). Let $e = y - X\hat{\beta}$ be
the regression residuals. Let e^* be a *resample* of the regression
residuals. More explicitly, e^* is a $T \times 1$ random vector defined
as follows: for $t = 1, 2, \dots, T$, e_t^* is i.i.d. and equals e_j (for some
$j = 1, 2, \dots, T$) with probability $1/T$. That is, each e_t^* is one
of the original regression residuals, chosen randomly with equal

probability. A typical value of e^* will be some rearrangement of
the original residuals but (and this is very important) with some
repetitions and some omissions. Next define $y^* = X\hat{\beta} + e^*$. Next
calculate MC^* to be the maximum value of the F statistic for
this 'new' data set. Repeat this process 100 times (for example)
to get a bootstrap sample $MC_1^*, MC_2^*, \ldots, MC_{100}^*$. Arrange these
values in increasing order and let $MC_{(95)}^*$ be the fifth largest value.
Then a level 5% test for the null rejects for $MC > MC_{(95)}^*$.

4. Show that

$$\int \exp\left(-\|y - X\beta\|^2/2\sigma^2\right)\, d\beta = (2\pi)^{-K/2}\, \det(X'X)^{-1/2}.$$

$$\int \sigma^{-2p} \exp -\frac{S}{2\sigma^2}\, \frac{d\sigma^2}{\sigma^2} = S^{-p}\Gamma(p).$$

Use these to calculate $P(Y|H_j) = \int\int l_j(Y)\, d\beta\, (d\sigma^2/\sigma^2)$ and
prove lemma 10.1

5. Suppose $Z \sim N(\beta, \Sigma)$, where Σ is a known covariance matrix.
We wish to test $R\beta = 0$, where R is an $m \times k$ matrix of full rank
$m \leq k$.

   (a) Show that the best invariant test can be based on $X = RZ$.
   Hint: introduce $(k - m) \times k$ matrix T such that rows of R
   and T together span $\mathbb{R}^k$. Let $Y = TZ$, and consider testing
   based on (X, Y) a one-to-one transform of Z. Show that Y
   can be eliminated by invariance considerations as the null
   hypothesis places no restrictions on its mean.

   (b) Show that a best invariant test based on X rejects for large
   values of the statistic $X'(R\Sigma R')^{-1}X$, which has a chi-square
   density under the null.

   (c) Specialize this general framework to get the formula in Eq.
   (10.3).

   (d) Derive the likelihood ratio statistic for this test. Show that
   in the special case of the previous question, it also coincides
   with the statistic in Eq. (10.3).

6. Given independent random variables X, Y with distributions $X \sim \Gamma(p_1, \lambda)$ and $Y \sim \Gamma(p_2, \lambda)$, show that $R = X/(X + Y)$ has a beta density and is independent of $S = (X + Y)$. Recall that $X \sim \Gamma(p, \lambda)$ means that

$$f^X(x) = \frac{\lambda^p}{\Gamma(p)} x^{p-1} \exp(-\lambda x).$$

Part III

Asymptotic Theory

Courtiers of Sultan Mahmood of Ghazna were jealous of the Sultan's favorite, a slave named Ayaz. Aware of this, Mahmood arranged a test to demonstrate the superiority of Ayaz. He had the largest and most precious gem in his treasury brought out and ordered his courtiers to crush it with a hammer. None of them could bring themselves to destroy the precious jewel, and excused themselves from the job. At his turn, Ayaz did not hesitate but struck a strong blow and destroyed the stone. In response to remonstrations of the courtiers, he said that it was necessary to break either the order of the Sultan or the stone, and the former was much more precious.

Chapter 11

Consistency of Tests and Estimators: Direct Methods

One of the problems in teaching asymptotic theory is that there are many different concepts of convergence for random variables. We have tried without much success to limit the number of concepts of convergence used. When it comes to the Law of Large Numbers, essential for consistency results, an important tool of wide applicability is quadratic mean convergence. This is easily understood, as it merely involves the convergence of the variance to zero. Furthermore, it is quite general and flexible: we can deal with arbitrarily complex covariance structures. Unfortunately, it is not quite sufficient for the needs of the econometrician because of one major defect. If $\hat{\theta}_n$ converges to θ in quadratic mean, it does not follow that $f(\hat{\theta}_n)$ converges to $f(\theta)$ for all contin-

uous functions f. This is why (weak) consistency is defined in terms
of convergence in probability, which is implied by, but does not imply quadratic mean convergence. In practice, we prove convergence in
probability by proving quadratic mean convergence for some suitable
sequence, and then use continuous functions to conclude consistency
for a wide range of different types of estimators. It is the author's
impression that strong consistency, based on almost sure convergence
or convergence with probability one, can safely be left out of elementary asymptotic theory for econometrics. Nonetheless, we have put in
a section on this concept so as to enable the student to approach the
literature.

The main idea behind proving consistency of a sequence of estimators, e.g., $\hat{\theta}_n$ is simple. We must show that $\mathbb{E}\hat{\theta}_n = \theta_n$ converges to
the true parameter θ_0 (this part is trivial for unbiased estimators) and
also that $\mathbb{V}ar(\hat{\theta}_n)$ converges to zero. This is typically easy to verify for
estimators which are linear functions of errors. For more complicated
functions of the errors we look for the simpler building blocks. If we
can find sequences X_n, Y_n, Z_n which converge to limiting values a, b, c
than a continuous function $f(X_n, Y_{,n}, Z_n)$ must converge to $f(a, b, c)$.
Together these techniques can handle many but not all issues of consistency. Those which remain can often be handled by techniques described in the next chapter.

11.1   Quadratic Mean Convergence

For the real numbers, there is essentially only one useful way to define
convergence. However, there are many different ways of defining convergence for random variables. One of the simplest definitions is the
following.

Definition 11.1 *A sequence of random variables X_n converges to random variable X in* quadratic mean *if*

$$\lim_{n \to \infty} \mathbb{E}\|X_n - X\|^2 = 0.$$

We can write this as **Qlim**$_{n \to \infty} X_n = X$ *and also* $X_n \overset{qm}{\to} X$.

As a simple example, suppose $X_n \sim N(0, \sigma_n^2)$. It is easily verified that X_n converges to 0 in quadratic mean if and only if $\lim_{n\to\infty} \sigma_n^2 = 0$.

Quadratic mean convergence is very useful in verifying many *laws of large numbers*. Roughly speaking, a law of large numbers is a statement showing that averages of a sequence of random variables converge to their expected value. Under the hypothesis that variances are bounded, one such law is easily proven:

Theorem 11.1 (QM LLN) *Let X_n be a sequence of uncorrelated random variables such that $\mathbb{E}X_n = 0$ and $\mathbb{E}\|X_n\|^2 \leq M < \infty$ for all n. Then the averages $S_n = (1/n)\sum_{i=1}^{n} X_i$ converge to zero in quadratic mean.*

Proof: The following computation is simpler for real (instead of vector) valued X_j:

$$
\begin{aligned}
\mathbb{E}\|S_n\|^2 &= \mathbb{E}S_n'S_n = \mathbb{E}\,\mathbf{tr}\,S_nS_n' = \mathbb{C}ov(S_n) \\
&= \frac{1}{n^2}\,\mathbf{tr}\left(\sum_{i=1}^{n}\sum_{j=1}^{n}\mathbb{C}ov(X_i, X_j)\right) = \frac{1}{n^2}\sum_{i=1}^{n}\mathbf{tr}\,\mathbb{E}X_iX_i' \\
&= \frac{1}{n^2}\sum_{i=1}^{n}\mathbb{E}\|X_i\|^2 \leq \frac{1}{n^2}\sum_{i=1}^{n}(M) = \frac{nM}{n^2} = \frac{M}{n}.
\end{aligned}
$$

Obviously, as n goes to infinity, M/n converges to zero.

The basic idea here can be substantially generalized. It should be clear that if, instead of being zero, the covariances die away to zero for distant indices i and j, the same type of proof will apply. Of course if the covariances are not bounded, then $\mathbb{C}ov\,S_n = (1/n^2)\sum_i\sum_j\mathbb{C}ov(X_i, X_j)$ is the sum of n^2 terms divided by n^2 and hence will not converge to zero.

11.2 Consistency of OLS

Laws of large numbers are used in econometrics primarily to prove *consistency* of estimators: if a sequence of estimators $\hat{\theta}_T$ converges to some value θ in quadratic mean, we say that the estimator is quadratic mean consistent for θ.

Write Y_T for the vector of observations $(y_1, y_2, \ldots, y_t)$ generated by the linear regression $Y_T = X_T\beta + \epsilon_T$. Here X_T is the $T \times K$ *design matrix*, β is the $K \times 1$ vector of parameters, and ϵ_T is the $T \times 1$ vector of errors. We assume that the errors satisfy $\mathbb{E}\epsilon_t = 0$ and $\mathbb{C}ov(\epsilon_T) = \sigma^2 \mathbf{I}_T$. The maximum likelihood estimator of β is just $\hat{\beta}_T = (X'_T X_T)^{-1} X'_T Y_T$. Conditions for the consistency of this estimator are given below:

Theorem 11.2 *The MLE $\hat{\beta}_T$ is quadratic mean consistent if and only if the smallest eigenvalue of the matrix $X'_T X_T$ goes to infinity as T goes to infinity.*

Remark: The condition that the smallest eigenvalue of $X'_T X_T$ goes to infinity can be expressed in many alternative equivalent forms. See exercise 2. This same condition also implies almost sure convergence of the estimators. See Theorem 11.9 below.

Proof: $\hat{\beta}_T$ is quadratic mean consistent if and only if $\mathbb{E}\|\hat{\beta}_T - \beta\|^2$ converges to zero as T goes to infinity. In the proof we will omit the subscript T (on X, Y, ϵ, and $\hat{\beta}$) for notational convenience. Recall that $\hat{\beta} \sim N(\beta, \sigma^2(X'X)^{-1})$. We use the standard trick to calculate this expected value. Note that $\mathbb{E}\|\hat{\beta}-\beta\|^2 = \mathbf{tr}\,\mathbb{E}(\hat{\beta}-\beta)(\hat{\beta}-\beta)' = \mathbf{tr}\,\mathbb{C}ov(\hat{\beta})$.

Since $\mathbb{C}ov(\hat{\beta}) = \sigma^2(X'X)^{-1}$, $\hat{\beta}$ is consistent if and only if $\mathbf{tr}(X'X)^{-1}$ goes to zero as T goes to infinity. Let $\lambda^1_T, \lambda^2_T, \ldots, \lambda^K_T$ be the K eigenvalues of $X'_T X_T$. It is easily established that the eigenvalues of A^{-1} are just the reciprocals of the eigenvalues of A. Recalling that the trace of a matrix is the sum of its eigenvalues, we get

$$\mathbf{tr}(X'_T X_T)^{-1} = \frac{1}{\lambda^1_T} + \frac{1}{\lambda^2_T} + \cdots + \frac{1}{\lambda^K_T}.$$

Thus the trace converges to zero if and only if all the eigenvalues converge to infinity.

This result on the consistency of OLS generalizes to the case where the error term ϵ has covariance Σ rather than $\sigma^2\mathbf{I}$ as assumed above. Even with a misspecified error covariance matrix, consistency holds as long as the largest eigenvalue of Σ remains bounded. See exercise 4 for a proof. This assumption on the error covariance structure allows for extremely complex patterns of covariance among the errors. In particular, the standard ARMA process errors do satisfy this condition, and hence the OLS is consistent whenever $\lambda_{\min}(X'X)$ goes to infinity for ARMA errors. To prove this, it is sufficient to show that the ARMA error covariance matrix has a bounded maximum eigenvalue. For a proof of this fact, see Theorems 5.1.2 and 6.1.2 of Amemiya (1985).

11.3   Convergence in Probability

While quadratic mean convergence is a useful tool, consistency of estimators is usually defined using convergence in probability, to be defined below. The single most important reason for this is that if $\hat{\theta}_n$ converge to θ in probability and f is a continuous function, then $f(\hat{\theta}_n)$ converges to $f(\theta)$ in probability; this property does not hold for quadratic mean convergence. We introduce the basic definitions after presenting an example.

Example 11.1:   Consider a sequence of random variables $X_n = 0$ with probability $1 - 1/n$ and $X_n = n^{1/4}$ with probability $1/n$. Then $\mathbb{E}X_n^2 = n^{1/2} \times n = 1/\sqrt{n} \to 0$. Thus X_n converges to 0 in quadratic mean. Consider however, $Z_n = X_n^2$. Then we have $\mathbb{E}Z_n^2 = n/n = 1$ for all n. Thus X_n converges to 0 but X_n^2 fails to converge to zero in quadratic mean.

This shows that continuous transforms of quadratic mean convergent sequences need not be convergent. Also note that even though Z_n does not converge to 0 in quadratic mean, for large n the probability that $Z_n = 0$ is arbitrarily close to one. We now define a mode of convergence which is based on probabilities of closeness.

Definition 11.2 *A sequence X_n of random variables converges* in probability *to X if for each $\epsilon > 0$*

$$lim_{n\to\infty}\mathbb{P}(\|X_n - X\| > \epsilon) = 0.$$

In this case, we also write **plim**$_{n\to\infty} X_n = X$.

It is now easily verified that the sequence $Z_n = X_n^2$ considered in Example 11.1 does converge in probability to zero, even though it does not in quadratic mean. There is a close relationship between convergence in probability and quadratic mean convergence. Quadratic mean convergence implies convergence in probability, as shown below in Theorem 11.3. The converse holds provided that we exclude very large values of the random variable which occur with very low probabilities, as shown in Corollary 11.3.1. Thus Example 11.1 is the only type of example where the two concepts are different.

Theorem 11.3 *Suppose X_n converges to X in quadratic mean. Then X_n also converges to X in probability.*

Proof: Let $X \geq 0$ be a positive random variable. Define a new random variable Z as follows: $Z = 0$ if $0 \leq X < 1$ and $Z = 1$ if $X \geq 1$. Since $X \geq Z$ by the way it is constructed, it follows that $\mathbb{E}X \geq \mathbb{E}Z = \mathbb{P}(X \geq 1)$. This immediately implies Chebyshev's inequality:

$$\mathbb{P}(\|X\| > k) = \mathbb{P}\left(\frac{\|X\|^2}{k^2} > 1\right) \leq \frac{\mathbb{E}\|X\|^2}{k^2}.$$

Now fix $\epsilon > 0$ and note that

$$\mathbb{P}(\|X_n - X\| > \epsilon) = \mathbb{P}(\frac{\|X_n - X\|^2}{\epsilon^2} > 1) \leq \frac{\mathbb{E}|X_n - X|^2}{\epsilon^2}.$$

If X_n converge to X in quadratic mean then the last term converges to zero, so that convergence in probability also holds.

Example 11.1 shows that the converse is not true; sequences can converge in probability without converging in quadratic mean. Nonetheless, convergence in probability is conceptually quite close to quadratic mean convergence. If X_n converge in probability to X and $\mathbb{E}|X_n|^3$ remains bounded for all n, or if the sequence of differences $|X_n - X|$ is bounded by a random variable for which the second moment exists, then quadratic mean convergence must also hold. For these general results see Theorem 4.1.4 of Chung (1974). We present here a simpler result which shows the close connection between the two modes of convergence. Define the truncation $T_M(X)$ by $T_M(X) = X$ if $|X| < M$ and $T_M(X) = 0$ if $|X| \geq M$. Note that for large M, $T_M(X)$ is the same as X with very high probability. We then have:

Corollary 11.3.1 (Qlim and plim) *A sequence X_n converges to X in probability if and only if for all $M > 0$, the truncations $T_M(X_n)$ converge in quadratic mean to $T_M(X)$.*

The proof is given in exercise 5. The significance of this corollary is that convergence in probability can frequently be checked by calculating variances and seeing if they go to zero. If the original sequence of random variables is bounded, this condition is necessary and sufficient. If not, we can always get a necessary and sufficient condition by using truncations of the random variables.

Using Theorem 11.3, we can easily convert the quadratic mean law of large numbers to a *weak* law of large numbers — weak is used in this context to refer to convergence in probability. The terms 'consistency' or 'weak consistency' will refer to convergence of estimators to some true value in probability. An important property of convergence in probability, that it is preserved by continuous transforms, is stated below.

Theorem 11.4 *Suppose X_n is a (possibly vector valued) sequence of random variables converging in probability to X. If f is a continuous function then $f(X_n)$ converges to $f(X)$ in probability.*

The somewhat tedious proof is omitted. The diligent reader is referred to exercise 10 of Section 4.1 of Chung (1974). As a special case, we have the following result on consistency, sometimes referred to as Slutsky's Theorem.

Corollary 11.4.1 (Slutsky's Theorem) *If $\hat{\theta}_n$ is a sequence of estimators consistent for θ and f is a function continuous at θ, then $f(\hat{\theta}_n)$ is consistent for $f(\theta)$.*

Although this result follows from Theorem 11.4, it is also easily established directly (see exercise 1). This result can be used to conclude that if we have a consistent sequence of estimators $\hat{\rho}_t$ for ρ, then (for example) the powers ρ^j can be consistently estimated using $\hat{\rho}_t^j$. Theorem 11.4 also has the following important implication:

Corollary 11.4.2 *If sequences of random variables X_n and Y_n converge in probability to X and Y, then $X_n + Y_n$, $X_n Y_n$ and X_n/Y_n converge to $X + Y$, XY and X/Y (provided $\mathbb{P}(Y \neq 0) = 1$) respectively.*

Proof: Note that the ordered pair (X_n, Y_n) converges in probability to (X, Y) since

$$\mathbb{P}(\|(X_n, Y_n) - (X, Y)\| > \epsilon) \le \mathbb{P}(|X_n - X| > \epsilon/2) + \mathbb{P}(|Y_n - Y| > \epsilon/2).$$

The corollary follows from noting that the functions $f_1(x, y) = x + y$, $f_2(x, y) = xy$ and $f_3(x, y) = x/y$ are continuous; for f_3 we need to make the restriction $y \ne 0$ for continuity.

An application of this corollary to a consistency proof in a simple dynamic regression model is given in exercise 13.

11.4   Consistency of Tests

Roughly speaking, a hypothesis test is called *consistent* if it takes the right decision asymptotically. Below we give a definition somewhat more general than usual.

Definition 11.3 *Let T_n be a sequence of tests of levels α_n of the null hypothesis $H_0 : \theta \in \Theta_0$ versus the alternative $H_1 : \theta \in \Theta_1$. We say that T_n is consistent at levels α_n for the sequence of alternatives $\theta_n \in \Theta_1$ if $\mathbb{P}(T_n$ rejects $H_0 | \theta_n)$ converges to 1.*

The usual definition fixes the alternative $\theta_n \equiv \theta$ and also the levels $\alpha_n \equiv \alpha$ for all n. The more general definition given here is convenient for many purposes. For example, Andrews (1986) defines a test to be *completely consistent* if the above definition holds for some sequence α_n converging to zero, with the sequence of alternatives θ_n fixed (so that $\theta_n \equiv \theta^*$ for all n) at any $\theta^* \in \Theta_1$. As the sample size increases, it is reasonable to reduce the probability of type I error to zero; see Leamer (1978, pp 104-106) for a Bayesian style argument, and Andrews (1986) for a frequentist argument. Another way to assess tests is to send the alternative θ_n towards the null hypothesis. This makes the testing problem harder, since nearer values are more difficult to discriminate against. If the alternative goes towards the null slowly enough, consistency will still hold and the power of the test will go asymptotically to one. We illustrate some of these concepts with reference to the F test for linear restrictions.

11.4.1 Consistency of the F Test

Almost every symbol to follow will depend on the sample size T, but it will be notationally convenient to suppress this dependence. Consider a regression model $y = \alpha 1 + X\beta + \epsilon$, where y, ϵ are $T \times 1$, X is $T \times K(T)$ nonstochastic matrix of full rank, β is $K(T) \times 1$, α is the unknown scalar constant, and 1 is a $T \times 1$ vector of ones. It will be convenient to assume that the variables in X have been centered so that $X'1 = 0$. Introduce a $T \times (T - (K(T) + 1))$ matrix Z of full rank orthogonal to X and 1. For any full rank matrix M, define $\Pi_M = M(M'M)^{-1}M'$ to be the matrix of the projection into the vector space spanned by the columns of M. We will also use the notation $\hat{y}_M$ for $\Pi_M(y) = M(M'M)^{-1}M'y$.

In the regression model just introduced, we wish to test the null hypothesis that $\beta = 0$, or equivalently, that the regressors X play no role in determining the mean of the dependent random variable y. The standard F statistic for testing this hypothesis can be written in the revealing form:

$$F = \frac{\|\Pi_X(y)\|^2 / dim(X)}{\|\Pi_Z(y)\|^2 / dim(Z)} = \frac{\|\hat{y}_X\|^2 / dim(X)}{\|\hat{y}_Z\|^2 / dim(Z)} \tag{11.1}$$

See exercise 14. Note that $dim(X) = K(T)$ and $dim(Z) = T - (K(T) + 1)$. Thus the F statistic compares the average projection of y on X with the average projection of y on Z (orthogonal to X and the constant 1), averaging being done with respect to the dimensions of the two vector spaces. We discuss conditions under which the F test rejects the null hypothesis with probability one in large samples.

Define $\mu = \mathbb{E}y$ and $\Sigma = \mathbb{C}ov(y)$. If the regression model is correctly specified then $\mu = \alpha + X\beta$ and $\mathbb{C}ov(\epsilon) = \Sigma$, but we allow for misspecification by not assuming any particular relation between X and μ. We will make three assumptions on the stochastic structure of the vector y. It will be clear from the proof that the result to follow is valid in much greater generality, but these assumptions yield an easy proof.

[A] Either y is normal, so that $y \sim N(\mu, \Sigma)$ or components of y are independent, so that Σ is diagonal.

[B] $\limsup_{T \to \infty} \lambda_{\max}(\Sigma) < \infty$.

[C] $\liminf_{T\to\infty} \lambda_{\min}(\Sigma) > 0$.

Under these conditions the following result is easily obtained

Theorem 11.5 (Consistency of F test) *Suppose the level of the F test is fixed at some α greater than 0. If y satisfies [A], [B], and [C] above then the probability of rejecting the null goes to one (i.e., the test is consistent) asymptotically if and only if*

$$\lim_{T\to\infty} \frac{\|\hat{\mu}_X\|^2 / \mathbf{dim}(X)}{1 + \|\hat{\mu}_Z\|^2 / \mathbf{dim}(Z)} = +\infty.$$

Remark: This result shows that as long as the averaged projection of the mean of the dependent variables on the space spanned by columns of X is large relative to the projection on Z, the F test will reject the null with high probability. Equivalently, the test compares the linear relationship of y with X to that of y with Z. The proof is given in exercise 17.

11.4.2   An Inconsistent Lagrange Multiplier Test

Using the characterization for the consistency of the F test developed above, we will show that a natural testing principle leads to an inconsistent test. Suppose that $y_t = f(\alpha + x_t\beta) + \epsilon_t$, where f is some smooth function, and x_t is $k \times 1$ vector of nonconstant regressors at time period t. This is a typical nonlinear regression model. Consider testing the null hypothesis $\beta = 0$ as before. One testing principle suggests looking at the Lagrange multiplier for the constraint imposed by the null hypothesis; as the Lagrange multiplier is the shadow price of the constraint, it measures by how much the hypothesis is violated. Assume $\epsilon \sim N(0, \sigma^2 \mathbf{I}_T)$ and maximize the log likelihood of the model subject to the constraint that $\beta = 0$. Formulate the Lagrangean

$$\mathcal{L} = -\frac{T}{2} \log(2\pi\sigma^2) - \frac{1}{2\sigma^2} \sum_{t=1}^{T} \{y_t - f(\alpha + x_t\beta)\}^2 + \lambda'\beta.$$

We can simplify the problem by assuming, without essential loss of generality, that $X'\mathbb{1} = 0$. Let $\tilde{\alpha}$ be the ML estimate of α subject to the constraint that $\beta = 0$. Differentiating with respect to β yields the first order condition

$$f'(\tilde{\alpha}) \sum_{t=1}^{T} x_t(y_t - f(\tilde{\alpha}) + \lambda = 0,$$

so that $-\lambda = f'(\tilde{\alpha}) \sum_{t=1}^{T} x_t(y_t - f(\tilde{\alpha}) = f'(\tilde{\alpha}) X'y$.

The Lagrange multiplier (LM) test involves testing to see if λ is significantly different from 0. Under the null hypothesis we have $y \sim N(f(\alpha)\mathbb{1}, \sigma^2 \mathbf{I}_T)$, so that $\lambda = f'(\tilde{\alpha}) X'y \sim N(0, (f'(\tilde{\alpha}))^2 \sigma^2 (X'X))$. A natural way to test if λ is significantly different from 0 is to look at $LM = \lambda' \mathbb{C}ov(\lambda)^{-1} \lambda$; if λ is approximately normal, and the covariance matrix is known then this will have a χ^2_K density. We reject the null for significantly large values of this statistic. Since the covariance matrix is unknown, we must replace it by an estimate, getting

$$\widehat{LM} = \frac{y' X (X'X)^{-1} X'y}{f'(\tilde{\alpha})^2 \hat{\sigma}^2}.$$

If use the standard unbiased regression estimator $\hat{\sigma}^2 = \|y - \hat{\alpha}\mathbb{1} - X\hat{\beta}\|^2 / (T - (K + 1))$, then except for the term $f'(\tilde{\alpha})^2$, this is exactly in the form of the F statistic discussed above. Assuming that $f'(\tilde{\alpha})^2$ converges to a strictly positive quantity under the null, the conditions for the consistency of this F test are exactly the same as those given in Theorem 11.5. Thus it appears that the F test is valid for general nonlinear relationships. However, the apparent generality of the LM test in the present context is an illusion. The conditions for consistency, and the power of the LM test, depend solely on the *linear* relationship between the regressors X and the mean of y.

In particular, suppose that the true relationship between y and x is $y = (\alpha + \beta x)^2 + \epsilon$. This is a model of the form $y = f(\alpha + x\beta)$ with $f(x) = x^2$. The LM principle discussed above suggests that an F test for the standard linear regression of y on x also tests for the presence of a nonlinear relationship. However, suppose that α and β are such that $\mathbb{C}ov(x, 2\alpha\beta x + \beta^2 x^2) = 0$. Then it is easily verified that the projection of $\mathbb{E}y = (\alpha + \beta x)^2$ on x will be asymptotically zero, so that the F test will accept the null hypothesis that there is no relationship between y and x (more precisely, the power of the test will be equal to its level). Thus the test is inconsistent for such parameter values. Of course, for most parameter values α, β the covariance will be nonzero and hence the F test will have some power against such parameter values. However, it is wrong to interpret this as meaning that the F test detects nonlinear relationships. Rather, nonlinear functions $f(x)$ are detected to the extent that they are correlated with x itself.

The Breusch-Pagan test for heteroskedasticity is similar to, though somewhat more complex than, the LM test above. Let e_t^2 be the residuals from a regression. If the errors are heteroskedastic, with $\mathbb{V}ar(\epsilon_t) = \sigma_t^2$, it will be the case (under some assumptions) that $\mathbb{E}e_t^2$ is approximately σ_t^2 in large samples. To detect if the variances are explained by some regressors Z, we could therefore use the F test in a regression of $y = e^2$ on Z and a constant. This will do an adequate job of detecting a linear relationship between the variances and the regressors. Presented as an LM test, it appears that the test detects arbitrary relationships between the regressors Z and the variances. However, this is an illusion, and the test works only for linear relationships; see also Dutta and Zaman (1990).

11.5 Heteroskedasticity-Consistent Covariance Estimates

The result to be presented, due to Eicker (1966a), is worthy of being regarded as the eighth wonder of the world. In the regression model $y = X\beta + \epsilon$, if $\mathbb{C}ov(\epsilon) = \Sigma$ then the the covariance matrix of the OLS estimate is easily seen to be $\mathbb{C}ov(\hat{\beta}) = (X'X)^{-1}(X'\Sigma X)(X'X)^{-1}$. In general, heteroskedasticity is the assumption that Σ is some diagonal matrix (instead of $\sigma^2 \mathbf{I}$, the usual, homoskedastic case). Since Σ is $T \times T$ matrix with T unknown parameters in the case of heteroskedasticity, it is rather surprising that reasonable estimates of $\mathbb{C}ov(\hat{\beta})$ can be found. The explanation lies in the fact that the matrix Σ with T unknown parameters is not estimated but rather a fixed $K \times K$ matrix is estimated. Elements of the $K \times K$ covariance matrix of $\hat{\beta}$ can be regarded as various types of averages of the sequence of true variances. These averages can be estimated even when the individual variances cannot. As a simple example of this process, suppose that X_i equals 1 with probability p_i and 0 with probability $(1 - p_i)$. It is clear that observing an independent sequence X_i we have no hope of estimating p_i, since we have only one observation relating to each p_i. However, it is possible to estimate the average value of p_i since it is easy to check that

$$\lim_{n \to \infty} \frac{1}{n} \sum_{j=1}^{n} X_j = \lim_{n \to \infty} \frac{1}{n} \sum_{j=1}^{n} p_j,$$

provided that the limit on the RHS exists.

We will follow the notation of Eicker (1966a) for the most part. Given a positive definite matrix M, let $M^{1/2}$ denote a symmetric positive definite matrix satisfying $(M^{1/2})(M^{1/2}) = M$. Suppose we observe $y_t = x_{t1}\beta_1 + \cdots + x_{tK}\beta_K + \epsilon_t$. for $t = 1, 2, \ldots, T$. We will consider the asymptotic situation where T approaches infinity. In matrix notation, we can write this as $Y_T = X_T\beta + \epsilon(T)$, where

$$Y_T = \begin{pmatrix} y_1 \\ y_2 \\ \vdots \\ y_T \end{pmatrix}, X_T = \begin{pmatrix} x_{11} & x_{12} & \cdots & x_{1K} \\ x_{21} & x_{22} & \cdots & x_{2K} \\ \vdots & \vdots & & \vdots \\ x_{T1} & x_{T2} & \cdots & x_{TK} \end{pmatrix}, \epsilon(T) = \begin{pmatrix} \epsilon_1 \\ \epsilon_2 \\ \vdots \\ \epsilon_T \end{pmatrix}.$$

Then $\hat{\beta}_T = (X_T'X_T)^{-1}X_T'Y_T$. We will use $x_t = (x_{t1}, \ldots, x_{tK})$ to refer to the t-th row of the matrix X_T.

Instead of estimating the covariance matrix of $\hat{\beta} - \beta$ (which converges to zero) it is convenient to define and estimate D_T^2 to be the covariance matrix of $\eta_T = (X_T'X_T)^{1/2}(\hat{\beta}_T - \beta)$. It is easily checked that

$$D_T^2 = (X_T'X_T)^{-1/2}(X_T'\Sigma_T X_T)(X_T'X_T)^{-1/2}.$$

The matrix D_T involves the unknown (and since the parameters are equal in number to the observations, the unknowable) covariance matrix of the errors Σ_T. Our goal in this section is to provide a reasonable estimator for D_T. Now $\mathbb{E}\epsilon_t^2 = \sigma_t^2$ so it is reasonable to estimate σ_t^2 by ϵ_T^2. In turn, ϵ_t is not observable, so we replace it by the residual $e_{Tt} = y_t - x_t\hat{\beta}_T$. Define S_T to be a diagonal matrix with (t, t) entry e_{Tt}^2. Eicker's result is that D_T can be estimated by C_T defined by replacing Σ_T by S_T as follows:

$$C_T^2 = (X_T'X_T)^{-1/2}(X_T'S_T X_T)(X_T'X_T)^{-1/2}.$$

This holds even though S_T is not a consistent estimator of Σ_T.

A slight adaptation of Eicker's (1966a) results is given below. The proof is given in exercise 18.

Theorem 11.6 (Eicker) *Suppose there exist m, M satisfying $0 < m \leq M < \infty$ and for all t $m < \mathbb{E}|\epsilon_t|^2$ and $\mathbb{E}|\epsilon_t|^3 < M$. Further suppose there exists M^* such that for all t and T, $x_t'(X_T'X_T)^{-1}x_t < M^*$. Then $C_T^{-1}D_T$ converges to $\mathbf{I}_K$.*

Remarks: Several aspects of this results are worth discussing further. These are listed below.

1. *Practical Consequence:* We know that $D_T^{-1}\eta_T$ has an identity co-variance, and hence η_T has covariance D_T^2. Since D_T involves the unknown matrix Σ_T, this result cannot be used to estimate the covariance matrix of η_T. The theorem allows us to conclude that $(C_T^{-1}D_T)(D_T^{-1}\eta_T) = C_T^{-1}\eta_T$ has identity covariance asymptotically. Hence a large sample approximation to the covariance of η_T is C_T^2. This last matrix involves no unknown parameters and hence provides a usable estimate for the covariance of least squares.

2. *Leverage:* The quantities $x_t'(X_T'X_T)^{-1}x_t = h_{tt}$ are the diagonal entries of the 'hat' matrix $H_T = X_T(X_T'X_T)^{-1}X_T'$. Sometimes h_{tt} is referred to as the *leverage* of the t-th observation. See Section 5.8 for further details about leverage. These play an important role in all aspects of asymptotic behavior for the least squares estimators. The theorem above shows that bounded third moments for the errors and a lower bound for the variances, together with bounded leverage permit us to estimate the asymptotic covariance matrix of the least squares estimators. In a later chapter we will show that the stronger condition that the maximum leverage should go to zero is needed for asymptotic normality of the OLS estimate $\hat{\beta}$.

3. *Variance Bounds:* It should be clear that if variances are unbounded, the OLS need not converge, and this will certainly cause difficulties in estimating its covariance. It may not be clear why a lower bound on the variances is needed. If in fact the variances

are close to zero, then the OLS can become exact — in the polar case, if the variance is zero, the OLS becomes exact in finite samples. In this case the covariance matrix is singular, and this causes difficulties in attempts to estimate it.

4. *Necessary and Sufficient Conditions:* These conditions are not too far from being necessary and sufficient. One can replace the bound on the third moment by a bound on the $2 + \epsilon$ moment, where ϵ is any positive number. This can be weakened further to a uniform integrability condition on the second moment itself. This last condition is the most general condition possible, and is given in this form by Eicker. In practice, the author does not know of applications where it would be safe to assume uniform integrability of second moments but not uniform boundedness of third moments. Thus the additional complications to prove the necessary and sufficient results do not appear worthwhile in the present context. However, it is important to know that the conditions given are reasonably sharp and that we can expect the result to fail if they are substantially violated.

11.6 Almost Sure Convergence

We have introduced so far two closely related concepts of convergence. Convergence in probability is equivalent to the convergence in quadratic mean of truncations of a sequence of random variables. Almost sure convergence is more demanding than convergence in probability or quadratic mean convergence. Intuition for the difference is most easily developed in the context of a special case. Consider a sequence of independent random variables X_n such that $\mathbb{P}(X_n = 0) = 1 - p_n$ and $\mathbb{P}(X_n = 1) = p_n$. A sequence of observations on X_n is then just a sequence of 0's and 1's. When does the sequence X_n converge to 0 in probability? A little reflection shows that this happens when $\lim_{n \to \infty} p_n = 0$. Intuitively, the meaning of this condition is that for large n, the chances are very high that X_n will be zero. Almost sure convergence demands more of the sequence, namely that the chances should be very high that $X_n, X_{n+1}, X_{n+2}, \ldots$ should *all* equal zero with very high probability. To see that there is a difference between these two conditions, consider the following deterministic sequence: $(0, 1, 0, 0, 1, 0, 0, 0, 1, 0, 0, 0, 0, 1 \ldots)$. The occurrence of 0's and 1's is predictable on this sequence, but let us pretend that we lose count after a while so that we do not know exactly where we are in this sequence. For large n, the probability that the next term is 1 is very small, and this probability keeps decreasing the further we go. Thus such a sequence converges in probability to 0. However, if we consider whether or not a 1 will ever occur from now on, the answer is that 1's will occur arbitrarily far out in the sequence, preventing the sequence from converging to 0 almost surely.

To make the situation more vivid, suppose that p_n is the probability of a devastating accident, a nuclear explosion at a power plant. If the sequence goes to zero, people living in the neighborhood can feel safe when they wake up in the morning about surviving that particular day. As time progresses, they are entitled to feel safer on each day that no explosion will occur on that particular day. However, if we take a longer view and consider whether or not an explosion will ever occur from now on, then we need to consider the almost sure convergence property. It is not enough that the probability of an explosion on any particular day is low, since the number of days to eternity is very long, and a rare event can happen if the trials are repeated for a very long time. Fix n and consider the event E that an explosion never occurs from now on. The probability of no explosion at time n is $1 - p_n$ and these events are independent so the probability of E is

$$\mathbb{P}(E) = \prod_{j=n}^{\infty}(1 - p_j) \approx 1 - \sum_{j=n}^{\infty} p_j,$$

where the approximation is valid if p_n are all very small numbers. This shows heuristically that for almost sure convergence, we need the condition that $\sum_{j=n}^{\infty} p_n$ should go to zero as n goes to infinity. This is equivalent to the condition that the sum $\sum_{j=1}^{\infty} p_j < \infty$ should be finite. This is stronger than the requirement that p_j goes to zero. For example, if $p_j = 1/j$ then p_j goes to zero but $\sum_{j=1}^{n}(1/j) \approx \log(n) \to \infty$ so the sum diverges to infinity. We now give a formal definition of almost sure convergence, which is also called convergence with probability 1.

Definition 11.4 *Let X_n be a sequence of random variables. Define the*

*event $E_n(\epsilon)$ to occur whenever $|X_n - X| < \epsilon$. Define $E_n^*(\epsilon) = \cap_{j=n}^{\infty} E_j(\epsilon)$
to occur whenever all the events $E_n(\epsilon), E_{n+1}(\epsilon), \ldots$ occur. We will say
that X_n converges* almost surely *(and also with probability one) to X
if $\lim_{n \to \infty} \mathbb{P}(E_n^*(\epsilon)) = 1$ for all $\epsilon > 0$.*

Note that convergence in probability requires $\mathbb{P}(E_n(\epsilon))$ to converge
to 1 for all ϵ. Since $E_n^*(\epsilon)$ is a subset of $E_n(\epsilon)$ it is necessarily the case
that $\mathbb{P}(E_n(\epsilon))$ converges to 1 whenever $\mathbb{P}(E_n^*(\epsilon))$ does. It follows immediately almost sure convergence implies convergence in probability.
Almost sure convergence does not imply quadratic mean convergence,
essentially for the same reasons that convergence in probability does
not imply quadratic mean convergence. The moments of the sequence
may fail to exist or may go off to infinity because with a very tiny
probability, the random variables may take a huge value. Both convergence in probability and almost sure convergence ignore such rare
events but they make a big difference for quadratic mean convergence.
Since almost sure convergence implies convergence in probability, the
conditions of Theorem 11.3 can also be used to get quadratic mean
convergence starting from almost sure convergence.

11.6.1 Some Strong Laws

A law of large numbers which asserts that averages of some sequence
of random variables converge almost surely to their mean is called a
Strong Law of Large Numbers (SLLN). We list a few important strong
laws below. The first law, valid for i.i.d. random variables, is very
elegant since it has no assumptions other than the existence of the
expected value.

Theorem 11.7 (Kolmogorov's i.i.d. Strong Law) *If X_i is an
i.i.d. sequence of random variables, and $\mathbb{E}|X_1| < \infty$, then the aver-
ages $A_n = (1/n) \sum_{i=1}^{n} X_i$ converge almost surely to $\mathbb{E}X_1$.*

For a proof see Theorem 5.4.2 of Chung (1974). An excellent reference which is comprehensive, definitive, and clearly written, for all
questions related to convergence of sums of independent random variables is Gnedenko and Kolmogorov (1951). It is worth noting that the

assumption that $\mathbb{E}|X_1| < \infty$ is necessary to ensure the $\mathbb{E}X_1$ is well defined. Thus, the sole condition that the mean is well defined is enough for the SLLN to hold in the i.i.d. case.

While Kolmogorov's Law is elegant, the i.i.d. case is rare in econometrics, and more general laws are needed. Chung's Strong Law is powerful, flexible, and sufficient to handle most applications. For a proof, see the remarks following the Corollary to Theorem 5.4.1 in Chung (1974).

Theorem 11.8 (Chung's SLLN) *Let X_i be an independent sequence of random variables with finite variances. Define $S_n = \sum_{i=1}^{n}(X_i - \mathbb{E}X_i)$. Let $v_n^2 = \mathbb{V}ar(S_n)$ and suppose that $\lim_{n\to\infty} v_n^2 = \infty$. Then*

$$\lim_{n\to\infty} \frac{S_n}{v_n \log(v_n)} = 0, \quad almost\ surely.$$

For example, suppose that the variances of all the X_i are bounded above by M and also that $\mathbb{V}ar(S_n)$ goes to infinity . Then the theorem asserts that $S_n/\sqrt{n}\log(n)$ goes to zero. It is easy to see that $S_n/\sqrt{n}$ does not go to zero so that Chung's law gives us the precise rate of growth needed to make S_n converge to zero.

11.6.2   Strong Consistency

We have defined weak consistency of a sequence of estimators in terms of convergence in probability. Strong consistency is defined in terms of almost sure convergence.

Definition 11.5 (Strong Consistency) *A sequence of estimators $\hat{\theta}_T$ is said to be* strongly consistent *for θ if it converges almost surely to θ.*

There have been some discussions in the literature on whether strong consistency or weak consistency are more relevant to applications. From the heuristic discussions given above, it should be clear that strong consistency ensures with high probability that once an estimator gets close to the true value, it never strays from it subsequently. Weak consistency ensures that on any given (fixed) large sample size the estimator is close to the truth with high probability. For one-time applications on a fixed data set, weak consistency is clearly the appropriate concept. With large data sets and the need for repeated estimation in a stable environment strong consistency may be suitable. It seems clear to the author that weak consistency is perfectly satisfactory for most econometric applications. In most situations the environment is unstable and the parameters change from time to time so that attention to long run performance does not appear worthwhile. Nonetheless, it is at least of theoretical interest to know whether strong or weak consistency holds for a particular sequence of estimators. Some important results on strong consistency are presented below without proof. The following result is due to Lai *et al.* (1978).

Theorem 11.9 (Lai, Robbins, and Wei) *Suppose* $y_t = x_t'\beta + \epsilon_t$ *where the regressors* x_t *are nonstochastic and the errors* ϵ_t *are independent. If the variances of the errors are bounded then*

$$\lim_{T\to\infty} \lambda_{min}(X_T'X_T) = \lim_{T\to\infty} \lambda_{min} \sum_{t=1}^{T} x_t x_t' = \infty,$$

is necessary and sufficient for the strong consistency of least squares.

The condition placed on the regressors by this result is the same as that obtained earlier and required for weak consistency. However the errors are required to be independent, whereas weaker conditions suffice for weak consistency. In fact the errors do not need to be independent; the hypothesis can be weakened to the assumption that the errors form an *innovations process*, as defined in the next section. This hypothesis still places strong restrictions on the errors compared to the case of arbitrary covariance structure. The theorem above is valid for nonstochastic regressors. In the next section we examine the consequences of stochastic regressors for consistency results.

11.7 Consistency in Dynamic Models

In dynamic models, consistency of least squares requires stronger conditions than $\lambda_{min}(X'X)$ going to infinity. Below we present the most general results available for consistency of estimators in linear dynamic models. These results are close to being necessary and sufficient in the context presented, and have been recently obtained by Lai and Wei. The proofs are somewhat long and involved, and hence omitted here.

A linear dynamic model has the form

$$y_t = \alpha_1 y_{t-1} + \cdots + \alpha_p y_{t-p} + \beta_1 z_{t1} + \cdots + \beta_k z_{tk} + \epsilon_t. \qquad (11.2)$$

Here y_{t-j} are lagged values of the dependent variable, and z's are the regressors, which may also include lagged values. As discussed in Section 8.3, we may assume without loss of generality that if the model is not misspecified then

$$E(\epsilon_t \,|\, \epsilon_{t-1}, \epsilon_{t-2}, \ldots,) = 0.$$

If this does not hold, past errors contain information about the present error. Past errors are functions of observables, and these functions when incorporated into the model will eliminate such dependence. Thus the assumption that the errors form an innovations process (or, equivalently, a martingale difference sequence) can be made without loss of generality for a correctly specified model.

Theorem 11.10 (Lai and Wei) *Consider the regression model $y_t = x_t' \beta + \epsilon_t$ where the regressors x_t are stochastic and the errors ϵ_t form an innovations process. If the variances of the errors are bounded then the condition*

$$\lim_{T \to \infty} \frac{\log(\lambda_{max}(X_T' X_T))}{\lambda_{min}(X_T' X_T)} = 0,$$

is sufficient for the strong consistency of least squares.

For a proof, see Lai and Wei (1982a). This condition is also close to being necessary. Lai and Wei produce an example with stochastic regressors where this condition is slightly violated and the strong consistency fails. Use of this result permits proof of consistency of OLS in dynamic models under the most general conditions currently available. The details of the proof are somewhat involved and will not be presented here. To apply this result to particular dynamic models, we need to be able to estimate λ_{max} and λ_{min} for the matrix $X'X$ arising in dynamic models. Various theorems permitting one to estimate these eigenvalues and apply these results are developed by Lai and Wei (1982b). We will now present some of these results.

Every k-th order autoregressive dynamic model such as Eq. (11.2) can be represented as a first order vector autoregressive model. This representation is very convenient for many purposes, including the derivation of conditions for stationarity, which we need to discuss consistency of estimators in such models. Define the following vectors and matrices

$$
Y_t = \begin{pmatrix} y_t \\ y_{t-1} \\ \vdots \\ y_{t-(p-1)} \end{pmatrix} \quad Z_t = \begin{pmatrix} z_{t1} \\ z_{t2} \\ \vdots \\ z_{tk} \end{pmatrix} \quad A = \begin{pmatrix} \alpha_1 & \alpha_2 & \alpha_3 & \cdots & \alpha_p \\ 1 & 0 & 0 & \cdots & 0 \\ 0 & 1 & 0 & \cdots & 0 \\ & \vdots & & & \vdots \\ 0 & 0 & 0 & \cdots & 0 \end{pmatrix} \quad E_t = \begin{pmatrix} \epsilon_t \\ 0 \\ \vdots \\ 0 \end{pmatrix}.
$$

With these definitions, Eq. (11.2) can be written in matrix form as

$$
Y_t = AY_{t-1} + \beta'Z_t + E_t.
$$

Repeatedly substituting for Y_{t-1} leads to the equation:

$$
Y_t = \sum_{j=0}^{K} A^j \left(\beta'Z_{t-j} + E_{t-j} \right) + A^{K+1}Y_{t-(K+1)}.
$$

Many physical and economic systems satisfy the condition that the effects of the distant past should be negligible on the present. On the present equation, this translates to the requirement that the term $A^{K+1}Y_{t-(K+1)}$ should be small for large K. This is equivalent to requiring that all eigenvalues of A should be strictly less than one in magnitude. Now the eigenvalues of A are the roots of the determinantal equation which can be written explicitly due to the simple form of A as follows (see exercises 10 and 20):

$$
det(A - \lambda\mathbf{I}) = \lambda^p - \alpha_1\lambda^{p-1} - \alpha_2\lambda^{p-2} - \cdots - \alpha_{p-1}\lambda - \alpha_p = 0. \quad (11.3)
$$

We present without proof some results taken from Lai and Wei (1982b), applying to successively more complicated situations. While all require roots of Eq. (11.3) to be less than or equal to one, many of them hold for unit roots, a case of considerable importance in econometrics.

1. Suppose that there are no exogenous variables (i.e., $k = 0$ and z's do not enter) in Eq. (11.2). Suppose that all roots λ_i of Eq. (11.3) are less than *or equal to* one in magnitude. Also suppose that the variances of the innovations are bounded above. Then X_T, the standard matrix of regressors consisting solely of the lagged values of y, satisfies the hypotheses of Theorem 11.10. Thus OLS estimates of the model parameters are consistent.

2. Next consider the case where nonstochastic (or ancillary) variables z_j enter the equation, but all roots of Eq. (11.3) are strictly smaller than unity. The variances of the innovations must be bounded above as before. Let Z_T be the $T \times K$ matrix of the exogenous regressors and assume it satisfies the following conditions:

$$\liminf_{T \to \infty} \frac{1}{T} \lambda_{min}(Z_T' Z_T) > 0$$

$$\limsup_{T \to \infty} \frac{1}{T} \, \mathbf{tr}(Z_T' Z_T) < \infty.$$

Then the hypotheses of Theorem 11.10 hold and OLS is consistent.

3. The final case allows for both unit roots and exogenous variables. In this case we make the stronger assumption that for some $\delta > 0$,

$$\limsup_{t \to \infty} t^{-\delta} \|z_t.\|^2 < \infty.$$

Further, we must assume that the variance of the innovations $\mathbb{V}ar(\epsilon_t | \mathcal{I}_{t-1})$ is bounded above by some constant $M < \infty$ and below by some constant $m > 0$. Again the consistency of OLS estimates follows.

There remains the case of some roots larger than unity, which is not covered by the results of Lai and Wei described above. Gourieroux *et al.* (1987) have obtained results on cases where all regressors are nonstationary and trending. This corresponds to the case where all roots are bigger than one. The basic result is that OLS is consistent in this case. This still leaves open the case where some roots are less and others are larger than one.

11.8 Exercises

1. While the general result that continuous transforms preserve convergence in probability is somewhat tedious, the corollary which yields consistency is quite easy to prove directly, using the properties of continuity. Provide a proof of Corollary 11.4.1.

2. For $T = 1, 2, \ldots$ let X_T be a sequence of $T \times K$ matrices. Show that all the following conditions are equivalent:

   (a) The smallest eigenvalue of $X_T'X_T$ goes to infinity.

   (b) All entries of $X_T'X_T$ go to infinity.

   (c) All diagonal entries of $X_T'X_T$ go to infinity.

   (d) The largest eigenvalue of $(X_T'X_T)^{-1}$ goes to zero.

   (e) All diagonal entries of $(X_T'X_T)^{-1}$ go to zero.

   (f) All entries of $(X_T'X_T)^{-1}$ go to zero.

   See Theorem 3.5.1 of Amemiya (1985).

3. [Amemiya's Lemma] Show that if A and B are positive definite, $\text{tr}(AB) \leq \text{tr}(A)\lambda_{\max}(B)$. Hint: diagonalize A. If that fails, look up Lemma 6.1.1 of Amemiya (1985).

4. Consider a regression model $y = X\beta + \epsilon$, where $\mathbb{E}\epsilon = 0$ and $\text{Cov}(\epsilon) = \Sigma$. Show that the OLS $\hat{\beta} = (X'X)^{-1}X'y$ is consistent provided that $\lambda_{\min}(X'X) \to \infty$ and $\lambda_{\max}(\Sigma) \leq M < \infty$. You will need to use Amemiya's Lemma of the previous exercise.

5. Show that if X_n is a bounded sequence of random variables (that is, for some M, $\mathbb{P}(|X_n| < M) = 1$ for all n), and X_n converges to

X in probability, then X_n also converges to X in quadratic mean. Use this to prove Corollary 11.3.1.

6. Consider the regression of y on a trend: $y_t = a + bt + \epsilon_t$, with $\sigma_t^2 = \text{Var}(\epsilon_t)$.

    (a) First suppose $a = 0$, and $\sigma_t^2 = t^\alpha$. For what values of α is the OLS estimate of b consistent?

    (b) Now suppose $a \neq 0$. Show that the estimate of b is still consistent under the same conditions on α, but the estimate of a is not consistent.

7. The GLS estimator of β is $\hat{\beta}_T = (X_T' \Sigma_T^{-1} X_T)^{-1} X_T' \Sigma_T^{-1} y_T$. If the covariance matrix of the error ϵ_T is Σ_T, find conditions guaranteeing the consistency of this estimator.

8. Suppose $X_n = 0$ with probability $1 - (1/n)$ and $X_n = n^\alpha$ with probability $1/n$. For what values of α does X_n converge to zero in (a) quadratic mean, and (b) probability. Answer (a) and (b) also for X_n^2.

9. The level of radioactivity decays exponentially. If A is the quantity of uranium and y_t is the level of radioactivity measured by a Geiger counter at time t, the following equation holds:

$$y_t = Ae^{-t} + \epsilon_t.$$

If the errors are i.i.d. $N(0, \sigma^2)$, can one measure the constant A arbitrarily closely by taking sufficient readings (i.e., if t approaches infinity)?

10. Show that for any matrix A, A^K goes to zero as K approaches infinity if and only if all eigenvalues of A are less than 1 in magnitude. Hint: if $A = Q\Lambda Q^{-1}$ then $A^K = Q\Lambda^K Q^{-1}$. Every matrix with distinct eigenvalues can be diagonalized like this. Ignore the case of multiple eigenvalues unless you know about the Jordan canonical form.

11. Suppose $y_t = \beta x_t + \epsilon_t$, where $\epsilon_t \sim N(0, \sigma_t^2)$. Construct a sequence x_t and σ_t for which the OLS estimator is not consistent, but the GLS estimator is consistent.

12. Suppose $Y \sim N(0, \Sigma)$, so that $\mathbb{E}Y_a Y_b = \Sigma_{a,b}$. Using the moment generating function for the normal density, $\mathbb{E}\exp(\theta'Y) = \exp\{(1/2)\theta'\Sigma\theta\}$, show that

$$\mathbb{E}(Y_a Y_b Y_c Y_d) = \Sigma_{a,b}\Sigma_{c,d} + \Sigma_{a,c}\Sigma_{b,d} + \Sigma_{a,d}\Sigma_{b,c}$$

Hint: the tedious task of differentiation the MGF is substantially simplified if we first use $\exp(x) = 1 + x + x^2/2! + \cdots$ and then differentiate.

13. Consider the first order autoregressive model:

$$y_t = \alpha y_{t-1} + \epsilon_t.$$

Suppose the equation holds for all $t = 0, \pm 1, \pm 2, \ldots$, with $\epsilon_t \overset{i.i.d}{\sim} N(0, \sigma^2)$. We say that y is a stationary autoregressive process if $\alpha < 1$. Suppose we have observations on y_t for $t = 0, 1, \ldots, T$. The least squares estimator for α is

$$\hat{\alpha}_T = \frac{\sum_{t=1}^{T} y_t y_{t-1}}{\sum_{t=1}^{T} y_{t-1}^2} = \frac{(1/T)\sum_{t=1}^{T} y_t y_{t-1}}{(1/T)\sum_{t=1}^{T} y_{t-1}^2} \equiv \frac{N_T}{D_T}.$$

In this exercise our goal is to prove the consistency of this estimator.

(a) Show that $\mathbb{E}y_i y_j = \alpha^{|j-i|}\sigma^2/(1-\alpha)^2$.

(b) Show that

$$\mathbb{E}(N_T) = \frac{1}{T}\sum_{t=1}^{T} \mathbb{E}y_t y_{t-1} = \frac{\alpha\sigma^2}{1-\alpha^2} \equiv \overline{N}$$

$$\mathbb{E}(D_T) = \frac{1}{T}\sum_{t=1}^{T} \mathbb{E}y_{t-1}^2 = \frac{\sigma^2}{1-\alpha^2} \equiv \overline{D}.$$

(c) Use the formula of exercise 12 to establish that

$$\lim_{T\to\infty} \text{Var}(N_T) = \lim_{T\to\infty} \frac{1}{T^2} \sum_{j=1}^{T} \sum_{k=1}^{T} \mathbb{E} Y_j Y_{j-1} Y_k Y_{k-1} = 0$$

$$\lim_{T\to\infty} \text{Var}(D_T) = \lim_{T\to\infty} \frac{1}{T^2} \sum_{j=1}^{T} \sum_{k=1}^{T} \mathbb{E} Y_j^2 Y_k^2 = 0.$$

Show that this implies the consistency of the estimator $\hat{\alpha}_T$.

(d) Derive the ML estimator $\tilde{\alpha}_T$ for α. Can you show that $|\tilde{\alpha}_T - \hat{\alpha}_T|$ converges to zero in probability? What does this imply about the consistency of the ML?

14. Show that the F statistic for testing $\beta = 0$ is exactly as given in Eq. (11.1).

15. Suppose $\epsilon \sim N(0, \Sigma)$. If P is any symmetric matrix then

$$\text{Var}(\epsilon'P\epsilon) = 2\,\textbf{tr}(P\Sigma P\Sigma).$$

16. Suppose ϵ is i.n.i.d. and Σ is the diagonal matrix of variances. For any matrix P, we have

$$\text{Var}\,\epsilon'P\epsilon = \sum_{i=1}^{T} \text{Var}(P_{ii}\epsilon_i^2) + 2 \sum_{i=1}^{i=T} \sum_{1\le j\ne i}^{j=T} \Sigma_{ii}\Sigma_{jj} P_{ij}^2.$$

17. Prove Theorem 11.5 using the following steps. First define

$$N_T = \frac{\|\Pi_X(y)\|^2 / \textbf{dim}(X)}{(1 + \|\Pi_Z(\mu)\|^2)/ \textbf{dim}(Z)}$$

$$D_T = \frac{\|\Pi_Z(y)\|^2 / \textbf{dim}(Z)}{(1 + \|\Pi_Z(\mu)\|^2)/ \textbf{dim}(Z)}$$

and note that $F = N/D$.

(a) Show that the F test is consistent if and only if the F statistic $F_T = N_T/D_T$ converges to $+\infty$ in probability. This means that for any constant C, $\mathbb{P}(F > C)$ converges to unity.

(b) Show that $\mathbb{E}D_T$ converges to a strictly positive quantity and the variance $\mathbb{V}ar(D_T)$ converges to 0. Thus the convergence of F to $+\infty$ is equivalent to the convergence of N to ∞.

(c) Show that $\mathbb{V}ar(N)$ goes to $+\infty$ and also $\mathbb{E}N/\sqrt{\mathbb{V}ar(N)}$ also converges to $+\infty$.

(d) If X_T is a sequence of random variable for which $\mathbb{V}ar(X)$ and $\mathbb{E}X/\sqrt{\mathbb{V}ar(X)}$ converge to $+\infty$ then X_T converges to $+\infty$.

18. (Eicker's HCCE) Prove Theorem 11.6 using the sequence of steps below:

(a) Show that
$$\lim_{T \to \infty} \left(D_T^{-1} C_T^2 D_T^{-1} - \mathbf{I}_K\right) = 0. \qquad (11.4)$$

(b) Show that if $\lim_{T \to \infty} M_T = 0$, and N_T is a *bounded* sequence of matrices (that is, for all T the largest eigenvalue satisfies $\lambda_{max}(N_T) \le C < \infty$), then $\lim_{T \to \infty} N_T M_T = 0$.

(c) Show that both D_T and D_T^{-1} are bounded sequences of matrices.

(d) Using the previous steps, multiplying Eq. (11.4) from left by D_T and from right by D_T^{-1} yields $\lim_{T \to \infty} C_T^2 D_T^{-2} - \mathbf{I}_K = 0$ or equivalently, $\lim_{T \to \infty} C_T^2 D_T^{-2} = \mathbf{I}_K$.

(e) To complete the proof, show that if a sequence of positive definite matrices M_T converges to $\mathbf{I}$ then so does $M_T^{1/2}$. Hint: this follows directly from the continuity principle — convergence in probability is preserved by continuous transforms.

19. Suppose X_t are i.i.d. random variables with $\mathbb{E}X_t^4 = M < \infty$. Show that the least squares estimator of β in the model $y_t = X_t'\beta + \epsilon_t$ is consistent if $\mathbb{C}ov(X_t, \epsilon_t) = 0$ and $\mathbb{E}\epsilon_t^2 = M' < \infty$. Calculate the probability limit of the least squares estimator in the case that the covariance is not zero.

20. Prove that the determinant of the *companion* matrix A defined in the last section is given by Eq. (11.3). Hint: form the matrix $A - \lambda \mathbf{I}$. Multiply the first column by λ and add to the second.

This operation does not change the determinant (why?). Do the same thing to the second and subsequent columns in sequence. The determinant of the matrix you will get is just the last entry in the first row, and it is zero if and only if Eq. (11.3) is zero.

His declarations of undying love and praises of her beauty appeared to have no effect on her. Finally, she turned and said to him, "I am nothing. Look, there goes one who is truly beautiful". When he turned to look, he received a violent slap. "Your inconstancy gives the lie to your empty words and meaningless flattery", she said.

Chapter 12

Consistency of Estimators: Indirect Methods

The previous chapter gave a straightforward and down-to-earth approach to consistency of tests and estimators. In the recent past some dazzling space-age techniques have been developed to achieve these goals. It is impossible to present the rather heavy theoretical machinery that lies behind these techniques in this text. However, the end results are of great practical value and importance, since the results work under fairly general conditions and with relatively easily verified hypotheses. Our goal in this chapter is to give a users guide to some fairly advanced techniques which can be used to prove consistency of tests, estimators, and other objects of statistical interest.

The main idea behind one important technique is very simple; if we forego the technical conditions, it is very easy to use. Suppose $\hat{\theta}_n$ is defined to be an estimator which maximizes a random function $S_n(\theta) = (1/n) \sum_{i=1}^{n} \rho(X_i, \theta)$. For example, the maximum likelihood, OLS, as well as many other estimators can be put into this form. In large samples $S_n(\theta)$ should be close to $\mathbb{E}S_n(\theta)$. Let θ_n be the value which maximizes $\mathbb{E}S_n$. We may hope that $\hat{\theta}_n$ should be close to θ_n.

Now if θ_n converge to some limit θ_0, then $\hat{\theta}_n$ should also converge to θ_0. Thus the problem of convergence of the random variables $\hat{\theta}_n$ maximizing random functions $S_n(\theta)$ can be converted to a problem of a deterministic sequence θ_n which maximizes a deterministic function $\mathbb{E}S_n$. Making this idea precise and correct requires filling in several details which is what is undertaken in this chapter.

12.1 Convergence of Maxima: Deterministic Case

Given a sequence of deterministic functions $f_n(\theta)$, what can we say about the convergence of their maxima? Understanding the answer to this question is a key to the more complex case regarding consistency of estimators defined by a maximum property, such as the maximum likelihood estimator.

We will use the notation $\theta_n = \mathrm{argmax} f_n(\cdot)$ to mean that for all θ, $f(\theta_n) \geq f(\theta)$. We would like to find condition on the sequence of functions f_n such that $\theta_n = \mathrm{argmax} f_n(\cdot)$ converge to some fixed value θ_0. A useful condition involves assuming that f_n themselves converge to a limit function f and that $\theta_0 = \mathrm{argmax} f$. Such a condition can be made to work with suitable additional assumptions.

A sequence $f_n(\theta)$ is said to converge pointwise to a limit function f if for each θ $\lim_{n\to\infty} f_n(\theta) = f(\theta)$. First we note that pointwise convergence is not enough for the kind of result that we need. This can be seen from Figure 12.1 (A), which plots the functions $f_n(x) = e^{-x} + \phi(x-n)$, where ϕ is the normal density. For fixed x as n approaches infinity $\phi(x - n)$ goes to zero, so that $f_n(x)$ converges to $f_\infty(x) = e^{-x}$. However, it is easily seen that $f_n(x)$ has a maximum at $x = n$, while $f_\infty(x)$ has a maximum at $x = 0$ for $x \geq 0$. The difficulty with pointwise convergence, as can be seen from Figure 12.1 (A), is that the maximum gap between the functions f_n and the limit function f need not decrease with large n. This means that the shape of f_n for large n does not necessarily conform to the shape of the limiting function. We can strengthen the notion of convergence to get the desired result as follows.

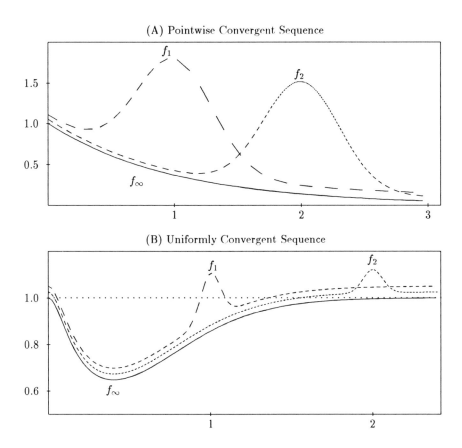

Figure 12.1: Convergent Functions with Divergent Maxima

Definition 12.1 (Uniform Convergence) *A sequence of functions* $f_n : \Theta \to \mathbf{R}$ *converges* uniformly *to limit function* $f : \Theta \to \mathbf{R}$ *if*

$$\lim_{n \to \infty} \sup_{\theta \in \Theta} |f_n(\theta) - f(\theta)| = 0.$$

Uniform convergence requires that the maximum gap between the sequence and the limit function should go to zero. This forces the sequence to conform in shape to the limit function for large n. Roughly speaking, it is true that if f_n converge uniformly to a limit function f and the limit function has a unique maximum θ_0, then maxima of f_n will also converge to θ_0. However, one has to be careful about ruling out the subtle phenomenon depicted in Figure 12.1 (B). Here the limit function is $f_\infty(x) = (1 + x^2 e^{-x})^{-1}$ which has a unique maximum at $x = 0$, with $f_\infty(0) = 1$. However, as $x \to \infty$, $f_\infty(x)$ converges to 1 from below. If we define a sequence $f_n(x)$ by $f_n(x) = f_\infty(x) + (1/n)\phi(x - n)$, then f_n will have a unique global maximum at $x = n$ but also converge uniformly to f_∞. The problem arises because, in effect, f_∞ has another global maximum at $x = +\infty$. This kind of phenomenon is ruled out as follows. We first define the concept of a sequence of approximate maxima. Unlike maxima, which may fail to exist, approximate maxima always exist for all functions.

Definition 12.2 *Let $f_n : \Theta \to \mathbf{R}$ be any sequence of bounded functions. A sequence of points $\theta_n \in \Theta$ is called* a sequence of approximate maxima *of f_n if*

$$\lim_{n \to \infty} \sup_{\theta \in \Theta} f_n(\theta) - f_n(\theta_n) = 0$$

Thus for large n, sequences of approximate maxima θ_n come closer and closer to being a maximizing value for f_n. We can also apply this definition to a single function f and say that θ_n is a sequence of approximate maxima of f if $\sup_{\theta \in \Theta} f(\theta) - f(\theta_n)$ converges to zero. For many reasons, it is technically very advantageous to study convergence of approximate maxima rather than exact maxima. The following definition ensures that all sequences of approximate maxima converge to the same limit.[1]

Definition 12.3 *A point $\theta_0 \in \Theta$ is called a* Strong Maximum *of the sequence f_n if all sequences of approximate maxima of f_n converge to*

---

[1] The definition of *strong maximum* is closely related to the concept of *identifiably unique* maxima, introduced by Domowitz and White (1982). To be somewhat more precise, the condition of identifiable uniqueness requires that the distance between a sequence of exact maxima and any sequence of approximate maxima should converge to zero. Our definition has the advantage of not requiring the existence of exact maxima. See also Pötscher and Prucha (1991a) for a survey of this and related concepts.

θ_0. *As a special case, θ_0 is a strong maximum of the single function f, if this property holds for the sequence $f_n \equiv f$.*

In order that maxima of sequences of functions convergent to f should converge to a maximum of f, f has to have a strong maximum. That is, all sequences of approximate maxima of f must converge to θ_0. It follows immediately from the definition that θ_0 must be a unique maximum of f. But the condition is somewhat stronger than that. Note that the counterexample of Figure 12.1 (B) arises because 0 is not a strong maximum of f. If we take $x_n = n$ then $\lim_{n\to\infty} f(x_n) = 1 = \sup_x f(x)$. Thus x_n is a sequence of approximate maxima of f which fails to converge to the unique maximum 0.

Theorem 12.1 *Suppose θ_0 is a strong maximum of f and f_n converge to f uniformly. Then any sequence of approximate maxima θ_n of f_n converges to θ_0.*

Remark: For a stronger version of this theorem, see exercise 2.
Proof: Since θ_0 is a strong maximum of f, to prove the theorem it suffices to show that θ_n is a sequence of approximate maxima of f. This can be achieved as follows.

$$\lim_{n\to\infty}\sup_{\theta\in\Theta} f(\theta) - f(\theta_n) \leq \lim_{n\to\infty}\left\{\left(\sup_{\theta\in\Theta} f(\theta) - \sup_{\theta\in\Theta} f_n(\theta)\right)\right.$$
$$+ \left.\left(\sup_{\theta\in\Theta} f_n(\theta) - f_n(\theta_n)\right) + (f_n(\theta_n) - f(\theta_n))\right\}$$
$$= A + B + C.$$

The middle term B goes to zero by the hypothesis that θ_n is a sequence of approximate maxima of f_n. Uniform convergence of f_n to f immediately implies that C goes to zero since $|f_n(\theta_n) - f(\theta_n)| \leq \sup_{\theta\in\Theta} |f_n(\theta) - f(\theta)| \to 0$. The following lemma, applied to the set $S = \Theta$ shows that A goes to zero and completes the proof.

Lemma 12.1 *If f_n converge uniformly to f then for any subset S of Θ,*
$$\lim_{n\to\infty}\sup_{\theta\in S} f_n(\theta) = \sup_{\theta\in S} f(\theta).$$

Proof: Let ψ_n be a sequence of approximate maxima of f in the set A, so that $\lim_{n\to\infty} f(\psi_n) = \sup_{\theta\in A} f(\theta)$. Then $\sup_{\theta\in\Theta} f_n(\theta) \geq f_n(\psi_n)$ and $|f_n(\psi_n) - f(\psi_n)| \to 0$, so that $\lim_{n\to\infty} \sup_{\theta\in A} f_n(\theta) \geq \sup_{\theta\in A} f(\theta)$. Next let ψ'_n be a sequence of approximate maxima of f_n over the set A, so that $\lim_{n\to\infty} \sup_{\theta\in A} f_n(\theta) - f_n(\psi_n) = 0$. Since $|f_n(\psi_n) - f(\psi_n)| \to 0$ and $f(\psi_n) \leq \sup_{\theta\in A} f(\theta)$, substitutions yield the following reverse inequality, proving the lemma.

$$\lim_{n\to\infty} \sup_{\theta\in A} f_n(\theta) - \sup_{\theta\in A} f(\theta) \leq 0.$$

It may be difficult to verify whether a given maximum θ_0 is strong or not. Fortunately, when the set Θ is compact, every unique maximum is automatically a strong maximum. This simplifies the task substantially. The proof of the following useful lemma, based on elementary properties of compact (i.e., closed and bounded) sets, is given in exercise 1.

Lemma 12.2 *Suppose Θ is compact and $f : \Theta \to \mathbf{R}$ is a continuous function. If θ_0 is the unique global maximum of f on Θ then θ_0 is a strong maximum.*

12.2   Uniform Laws of Large Numbers

Let y_i be a sequence of observations, and $\rho_n(y_i, \theta)$ a sequence of functions. $\hat{\theta}_n$ is called a sequence of M-estimators (where M stands for maximum) if $\hat{\theta}_n$ maximizes the function $S_n(\theta) = \sum_{i=1}^n \rho_n(y_i, \theta)$. Note that since this function depends on the random variables y_i, it is a random function. For each fixed θ, we can expect that $S_n(\theta)/n$ will converge to $\mathbb{E}(S_n(\theta)/n)$ under conditions guaranteeing that a law of large numbers holds. This means that pointwise convergence of the random functions to a limit function holds. If we can strengthen the pointwise convergence to uniform convergence, and if we can show that the limit function has a strong maximum, we can deduce the consistency of the estimators using arguments sketched in the previous section. Results which prove that $S_n(\theta)/n$ converges to $\mathbb{E}S_n(\theta)/n$ for a fixed θ are referred to as Laws of Large Numbers (or LLNs). A Uniform Laws of Large Numbers (ULLN) shows that $\sup_\theta |S_n(\theta) - \mathbb{E}S_n(\theta)/n|$ converges to zero under suitable hypothesis. The convergence can be *weak*, that is, in probability, or *strong*, that is, almost surely. We will present some ULLNs for the case of identically distributed random variables. This case is of limited applicability in econometric applications, where heterogeneity of observations and errors is the rule rather than the exception. However, the identical distribution assumptions gives fairly powerful results under minimal assumptions.

12.2.1   IID Case: Compact Parameter Space

Let Z_i be a sequence of random variables taking values in some set $\mathcal{Z}$. Let Θ be an index set (typically the parameter space or some subset of it in applications). Let $\rho_n(z, \theta)$ be a real-valued function of $z \in \mathcal{Z}$ and $\theta \in \Theta$. We will consider the question of uniform convergence for sums of the type $S_n(\theta) = \sum_{i=1}^n \rho_n(Z_i, \theta)$. Our first result is a slight variant of Wald (1949). The proof is straightforward, but requires more analysis than we are assuming as background, and hence is omitted.

Theorem 12.2 *Suppose Z_i is an i.i.d. sequence of random variables. Suppose that ρ is dominated: that is,*

$$\mathbb{E} \sup_{\theta \in \Theta} \rho(Z_i, \theta) < \infty. \tag{12.1}$$

Suppose that for each z, $\rho(z, \theta)$ is a continuous function of θ and Θ is compact. If θ_0 is a unique maximum of $\mathbb{E}\rho(Z_1, \theta)$ then all sequences $\hat{\theta}_n$ of maxima of $S_n(\theta) = (1/n) \sum_{i=1}^n \rho(Z_i, \theta)$ converge to θ_0 with probability one and also in quadratic mean.

Remarks: If we also assume that $\mathbb{E}\inf_\theta \rho(Z_1, \theta) > -\infty$ then we get a ULLN that $S_n(\theta) - \mathbb{E}\rho(Z_1, \theta)$ converges uniformly to zero. However, the condition given is enough to ensure that maxima of S_n converge to the unique maximum of $\mathbb{E}\rho(Z_1, \theta)$. With the possible exception of the Glivenko-Cantelli theorem discussed later, this is the simplest and most elegant ULLN available. Both the i.i.d. assumption and the compactness of Θ are strong hypotheses and restrict the domain of application. The result can be extended to noncompact Θ at the price of some additional complexity in formulation, as discussed in the following section.

12.2.2   IID Case: Noncompact Parameter Space

We briefly sketch three results on consistency for i.i.d. random variables with noncompact parameter spaces. We will not discuss the somewhat technical regularity conditions required for the validity of these results, except to indicate that the three are listed in decreasing order of generality (i.e., the third result requires the strongest hypotheses). Let Z_i be a sequence of i.i.d. random variables, and define $\hat{\theta}_n$ to be a sequence of approximate maxima of $S_n(\theta) = (1/n)\sum_{i=1}^{n} \rho(Z_i, \theta)$. Suppose that $\rho(x, \theta)$ is continuous[2] for all $\theta \in \Theta$, where Θ is the possibly noncompact parameter space. According to Perlman (1972), all such sequences of approximate maxima converge almost surely to a strong maximum θ_0 of $\mathbb{E}\rho(Z_1, \theta)$ if for every neighborhood $N(\theta_0)$ of θ_0, it is true that for some integer $n > 0$:

$$\mathbb{E} \sup_{\psi \notin N(\theta_0)} \sum_{i=1}^{n} \rho(Z_i, \psi) < \infty.$$

Note that this is very close to the dominance condition for the case $n = 1$.

Perlman's condition above is in fact necessary and sufficient, but is frequently difficult to verify in applications. An alternative sufficient condition which is sometimes more convenient is given by Zaman (1991). For integers $n = 1, 2, \ldots$, define the expectation of the truncation $\rho_n^* = \mathbb{E}\max(\rho(Z_1, \theta), -n)$. If ρ is dominated (that is, it satisfies Eq. (12.1) above), then all sequences of approximate maxima $\hat{\theta}_n$ of $S_n(\theta)$ converge to θ_0 if and only if all sequences of the approximate maxima of ρ_n^* also converge to θ_0.

The last result is useful in the cases where dominance fails to hold. Define

$$\psi(Z_i, \theta) = (\rho(Z_i, \theta) - \rho(Z_i, \theta_0))/\mathbb{E}\rho(Z_1, \theta),$$

and let $\psi_n^*(\theta) = \mathbb{E}\max(\psi(Z_1, \theta), -n)$. Zaman (1991) shows that $\hat{\theta}_n$ converge almost surely to θ_0 if and only if ψ is dominated *and* all sequences of approximate maxima of $\psi^*(\theta)$ converge to θ_0.

---

[2]Continuity can be weakened to upper semicontinuity; if θ_n converge to θ, and upper semicontinuous function ρ satisfies $\limsup_{n \to \infty} \rho(\theta_n) \leq \rho(\theta)$.

12.2.3 Dependent but Identical Distributions

The results cited above require independence of the random variables as well as identical distribution and conclude strong convergence. Our next result only requires identical distributions and dispenses with independence. The price of this relaxation is that we can only conclude weak convergence for the sequence of estimators.

Theorem 12.3 (Taylor) *Suppose X_i is a sequence of random variables with identical distributions. Further suppose that for each θ $S_n(\theta) = (1/n)\sum_{j=1}^{n} \rho(X_j, \theta)$ converges in probability to $\mathbb{E}\rho(X_1, \theta)$. Then the convergence is uniform in probability. That is*

$$\mathbf{plim}_{n\to\infty} \sup_{\theta\in\Theta} |S_n(\theta) - \mathbb{E}\rho(X_1, \theta)| = 0.$$

This holds provided either Θ is compact or the collection of functions $\rho(x, \cdot)$ spans a separable set in the uniform topology.

Proof: See Taylor (1978, Theorem 5.1.1).

12.2.4 Dependent Heterogeneous Variables

One frequently needs to deal with nonindependent and nonidentically distributed random variables in econometrics. Unfortunately, this field is vast, deep, and complex. It is our hope that armed with the basic concepts which have been developed above in the simplest and most transparent cases, the reader will be able to follow the complicated extensions to various types of dynamic models, allowing for dependence and heterogeneity. Pötscher and Prucha (1991a, 1994) have given useful surveys of the extensive literature. We discuss below just one result, a 'generic' ULLN, because it is very flexible and has assumptions that are relatively easy to verify.

Suppose X_i is a sequence of random variables taking values in sample space $\mathcal{X}$. Let Θ be a compact parameter space, and $\rho_{ni} : \mathcal{X} \times \Theta \to \mathbf{R}$ a sequence of criterion functions. A key feature of the generic results, the reason for the name, is the hypothesis that for each $\theta \in \Theta$,

$$\mathbf{plim}_{n\to\infty} \frac{1}{n} \sum_{i=1}^{n} \rho_{ni}(X_i, \theta) - \mathbb{E}\rho_{ni}(X_i, \theta) = 0.$$

This is a 'high-level' assumption, in the sense that verifying it will generally involve applying some other theorem. It is convenient to make such an assumption, since the LLN being assumed may be verifiable under large numbers of different kinds of hypothesis — indeed it is theoretically possible that different values of θ may require different approaches. Assuming that we get can somehow get past this hurdle of proving pointwise convergence to zero, the generic ULLNs address the issue of what additional conditions are needed to convert the pointwise convergence to uniform convergence. Extensive catalogs of conditions which will do the job are given by Pötscher and Prucha (1994).

We present below one of the simplest conditions, due to Andrews (1992). Suppose there exists a continuous function h satisfying $h(0) = 0$ and functions B_{nj} such that (i) for any $\theta, \theta' \in \Theta$, $|\rho_{nj}(X_j, \theta) - \rho_{nj}(X_j, \theta')| \leq B_{nj}(X_j)h(\|\theta - \theta'\|)$ and (ii) $\sup_{n \geq 1}(1/n)\sum_{j=1}^{n} B_{nj}(X_j) < \infty$. Then the average $S_n(\theta) = (1/n)\sum_{j=1}^{n} \rho_{nj}(X_n, \theta)$ satisfy the following ULLN:

$$\boldsymbol{plim}_{n \to \infty} \sup_{\theta \in \Theta} |S_n(\theta) - \mathbb{E}S_n(\theta)| = 0.$$

Intuitively, if the functions converge pointwise, and the derivatives are wellbehaved, then the functions must converge uniformly. Many variants of this type of result exist. These results and further references can be found in the papers cited earlier.

12.3   Consistency of M-Estimators

Having developed several ULLNs, we will now give some examples of how they are applied to prove consistency of estimators. When an estimator can be written explicitly, it is frequently possible to calculate its mean and variance, at least approximately for large samples, using various results in asymptotic theory. In such cases, one can evaluate consistency by checking whether the asymptotic mean converges to the right value, and the asymptotic variance goes to zero. In other cases an estimator may be defined indirectly by various conditions. In such cases, one may not have an explicit expression for the estimator. Even when an explicit expression can be found, it may be difficult to calculate its variance. In such cases, the following results, which establish consistency of certain broad classes of estimators, are very useful. In particular, we will cover the case of maximum likelihood, method of moments, and also Bayes estimators.

12.3.1 Maximum Likelihood

Suppose $y_i \overset{i.i.d.}{\sim} f^Y(y, \theta_0)$, where $\theta_0 \in \Theta$. The log likelihood is just the (random) function S_n defined below:

$$S_n(y_1, y_2, \ldots, y_n, \theta) = \sum_{i=1}^{n} \log f^Y(y_i, \theta).$$

The maximum likelihood estimator $\hat{\theta}_n$ maximizes $S_n(., \theta)$ over all $\theta \in \Theta$, the parameter space. Using the uniform strong law in Theorem 12.2, we can conclude that this estimator is strongly consistent. The needed assumptions will be formulated in terms of the function $\rho(y, \theta)$ defined as $\rho(y, \theta) \equiv \log(f^Y(y, \theta)/f^Y(y, \theta_0))$. Also define $\rho^*(\theta) = \mathbb{E}\rho(Y_1, \theta)$. We will also need the following definition.

Definition 12.4 *Suppose the set of possible densities of Y is given by $f(y, \theta)$ for $\theta \in \Theta$. The parameter θ is called* identified *if for any two parameters θ_1 and θ_2 such that $\theta_1 \neq \theta_2$, there is some set A such that $\mathbf{P}(Y \in A|\theta_1) \neq \mathbf{P}(Y \in A|\theta_2)$.*

The condition of identification ensures that there is an observable difference between any two parameters and is an essential hypothesis for the consistency result given below.

Theorem 12.4 (Wald) *Let Y_i be an i.i.d. sequence of random variable with common density $f(y, \theta_0)$ and suppose $\hat{\theta}_n$ is a sequence of approximate maxima of the joint likelihood $l_n(\theta) = \prod_{i=1}^{n} f(Y_i, \theta)$ for $\theta \in \Theta$ a compact parameter space. Then $\hat{\theta}_n$ converges almost surely to θ_0 provided that θ is identified and the log likelihood is dominated.*

Proof: This follows from Theorem 12.2 above, after setting $\rho(x, \theta) = \log f(x, \theta)$ and noting that $\hat{\theta}_n$ maximizes $(1/n) \sum_{i=1}^{n} \rho(Y_i, \theta)$. Note that dominance holds trivially if, as is usually is the case, the densities

$f(y, \theta)$ are all bounded above: that is if $\sup_y f(y, \theta) \leq M < \infty$ for all θ. Lemma 12.3, which shows that θ_0 is always a unique maximum of $\mathbb{E}_\rho(Y_1, \theta)$, is usually referred to as the information inequality. Since unique maxima are necessarily strong maxima on compact sets, this establishes the consistency of maximum likelihood.

Lemma 12.3 (Information Inequality) *For all θ such that $f^Y(y, \theta)$ is a different density from $f^Y(y, \theta_0)$, we have the inequality*

$$\mathbb{E}\log f^Y(y_1, \theta) < \mathbb{E}\log f^Y(y_1, \theta_0).$$

Proof: By Jensen's inequality (see exercise 5), $\mathbb{E}\log Z < \log(\mathbb{E}Z)$ whenever Z is a nondegenerate random variable. Thus we can bound the difference $\Delta(\theta) = \mathbb{E}(\log f^Y(Y_1, \theta)) - \log f^Y(Y_1, \theta_0)$ as follows:

$$
\begin{aligned}
\Delta(\theta) &= \mathbb{E}\log\frac{f^Y(Y_1, \theta)}{f^Y(y_1, \theta_0)} < \log\mathbb{E}\frac{f^Y(Y_1, \theta)}{f^Y(Y_1, \theta_0)} \\
&= \log\int\frac{f^Y(y_1, \theta)}{f^Y(y_1, \theta_0}f^Y(y_1, \theta_0)dy = \log\int f^Y(y, \theta)dy = \log 1 = 0.
\end{aligned}
$$

This completes the proof of the lemma and the theorem.

This result can be extended to the case of noncompact parameter spaces by using the results of Perlman (1972) and Zaman (1991) discussed above. General results covering the consistency of ML estimators in the i.n.i.d. case and valid for noncompact parameter spaces are given by Ibragimov and Has'minskii (1981, Theorems 4.4, 5.1, and 5.3). The conditions are a bit involved and hence not presented here.

12.3.2   Method of Moments

Historically, this is the first general method for construction of estimators, and was suggested by Karl Pearson. Following an acrimonious debate between Pearson and Fisher, it fell into disfavor after Fisher showed that his favorite, the ML estimation procedure, produced asymptotically efficient estimates. Recently it has been realized that the efficiency of ML is tied to delicate and unverifiable assumptions regarding the class of parametric densities, while the method of moments can produce more robust estimators in many cases.

Suppose that X_i are i.i.d. with common distribution function $F(x,\theta)$, where the θ is an element of the parameter space $\Theta \subset \mathbf{R}^K$. Assume that the distribution F has K finite absolute moments. For $k = 1, 2, \ldots, K$ the moments $M_k(\theta) = \mathbb{E}X_1^k$ will be functions of the parameters θ. Let $\hat{M}_{k,n} = (1/n) \sum_{i=1}^{n} X_i^k$ be the estimated sample moments. The MOM estimates are defined as solutions of the K equations

$$\hat{M}_{k,n} = M_k(\theta) \quad \text{for } k = 1, \ldots, K. \qquad (12.2)$$

Under relatively simple conditions, such estimates are consistent.

Theorem 12.5 *Suppose that there is a unique solution to the system of Eq.s (12.2) with probability approaching 1 as n goes to infinity. Suppose the functions $M_k(\theta)$ are continuously differentiable and the Jacobian matrix $J_{ij} = \partial M_i(\theta)/\partial\theta_j$ has a nonzero determinant over the entire parameter space Θ. Then the method of moments estimates converge to the true parameter values.*

Proof: Under the hypotheses, the map $M : \Theta \to U \subset \mathbf{R}^K$ defined by $M(\theta) = (M_1(\theta), \ldots, M_K(\theta))$ is continuous and locally one-to-one. Suppose that $\theta_0 \in \Theta$ is the true parameter and let $u_0 = M(\theta_0) \in U$ be the image of θ_0 under the mapping M, and find a neighborhood N of u_0 over which the map M is invertible. By Kolmogorov's SLLN, the existence of the moments is sufficient to guarantee convergence of the sample moments $\hat{M}_{k,n}$ to the population moments $M_k(\theta_0) = \mathbb{E}X_1^k$, so that for large n, $\hat{M}_{k,n}$ must belong to the neighborhood N. Now the convergence of the moments and the invertibility of the map guarantees the consistency of the estimates.

As a practical matter, it requires very high sample sizes to accurately estimate the higher moments. Thus the original method of moments is not recommended in cases where there are more than one or two parameters. A substantial generalization of the method of moments can be made by choosing *any* K functions $h_k(X_1, \ldots, X_n, \theta)$ and declaring the GMM (*Generalized Method of Moments*) estimator to be the solutions $\hat{\theta}_n$ to the equations $h_k(X_1, \ldots, X_n, \theta) = 0$ for $k = 1, 2, \ldots, K$. This is in fact more or less a generalization of the concept of M-estimators defined above as maxima of arbitrary functions. If the ρ function is differentiable, then maxima will satisfy the

K first order conditions $\partial\rho/\partial\theta_i = 0$. Thus, by defining h_k to be the partial derivatives of ρ, we can convert M-estimators into GMM estimators. The theory of consistency of GMM estimators is again based on the use of the uniform laws of large numbers. If, for example, $h_k(X_1, X_2, \ldots, X_n, \theta) = (1/n)\sum_{i=1}^{n} h_k^*(X_i, \theta)$ then h_n will, under suitable assumptions, converge uniformly to $H_k^*(\theta) = \mathbb{E}h_k^*(X_1, \theta)$, and the GMM estimators will converge to roots of the (deterministic) system of equations $H_k(\theta) = 0$ for $k = 1, 2, \ldots, K$. See Hansen (1982) for details.

12.3.3   Bayes Estimators

The method of Bayes estimation produces consistent estimators under conditions weaker than those required by other methods. Suppose that X_i are i.i.d. random variables taking values in sample space $\mathcal{X} \subset \mathbf{R}^p$ with common densities $f(x, \theta)$ where θ belongs to the parameter space Θ. Introduce a measure of distance on Θ by defining $d(\theta, \theta') = \int_{\mathcal{X}} |f(x, \theta) - f(x, \theta')|\, dx$. Let $\pi(\theta)$ be a prior probability density on Θ.

Definition 12.5 *The prior $\pi(\theta)$ is called a* dogmatic *prior if for some open subset U of Θ, $\pi(U) \equiv \int_{\theta \in U} \pi(\theta)\, d\theta = 0$.*

To avoid inconsistency, it is essential to rule out dogmatic priors. This is because it is easily verified that if the prior measure of a set is zero, then its posterior measure is also zero. Thus if the true parameter lies in the interior of a set assigned zero probability a priori, no amount of data will cause that probability to change, and hence the Bayes procedure will be inconsistent. Provided that we use nondogmatic priors, Bayes procedures are consistent under very weak conditions. Our first result shows that whenever the ML estimates are consistent, Bayes estimators are also consistent:

Theorem 12.6 (Strasser) *Suppose that $\pi(\theta)$ is not a dogmatic prior and also that $H(\sigma) = \int_{\mathcal{X}} f(x, \sigma)f(x, \theta)\, dx$ is a continuous function of σ for each θ. If sequences of ML estimators based on i.i.d. observations with common density $f(x, \theta_0)$ are consistent for θ_0, then for every open neighborhood N of θ_0, the posterior probability of N converges to 1 in large samples*

See Strasser (1981) for proof. The converse is not true, and there exist many situations where Bayes procedures are consistent while the ML is not. Ibragimov and Has'minskii (1981) and Le Cam and Yang (1990) present general results on consistency of Bayes estimates. Because the conditions given are somewhat cumbersome, we do not present these results here. A comparison of the conditions needed for consistency of Bayes and ML shows that Bayes estimators require weaker hypotheses for consistency. In finite dimensions, it is more or less true that all nondogmatic priors lead to consistent estimators. With infinite dimensional parameter spaces, the situation is more complicated and natural examples of inconsistency of Bayes estimators arise. See Le Cam and Yang (1990) for examples and references. In the next section we give an example of an inconsistent ML estimator.

12.3.4 An Inconsistent ML Estimator

Although ML estimators are typically consistent, there do exist situations where this is not true. One important example, the case of a mixture of normal densities, is discussed in the present section. Another example is given in exercise 8.

Suppose X_i has the following density:

$$f^{X_i}(x_i) = \frac{\alpha}{\sqrt{2\pi\sigma_1^2}} \exp\left(-\frac{(x_i - \mu_1)^2}{2\sigma_1^2}\right) + \frac{1-\alpha}{\sqrt{2\pi\sigma_2^2}} \exp\left(-\frac{(x_i - \mu_2)^2}{2\sigma_2^2}\right)$$

That is, with probability $\alpha \in [0,1]$, $X_i \sim N(\mu_1, \sigma_1^2)$ and with probability $1 - \alpha$, $X_i \sim N(\mu_2, \sigma_2^2)$. In order to identify the model, we need to make some restriction such as $\mu_1 > \mu_2$. We show below that the ML estimators will typically be inconsistent in such mixture models.

Consider maximizing the joint density, which is just $f(X_1, \ldots, X_n)$ $= \prod_{i=1}^{n} f^{X_i}(x_i)$ with respect to the parameters. Consider the effect of sending σ_1 to zero. In standard models, the term $\exp((X_j - \mu_1)^2/2\sigma_1^2)$ declines to zero exponentially fast, and no problems occur. However, if we set $\hat{\mu}_1 = X_j$ for any j, this knocks out the exponential term, and so the $\alpha/\sqrt{2\pi}\sigma_1$ term goes to infinity. The other terms cannot lend a hand in preventing this blowup of the likelihood function since they are of the form $\alpha f_1 + (1 - \alpha)f_2$; while the f_1 terms dutifully go to zero exponentially fast for all other observations, the f_2 spoils the show by preventing the density from going to zero. Thus the likelihood is maximized to be $+\infty$ by setting $\sigma_1 = 0$ and μ_1 equal to any of the observations (and similarly for σ_2 and μ_2 also). This holds for any number of observations and implies that the ML is inconsistent. It is easily checked that the dominance condition (Eq. 12.1) fails in this example.

12.4   Case of Trending Regressors

It has been realized by several authors that the method of proof of consistency developed above is deficient in econometric applications where regressors have some type of trend. The results of the present section can be applied whenever the regressors are ancillary – this means that we can condition on the observed values of the regressors and treat them as constants. In particular, these results allow for arbitrary trends in the regressors. The approach described is due to Wu (1981), as extended by Zaman (1989). For a different approach to the same problem, see Andrews and McDermott (1993).

Let y_i be a sequence of independent variables taking values in the set $\mathcal{Y}$. Let Θ be a compact subset of $\mathbf{R}^K$. Let $\rho_t : \mathcal{Y} \times \Theta \to \mathbf{R}$ be a sequence of functions. Suppose estimators $\hat{\theta}_T$ maximize the random function $S_T(\theta) = \sum_{t=1}^{T} \rho_t(y_t, \theta)$. Note that $\rho_t(y_t, \theta)$ is allowed to depend on t so that $\rho_t(y_t, \theta) = \rho(y_t, x_t, \theta)$ is possible. Let θ_T be a maximizing value for $\mathbb{E}S_T$ and suppose that θ_T converges to some θ_0. The theorem below gives conditions under which $\hat{\theta}_T$ converges to θ_0. Fix some element $\theta_0 \in \Theta$. For any function $f : \Theta \to \mathbf{R}$, we define the Lipschitz norm of f by

$$\|f\|_{Lip}^2 = \|f(\theta_0)\|^2 + \int_\Theta \|\sum_{i=1}^K \frac{\partial f}{\partial \theta_i}\|^2 \, d\theta_1 d\theta_2 \cdots d\theta_K.$$

Theorem 12.7 (Wu-Zaman) *Assume that for some $\theta_0 \in \Theta$, $\mathbb{E}S_T(\theta_0) = 0$ for all T. Further assume that for some sequence of constants λ_T increasing to infinity*

$$\mathbb{E}S_T(\theta) \geq \lambda_T \|\theta - \theta_0\|. \tag{12.3}$$

The difference $|\hat{\theta}_T - \theta_0|$ converges to zero almost surely, and in quadratic mean, if

$$\sum_{t=1}^{\infty} \mathbb{E}\|\frac{\rho_t(y_t, \theta) - \mathbb{E}\rho_t(y_t, \theta)}{\mathbb{E}S_T(\theta)}\|_{Lip}^2 < \infty.$$

Remarks: The conditions imply that the ratio of S_T to $\mathbf{ES_T}$ converges uniformly to 1 for all θ outside a neighborhood of θ_0. It turns out to be somewhat more general to use the uniform convergence of the ratio to 1 rather than the uniform convergence of the difference to zero. See exercise 4 for a deterministic example of the phenomena. The Lipschitz norm can be replaced by a variation norm, which yields sharper results in some cases; however, it is harder to compute. For details see Zaman (1989).

12.4.1 Nonlinear Least Squares

The (nonlinear) least squares estimator is a special case of the L_p estimator described above with $p = 2$. Wu (1981) has given several examples of interest. The condition (Eq. 12.3) has been stated in a form which makes it easier to verify in applications. A slightly more general form is actually a necessary condition for the existence of a weakly consistent estimator. This result of Wu is given below:

Theorem 12.8 (Wu) *Consider the model $y_i = f_i(\theta) + \epsilon_i$, where ϵ_i are i.i.d. with common density g satisfying $g > 0$ and $\int(g'/g) < \infty$. There exists a weakly consistent estimator for θ if and only if for all θ, θ' such that $\theta \neq \theta'$,*

$$\lim_{n \to \infty} \sum_{i=1}^{n}(f_i(\theta) - f_i(\theta'))^2 = \infty.$$

Note that the hypothesis (Eq. 12.3) is sufficient but not necessary for the condition above. This condition guarantees that enough information about θ accumulates to permit a consistent estimator to exist. If it is violated then no consistent estimator can exist. This same condition is also enough for the consistency of least squares so that if any consistent estimator exists then least squares will also be consistent.

One example of the use of this condition is given below. For another, see exercise 10.

Consider the power curve model $y_t = (t + \theta)^d + \epsilon_t$, where d is a known constant. According to the necessary condition given by the theorem above, we need

$$\sum_{t=1}^{\infty}((t + \theta)^d - (t + \theta')^d) \to \infty$$

for consistent estimators to exist. This holds if and only if $d > 1/2$. In this case the consistency of least squares is easily established by using the Wu-Zaman theorem above.

12.5   Glivenko-Cantelli and Bootstrap

The Glivenko-Cantelli theorem is of fundamental importance in theory and practice. Let $X_1, X_2, X_3, \ldots$ be an i.i.d. sequence of random variables with common cumulative distribution function F. Suppose we observe $X_1 = x_1, \ldots, X_n = x_n$. The empirical distribution function $\hat{F}_n$ is defined to be the cumulative distribution function of a random variable which takes the values $x_1, \ldots, x_n$ each with equal probability. Let $I\{x \geq t\}$ be the indicator function taking value 1 for $x \geq t$ and 0 for $x < t$. Then the empirical CDF can be written as

$$\hat{F}_n(x) = \frac{1}{n}\sum_{j=1}^{n} I\{x \geq X_j\}.$$

As an illustration, Figure 12.2 plots the empirical CDF of a sample of 10 random $N(0,1)$ observations against the true CDF. The function $\hat{F}_n$ is a natural estimate of the CDF of the random variables X_i. Each of the observed values is assigned equal probabilities as one of the possible outcomes of X. If it is known in advance that the random variable is continuous then a version of $\hat{F}_n$ modified to be continuous may be superior in some (but not all) contexts. The Glivenko-Cantelli Theorem asserts that $\hat{F}_n$ converges uniformly to the true distribution F. This is of fundamental importance in probability and statistics. It is also quite easy to prove:

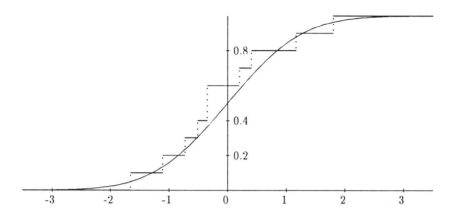

Figure 12.2: Empirical and Actual Normal CDF

Theorem 12.9 (Glivenko-Cantelli) *The maximum difference between the empirical and the actual CDF,* $\sup_{x \in \mathbf{R}} |\hat{F}_n(x) - F(x)|$, *converges to zero almost surely as n goes to* ∞.

Proof: Fix a particular value of x, e.g., $x = x_0$. Then $\hat{F}_n(x_0)$ is the average of i.i.d. Binomial random variables and converges to $F(x_0)$ with probability one according to Theorem 11.7. Now choose $x_0, x_1, x_2, \ldots, x_{99}, x_{100}$ so that $F(x_0) = 0, F(x_1) = 0.01, F(x_2) = 0.02,$ $\ldots, F(x_{99}) = 0.99, F(x_{100}) = 1$. We allow for $x_0 = -\infty$ and $x_{100} = +\infty$. With probability one $\hat{F}_n(x_j)$ converges to $F(x_j)$ for $j = 0, 1, 2, \ldots, 99, 100$. We will show that if $F_n(x_j)$ and $\hat{F}_n(x_j)$ are close, they are also close at all points between x_j. For any x there is an x_j such that $x_{j-1} \leq x \leq x_j$, so that

$$|F(x) - \hat{F}_n(x)| \leq |F(x) - F(x_j) + F(x_j) - \hat{F}_n(x_j) + \hat{F}_n(x_j) - \hat{F}_n(x)|$$

$$\leq |F(x) - F(x_j)| + |F(x_j) - \hat{F}_n(x_j)| + |\hat{F}_n(x_j) - \hat{F}_n(x)|.$$

The first term is less than or equal to 0.01, the second is convergent to zero, and the third can be bounded as follows:

$$\hat{F}_n(x_j) - \hat{F}_n(x) \leq \hat{F}_n(x_j) - \hat{F}_n(x_{j-1}) \to F(x_j) - F(x_{j-1}) = 0.01.$$

It follows that with probability one for large n $F(x) - \hat{F}_n(x)$ is less then or equal to 0.03. By increasing the subdivisions, we can show by the same argument that the maximum difference of $F(x) - \hat{F}_n(x)$ over all x must be smaller than any fixed positive real number for large n with probability one. The only number smaller than all fixed positive real numbers is zero, so it follows that the maximum difference must converge to zero with probability one.

The basic bootstrap procedure in the simplest setting works as follows. Suppose $X_1, X_2, \ldots$, are i.i.d. observations with common distribution function F. Suppose we wish to estimate some property of the distribution function F, such as the mean, variance, or even some arbitrary complex function of F, e.g., $\theta(F)$. Since $\hat{F}_n$ is an estimate of F, $\theta(\hat{F}_n)$ provides an estimate of $\theta(F)$. This simple idea works very well in extremely general settings. Two basic conditions are needed for it to work. One is that $\hat{F}_n$ must converge to the true distribution. This is guaranteed by the Glivenko-Cantelli theorem in the simple i.i.d. setting under discussion but applications frequently require more complex estimates for distributions, and consistency must be checked. The second condition needed is that $\theta(F)$ must be continuous in a suitable sense, so that when $\hat{F}_n$ is (uniformly) close to F, $\theta(\hat{F}_n)$ should also be close to $\theta(F)$. We give several examples to clarify the ideas.

Suppose $X_1, \ldots, X_n$ is an i.i.d. sample from a common unknown distribution F, and we would like to study the properties of the estimator $M(X_1, \ldots, X_n)$, for example the median of the observed values. If we know the underlying distribution F, it is easy to calculate the distribution of $M(X_1, \ldots, X_n)$. Theoretical formulas are available which permit calculation of the exact distribution of the median M from the known distribution F. Even if theoretical formulas are not available (typically the case for more complex functions of the data), one always has the Monte Carlo method available: generate a lot of samples from the true distribution and calculate medians for each sample. Plot the histogram of the median. This will converge to the density of the median asymptotically. Typically we are not interested in the full density of any given estimator but just some characteristics. For example, we may want to know the variance of some complicated estimator proposed. Again there is no problem if we know the true distribution.

Just generate 1000 i.i.d. samples from the true distribution and calculate the estimator for each sample. Now we have a sample of 1000 values of the estimator and the sample mean and variance will provide a reasonably accurate estimate of the true mean and variance.

The problem with the Monte Carlo procedure for calculating variances of arbitrarily complex estimators is that the true distribution is never known. The solution, which seems obvious in retrospect, has nonetheless been compared to alchemy and given the improbable name 'bootstrap.' Simply substitute the estimated $\hat{F}_n$ for F and pretend that this is the true distribution. That is, the bootstrap is merely a Monte Carlo estimate, with the unknown true distribution replaced by a sample estimate of it. It is clear from the Glivenko-Cantelli theorem that this should give good results in large samples. Substantial theoretical research shows that it provides relatively good finite sample approximations under a wide range of conditions. The theory of bootstrapping is discussed in greater detail in Chapter 11. Here we explore a few simple examples in a more elementary way.

12.6 Bootstrapping a Regression Model

We first spell out the workings of the bootstrap for a standard regression model, $y_t = x_t'\beta + \epsilon_t$, where ϵ_t are i.i.d. according to some common distribution F. Let X_T be the $T \times K$ matrix of regressors and Y_T be the $T \times 1$ vector of observations on the dependent variable. Our goal is to determine the distribution of the OLS estimator $\hat{\beta}_T = (X_T'X_T)^{-1}X_T'Y_T$. The bootstrap is very similar to the Monte Carlo method discussed earlier. To have clarity about the similarity and differences we first discuss how we would approach this problem via Monte Carlo.

In this problem there are two unknowns: the parameter β and the distribution F of ϵ. Monte Carlo methods require both of these two be known. Suppose we fix some value of β, e.g., $\beta = \beta_0$ and also assume F is $N(0, \sigma^2)$ for some known value of σ^2, e.g., $\sigma = \sigma_0$. Under these assumptions we can do Monte-Carlo to find out the distribution of $\hat{\beta}$. The procedure would be as follows:

Step 1: Generate a random sample $\epsilon_1^m, \ldots, \epsilon_T^m$.

Step 2: Calculate $y_t^m = x_t'\beta + \epsilon_t^m$.

Step 3: Calculate the OLS estimate $\hat{\beta}_T^m$ using the y^m.

This process generates one random variable $\hat{\beta}^m$. By repeating it a 1000 times, we can generate 1000 random variables which will have exactly the same distribution as the OLS estimator *given that* $\beta = \beta_0$ *and* $F = N(0,\sigma_0^2)$. Any information we want about this distribution can be obtained from the random sample. For example, we could estimate the mean, the variance, and also the general shape of the density from this sample. Of course this is unnecessary since the density can be computed from theoretical considerations. In the case under discussion, we know that $\hat{\beta} \sim N(\beta_0, \sigma_0^2 (X'X)^{-1})$. Here and subsequently the subscript T for the sample size will be suppressed for notational convenience.

How should we proceed when β and F are unknown? An obvious answer is to replace them by estimated values and then use the Monte Carlo methodology. The OLS estimator $\hat{\beta}$ readily suggests itself as an estimate for β. The estimation of the distribution is not so obvious. There are two ways to proceed. We can *assume* that the distribution is Normal, so that $F \sim N(0,\sigma^2)$. Then we can estimate F by $N(0,\hat{\sigma}^2)$, where $\hat{\sigma}^2$ is the usual estimate of the variance. Now F and β are fixed so the Monte Carlo method can be used. Since we made an assumption which reduced F to a parametric family, this method is called the *parametric bootstrap*.

For the nonparametric bootstrap, we make no assumptions about F but proceed to estimate it directly via the Glivenko-Cantelli theorem. If ϵ_t were directly observed, we would use the empirical distribution function of ϵ_t to estimate F. Since they are not observed, we can use the residuals $e_t = y_t - x_t'\hat{\beta}$ as a substitute. This leads to the right estimate, as discussed in exercise 12. If we estimate the distribution function F by the empirical distribution function of e_t we get the nonparametric bootstrap. We now run a Monte Carlo exactly as described above to get a distribution for the OLS estimator. The only difference arises in the first step, where we must obtain random variables $e_1^*, \ldots, e_T^*$ from the estimated distribution $\hat{F}$. Since the estimated distribution is just the empirical distribution of the observed residuals $e_1, \ldots, e_T$, the random variables e_i^* are generated as follows: Each e_i^* is set equal to one of the

T observed residuals, where all the residuals are equally likely. The resulting residuals $e_1^*, \ldots, e_T^*$ are known as a *resample* of the original residuals. This is because the new (starred) residuals simply replicate in some random order the original residuals. Note, however, that there will typically be replications and omissions. That is some of the original residual may appear several times, while others may not appear at all in the resample. This process of resampling is fundamental to the bootstrap and responsible for its name. We have only one sample of the residuals and yet we use it to generate many new samples, effectively pulling ourselves up by the bootstraps.

Unfortunately, for the regression problem, this appears to be an exercise in futility. All methods lead to the same answer. To see this, let $\beta^* = (X'X)^{-1}X'Y^*$, where $Y^* = X\hat{\beta} + \epsilon^*$ is generated from a resample of the residuals and the estimated β. Then $\mathbb{E}\beta^* = \hat{\beta} + (X'X)^{-1}X'\epsilon^*$. Now the vectors of the resample are i.i.d. and have mean 0, so it is immediate that $\mathbb{E}\beta^* = \hat{\beta}$. Our estimate of the mean of β is $\hat{\beta}$. What about the covariance matrix? Note that ϵ_i^* are i.i.d. and have mean zero and variance equal to the sampling variance of the residuals: $\mathbb{V}ar(\epsilon_i^*) = (1/T) \sum_{t=1}^{T} e_t^2 = s^2$. It follows that $\mathbb{C}ov(\epsilon^*) = s^2 \mathbf{I}_T$ and hence that $\mathbb{C}ov(\beta^*) = s^2(X'X)^{-1}$. This differs slightly from the standard estimate in that $s^2 = (1/T) \sum_{t=1}^{T} e_t^2$, while the usual estimate for the covariance matrix of β uses $1/(T-K)$ instead of $1/T$ to get an unbiased estimate of the variance.

However, appearances are deceiving. We will see later that the bootstrap typically provides a better approximation to the true density of the OLS estimator than the asymptotic normal approximation. However, this fact is not reflected by the mean and the standard deviation which are the same. The differences arise because skewness or lack of symmetry in the true distribution are captured by the bootstrap, and not by the asymptotic normal distribution.

12.7 Bootstrap Failure: Discontinuity

There are two major types of situations where the bootstrap fails to work. One, which is relatively rare, is where continuity breaks down. That is, we estimate the true distribution $F(x)$ by $\hat{F}$, and then do calculations on the basis of the known $\hat{F}$. However, if the quantity of interest is such that its value at $F(x)$ is not well approximated by its value at $\hat{F}$ (due to a failure of continuity), then bootstrapping will fail. We illustrate this type of failure by an example below. The second type of failure, due to heterogeneity, is discussed in the next section.

Consider a first order autoregressive process, with $y_t = \rho y_{t-1} + \epsilon_t$, for $t = 1, 2, \ldots, T$ where $\epsilon_t \overset{i.i.d.}{\sim} N(0, \sigma^2)$. Assume y_0 is a fixed constant. We wish to determine the distribution of the least squares estimator $\hat{\rho} = (\sum_{t=1}^{T} y_t y_{t-1})/(\sum_{t=1}^{T} y_{t-1}^2)$. According to asymptotic theory, $\hat{\rho} \sim N(\rho, \sigma^2/(\sum_{t=1}^{T} y_{t-1}^2))$. This is exactly the formula we would get for the distribution of $\hat{\rho}$ if we pretended that the regressors were nonstochastic (or ancillary) and applied the classical theory. In dynamic models, this formula is no longer the exact distribution of $\hat{\rho}$ but it remain asymptotically valid for stable models, i.e. those models for which $\rho < 1$.

Thus asymptotic theory provides us with an approximation to the distribution of $\hat{\rho}$, namely

$$\frac{\hat{\rho}}{\sqrt{\sum_{t=1}^{T} y_{t-1}^2}} \approx N(\rho, \sigma^2). \tag{12.4}$$

This approximation is valid for all $|\rho| < 1$. The asymptotic theory is nonstandard when $|\rho| > 1$ in the following sense. In typical cases, the variance of an estimator divided by the square of the sample size converges to an asymptotic distribution. For $|\rho| < 1$ the standard deviation $\sqrt{\sum_{t=1}^{T} y_{t-1}^2}$ is approximately a constant multiple of $\sqrt{T}$. For $\rho > 1$ the standard deviation goes to zero much faster than $T^{-1/2}$. However, the asymptotic approximation given by Eq. (12.4) remains valid. Indeed, empirical studies show that it becomes more accurate more quickly (i.e., at smaller sample sizes). However there is a problem when $\rho = 1$, the so-called *unit root* case. In this case, the approximation of Eq. (12.4) is no longer valid, and the distribution of $\hat{\rho}$ is quite different. Thus there is a discontinuity in the distribution of $\hat{\rho}$ at $\rho = 1$. This discontinuity leads to failure of the bootstrap — the procedure does not accurately estimate the distribution of $\hat{\rho}$ when $\rho = 1$. One might expect the failure at 1 to cast shadows on near unit-root cases.

Somewhat surprisingly, the bootstrap continues to work well even for ρ very close to one. Some intuition for why this is so is offered later in Section 14.5.3, where we study bootstrapping an autoregressive model in some detail.

It is worth noting that it is discontinuities in the distribution of the statistic as a function of the parameter values which adversely affect the bootstrap. If statistic $T(X)$ is a discontinuous function of the data, this will not affect the bootstrap as long as the underlying distribution of T remains continuous. See Peracchi (1992, Section 5) for an example and discussion.

12.8 Bootstrap Failure: Heterogeneity

In resampling, we assume that the objects being resampled are i.i.d. If in fact these objects come from a heterogeneous population, the bootstrap will fail to give good results. This kind of failure is relatively frequent in econometric applications. We give just one example here.

Suppose we are interested in explaining a random variable y_t as a function of a second $K \times 1$ vector of random variable x_t. More precisely, we are interested in estimating the conditional density of y given x. If it is possible to assume that $\mathbb{E}(y|x) = \beta'x$, we can get a regression model by defining $\epsilon_t = y_t - \mathbb{E}(y_t|x_t)$ so that $y_t = \beta'x_t + \epsilon_t$. More generally, it is possible to use a linear regression model if $\mathbb{E}(y|x) = \sum_{i=1}^{k} \beta_i f_i(x)$ where $f_i(x)$ are arbitrary known functions of x. In this way of formulating the model, it is clear that just as the conditional mean of y is a function of x, so we should expect the conditional variance of y given x to be a function of x. Thus it would not be reasonable to assume that ϵ are homogeneous errors as is usually assumed. If, as will typically be the case in a situation like this, the error variance does depend on x, the standard bootstrapping method will fail to give good results.

To be more precise, let $\hat{\beta} = (X'X)^{-1}X'y$ be the usual OLS estimator. Let $e_t = y_t - \hat{\beta}'x_t$. Let e_t^* be a resample of the residuals, and generate new values for the dependent variable y using $y_t^* = \hat{\beta}'x_t + e_t^*$. Define the bootstrap estimate $\beta^* = (X'X)^{-1}X'y^*$. If we generate many, many values for β^* using Monte Carlo methods and compare the distribution of β^* with the true distribution of $\hat{\beta}$, we will find that these distributions are different. Our resampling technique is based on the implicit assumption that the e_t are i.i.d., but this is violated because the variances of the residuals are functions of x_t.

In this particular case, we can solve the problem by resampling data points instead of resampling residuals. A resample of the data consists of values $(y_1^*, x_1^*), \ldots, (y_T^*, x_T^*)$, where each (y_j^*, x_j^*) is chosen at random to equal one of the original T data points (y_t, x_t) with equal probability. In this way of resampling, the error at the t-th observation remains attached to the t-th observation, and so the heterogeneity of the regression residuals is properly accounted for by the bootstrapping scheme.

In the model described in this section, the errors are heteroskedastic and the error variances are related to the regressors. The bootstrapping scheme provides an estimator for the covariance matrix of the OLS estimates which appropriately adjusts for the heteroskedasticity. We have already learned another method of adjusting the OLS covariance for heteroskedasticity. This is based on the Eicker estimator for the covariance matrix. While the formulas and techniques for the two are quite different, both of these techniques give similar results in regression models where the regressors are 'balanced' — see Section 1.4 for a definition of balance. Efron (1982, Section 3.8) shows that Eicker's estimator for covariance under heteroskedasticity is an approximation to the bootstrap estimate.

The Eicker-White estimator for covariance under heteroskedasticity is asymptotically valid under conditions weaker than those required for the bootstrap. Our bootstrap technique requires (y_t, x_t) to be i.i.d. with some common joint distribution, and fails to work when this assumption does not hold. In particular, if we have some deterministic sequence of regressors, such as $x_t = t$, and the error variances are also related to x_t, the bootstrap technique does not give good estimates for the covariance matrix of the OLS. Theoretically the Eicker estimator continues to work in this situation, but practically it has been discovered that the relatively large sample sizes are required to eliminate biases in estimation of the covariance matrix under heteroskedasticity, especially when the regressors are not balanced.

12.9 Exercises

1. Let $f : [a, b] \to \mathbf{R}$ be a continuous function with a unique maximum at $x_0 \in [a, b]$. Show that x_0 is a strong maximum of f. Hint: from a sequence of approximate maxima, we can extract a convergent subsequence which must converge to the unique maximum.

2. Suppose g_n is a sequence of functions with a strong maximum at θ_0. Show that if $f_n - g_n$ converges uniformly to zero then all sequences of approximate maxima of f_n must also converge to θ_0. Construct a sequence g_n which has a strong maximum even though the sequence does not converge to any limit function, to see that this is stronger than Theorem 12.1.

3. Suppose θ_0 is a unique maximum of f on Θ and $f(\theta_0) > 0$. Suppose we can find a compact subset K of Θ such that for all $\theta' \notin K$, $f(\theta') < 0$. Then θ_0 must be a strong maximum of f.

4. Let $f(x) = -x^2$ so that f has a unique maximum at 0. Consider $f_n(x) = ((1/n) - 1)x^2$. Show that $\sup_x |f_n(x) - f(x)|$ is infinite for every n so that f_n does not converge uniformly to f. On the other hand f/f_n does converge uniformly to 1. Show that this last condition is enough to conclude that maxima of f_n converge to a maximum of f.

5. If $f : \mathbf{R} \to \mathbf{R}$ is strictly convex then for any $x_0 \in \mathbf{R}$ we can find a line $l(x) = a + bx$ such that (i) $l(x_0) = f(x_0)$ and (ii) $l(x) < f(x)$ if $x \neq x_0$. Use this to prove Jensen's inequality: $Ef(X) > f(EX)$ for strictly convex f provided that X is not degenerate. Note that X is degenerate if and only if $X = EX$ with probability 1.

6. Suppose $y_t = \alpha_0 x_t + \epsilon_t$ where $x_t = +1$ with probability $1/2$ and $x_t = -1$ with probability $1/2$. Suppose that ϵ_t are i.i.d. Cauchy random variables, with common density $f(\epsilon) = \{\pi(1 + \epsilon^2)\}^{-1}$.

   (a) Show that OLS estimates of α are not consistent. Hint: first show the averages of Cauchy random variables have the same Cauchy density. This follows easily using the characteristic

function of the Cauchy density; see exercise 13.2. Calculate
the variance of OLS conditional on the observed values of x_t
and show that this does not go to zero. .

(b) Assume, for simplicity, that $|\alpha| < K$ and show that LAD
estimates are consistent. Hint: it is necessary to define
$\rho((x,e),\alpha) = |e - (\alpha - \alpha_0)x| - |e|$ to ensure that assumption
B holds.

7. This exercise should be compared with 11.13. Suppose y_t is generated by the first order autoregressive process,

$$y_t = \alpha_0 y_{t-1} + \epsilon_t.$$

Suppose the equation holds for all $t = 0, \pm 1, \pm 2, \ldots,$ with
$\epsilon_t \overset{i.i.d}{\sim} N(0, \sigma^2)$. We say that y is a stationary autoregressive
process if $\alpha_0 < 1$. Suppose we have observations on y_t for
$t = 0, 1, \ldots, T$. The least squares estimator for α is the value
$\hat{\alpha}_T$ which minimizes $S_T(y_T, \ldots, y_0, \theta) = (1/T) \sum_{t=1}^{T} \rho(y_t, y_{t-1}, \alpha)$,
where $\rho(y_t, y_{t-1}, \alpha) = (y_t - \alpha y_{t-1})^2$.

(a) Show that $\rho(y_t, y_{t-1}, \alpha) = (\epsilon_t + (\alpha_0 - \alpha)y_{t-1})^2$. Also show
that this has identical distribution for all t (Hint: this follows
from the stationarity of y_t).

(b) Show that for each fixed $\alpha \in (-1, 1)$, $S_T(\ldots, \alpha)$ converges
in probability to $\mathbb{E}\rho(y_1, y_0, \alpha)$.

(c) Show that the above convergence must be uniform (in probability) Hint: this follows from Taylor's ULLN (Theorem 12.3
above).

(d) Show that the limit function has a strong maximum at α_0
and conclude consistency of OLS.

8. For $i = 1, 2, \ldots,$ (X_i, Y_i) are independent bivariate normal with
mean vector (μ_i, μ_i) and covariance matrix $\sigma^2 I_2$. Show that the
ML estimate of σ^2 is inconsistent, but there does exist a consistent
estimator of σ^2. This problem has an infinite number of nuisance
parameters and hence does not satisfy standard conditions for
consistency. Hint: first note that the ML of μ_i is $\hat{\mu}_i = (X_i + Y_i)/2$.

9. Consider the following dynamic model:

$$y_t = \alpha_1 y_{t-1} + \cdots + \alpha_p y_{t-p} + \beta_1 x_{t1} + \cdots + \beta_k x_{tk} + \epsilon_t.$$

We wish to estimate the parameters α_i and β_i by least squares. Define $\rho_t(y, x, \alpha, \beta) = (y_t - \sum_{i=1}^p \alpha_i y_{t-i} - \sum_{j=1}^k \beta_k x_{tk})^2$, and $S_T = (1/T)\sum_{t=1}^T \rho_t(y, x, \alpha, \beta)$. Assume that $(y_t, \ldots, y_{t-p}, x_{t1}, \ldots, x_{tk})$ is a stationary and ergodic vector sequence. Then the ergodic law of large numbers permits us to conclude that S_n converges for each α, β to $\mathbb{E}S_n$. Now from Taylor's Weak Law of Large Numbers, since the ρ_t have identical distributions, we can conclude that S_n converges to $\mathbb{E}S_n$ uniformly over any compact set of α, β.

10. Suppose that $y_t = \exp(-\alpha t) + \epsilon_t$; that is, y_t decays exponentially at rate α plus some noise. Show that for any α, α',

$$\sum_{t=1}^{\infty} (\exp(-\alpha t) - \exp(-\alpha' t))^2 < \infty$$

Conclude that α cannot be consistently estimated.

11. Consider the power curve model $y_t = \theta_1 t^{-\theta_2} + \epsilon_t$. Use Wu's result on existence of consistent estimators to decide for which values of θ_1, θ_2 this model can be estimated consistently.

12. To show that the empirical CDF based on estimated residuals converges to the same value as the one based on the true residuals, it is enough to prove that $\mathbf{P}(e_t \le y)$ converges to $\mathbf{P}(\epsilon_t \le y)$ when t is large. Intuitively, this is plausible because as $\hat{\beta}$ converges to β so $y_t - x_t'\hat{\beta} = e_t$ converges to $y_t - x_t'\beta = \epsilon_t$. Formally, show that $e_t = \epsilon_t + x_t'(X'X)^{-1}X'\epsilon = \epsilon_t + Z_t$. Next show that Z_t converges to zero in quadratic mean. Conclude that $\mathbf{P}(e_t \le y)$ converges to $\mathbf{P}(\epsilon_t \le y)$ under the same hypothesis needed for consistency of least squares, namely that the smallest eigenvalue of $(X'X)$ converges to infinity.

13. Consider a regression equation $y_t = x_t \beta + \epsilon_t$, where x_t is a $1 \times K$ vector of constants and β is a $K \times 1$ vector of parameters. Let $e = (e_1, e_2, \ldots, e_T)'$ be the vector of residuals from OLS; $e = y - X\hat{\beta}$,

where $\hat{\beta} = (X'X)^{-1}X'y$. Let $e_1^*, \ldots, e_T^*$ be i.i.d. observation, each being a *resample* of the residuals. This means that for any j, e_j^* equals one of the e_t with each having equal probability $1/T$. Show that $\mathbb{E}\sum_{t=1}^{T} x_t e_t^* = 0$ and that $Cov(\sum_{t=1}^{T} x_t e_t^*) = Var(e_1^*)(X'X)$. Let $\beta^* = (X'X)^{-1}X'y^*$, where $y^* = X\hat{\beta} + e^*$. Consider the original observations y, X and the unobserved errors ϵ as fixed, and let E^* and Cov^* denote expectation and covariance with respect to the randomness induced by the resampling. We will show below that bootstrapping leads to standard estimates of the covariance for $\hat{\beta}$.

(a) Show that $\mathbb{E}\beta^* = \hat{\beta}$.

(b) Show that $Cov^* \beta^* = \nu^2 (X'X)^{-1}$, where ν^2 is the variance of the e^*.

(c) Let β_j^* for $j = 1, \ldots, J$ be a bootstrap sample. Show that $\overline{\beta^*} = (1/J)\sum_{j=1}^{J} \beta_j^*$ will converge to $\hat{\beta}$ and $\frac{1}{J}\sum_{j=1}^{J}(\beta^* - \overline{\beta^*})(\beta^* - \overline{\beta^*})'$ will converge to $c\hat{\sigma}^2(X'X)^{-1}$, where c is a constant close to 1.

The generosity of Hatim Tai was famous in the land, as was his Arabian horse. A king decided to test his generosity be sending him messengers with a request that he be given the horse. Hatim welcomed the guests and had a meal prepared for them. Over dinner, the messengers explained that their mission was to obtain his horse as a gift for the king.

'I am truly sorry, and I wish you had explained your mission earlier,' said Hatim. 'You see, it happened that I had nothing suitable to serve you on your arrival, so I ordered that the horse be killed, and it is his meat we are now eating.'

Chapter 13

Asymptotic Distributions

The tools needed to calculate asymptotic distributions of estimators are quite straightforward. Suppose we can verify that $\hat{\theta}_n$ converge to θ_0. Suppose we can also calculate the covariance matrix (maybe only approximately) $C_n = \text{Cov}(\hat{\theta}_n)$. If regularity conditions hold then the large sample distribution of $\hat{\theta}_n$ can be approximated by $N(\theta_0, C_n)$. For linear functions of errors, the regularity conditions are quite easy. For a more complex function, we use the continuous mapping theorem. We break it down to simpler constituents, e.g., X_n, Y_n, Z_n converging in distribution to X, Y, Z. Then a continuous function $f(X_n, Y_n, Z_n)$ will converge in distribution to $f(X, Y, Z)$. Using these methods we can calculate asymptotic distributions for many estimators and test statistics.

Estimators or test statistics which are implicitly defined by maximizing properties require a variant technique. We use a Taylor's expansion of the first order conditions for the maximum to approximate the estimator via a function of sums of random variables. Once this is done, standard techniques based on the CLT and the continuity principle become available. Thus the essence of the method is quite simple. Unfortunately, the regularity conditions needed to ensure that the remainder terms remain small are complex. The following presentation is geared towards saving some trees by skimping somewhat on the rigor.

The text covers only the finite dimensional CLT. The exclusion of the infinite dimensional case, the functional central limit theorem, constitutes a major gap of coverage in the present text. Many unsolved distributional problems have fallen to this powerful approach. See the review by Andrews (1994) for a survey and many examples of applications. Fortunately, McCabe and Tremayne (1993) have provided an excellent, detailed, and readable book-length exposition of this class of methods.

13.1 Convergence in Distribution

Central limit theorems gives us a way of approximating distributions for a large class of statistics and estimators. Such large sample approximations frequently provide an easy way to compare the properties of various estimators and test statistics. These approximations are also used to provide significance levels of tests, and confidence intervals for estimates, in cases where the exact distributions are not available. Ways of obtaining better approximations are discussed in the following chapter.

In order to state central limit theorems, we must introduce the concept of 'convergence in distribution,' also known as 'weak convergence.' There are many equivalent definitions. The following somewhat abstract definition is now the preferred one for several reasons:

Definition 13.1 (Weak Convergence) *Suppose X_n is a sequence of random variable taking values in space $\mathcal{X}$. If there exists a random variable X such that for every real valued bounded continuous function $f : \mathcal{X} \to \mathbf{R}$, $\lim_{n \to \infty} \mathbb{E}f(X_n) = \mathbb{E}f(X)$ then X_n converges in distri-*

bution *(or* converges weakly*) to* X. *We write* $\lim_{n\to\infty} X_n \overset{\mathcal{D}}{=} X$ *or* $X_n \overset{\mathcal{D}}{\to} X$.

Below we list some properties of weak convergence without proof.

Theorem 13.1 (Properties of Weak Convergence) *Suppose* X_n *converge weakly to* X.

A *For any continuous function* g, $g(X_n)$ *converges weakly to* $g(X)$.

B *Let* $F_n(x) = \mathbf{P}(X_n \leq x)$ *and* $F(x) = \mathbf{P}(X \leq x)$ *be the CDF of* X_n *and* X. *If* $F(x)$ *is continuous at* x, *then* $\lim_{n\to\infty} F_n(x) = F(x)$.

C *If for some* $\delta > 0$ *and* $M < \infty$, $\mathbb{E}|X_n|^{K+\delta} < M$ *for all* n, *then* $\lim_{n\to\infty} \mathbb{E}|X_n|^K = \mathbb{E}|X|^K$.

D *If moment generating functions of* X_n *and* X *are welldefined over an interval around 0 then the moment generating function of* X_n *converges uniformly to the moment generating function of* X *on that interval.*

Conversely, any of the following properties imply the weak convergence of X_n *to* X:

E *The moment generating functions of* X_n *converge to the moment generating function of* X *on an interval.*

F *If* f *is a bounded, infinitely differentiable function with compact support,* $\mathbb{E}f(X_n)$ *converges to* $\mathbb{E}f(X)$.

G *The density* $f_n(x)$ *of* X_n *converges to* $f(x)$ *the density of* X.

H *If* $\lambda : \mathcal{X} \to \mathbf{R}$ *is linear, then* $\lambda(X_n)$ *converges weakly to* $\lambda(X)$

Remarks: Several remarks on these properties of weak convergence are in order. These are listed below in correspondence with the properties.

A This extremely convenient property is called the *continuity principle*. Knowledge of weak convergence of simple functions can,

by the means of continuous mappings, furnish us with information on the behavior of very complex function of random variables. The requirement of continuity for f can be weakened; see
Chapter 4.2 of Pollard (1984) for a more general result. Since
$g_\theta(x) = \exp(i\theta'x)$ is a bounded continuous functions, it follows
that characteristic function of X_n, $f_n(\theta) = \mathbb{E}\exp(i\theta X_n)$, converges to $f(\theta) = \mathbb{E}\exp(i\theta X)$.

B This is the usual definition of weak convergence. However it does
not generalize readily to abstract vector spaces. Note that the
CDFs are only required to converge at points of continuity.

C Note that the definition of weak convergence, with $\mathbb{E}f(X_n)$ converging to $\mathbb{E}X$, says *nothing* about the convergence of moments
of X_n to that of X. This is because the functions $f(x) = x^p$ are
not bounded. Additional conditions which yield convergence of
moments are given here.

D,E Moment generating functions need not exist, but if they do, then
weak convergence implies and is implied by convergence of the
corresponding moment generating functions. A more satisfactory result is obtained by considering the characteristic func-
tions $\psi_n(\theta) = \mathbb{E}\exp(i\theta X_n)$ which are always well defined for
all random variables. In this case, weak convergence is equiv-
alent to the convergence of the characteristic functions $\psi_n(\theta)$ to
$\psi(\theta) = \mathbb{E}\exp(i\theta X)$. See Gnedenko and Kolmogorov (1956) Theorem 1 of §13 for a precise statement and proof.

F A collection of functions $\mathcal{F}$ is called a *convergence determining
class* if the convergence of $\mathbb{E}f(X_n)$ to $\mathbb{E}f(X)$ for all $f \in \mathcal{F}$ guarantees that X_n converges weakly to X. By the definition of weak
convergence, the set of all bounded continuous functions is a convergence determining class. This proposition gives a smaller convergence determining class. According to the previous remark,
the set of function $f_\theta(x) = \exp(i\theta'x)$ indexed by $\theta \in \mathbf{R}^k$ is also
convergence determining.

G Interestingly, weak convergence *does not* imply the convergence
of the density functions (although the distribution functions do

converge, as claimed in [C]). However, if density functions do converge, then weak convergence must hold. Additional conditions which guarantee convergence of the probability density functions are listed in Gnedenko and Kolmogorov (1956) §46.

H   This is called the 'Cramer-Wold' device. It is used to reduce a high-dimensional problem to a one-dimensional one. For example, if X_n and X are vector valued, $\lambda(X_n)$ is a real valued random variable. If we can prove weak convergence of $\lambda(X_n)$ to $\lambda(X)$ for every linear function λ, using one-dimensional techniques or theorems available, we can conclude the weak convergence of the vector X_n to X.

13.2   Cumulant Generating Functions

Valuable tools in obtaining central limit theorems are the *cumulants* and the *cumulant generating function* of a distribution. The extra investment of effort over and above that required by the approach via the moment generating function is handsomely repaid by extra insight into asymptotic behavior of sums of random variables[1] Recall that the moment generating function (MGF) $\mathcal{L}_X(\theta) = \mathbb{E}\exp(\theta X)$ can be expanded as $\mathcal{L}_X(\theta) = 1 + \sum_{j=1}^{\infty} \mu'_j \theta^j / j!$, where $\mu'_j = \mathbb{E}X^j$ is the j-th raw moment of X. From the series expansion, it is clear that the j-th raw moment can be evaluated by differentiating the MGF j times and evaluating at 0:

$$\mu'_j = \mathbb{E}X^j = \mathcal{L}_X^{(j)}(0) \equiv \frac{\partial^j}{(\partial \theta)^j}\mathcal{L}_X(\theta)\bigg|_{\theta=0}.$$

The *cumulant generating function* or CGF $K(\theta, X)$ of X is defined to be the log of the moment generating function, so that $K(\theta, X) = \log(\mathcal{L}_X(\theta))$. The j-th cumulant κ_j of X is defined to be $\kappa_j(X) \equiv K^{(j)}(0, X)$. Exactly as differentiating the MGF and evaluating at 0

---

[1]It is substantially more useful to work with the *cumulant function* defined as the log of the characteristic function, since this avoids unnecessary assumptions regarding the existence of all moments. Nonetheless, we have not taken this route to avoid *complexity*.

yields raw moments, so differentiation the CGF(=log MGF) and evaluating at 0 yields cumulants.

If $X \sim N(\mu, \sigma^2)$ then the MGF of X is $\mathcal{L}_X(\theta) = \exp(\mu\theta + \sigma^2\theta^2/2)$,
and hence the CGF is $K(\theta, X) = \mu\theta + \sigma^2\theta^2/2$. From this it is easily
seen that $\kappa_1(X) = \mu$, $\kappa_2(X) = \sigma^2$ and $\kappa_j(X) = 0$ for all $j > 2$. Thus
the third and higher order cumulants of a distribution can be seen as
measures of its difference from the normal distribution. The widely
used Jarque-Bera (1987) test (see Section 8.2) for normality simply
tests a suitable linear combination of the third and fourth cumulants
for being significantly different from zero. Several properties of the
cumulant generating functions which make them extremely convenient
for use in asymptotic theory dealing with sums are listed below.

Theorem 13.2 (Properties of CGF) *Let $K(\theta, X) = \log \mathbb{E}\exp(\theta X)$
be the CGF of random variable X. Assume it is well defined for all
$\theta \in (-a, a)$ for some $a > 0$. Then the CGF has the following properties:*

$$K(\theta, aX + b) = \theta b + K(a\theta, X)$$

$$K\left(\theta, \sum_{i=1}^{n} X_i\right) = \sum_{i=1}^{n} K(\theta, X_i)$$

$$K(\theta, X) = \sum_{j=1}^{\infty} \kappa_j(X)\theta^j/j!$$

*It follows that $\kappa_j(aX) = a^j\kappa_j(X)$ and that $\kappa_j(\sum_{i=1}^{n} X_i) = \sum_{i=1}^{n} \kappa_j(X_i)$,
so that the cumulants cumulate in sums. Also $\kappa_j(X + b) = \kappa_j(X)$ if
$j > 1$, while $\kappa_1(X + b) = \kappa_1(X) + b$.*

The proof is straightforward and left to the reader. We now turn
to the relationship between the cumulants and the central moments
$\mu_j(X) = \mathbb{E}(X - \mu)^j$. To study this, suppose that $\mu = \mathbb{E}X$ and
let $Y = X - \mu$ be the centered variable. Note that (*) $\mathcal{L}_Y(\theta) =
\exp(-\theta\mu)\mathcal{L}_X(\theta)$, and that the raw moments of Y are the centered mo-
ments of X: $\mu'_j(Y) = \mu_j(X)$. Multiplying both sides of (*) by $e^{\theta\mu}$ and
taking logs yields the following expression for the cumulant generating
function of X:

$$K(\theta, X) = \theta\mu + \log(\mathcal{L}_Y(\theta)) = \theta\mu + \log\left(1 + \sum_{j=1}^{\infty} \mu_j(X)\frac{\theta^j}{j!}\right).$$

Using the definition of cumulants $\kappa_j(X) = K^{(j)}(0, X)$ and a little differentiation yields the following relationships:

$$\kappa_1 = \mu = \mathbb{E}X, \; \kappa_2 = \mu_2 = \mathrm{Var}(X), \kappa_3 = \mu_3, \kappa_4 = \mu_4 - 3\mu_2^2.$$

See also exercise 2 on page 384. It is frequently convenient to calculate the cumulants for a standardized variable, to eliminate dependence on the units of measurement. If σ is the standard deviation of X, it is easily seen that $\kappa_j((X - \mu)/\sigma) = \kappa_j(X)/\sigma^j$ for $j > 2$, while $\kappa_1 = 0$ and $\kappa_2 = 1$ for the standardized variable. The third cumulant of the standardized variable (i.e. $\kappa_3(X)/\sigma^3 = \kappa_3(X)/\kappa_2^{3/2}(X)$ is called the *skewness* and the fourth cumulant $\kappa_4(X)/\sigma^4 = \kappa_4(X)/\kappa_2^2(X)$ is called the *kurtosis*. Since the third cumulant vanishes for all symmetric distributions, skewness is a measure of the lack of symmetry. Kurtosis is a measure of the heaviness of the tails of the distribution in comparison with the tail of the normal — it is standardized so that the normal has kurtosis 0, while heavier tailed distributions have positive and lighter tailed distributions have negative kurtosis. For example, a Gamma density has tails going to zero at the rate $\exp(-\lambda x)$, which is slower than $\exp(-x^2/2)$ of the normal, so it is *leptokurtic*, another name for positive kurtosis. The uniform density has no tail at all, and hence is *platykurtic*, another name for negative kurtosis.

13.3 Central Limit Theorems

If X_i is an i.i.d. sequence of random variables, we obtain a strong law of large numbers under the sole assumption that $\mathbb{E}X_1$ exists (more precisely: $\mathbb{E}|X_1| < \infty$, see Theorem 11.7). Similarly, for an i.i.d. sequence, the central limit theorem takes a particularly simple form. If $\mathbb{E}X_i = \mu$ and $\mathrm{Var}(X_i) = \sigma^2$ then it can be concluded that

$$\lim_{n\to\infty} \frac{1}{\sqrt{n}} \sum_{j=1}^{n} (X_i - \mu) \overset{D}{=} N(0, \sigma^2).$$

Somewhat stronger hypotheses are needed in the case of i.n.i.d. variables. An essential hypothesis in all versions of the central limit theorem is that each individual term in the sequence should be small in relation to the sum. This is necessary since if any of the terms is large, then its density (which is not assumed normal) will contribute to the asymptotic density, thereby preventing the asymptotic distribution from being normal. For i.i.d. sequences each term is symmetric and hence has a proportional influence of $(1/n)$ in the sum. In the i.n.i.d. case, the central limit theorem takes the following form:

Theorem 13.3 (Lyapunov) *Suppose that X_i are independent with $\mu_i = \mathbb{E}X_i$ and $\sigma_i^2 = Var(X_i)$. Assume that*

$$\lim_{n \to \infty} \frac{\sum_{j=1}^{n} \mathbb{E}|X_i - \mathbb{E}X_i|^3}{\left(\sum_{j=1}^{n} \sigma^2\right)^{3/2}} = 0. \tag{13.1}$$

Then

$$\lim_{n \to \infty} \frac{\sum_{j=1}^{n}(X_j - \mu_j)}{\left(\sum_{j=1}^{n} \sigma_j^2\right)^{1/2}} \overset{D}{=} N(0, 1).$$

Remarks: The hypothesis of this theorem is somewhat stronger than necessary. The number '3' in the numerator and denominator of Eq. (1.1) can be replaced by '2+ϵ' for any $\epsilon > 0$. Even weaker is the Lindeberg-Feller condition, which is more or less the best possible. However, the above hypothesis is enough for nearly all applications. For a lucid and detailed discussion of the central limit theorem and all variations of hypotheses possible in the case of independent variables, the reader is referred to the excellent monograph of Gnedenko and Kolmogorov (1954). The relation between Eq. (1.1) and the condition of negligibility of individual terms is clarified in exercise 4.

Proof: The argument to be given lacks rigor because of the sloppy treatment of the remainder term. See exercise 3 for the missing details. The odd-looking hypothesis of Eq. (1.1), and the theorem itself, fall right out of an analysis via the cumulant generating function. Define $\sigma_n^2 = \sum_{i=1}^{n} \sigma_i^2$. Let $S_n^* = \sum_{i=1}^{n}(X_i - \mu_i)/\sigma_n$ be the *standardized* sum; here and elsewhere, standardization of a random variable means subtracting the mean and dividing be the standard deviation. We will show that the cumulant generating function of S_n^* converges to $\theta^2/2$

which is the CGF of the $N(0,1)$ density. Using Theorem 1.2, we get

$$K(\theta, S_n^*) = K\left(\theta/\sigma_n, \sum_{i=1}^{n}(X_i - \mu_i)\right) = \sum_{i=1}^{n} K(\theta/\sigma_n, X_i - \mu_i)$$

$$= \sum_{i=1}^{n} -\mu_i\theta + \left(\frac{\kappa_1(X_i)\theta}{\sigma_n} + \frac{\kappa_2(X_i)\theta^2}{2!\sigma_n^2} + +\frac{\kappa_3(X_i)\theta^3}{3!\sigma_n^3} + \cdots\right).$$

Now recall that $\kappa_1(X_i) = \mathbb{E}X_i = \mu_i$, $\kappa_2(X) = \mathbb{V}ar(X)$ and $\kappa_3(X) = \mathbb{E}(X - \mathbb{E}X)^3$. The first term, corresponding to θ, is 0, and the second term corresponding to $\theta^2/2$ is 1, since the variables have been standardized to have mean 0 and variance 1. The coefficient of the third term θ^3 is $\sum_{i=1}^{n}\kappa_3(X_i)/\sigma_n^3$. In order to make sure that this and all higher terms go to zero, we have to replace $\kappa_3(X_i)$ by the absolute third cumulant, $\mathbb{E}|X_i - \mathbb{E}X_i|^3$. Upon doing this, we see that the hypothesis of the Lyapunov theorem is precisely that the coefficient of the third term goes to zero. It follows that the CGF converges to $\theta^2/2$ asymptotically, proving the theorem. For a rigorous treatment of the remainder, see exercise 3.

In most applications, especially in econometrics, a slightly more general result is needed than the Lyapunov central limit theorem given above. Usually this is stated in terms of a double array of random variables. The version given below, due to Eicker (1966b), is more directly adapted to econometric applications. Let ϵ be a $T \times 1$ vector of independent random variables. Assume $\mathbb{E}\epsilon = 0$ and let Σ be the (diagonal) $T \times T$ covariance matrix of ϵ. Let M be a $K \times T$ matrix of full rank, where K is fixed. We consider conditions under which $M\epsilon$ will be asymptotically normal as $T \to \infty$. As usual, the notation $M^{1/2}$ will refer to a positive semidefinite root of a positive semidefinite matrix.

Theorem 13.4 (Eicker) *Let $A = (M\Sigma M')^{-1/2}$ so that $\mathbb{C}ov(AM\epsilon) = \mathbf{I}$. If the following conditions hold then $AM\epsilon$ will be asymptotically $N(0, \mathbf{I}_K)$ as T goes to infinity.*

1. *For some constants b, B, $0 < b < \mathbb{V}ar(\epsilon_t) < B < \infty$ holds for all t.*

2. *There exists $B' < \infty$ such that for all t $\mathbb{E}|\epsilon_t|^3 < B'$.*

3. *The hat matrix $H = M'(MM')^{-1}M$ is asymptotically balanced;* *that is,*

$$\lim_{T \to \infty} \max_{1 \le t \le T} H_{tt} = 0.$$

Remarks: The assumption that the third moments are bounded can be weakened to the $2 + \epsilon$ moment. It can be further weakened to the uniform integrability of ϵ_t^2, and this last is a necessary and sufficient condition. Thus, roughly speaking, the theorem requires a hypothesis slightly stronger than that all variances are bounded above. It is worth noting that if the third absolute moments are bounded above, the variances are necessarily bounded above, so that the assumption that $\mathbb{V}ar(\epsilon_t) < B$ is redundant. The condition (3) is interpreted in terms of vanishing leverage for individual error terms in the following section. This theorem follows directly from Lyapunov's. A proof is sketched in exercise 5.

13.4   Asymptotic Distribution of OLS

We now apply Eicker's theorem to get asymptotic normality of the least squares estimators under the assumption of independent errors. Let $y = X\beta + \epsilon$, where y, ϵ are $T \times 1$, X is $T \times K$, and β is $K \times 1$. We want to study the asymptotic distribution of $\hat{\beta} = (X'X)^{-1}X'y$. All quantities involved (i.e. $y, X, \hat{\beta}, \epsilon$) depend on the sample size T, but it is convenient to suppress this dependence in the notation. We have seen that if $\epsilon \sim N(0, \sigma^2 I)$ then the distribution of $\hat{\beta}$ is $N(\beta, \sigma^2(X'X)^{-1})$. We will see that this result is fairly robust, in the sense that it will continue to be approximately true under substantially weaker assumptions. The asymptotic theory also provides information on the conditions which are needed for this result to hold, and when we can expect this result to fail.

Define $H = X(X'X)^{-1}X'$ to be the $T \times T$ projection matrix. H is sometimes called the *hat* matrix, since $Hy = \hat{y}$ so the action of H puts the hat on the y. Of critical importance are the diagonal entries H_{tt} of the hat matrix. H_{tt} was introduced in Chapter 5 as a measure of the *leverage* of the t-th observation. It is desirable for robustness that leverages be small. This is also a key condition for the asymptotic normality of the OLS estimator. As a finite sample counterpart to the asymptotically balanced condition introduced in the previous theorem, we will say that a matrix of regressors is *balanced* if $\max_t H_{tt}$ is small. Since we have not defined exactly what we mean by small, this is just a qualitative term. Note that $\sum_{t=1}^{T} H_{tt} = \text{tr}(H) = K$, and H is positive semidefinite so that $H_{tt} \geq 0$. It follows that $H_{tt} = K/T$ is the smallest possible value for the maximum diagonal entry.

Another condition needed for the asymptotic distribution of OLS to be valid is homoskedasticity, which refers to the equality of the variances of the error terms. The condition needed is actually substantially less restrictive:

Definition 13.2 (Generalized Homoskedasticity) *We will say that errors ϵ are* homoskedastic *relative to regressors X if $\text{Cov}(\epsilon) = \Sigma$ and for some constant $\overline{\sigma}^2$*

$$\lim_{T \to \infty} (X'X)^{-1/2} X' \left(\Sigma - \overline{\sigma}^2 \mathbf{I}_T \right) X(X'X)^{-1/2} = 0.$$

In intuitive terms, homoskedasticity of ϵ relative to the regressors X means that fluctuations in variances of the ϵ_t bear no relationship to the cross-products of the regressors. For example, when there is only one regressor x, generalized homoskedasticity relative to x calls for

$$\lim_{T \to \infty} \frac{\sum_{t=1}^{T} (\sigma_t^2 - \overline{\sigma}^2) x_t^2}{\sum_{t=1}^{T} x^2} = 0$$

This is very close to asking for the sample correlation between x^2 and the variances to be zero. For multiple regressors, the condition has a similar intuitive interpretation. Note that if $\sigma_t = a z_t$ where z is a random variable independent of x, then the errors will be heteroskedastic, but will still satisfy the condition of generalized homoskedasticity relative to x.

Theorem 13.5 (Asymptotic Normality of OLS) *Consider a regression model $y = X\beta + \epsilon$. Suppose that the errors ϵ_t are independent and homoskedastic relative to the regressors X. Suppose the third absolute moment of the errors is uniformly bounded: that is, for all t $\mathbb{E}|\epsilon_t|^3 < \infty$. Suppose the design matrix is asymptotically balanced. Then $(X'X)^{-1/2}(\hat{\beta} - \beta)$ converges in distribution to $N(0, \mathbf{I})$.*

The proof is straightforward; see exercise 6. The conditions given above are more or less necessary and sufficient for this result to hold. If the regressors are not balanced asymptotically, this means that individual observations will have influence on the final distribution of $\hat{\beta}$. To get normality, we can either exclude observations with high leverage (and possibly lose valuable information), or else assume (possibly incorrectly) that the high leverage observations are normal themselves. If the errors fail to be generalized homoskedastic, an asymptotic normality result still holds, but the asymptotic covariance matrix of OLS will be different from the simple $\sigma^2(X'X)^{-1}$. We turn to this case next.

13.4.1 OLS Covariance with Heteroskedasticity

If the error covariance matrix is Σ, it is easily seen that the covariance of $\hat{\beta} = \beta + (X'X)^{-1}X'\epsilon$ is the matrix $C = (X'X)^{-1}X'\Sigma X(X'X)^{-1}$. It can be shown that the OLS is asymptotically normal with covariance matrix C.

Theorem 13.6 (Eicker-White) *Suppose there exist constants b and B such that for all t, $0 < b < \mathbb{E}\epsilon_t^2$ and $\mathbb{E}|\epsilon_t^3| < B < \infty$. Also suppose that the largest diagonal element of $H = X(X'X)^{-1}X'$ goes to zero as T goes to ∞. Define $C = (X'X)^{-1}X'\Sigma X(X'X)^{-1}$. Then $C^{-1/2}(\hat{\beta} - \beta)$ converges in distribution to $N(0, \mathbf{I}_K)$.*

Remarks: We omit the proof, which follows directly from Eicker's Central Limit Theorem 1.4 above. Note the additional assumption of a lower bound on error variances, which is not needed in the case of homoskedasticity. The fact that $C^{-1/2}(\hat{\beta} - \beta)$ is approximately $N(0, \mathbf{I})$ is not directly useful in applications since the matrix C involves the unknown variances in Σ. However, we have shown earlier that $C^{1/2}$ can be consistently estimated in Theorem 11.6. There we found a matrix D involving no unknown parameters such that $DC_T^{1/2}$ converges to $\mathbf{I}_K$ in probability. This means that $DC^{1/2}C^{-1/2}(\hat{\beta} - \beta)$ also converges weakly to $N(0, \mathbf{I})$ and hence $\hat{\beta}$ is asymptotically normal with mean β and covariance $D^{-1}D'^{-1}$. This gives a usable asymptotic expression for the covariance of $\hat{\beta}$ in the presence of heteroskedasticity. Empirical/Monte Carlo investigation has shown that these asymptotics typically works well only in very large samples and in small samples the covariance matrix estimate is seriously biased. Useful proposals for improving the small performance by bootstrapping have been made and this is an active research area.

We have two possibilities for the asymptotic covariance of OLS. In the homoskedastic case, it is $(X'X)^{-1}X'(\hat{\sigma}^2\mathbf{I})X(X'X)^{-1}$, while in the heteroskedastic case it is $(X'X)^{-1}X'\hat{\Sigma}X(X'X)^{-1}$, where $\hat{\Sigma}_{tt} = e_t^2 = |y_t - \hat{\beta}'x_t|^2$. Since the heteroskedastic case includes the homoskedastic one as a special case, we could simply use the second covariance matrix estimate in all situations. The second estimate of the covariance is consistent in both cases (that is, whether or not homoskedasticity holds). However, since the Eicker-White estimate is poor in small to intermediate samples, it is worth making an initial test for heteroskedasticity to see if an adjustment for heteroskedasticity is called for. A straightforward test suggested by White (1980) is to compare the two estimators of the covariance. A version of this test is developed in exercise 7 on page 388. In particular, we could compare $X'\hat{\sigma}^2\mathbf{I}X$ with $X'\hat{\Sigma}X$ and check to see if these random variables are significantly different from each other. It is important to note that the null hypothesis for this test is generalized homoskedasticity relative to the regressors X. As indicated earlier, the errors could actually be heteroskedastic and have variances dependent on other random variable — as long as the determinants of the heteroskedasticity are uncorrelated with the products X_iX_j of any two regressors, generalized homoskedasticity will hold. We emphasize this point since there is a common misunderstanding regarding the test that it is effective for all types of heteroskedasticity. In fact it detects only that type of heteroskedasticity which has an effect on the covariance matrix of the OLS estimators.

13.5 Asymptotics in Dynamic Models

We have established the asymptotic normality of the least squares estimator in linear regression models with nonstochastic regressors in the previous section. An extensive survey of asymptotic distribution theory for linear and nonlinear dynamic models is given by Pötscher and Prucha (1991b). Below we give the bare essentials of the required results. The assumptions needed for asymptotic normality change somewhat in the case of stochastic regressors. With nonstochastic regressors we saw (in Theorem 1.5) that the diagonals of the hat matrix $X(X'X)^{-1}X'$ need to go to zero. However, this condition does not suffice for asymptotic normality with stochastic regressors. Instead we need the following new (stronger) conditions:

Definition 13.3 *A sequence of* $p \times p$ *stochastic matrices* M_n *for* $n = 1, 2 \ldots$ *is asymptotically* stable *if*

$$\underset{n \to \infty}{\boldsymbol{plim}}\, (\mathbb{E}M_n)^{-1} M_n = \mathbf{I}_p.$$

A sequence of random vectors $x_t \in \mathbf{R}^p$ *is asymptotically* stable *if the matrices* $M_n = \sum_{j=1}^{n} x_j x_j'$ *are stable. The sequence is asymptotically* balanced *if, in addition,*

$$\underset{n \to \infty}{\boldsymbol{plim}}\, \max_{1 \leq i \leq n} \| (\mathbb{E}M_n)^{-1} x_i \| = 0.$$

The following results on asymptotic normality of least squares in dynamic models were obtained by Lai and Wei:

Theorem 13.7 *In the regression model* $y_t = \beta' x_t + \epsilon_t$, *suppose the errors form an innovations process. Assume that the regressors (which may include lagged dependent variables and other stochastic regressors with lags) are stable and balanced. Assume that the variance of the innovations* ϵ_t *converges to some constant* σ^2 *asymptotically. Then the vector* $(X_T' X_T)^{-1/2} (\hat{\beta}_T - \beta)$ *converges in distribution to* $N(0, \sigma^2 \mathbf{I}_p)$.

The import of this theorem is that we may approximate the asymptotic distribution of $\hat{\beta}$ in the usual manner in dynamic models; that is $\hat{\beta} \sim N(\beta, \sigma^2 (X'X)^{-1})$ in large samples. Lai and Wei show by a counterexample that in models with unit roots, the condition for stability of the regressors fails. In such models, the OLS fails to have an asymptotic normal distribution (even though the diagonals of the hat matrix go to zero and Eicker's conditions hold). They also show that the usual estimator of the variance $\hat{\sigma}^2 = (1/T) \sum_{t=1}^{T} (y_t - x_t' \hat{\beta}_T)^2$ converges to σ^2 provided that

$$\lim_{T \to \infty} \frac{1}{T} \log(\lambda_{max}(X_T' X_T)) = 0.$$

This is a very weak condition, compared to usual requirements. If this condition holds, and the dynamic model does not have unit roots, then we can safely approximate the asymptotic covariance of $\hat{\beta} - \beta$ by $\hat{\sigma}^2 (X'X)^{-1}$ exactly as in the case of nonstochastic regressors.

For practical purposes, the above theorem is sufficient. It tells us when we can use standard asymptotic theory in dynamic models provided that the process has roots less then one. Things get complicated in situations where the process has unit roots or roots larger than one. Standard asymptotic theory does not hold, and nonnormal asymptotic distributions emerge. Problems in this area are usually handled using functional central limit theory, which we do not cover in this text.

13.6 Asymptotics for M-Estimators

If an explicit formula for an estimator is available, several techniques are available to determine its asymptotic distribution. There are many cases where the estimator is defined as a value maximizing a given function; such estimators are called M-estimators. Fortunately, one can determine an asymptotic distribution for such estimators without having an explicit formula. The key is to use a Taylor series expansion to represent an estimator as a function of sums of random variables. We first develop this representation.

If θ is a k-dimensional parameter, we will use $\nabla_\theta f(x, \theta)$ as notation for the $k \times 1$ vector of partial derivatives $\partial f / \partial \theta_i$ $i = 1, \ldots, k$ and $\nabla_\theta \nabla_\theta' f(x, \theta)$ as notation for the $k \times k$ matrix of second partial derivatives $\partial^2 f(x, \theta) / \partial \theta_i \partial \theta_j$.

Let X_i be a sequence of independent random variables taking values in sample space $\mathcal{X}$ and let $\Theta \in \mathbf{R}^k$ be the parameter space. For $x \in \mathcal{X}$ and $\theta \in \Theta$, suppose we have a function $\rho(x, \theta)$ which takes real values. Let $S_n(\theta) = \sum_{j=1}^n \nabla_\theta \rho(X_i, \theta)$ be the $k \times 1$ (random) vector of first derivatives of the criterion function. Let $M_n(\theta) = \sum_{i=1}^n \nabla_\theta \nabla_\theta' \rho(X_i, \theta)$ be the $k \times k$ (random) matrix of second partial derivatives of the criterion function. In order to obtain the basic representation, we will need to assume a condition related to, but stronger than, the stability condition introduced earlier. Let $B_n(\theta) = \mathbb{E} M_n(\theta)$. Stability of $M_n(\theta)$ requires that $B_n(\theta)^{-1} M_n(\theta)$ converges to $\mathbf{I}_k$. We will need to assume that this convergence is locally uniform in θ. An equivalent way of formulating the assumption is more transparently usable in the proof. We define $M_n(\theta)$ to be a *locally uniformly stable* sequence of matrices if for any consistent sequence $\hat{\theta}_n$ of estimators of θ, $B_n(\hat{\theta}_n)^{-1}M_n(\hat{\theta}_n)$ converges to $\mathbf{I}_k$ in probability. This condition is also equivalent to the *stochastic equicontinuity* of $M_n(\theta)$, discussed in detail in Pötscher and Prucha (1994). It is important to note that local uniform stability of $M_n(\theta)$ implies that the expectation $\mathbb{E}M_n(\theta)$ is continuous in θ, as is easily established.

Given this notation and terminology, we can get the basic representation result. For the remainder of this chapter, we will say A_n is approximately B_n and write $A_n \approx B_n$ whenever the difference $A_n - B_n$ converges to zero in probability.

Theorem 13.8 (Representation of M-estimators) *Let $\hat{\theta}_n$ be a sequence of maxima of the sums $\sum_{i=1}^n \rho(X_i, \theta)$. Suppose that $\hat{\theta}_n$ is consistent for θ_0, and θ_0 is in the interior of the parameter space Θ. Suppose the matrix of second partial derivatives of the criterion function $M_n(\theta) = \sum_{j=1}^n \nabla_\theta \nabla'_\theta \rho(X_j, \theta)$ is locally uniformly stable. Then $\hat{\theta}_n$ can be approximated as follows:*

$$\sqrt{n}(\hat{\theta}_n - \theta_0) \approx -\left(\frac{1}{n}\mathbb{E}M_n(\theta_0)\right)^{-1}\frac{1}{\sqrt{n}}S_n(\theta_0).$$

Remarks: The assumption of local uniform stability avoids the necessity of assuming existence of higher derivatives of ρ. It is also very close to being necessary and sufficient. Exercise 7 gives more transparent and more easily verifiable versions of this assumption. Since the hypothesis may be hard to verify, or fail, but the approximation may be valid, in practice one can test the approximation by empirical (for example, Monte Carlo) methods in cases of doubt, or simply assume the approximation holds, with little risk of going wrong.

Proof: Since $S_n(\theta)$ is the vector of first partial derivatives, and $\hat{\theta}_n$ is a maximum, we must have $S_n(\hat{\theta}_n) = 0$ by the first order conditions for maxima. Thus for some ξ on the line segment between θ_0 and $\hat{\theta}$ the following Taylor series representation must hold:

$$0 = S_n(\hat{\theta}_n) = S_n(\theta_0) + M_n(\xi)(\hat{\theta}_n - \theta_0),$$

so that (*) $M_n(\xi)(\hat{\theta}_n - \theta_0) = -S_n(\theta_0)$. Premultiply both sides of (*) by the product of the following three matrices:

$$E_1 E_2 E_3 \equiv (\mathbb{E}M_n(\theta_0))^{-1} \, \mathbb{E}M_n(\theta_0) \, (\mathbb{E}M_n(\xi))^{-1} \,.$$

Since ξ must converge to θ_0 (since $\hat{\theta}_n$ does), it follows from the continuity of $\mathbb{E}M_n(\theta)$ that $E_2 E_3 = \mathbb{E}M_n(\theta_0)(\mathbb{E}M_n(\xi))^{-1}$ is approximately the identity matrix, so that $E_1 E_2 E_3$ is approximately $E_1 = \mathbb{E}M_n(\theta_0)$. This is the term multiplying $S_n(\theta_0)$. Multiplying $\hat{\theta}_n - \theta_0$ is the product of four matrices $E_1 E_2 E_3 M_n(\xi)$. Now $E_3 M_n(\xi)$ converges to identity by the locally uniform stability of M_n. Since $E_1 E_2$ is the identity, the theorem follows directly.

From this representation, it is easy to get the asymptotic distributions of M-estimators. We note that the i.i.d. case is substantially simpler. If $\mathbb{E}\rho(X_1, \theta)$ is maximized at θ_0 then it will (typically) be the case that $\mathbb{E}S_n(\theta_0) = 0$ so that a CLT will yield the asymptotic distribution of $S_n(\theta_0)/\sqrt{n}$ to be $N(0, \mathbb{C}ov(\nabla_\theta \rho(X_1, \theta)))$. Also $(1/n)\mathbb{E}M_n(\theta_0)$ is a deterministic sequence which equals $\mathbb{E}\nabla_\theta\nabla'_\theta\rho(X_1, \theta)$. The more general form has been given here because the i.n.i.d. case is much more common in econometric applications.

13.7 The Case of Maximum Likelihood

We now use this representation to get the asymptotic distribution for the maximum likelihood estimator in the simplest case of i.i.d. observations. The *Fisher Score Function*, or just the *score function* is defined as the vector of partial derivatives $S(X, \theta) = \nabla_\theta \log f^X(x, \theta)$, for any random variable X with density $f^X(x, \theta)$. The asymptotic properties of the ML estimator fall out directly from the representation of Theorem 1.8 and certain properties of the score function which we develop below.

Lemma 13.1 (Properties of the Score Function) *Let* $S(X, \theta) = \nabla_\theta \log f^X(x, \theta)$, *and suppose the true density of* X *is* $f^X(x, \theta_0)$. *Then* $\mathbb{E}S(X, \theta_0) = 0$ *and* $\mathbb{E}\nabla_\theta \nabla'_\theta \log f^X(X, \theta) = -\mathbb{C}ov(S(X, \theta_0))$.

Proof: The proof of both assertions depends on interchanging the order of operation of differentiation with respect to θ and taking expectation with respect to X. This interchange is valid provided that the set of values for which $f^X(x, \theta) > 0$ is independent of θ and integrals on both sides of the interchange are welldefined and continuous. Note that $\nabla_\theta \log f^X(x, \theta_0) = (1/f^X(x, \theta_0)\nabla_\theta f^X(x, \theta_0$. It follows that

$$
\begin{aligned}
\mathbb{E}S(X, \theta) &= \mathbb{E}(1/f^X(x, \theta_0))\nabla_\theta(f^X(X, \theta_0) \\
&= \int \left\{(1/f^X(x, \theta_0))\nabla_\theta(f^X(X, \theta_0))\right\} f^X(x, \theta_0)\, dx \\
&= \int \nabla_\theta f^X(x, \theta_0)\, dx = \nabla_\theta \int f^X(x, \theta_0)\, dx = \nabla_\theta(1) = 0.
\end{aligned}
$$

The second equality is similar. We have

$$
\begin{aligned}
\mathbb{E}\nabla_\theta S(X, \theta) &= \mathbb{E}\nabla_\theta \frac{\nabla'_\theta(f^X(X, \theta_0))}{f^X(X, \theta_0} \\
&= \mathbb{E}\frac{\nabla_\theta \nabla'_\theta f^X(X, \theta_0)}{f^X(X, \theta_0)} - \frac{\left[\nabla_\theta f^X(X, \theta_0)\right]\left[\nabla'_\theta f^X(X, \theta_0)\right]}{\{f^X(X, \theta_0)\}^2}.
\end{aligned}
$$

By interchanging ∇_θ and ∇'_θ with $\mathbb{E}$, the first term is easily shown to be zero. The second term is easily seen to be equivalent to $\mathbb{E}SS' = \mathcal{C}ov\, S$. This proves the lemma.

The covariance of the score function plays an important role in asymptotic theory, and is called the *Information matrix* $\mathcal{I}(X, \theta)$. We can summarize the contents of the lemma as follows:

$$
\mathcal{I}(X, \theta) \equiv \mathbb{E}S(X, \theta)S'(X, \theta) = \mathcal{C}ov\, S(X, \theta) = -\mathbb{E}\nabla_\theta\nabla'_\theta \log f(X, \theta).
$$

Given a sequence of independent variables $X_1, X_2, \ldots$, we will use $X^{(n)}$ as notation for the first n, so that $X^{(n)} = (X_1, \ldots, X_n)$. It follows immediately from the definition of the score function that for independent variables $S(X^{(n)}, \theta) = \sum_{i=1}^n S(X_i, \theta)$ If the X_i are i.i.d. then $(1/\sqrt{n}S(X^{(n)}, \theta)$ must converge to $N(0, \mathcal{I}(X_1, \theta))$ by the CLT for i.i.d. case and the lemma above. We can now get the asymptotic distribution of the ML easily.

Theorem 13.9 *Suppose X_i are i.i.d. with density $f^X(x, \theta_0)$ and $\hat{\theta}_n$ maximizes the log likelihood $\sum_{j=1}^n \log f^X(X_j, \theta)$. Define $M_n(\theta) = \sum_{i=1}^n \nabla_\theta \nabla'_\theta \log f^X(X_i, \theta)$ and assume that for some neighborhood $N(\theta_0)$ and some integer k $\mathbb{E}\sup_{\psi \in N(\theta_0)} M_k(\psi) < \infty$. If $\hat{\theta}_n$ is consistent for θ_0, then $\sqrt{n}(\hat{\theta}_n - \hat{\theta})$ is asymptotically $N(0, \mathcal{I}_1^{-1}(X_1, \theta_0))$.*

Remarks: The assumptions made above are not quite enough to guarantee consistency of $\hat{\theta}_n$, and that is why consistency is directly assumed. If $\mathbb{E}S_n(\theta)$ has a unique 0 at θ_0, local consistency is ensured. For global consistency see Theorem 12.4 for a full list of conditions needed.

Proof: According to the representation given in Theorem 7,

$$\sqrt{n}(\hat{\theta}_n - \theta_0) \approx \left(\frac{1}{n}\mathbb{E}M_n(\theta_0)\right)^{-1}\frac{1}{\sqrt{n}}S_n(\theta_0).$$

As already discussed, $S_n(\theta_0)/\sqrt{n}$ converges to $N(0, \mathcal{I}(X_1, \theta_0))$. Next we have

$$\mathbb{E}M_n(\theta_0) = \sum_{j=1}^{n}\mathbb{E}\nabla_\theta\nabla'_\theta \log f^X(X_j, \theta_0) = -n\mathcal{I}(X_1, \theta_0).$$

Thus $(1/n)\mathbb{E}M_n(\theta_0) = -\mathcal{I}(X_1, \theta_0)$ and the theorem follows immediately using the continuity principle.

13.8   Likelihood Ratio Test

We have already used the likelihood ratio earlier. We will see in Chapter 15 that this test has several optimality properties asymptotically. It is always an invariant test and frequently (but not always) best invariant; see exercises 17 and 18 of Chapter 6, and also Chapter 8.8 of Lehmann (1986). When the finite sample distribution of the likelihood ratio statistic is known, or easily calculated, one obtains an exact test using this principle. When the distribution is not known, one is faced with the problem of selecting the constant c such that we will reject the null for LR$> c$. One of the traditional solutions to this problem is to calculate an asymptotic distribution for LR and use this approximate large sample distribution to find the constant c for which $\mathbf{P}(LR > c) = .05$ or some other chosen significance level under the null hypothesis. We derive below the asymptotic distribution of the LR in some standardized cases. We will study how to improve the accuracy of the asymptotic approximation via Bartlett corrections in the next chapter.

13.8.1   No Nuisance Parameters

We first present the asymptotic distribution of the LR for a simple null hypothesis in the case of no nuisance parameters. Suppose X_i is an i.i.d sequence of random vector with common density $f^X(x, \theta)$ for some $\theta \in \Theta$ where the parameter space Θ is an open subset of $\mathbf{R}^k$. We wish to test the null hypothesis $H_0 : \theta = \theta_0$ versus the alternative $H_1 : \theta \neq \theta_0$. As earlier, we will use the notation $S(X^{(n)}, \theta_0) = \sum_{i=1}^{n} \nabla_\theta \log f^X(X_i, \theta_0) = \nabla_\theta \log f(X^{(n)}, \theta)$ to mean the $k \times 1$ vector of partial derivatives of the joint density of n observations with respect to θ, evaluated at the point θ_0. We will also use $M_n(\theta) = \sum_{i=1}^{n} \nabla_\theta \nabla'_\theta \log f^X(X_i, \theta)$. Let $\hat{\theta}_n^{ML}$ maximize the log likelihood $\sum_{i=1}^{n} \log f^X(X_i, \theta)$, and note that the likelihood ratio test statistic is defined to be $LR \equiv f(X^{(n)}, \theta_0) / f(X^{(n)}, \hat{\theta}_n^{ML})$ so that $LR \leq 1$.

Theorem 13.10 *Suppose that $M_n(\theta)$ is locally uniformly stable. If the null hypothesis is true and the ML estimates $\hat{\theta}_n^{ML}$ are consistent then the statistic $\mathcal{L} = -2\log(LR)$ has an asymptotic chi-square distribution with k degrees of freedom under the null hypothesis.*

Proof: A somewhat careful treatment of the error terms has been given in the proof of Theorem 1.8 and in exercise 7. For the rest of this chapter we will Taylor with abandon and throw the remainders to the winds. Tayloring $\log f(X^{(n)}, \theta_0)$ around $\hat{\theta}_n^{ML}$ yields

$$\log f(X^{(n)}, \theta_0) \approx \log f(X^{(n)}, \hat{\theta}_n^{ML}) + (\theta_0 - \hat{\theta}_n^{ML})' S(X^{(n)}, \hat{\theta}_n^{ML})$$
$$+ \frac{1}{2}(\theta_0 - \hat{\theta}_n^{ML})' M_n(\theta_0)(\theta_0 - \hat{\theta}_n^{ML}).$$

Now the middle term on the RHS is zero since $S(X^{(n)}, \hat{\theta}_n^{ML}) = 0$ is the first order condition for $\hat{\theta}$ to maximize the likelihood. It follows immediately that in large samples $\mathcal{L} = -2\log(LR)$ satisfies:

$$\mathcal{L} \approx -\left\{ \sqrt{n}(\hat{\theta} - \theta_0) \right\}' \left(\frac{1}{n} M_n(\theta_0) \right) \left\{ \sqrt{n}(\hat{\theta} - \theta_0) \right\}.$$

If $X \sim N_k(0, \Sigma)$ then $X'\Sigma^{-1}X \sim \chi_k^2$. Now just use the fact that $\sqrt{n}(\hat{\theta} - \theta_0) \sim N(0, \mathcal{I}^{-1}(X_1, \theta_0))$, and that $(1/n)M_n(\theta_0)$ converges to $E\nabla\nabla'\log f(X, \theta) = \mathcal{I}(X_1, \theta_0)$ to get the desired conclusion.

13.8.2 Nuisance Parameters

The same asymptotic distribution for the LR test is valid in the presence of nuisance parameters, but the proof is more complex. As before, let $X_1, X_2, \ldots$ be i.i.d. observations with common density $f^X(x, \theta)$. We will write $f(X^{(n)}, \theta) = \prod_{i=1}^n f^X(X_i, \theta)$ for the joint density of $X^{(n)} = (X_1, \ldots, X_n)$. Partition θ as $\theta' = (\theta_1', \theta_2')$, where $\theta_1 \in \mathbf{R}^k$ and $\theta_2 \in \mathbf{R}^l$ are both vectors of parameters. Assume we wish to test the null hypothesis $H_0 : \theta_1 = \theta_1^0$ while θ_2 is unconstrained (and hence a nuisance parameter). Let $\hat{\theta} = (\hat{\theta}_1, \hat{\theta}_1)$ be the (unconstrained) maximum likelihood estimates of the parameters, and let $\tilde{\theta} = (\theta_1^0, \tilde{\theta}_2)$ be the ML estimates subject to the constraint imposed by the null hypothesis. The likelihood ratio statistic for testing H_0 can be written as

$$\log LR = \log f(X, \tilde{\theta}) - \log f(X, \hat{\theta}) \leq 1.$$

The following result shows that the asymptotic distribution for the likelihood ratio test remains the same in the presence of nuisance parameters.

Theorem 13.11 (LR:Asymptotic Distribution)
Assume $M(X^{(n)}, \theta) = \sum_{i=1}^n \nabla_\theta \nabla_\theta' \rho(X_i, \theta)$ is locally uniformly stable in a neighborhood of θ_0. Also suppose that both $\hat{\theta}$ and $\tilde{\theta}$ are consistent for θ_0 under the null hypothesis. Then in large samples $-2 \log LR$ is approximately chi-squared with k degrees of freedom under the null hypothesis.

Proof: For notational convenience, we drop the first argument $X^{(n)}$ of S, M. We will also drop the second argument when it equals the true parameter $\theta^0 = (\theta_1^0, \theta_2)$, so that M_{11} stands for $M_{11}(X^{(n)}, \theta^0)$. The line of reasoning adopted below is a slightly updated version of Roy (1957). For a proof along different lines see Stuart and Ord (1991, Sections 25.19 and 25.20). Partition the score function $S(\theta)$ and the matrix of second derivative $M(\theta)$ conformably with the parameters $\theta' = (\theta_1', \theta_2')$ as follows:

$$S(\theta) \equiv \begin{pmatrix} S_1(\theta) \\ S_2(\theta) \end{pmatrix}, M(\theta) = \begin{pmatrix} M_{11}(\theta) & M_{12}(\theta) \\ M_{21}(\theta) & M_{22}(\theta) \end{pmatrix}, \quad (13.2)$$

where $S_i(\theta) = \nabla_{\theta_i} \log f(X^{(n)}, \theta)$ and $M_{i,j}(\theta) = \nabla_{\theta_i} \nabla'_{\theta_j} \log f(X^{(n)}, \theta)$ for $i = 1, 2$ and $j = 1, 2$. Note that the unconstrained maximum likelihood estimate satisfies the first order conditions $S(X, \hat{\theta}) = 0$. The following lemma which represents the constrained estimates $\tilde{\theta}_2$ in terms of the unconstrained estimates $\hat{\theta}$ is the key to the proof.

Lemma 13.2 *The constrained MLE $\tilde{\theta}$ and the unconstrained MLE $\hat{\theta}$ satisfy the following relationship:*

$$\tilde{\theta}_2 - \theta_2 \approx (\hat{\theta}_2 - \theta_2) + M_{22}^{-1} M_{21}(\hat{\theta}_1 - \theta_1^0).$$

Proof: Expanding $S_2 = S_2(X^{(n)}, \theta^0)$ in a Taylor series around $\tilde{\theta} = (\theta_1^0, \tilde{\theta}_2)$, and noting that $S_2(\tilde{\theta}) = 0$, we get,

$$S_2 \approx \nabla_{\theta_2} S_2(\tilde{\theta})(\theta_2 - \tilde{\theta}_2) = M_{22}(\tilde{\theta})(\theta_2 - \tilde{\theta}_2). \qquad (13.3)$$

We can make a similar expansion of $S = S(X^{(n)}, \theta^0)$ around $\hat{\theta}$ to get

$$S \approx M(\hat{\theta})(\theta^0 - \hat{\theta}).$$

Using the partitioned form of M, this equation yields the following expression for S_2:

$$S_2 = M_{21}(\hat{\theta})(\theta_1^0 - \hat{\theta}_1) + M_{22}(\hat{\theta})(\theta_2 - \hat{\theta}_2). \qquad (13.4)$$

Now the local uniform stability of M guarantees that the error will be small if we replace $M(X^{(n)}, \hat{\theta})$ and $M(X^{(n)}, \tilde{\theta})$ by $M(X^{(n)}, \theta^0)$, provided that $\hat{\theta}$ and $\tilde{\theta}$ are consistent for θ^0 (as assumed). Making this substitution and combining Eqs.(1.3) and (1.4) yields the lemma.

With the help of Lemma 1.2, the asymptotic distribution of the LR is easily obtained. Expand $\log f = \log f(X^{(n)}, \theta^0)$ around $\hat{\theta}$ noting that the second term is zero due to the first order conditions satisfied by $\hat{\theta}$:

$$
\begin{aligned}
\log f &= \log f(\hat{\theta}) + \frac{1}{2}(\eta - \hat{\theta})' M(\hat{\theta})(\eta - \hat{\theta}) \\
&= \log f(\hat{\theta}) + \frac{1}{2}(\theta_1^0 - \hat{\theta}_1)' M_{11}(\hat{\theta})(\theta_1^0 - \hat{\theta}_1) \\
&\quad + (\theta_1^0 - \hat{\theta}_1)' M_{12}(X, \hat{\theta})(\theta_2 - \hat{\theta}_2) + \frac{1}{2}(\theta_2 - \hat{\theta}_2)' M_{22}(\hat{\theta})(\theta_2 - \hat{\theta}_2).
\end{aligned}
$$

We can similarly expand $\log f(X, \theta^0)$ around the constrained ML estimate $\tilde{\theta}$. Here too, the second term is zero because of the first order conditions satisfied by $\tilde{\theta}$:

$$
\begin{aligned}
\log f &= \log f(\tilde{\theta}) + \frac{1}{2}(\theta_2 - \tilde{\theta}_2)'M_{22}(\tilde{\theta})(\eta_2 - \tilde{\theta}_2) \\
&\quad \frac{1}{2}\left[(\eta_2 - \hat{\theta}_2) + M_{22}^{-1}M_{21}(\eta_1 - \hat{\theta}_1)\right]'M_{22}(X, \tilde{\theta}) \\
&\quad \left[(\eta_2 - \hat{\theta}_2) + M_{22}^{-1}M_{21}(\eta_1 - \hat{\theta}_1)\right]' \\
&= \frac{1}{2}(\theta_2 - \hat{\theta}_2)'M_{22}(\eta_2 - \hat{\theta}_2) + (\eta_1 - \hat{\theta}_1)'M_{12}(\eta_2 - \hat{\theta}_2) \\
&\quad + \frac{1}{2}(\eta_1 - \hat{\theta}_1)'M_{12}M_{22}^{-1}M_{21}(\eta_1 - \hat{\theta}_1).
\end{aligned}
$$

From these expansions, we get the following equation for the likelihood ratio, $\mathcal{L} = -2\log(LR)$:

$$
\mathcal{L} = -\left\{\sqrt{n}(\theta_1^0 - \hat{\theta}_1)\right\}'\left(\frac{1}{n}\left\{M_{11} - M_{12}M_{22}^{-1}M_{21}\right\}\right)\left\{\sqrt{n}(\theta_1^0 - \hat{\theta}_1)\right\}.
$$
(13.5)

To complete the proof, note that $\sqrt{n}(\hat{\theta} - \theta^0) \sim N(0, \mathcal{I}^{-1}(\eta))$. Also $(1/n)M(\theta^0)$ converges to $\mathcal{I}(\theta^0)$. Let $\mathcal{I}^{11}(\theta^0)$ be the top $k \times k$ block of the matrix $\mathcal{I}^{-1}(\theta^0)$. By the standard formulas for partitioned inverses, $\mathcal{I}^{11}(\eta) = \left(\mathcal{I}_{11} - \mathcal{I}_{12}\mathcal{I}_{22}^{-1}\mathcal{I}_{21}\right)^{-1}$, and M^{11} converges to $\mathcal{I}^{11}$. This implies the desired result.

13.9 The Wald Test and LM Test

The likelihood ratio test can be difficult to compute. Even when it is computable, its exact distribution can be intractable. Two alternatives, the Wald test and Lagrange multiplier test (more familiar to statisticians as Rao's Score test) can be derived as local approximations to the the LR test. Local approximations are held to be adequate because in large samples, parameters which are far from the null should easily be differentiated from the null. Thus only the local performance of the test should matter. This heuristic reasoning can be quantified via a measure called Pitman efficiency or Asymptotic Relative Efficiency, which evaluates test performance in a shrinking neighborhood of the null hypothesis. According to this measure, the LR has asymptotically equivalent performance to the Wald and LM tests. This appears plausible as both these tests are local approximations to the LR test. As discussed in Chapter 15, this local equivalence is valid up to the second order terms in asymptotic approximations. However, third order terms in the local approximations are different for the three tests. The picture which emerges is quite similar to the one which we studied in a particular special case in Chapter 6, with LR being superior to Wald, which is in turn superior to the LM test in terms of stringency. According to Efron's (1975) heuristic arguments, comparisons are likely to depend on the 'curvature' of the family of densities under study. The third order terms will be significant in highly curved families, and differences among the three tests will be negligible in less curved families. This issue has not been studied extensively. We will also see that the three types of tests can be compared in a nonlocal sense as well. In the nonlocal case, the LR appears superior to the other two tests. The object of this section is to present the first order theory according to which the Wald and the LM tests are locally equivalent to the LR.

    The Wald test is based directly on the unconstrained ML estimate $\hat{\theta} = (\hat{\theta}_1, \hat{\theta}_2)$. Under the null hypothesis $\theta_1 = \eta_1^0$. Thus if $\hat{\theta}_1$ is significantly different from η_1^0 we can reject the null hypothesis. To assess the significance of the difference, we can use the asymptotic distribution of $\hat{\theta}_1$, which is $N(\theta_1^0, \mathcal{I}^{11}(\eta))$. We can approximate $\mathcal{I}^{11}(\eta)$ by $\mathcal{I}^{1}1(\hat{\theta})$ to get the Wald statistic

$$W_1 = (\hat{\theta}_1 - \eta_1^0)' \mathcal{I}^{11}(\hat{\theta})(\hat{\theta}_1 - \eta_1^0).$$

The asymptotic covariance matrix can also be approximated in other ways and the question of which approximations are superior has been considered in the literature. Note that from Eq. (1.5), the Wald statistic is locally equivalent to $-2\log(LR)$. In fact, since the equivalence is based on a second order Taylor expansion, it will be exact if the log likelihood is quadratic in θ — in this case, the Taylor approximation is exact. This case occurs in the linear regression model, where the LR test for linear restrictions coincides with the Wald test and also with the LM test.

Whereas the Wald test is based on the unconstrained ML estimate $\hat{\theta}$, LM test (also known as the Aitchison-Silvey Lagrange multiplier test as well as Rao's score test) is based on the constrained ML estimate $\tilde{\theta}$. The score function $S(X, \theta)$ is zero when evaluated at $\hat{\theta}$. If the constraints are valid, then it should also be the case that $S(X, \tilde{\theta})$ should be approximately 0. Thus the extent to which $S(X, \tilde{\theta})$ differs from zero is also a measure of the extent to which the null hypothesis is violated by the data. This difference can also be assessed by look at the Lagrange multiplier associated with the constraints and testing it for being significantly different from zero. These two points of view turn out to lead to the same test statistics. Below we look directly at the score function and follow the original method of Rao.

Partitioning S as $S = (S_1, S_2)$ as in Eq. (1.2), note that $S_2(X, \tilde{\theta}) = 0$. To measure the difference of S and 0, it suffices to consider $S_1(X, \tilde{\theta})$. Now note that $Cov(S(X, \eta)) = \mathcal{I}^{-1}(\eta^0)$, so that $Cov(S_1(X, \eta) = \mathcal{I}^{11}(\eta^0)$. It follows that the significance of the difference between S_1 and 0 can be assessed using Rao's score statistic

$$LM = S_1(X, \tilde{\theta}) \left(\mathcal{I}^{11}(\tilde{\theta}) \right)^{-1} S_1(X, \tilde{\theta})$$

When the information matrix is block diagonal (so that $\mathcal{I}^{12} = 0$), the statistic takes a simpler form since $(\mathcal{I}^{11})^{-1} = \mathcal{I}_{11}$. In the general case, we have $(\mathcal{I}^{11})^{-1} = \mathcal{I}_{11} - \mathcal{I}_{12}\mathcal{I}_{22}^{-1}\mathcal{I}_{21}$. This test is also locally equivalent to the LR test and the Wald test. This can be seen as follows. It suffices to show that

$$- M^{11}(\tilde{\theta})(\hat{\theta}_1 - \theta_1^0) \approx S_1(\tilde{\theta}). \tag{13.6}$$

Once this is proven, it follows immediately from Eq. (1.5) that

$$-2 \log(LR) \approx -S_1'(\tilde{\theta}) \left(M^{11}(\tilde{\theta}) \right)^{-1} S_1(\tilde{\theta}),$$

which is the desired equivalence result.

To prove Eq. (1.6), expand $S(\tilde{\theta})$ around $\hat{\theta}$ via Taylor's formula as follows:

$$S(\tilde{\theta}) \approx S(\hat{\theta}) + M(\hat{\theta})(\tilde{\theta} - \hat{\theta}).$$

Note that $S(\hat{\theta}) = 0$. Using the partitioned form for M and $\tilde{\theta} - \hat{\theta}$, the first component S_1 can be written as

$$
\begin{aligned}
S_1(\tilde{\theta}) &\approx M_{11}(\hat{\theta})(\tilde{\theta}_1 - \hat{\theta}_1) + M_{12}(\hat{\theta})(\tilde{\theta}_2 - \hat{\theta}_2) \\
&= M_{11}(\hat{\theta})(\theta_1^0 - \hat{\theta}_1 + M_{12}(\hat{\theta})M_{22}^{-1}(\hat{\theta})M_{21}(\hat{\theta})(\hat{\theta}_1 - \theta_1) + \\
&= -M^{11}(\hat{\theta})(\hat{\theta}_1 - \theta_1^0).
\end{aligned}
$$

The third equality follows since $\tilde{\theta}_1 = \theta_1^0$, and from Lemma 1.2 after writing $\tilde{\theta}_2 - \hat{\theta}_2$ as $(\tilde{\theta}_2 - \theta_2) - (\hat{\theta}_2 - \theta_2)$. Now we use local uniform stability of M_n to conclude that $M_n(\hat{\theta})$ equals $M_n(\tilde{\theta})$, as needed for equivalence.

This demonstration only shows that the forms of the three statistics are equivalent under the null hypothesis. A slight extension of the argument allows us to show that they are also equivalent to second order terms at points close to the null hypothesis. Furthermore, their power curves coincide to the second order terms in a neighborhood of the null hypothesis. See Chapter 15 for more details and references.

13.10   Consistency of the LR Test

The asymptotic distribution theory developed permits us to prove the consistency of the LR test in a straightforward way. Suppose we have a situation where the ML estimators are consistent for the true parameter, and consider the asymptotic behavior of the likelihood ratio

$$
LR = \frac{f(X^{(n)}, \theta_1^0, \tilde{\theta}_2)}{f(X^{(n)}, \hat{\theta}_1, \hat{\theta}_2)}.
$$

Under the null hypothesis $\hat{\theta}_1$ will converge to θ_1^0 and $\tilde{\theta}_2$ and $\hat{\theta}_2$ will both converge to θ_2, the true value of the nuisance parameter. It follows that LR will converge to 1 asymptotically. Thus the critical value for the LR test will also converge to 1 asymptotically for any fixed level test.

Next suppose the null hypothesis is false, and the true parameter is $\theta^* = (\theta_1^*, \theta_2^*)$, where $\theta_1^* \neq \theta_1^0$. Under hypotheses ensuring the ULLN, $\tilde{\theta}_2$ will converge to a maximum η_2 of the function $f(\theta_2) = \mathbb{E} \log f(X_1, \theta_1^0, \theta_2)$. By the information inequality $\mathbb{E} \log f(X_1, \theta_1^0, \eta_2) =$

$\max_{\theta_2} \mathbb{E} \log f(X_1, \theta_1^0, \theta_2) < \mathbb{E} \log f(X_1, \theta_1^*, \theta_2^*)$. By applying the law of large numbers we can conclude that LR will converge to some number strictly less than 1 in the limit. It follows that the null hypothesis will be rejected with probability 1 in the limit when it is false, implying the consistency of the LR test.

When the parameter space is finite dimensional, the equivalence of the Lagrange multiplier and the Wald tests to the LR also implies their consistency. However, when the parameter space is infinite dimensional, equivalence does not hold, and the three tests behave differently. Note that the earlier example of an inconsistent LM test in Section 11.4.2 has an infinite dimensional parameter space — the set of all possible functions f is being considered.

13.11   Exercises

1. Let $X_n \sim \mathcal{U}(-1/n, 1/n)$; that is, X_n is uniformly distributed on the interval $(-1/n, 1/n)$. Show that X_n converge in distribution to the degenerate random variable X, where $\mathbf{P}(X = 0) = 1$. Show that the CDF $F_n(x)$ of X_n converges at all points *except* 0.

2. The characteristic function of a random variable X is defined as $\chi_X(\theta) = \mathbb{E} \exp(i\theta X)$, where $i = \sqrt{-1}$. When the moment generating function exists, then $\chi_X(\theta) = \mathcal{L}_X(i\theta)$. However the MGF is the expected value of an unbounded function and hence does not exist for many random variables. In contrast, since $\exp(i\theta) = \cos(\theta) + i\sin(\theta)$ is a bounded function, the characteristic function exists for all random variables.

    (a) Show that if $X \sim N(\mu, \sigma^2)$ then $\chi_X(\theta) = \exp(i\mu\theta - \sigma^2\theta^2/2$.

    (b) Show that if X is Cauchy, so that $f^X(x) = (\pi(1 + x^2))^{-1}$, then $\chi_X(\theta) = \exp(-|\theta|)$.

    (c) It can be proven that if the r-th moment of X exists then the characteristic function $\chi_X(\theta)$ is r times continuously differentiable in a neighborhood of $\theta = 0$. Conversely if the the r-th derivative of characteristic function exists at $\theta = 0$ then all even moments of X up to r exist. Also, if the moments exist, they can always be computed via the formula:

$$\mu_r \equiv \mathbb{E}X^r = i^{-r}\frac{\partial^j}{(\partial\theta)^j}\chi_X^{(r)}(\theta).$$

(See Theorems 6.4.1 and 6.4.2 of Chung (1974) for a proof.)
What can we conclude about the existence of moments of
the Cauchy distribution and the normal distribution from
the their characteristic functions?

3. A rigorous proof of Lyapunov's theorem can be based on the
characteristic function of the previous exercise.

  (a) The *cumulant function* of X is the log of the characteristic
function of X. Show that the cumulant can be obtained
from the cumulant function as follows:

$$\kappa_j = i^{-j}\frac{\partial^j}{(\partial\theta)^j}\log\chi_X(0).$$

Some care is needed in dealing with the logarithm of a complex function; see Chung (1974,p.199).

  (b) If the third absolute moment of X exists, then the cumulant
function has the following Taylor series expansion:

$$\log\mathbb{E}\exp(i\theta X) = i\kappa_1(X)\theta - \kappa_2(X)\theta^2/2 + \alpha(\mathbb{E}|X|^3)\theta^3/3!,$$

where α is some complex number satisfying $|\alpha| \leq 1$. See
Chung (1974,Theorem 6.4.2) for a proof. Show that this
yields a rigorous proof of Lyapunov's theorem.

4. An alternative form of Lyapunov's theorem brings out more
clearly the essential hypothesis that each individual term of the
sum must be negligible in relation to the whole.

  (a) Let w_{nj} be positive real numbers such that $\sum_{j=1}^{n} w_{nj}^2 = 1$. Show that $\sum_{j=1}^{n} w_{nj}^3$ converges to zero if and only if
$\max_{\{j:1\leq j\leq n\}} w_{nj}$ converges to zero.

(b) Let X_j be an i.i.d. sequence of random variable such that $\mathbb{E}X_i = 0$, $\mathbb{V}ar(X_i) = \sigma^2$, and $\mathbb{E}|X_i^3| = \gamma$. Let w_j be a positive sequence of weights. Show that

$$S_n = \frac{\sum_{j=1}^n w_j X_j}{\sum_{j=1}^n w_j^2}$$

converges to $N(0, \sigma^2)$ provided that

$$\lim_{n\to\infty} \max_{\{j:1\leq j\leq n\}} \frac{w_j^2}{\sum_{j=1} w_{nj}^2} = 0.$$

(c) Let X_j be an i.n.i.d. sequence satisfying Lyapunov's condition that $\left(\sum_{j=1}^n \mathbb{E}|X_j - \mathbb{E}X_j|^3\right) / \left(\sum_{j=1}^n \mathbb{V}ar(X_j)\right)^{3/2}$ converges to 0 and $n \to \infty$. In addition to implying that $\left(\sum_{j=1}^n (X_j - \mathbb{E}X_j)\right) / \sqrt{\sum_{j=1}^n \mathbb{V}ar(X_j)}$ converges to $N(0,1)$ the condition also implies that each X_j is negligible compared to the sum in the following sense:

$$\lim_{n\to\infty} \max_{\{j:1\leq j\leq n\}} \frac{\mathbb{V}ar(X_j)}{\sum_{i=1} \mathbb{V}ar(X_i)} = 0.$$

5. In this exercise, we will prove Theorem 1.4. Let $a = (a_1, \ldots, a_k)$ be a $k \times 1$ vector. To prove the theorem, we must show that $a'M_T \epsilon_T$ is asymptotically $N(0, a'\mathbf{I}_k a = a'a)$. Write $m_{i,j}$ for the (i, j) entry of the matrix M_T

   (a) Define $X_{T,j} = \sum_{i=1}^k a_i m_{i,j} \epsilon_j$ and $S_T = \sum_{t=1}^T X_{T,j}$. Show that $s_T = \text{Var } S_T = a'a$ and that $a'M_T \epsilon_T = S_T$.

   (b) Verify that $S_T / (\sqrt{a'a})$ converges to N(0,1), and conclude from this that Theorem 1.4 holds.

6. Show that Theorem 1.5 follows directly from Theorem 1.4.

7. This exercise gives more details about the conditions necessary to obtain the approximation given in Theorem 1.8.

(a) Suppose that for any i, j, k we have

$$E \sup_{\theta \in \Theta} \left| \frac{\partial^3}{\partial \theta_i \partial \theta_j \partial \theta_k} \rho(X_i, \theta) \right| < \infty.$$

By taking one more term in the Taylor expansion, show that the approximation in Theorem 1.8 must hold.

(b) The globally bounded third derivative assumption is rather extravagant. One can get by with locally bounded third derivatives. With more care, we can, using the ULLN for i.i.d. cases, show that the approximation is valid provided that for some integer k and some neighborhood $N(\theta_0)$ of the true value θ_0, it is true that

$$E \sum_{i=1}^{k} \sup_{\theta \in N(\theta_0)} \left| \frac{\partial^2 \rho(X_i, \theta)}{\partial \theta_i \partial \theta_j} \right| < \infty.$$

8. Partition $\mathcal{I}$ and $\mathcal{I}^{-1}$ conformably, so that

$$\mathcal{I}\mathcal{I}^{-1} = \begin{pmatrix} \mathcal{I}_{11} & \mathcal{I}_{12} \\ \mathcal{I}_{21} & \mathcal{I}_{22} \end{pmatrix} \begin{pmatrix} \mathcal{I}^{11} & \mathcal{I}^{12} \\ \mathcal{I}^{21} & \mathcal{I}^{22} \end{pmatrix} = \begin{pmatrix} \mathbf{I} & 0 \\ 0 & \mathbf{I} \end{pmatrix}.$$

Write out four equations in the four unknowns $\mathcal{I}^{11}, \mathcal{I}^{12}, \mathcal{I}^{21}, \mathcal{I}^{22}$ and solve in terms of the knowns $\mathcal{I}_{11}, \mathcal{I}_{12}, \mathcal{I}_{21}, \mathcal{I}_{22}$ to get the formula for the partitioned inverse.

9. Suppose that $\hat{\gamma}$ is an unbiased estimator of some function $\gamma(\theta)$. Define $T_n = \sum_{j=1}^{n} \partial \ln f^X(x, \theta) / \partial \gamma = (\partial \ln f / \partial \theta)(\partial \theta / \partial \gamma)$. Imitate the proof for the Cramer-Rao Lower Bound to find a lower bound for the variance of $\hat{\gamma}$.

*Khalifa Haroun Al-Rashid was well known for his gen-
erosity towards talent. A newcomer to his court planted
a needle at 50 feet, and threw a second needle in such
a way that it passed through the eyehole of the first needle. Haroun Al-Rashid ordered that the man should be
rewarded with 100 Dinars and also given 100 lashes. He
explained that the money was the reward for his extraordinary talent, and the lashes were the punishment for
the extraordinary amount of time he must have wasted in
learning this useless skill.*

Chapter 14

More Accurate Asymptotic Approximations

The central limit theorem provides us with approximations to distributions of estimators and test statistics. Unfortunately, in many cases,
these approximations are of poor quality and require impracticably
large sample sizes to yield desired levels of accuracy. Therefore this
chapter is devoted to finding superior approximations. It is worth noting that the theory of asymptotic expansions is an advanced topic.
However, our treatment below skips all the hard theory, and concen-
trates on getting useful formulas. For these, basically only two skills are
required: the ability to plod patiently through long algebraic formulas,
and the ability to do Taylor series expansions.

The basic Berry-Esseen Theorem tells us something about the accuracy of the CLT. Suppose that $X_1, \ldots, X_n$ are i.i.d. random variables
with $\mathbb{E}X_i = 0$, $\mathbb{E}X_i^2 = \sigma^2$, and $\mathbb{E}|X_i|^3 = \rho < \infty$. Let $\Phi(t) = \mathbf{P}(Z \le t)$,

where $Z \sim N(0,1)$ so that Φ is the CDF of a standard normal random variable. Let $F_n(t)$ be the CDF of $S_n = (\sum_{i=1}^n X_i)/(\sqrt{n}\sigma)$. According to the central limit theorem, $F_n(t)$ converges to $\Phi(t)$ for all t. The maximum difference between the limiting normal distribution and the true finite sample distribution can be bounded as follows:

$$\sup_t |F_n(t) - \Phi(t)| \leq \frac{0.7975\rho}{\sigma^3\sqrt{n}}.$$

For a proof see Bhattacharya and Rao (1976), page 110. Although we have cited the result for i.i.d. random variables, it can be generalized in many different directions.

As we can see from the Berry-Esseen theorem, the accuracy of the approximation is governed by the third absolute moment of the X_i, and goes to zero at the rate $1/\sqrt{n}$. Now $1/\sqrt{n}$ goes to zero rather slowly, and if the third absolute moment is large, we can expect the quality of the central limit approximation to be poor. The question arises as to whether we can obtain an improved approximation to the finite sample distribution $F_n(t)$ of the standardized sum S_n by utilizing information about higher moments of X_i.

14.1   Order of Approximation

First, we introduce the fundamental notations of *little* 'o' and *big* 'O' which measure order of approximation. Given two series x_n and y_n of numbers, if $|x_n - y_n|$ converges to zero, we can regard y_n as an asymptotic approximation to x_n: for large n the difference between the two will be small. Now suppose we want to assess the accuracy of the approximation. One way to do this is to ask how fast the difference $\epsilon_n = |x_n - y_n|$ goes to zero. Let c_n be another sequence of numbers going to zero. This sequence is used as our 'unit of measurement' — we measure how fast ϵ_n goes to zero in comparison with c_n. If ϵ_n/c_n remains bounded as n goes to infinity then we say that ϵ_n is of the same order as c_n and write $\epsilon_n = \mathcal{O}(c_n)$. This means that the two sequences ϵ_n and c_n converge to zero at the same rate. We also say that y_n approximates x_n to order $\mathcal{O}(c_n)$. If ϵ_n/c_n goes to 0, then we say that ϵ_n goes to zero faster than c_n and we write $\epsilon_n = o(c_n)$. In this case

y_n approximates x_n to order $o(c_n)$. It is immediate from the definitions that $\epsilon_n = \mathcal{O}(a_n \epsilon_n)$ for any *bounded* sequence of constants a_n. We also write that $\epsilon_n = o(1)$; that is, ϵ_n goes to zero faster than the constant sequence $c_n = 1$ (which does not go to zero at all).

In terms of these notations, the Berry-Esseen theorem tells us that the CLT provides an approximation to the distribution of $S_n = \sum_{i=1}^{n} X_i/\sqrt{n}$ which is accurate of order $\mathcal{O}(n^{-1/2})$. We will also say that this is a *first order* approximation. More generally, a k-th order approximation[1] is one of order $\mathcal{O}(n^{-k/2})$. We will now develop techniques which permit us to obtain more accurate (i.e., higher order) approximations.

14.2 The Edgeworth Expansion

We will now use asymptotic expansion techniques to produce large sample approximations to the distributions of sums of random variables which have higher order of accuracy than the usual normal asymptotic distribution. Such approximations have been found useful in a wide variety of contexts. It is reasonable to think of the central limit theorem as the most fundamental result of probability theory. The CLT allows us to say useful things about the distributions of sums of random variables when we know little about the individual terms. One important fact to note about CLT, asymptotic expansions, as well as nearly all asymptotic results, is that we frequently cannot tell at what sample sizes these results provide reliable guidance. This question is of extreme importance in applications, and can only be answered by experience or Monte Carlo studies over a limited range of circumstances. In situations where the CLT fails to provide adequate approximations in reasonable sample sizes, asymptotic expansions are frequently (but not always) helpful. Asymptotic expansions also suffer from the same difficulty: we do not know at what finite sample size they become reliable. However, they typically start working sooner than the CLT.

---

[1] Some authors refer to the normal approximation as 0-th order, and what we will call a k-th order approximation becomes a $(k-1)$-th order approximation.

14.2.1   Higher Order Approximation of the MGF

The central limit theorem is proven by showing that if X_i have mean 0 and variance 1, the MGF of the sum $Z_n = \sum_{i=1}^n X_i/\sqrt{n}$ is approximated by $\exp(\theta^2/2)$. A careful examination of the proof reveals that the order of approximation is $\mathcal{O}(n^{-1/2})$. It is quite straightforward to get higher order approximations by retaining more terms in an intermediate Taylor expansion in the course of the proof of the CLT. Keeping two additional terms yields the third order approximation given in the following theorem.

Theorem 14.1 *Let X_i be standardized i.i.d. variables so that $\mathbb{E}X_i = 0$ and $\mathbb{V}ar(X_i) = 1$. Let κ_3, κ_4 be the skewness and kurtosis of X_i. The moment generating function of $Z_n = \sum_{i=1}^n X_i/\sqrt{n}$ is approximated to order $\mathcal{O}(n^{-3/2})$ by*

$$\mathcal{L}_{Z_n}(\theta) = \exp\left(\frac{\theta^2}{2} + \frac{\kappa_3\theta^3}{6\sqrt{n}} + \frac{\kappa_4\theta^4}{24n}\right).$$

Proof: Use of the nice linearity properties of the CGF, as described in Theorem 1.2, lead directly to this formula. Note that

$$K(\theta, Z_n) = \sum_{i=1}^n K(\theta/\sqrt{n}, X_i) = nK(\theta/\sqrt{n}, X_1)$$

The last equality comes from the fact that the CGF depends only on the distribution of the X_i (and not the observed value), which is the same for all the X_i. The expansion of the CGF in Theorem 1.2 yields

$$nK(\theta/\sqrt{n}, X_1) \approx \sqrt{n}\kappa_1\theta + \kappa_2\frac{\theta^2}{2!} + \kappa_3\frac{\theta^3}{(3!)n^{1/2}} + \kappa_4\frac{\theta^4}{(4!)n} + \cdots.$$

Taking exponents yields the approximation of the theorem. The order of approximation is governed by the first neglected terms, which is of order $\mathcal{O}(n^{-3/2})$ so that this is a third order approximation as claimed.

Unfortunately, the MGF given in the theorem is not in a form which is convenient to 'invert' — that is, to calculate the density which gives rise to the MGF. However, it can easily be put into such a form by a little Tayloring, as follows:

Corollary 14.1.1 *Another third order accurate approximation to the moment generating function of* $Z_n = (1/\sqrt{n}) \sum_{i=1}^{n} X_i$ *is*

$$\mathcal{L}_{Z_n}(\theta) \approx \exp\left(\frac{\theta^2}{2}\right) \left(1 + \frac{\kappa_3\theta^3}{6\sqrt{n}} + \frac{3\kappa_4\theta^4 + \kappa_3^2\theta^6}{72n}\right) + \mathcal{O}(n^{-3/2}).$$

Proof: Write the MGF of Theorem 14.1 as

$$\mathcal{L}_{Z_n}(\theta) = \exp\left(\frac{\theta^2}{2}\right) \times \exp\left(\frac{\kappa_3\theta^3}{6\sqrt{n}} + \frac{\kappa_4\theta^4}{24n}\right)$$

Now expand the second exponential in its Taylor series and drop terms of order $n^{3/2}$ or higher to get the formula of the corollary.

The sloppy method of proof adopted restricts the validity of the results to random variables with moment generating functions. In fact these approximations hold in great generality provided that the fifth moment of X_i exists. Another hypothesis, called Cramer's condition, is also needed: the distribution of X_i should be nonlattice. A lattice distribution is one in which X takes values on an evenly spaced grid of points, such as the Binomial or the Poisson distribution. In this case the second order approximation is still valid, but the third order approximation requires some modifications. See, for example, Hall(1992,p.46) for further details. Since we do not deal with lattice distributions in the sequel, we ignore these refinements.

14.2.2 Inversion of the Approximate MGF

The reason the second form of the approximation to the MGF is easy to invert is clarified by the following lemma, which permits inversion of large classes of MGFs.

Lemma 14.1 *Suppose that h is the (bilateral) Laplace transform of the function f (that is, h is the m.g.f. of a random variable X with density f, in cases where f is a density):*

$$h(\theta) = \mathcal{L}_f(\theta) = \int_{-\infty}^{\infty} \exp(\theta x) f(x)\, dx.$$

Then $-\theta h(\theta)$ is the Laplace transform of $f'(x) = \partial f(x)/\partial x$.

Proof: This is easily shown via integration by parts. Letting $u(x) = e^{\theta x}$ and $dv(x) = f'(x)dx$, we have

$$\mathcal{L}_{f'}(\theta) = \int_{-\infty}^{\infty} u(x)dv(x) = u(x)v(x)\big|_{x=-\infty}^{\infty} - \int_{-\infty}^{\infty} v(x)u'(x)dx$$

$$= -\int \theta e^{\theta x} f(x)\, dx = -\theta \mathcal{L}_f \theta.$$

Suppose that $M(\theta)$ is the moment generating function of a known density f(x). This lemma permits us to calculate the density corresponding to any moment generating function of the type $M(\theta) \times (a_0 + a_1\theta + \cdots + a_k\theta^k)$. This density must be $a_0 f(x) - a_1 f^{(1)}(x) + \cdots + (-1)^k f^{(k)}(x)$. As a special case of great importance, let $\phi(x) = \exp(-x^2/2)/\sqrt{2\pi}$ be the standard normal density. Since $\exp(-\theta^2/2)$ is the Laplace transform of $\phi(x)$, it follows that the inverse transform of $\theta^j \exp(-\theta^2/2)$ is simply $(-1)^j \phi^{(j)}(x) = (-1)^j \partial^j \phi(x)/(\partial x)^j$. Using this, the third order approximation to the moment generating function given in Corollary 14.1.1 is easily inverted.

Theorem 14.2 (Edgeworth Density) *Suppose X_i are i.i.d. with mean 0 variance 1, and κ_3, κ_4, the skewness and kurtosis, respectively. A third order approximation to the density of $Z_n = (1/\sqrt{n})\sum_{i=1}^{n} X_i$ is*

$$f^{Z_n}(z) \approx \phi(z) + \frac{\kappa_3 (-1)^3 \phi^{(3)}(x)}{6\sqrt{n}} + \frac{3\kappa_4 (-1)^4 \phi^{(4)}(z) + \kappa_3^2 (-1)^6 \phi^{(6)}(z)}{72n}.$$

$$(14.1)$$

It is easily seen that the derivatives of $\phi(x)$ are simply polynomials in x times $\phi(x)$. It turns out to be convenient to define the *Hermite Polynomials* as follows via the identity $(-1)^j \phi^{(j)}(x) = H_j(x)\phi(x)$, or equivalently $H_j(x) \equiv (-1)^j \phi^{(j)}(x)/\phi(x)$. It is easily checked that $H_1(x) = x$, $H_2(x) = x^2 - 1$, $H_3(x) = x^3 - 3x$, $H_4(x) = x^4 - 6x^2 + 3$, $H_5(x) = x^5 - 10x^3 + 15x$, and $H_6(x) = x^6 - 15x^4 + 45x^2 - 15$. Using the Hermite polynomials, we can rewrite the third order approximation to the density of Z_n as

$$f^{Z_n}(z) \approx \phi(z)\left(1 + \frac{\kappa_3}{6\sqrt{n}}H_3(x) + \frac{3\kappa_4 H_4(z) + \kappa_3^2 H_6(z)}{72n}\right). \quad (14.2)$$

The expression for the Edgeworth density given in (14.1) can also be integrated to obtain an approximation for the cumulative distribution function of Z_n. Let $\Phi(z) = \int_{-\infty}^{z} \phi(x)dx$ be the c.d.f. for the standard normal. Note that $\int^{z} \phi^{(j)}(x)\,dx = \phi^{(j-1)}(z)$. Now integrate both sides of (14.1) to get

$$F^{Z_n}(z) \approx \Phi(z) - \phi(z)\left(\frac{\kappa_3}{6\sqrt{n}}H_2(z) + \frac{3\kappa_4 H_3(z) + \kappa_3^2 H_5(z)}{72n}\right). \quad (14.3)$$

14.2.3 Properties of Edgeworth Expansions

Users should be aware of several general features of the Edgeworth expansion. First, the Edgeworth expansion is not a probability density — that is, it does not integrate to one and can be negative. A variant sometimes called the Edgeworth-B approximation described later avoids this difficulty and is therefore used much more often in econometric applications. Also of importance is the fact that while the Edgeworth expansions are quite accurate near the center of the distribution, they frequently do poorly in the tails.

Let Z_i be i.i.d. χ_1^2 random variables and define $S_n = \sum_{i=1}^{n}(Z_i - 1)/\sqrt{2}$. Since S_n is a sum of i.i.d. random variables with mean 0 and variance 1, it is asymptotically $N(0,1)$. By calculating the third and fourth cumulant of S_n we can calculate Edgeworth approximations to the density of S_n. This is done in Figure 14.1, which plots the exact density of S_n, the asymptotic normal approximation, and also the first and second order Edgeworth approximations to the density of S_n. As can be seen in Figure 14.1, the Edgeworth approximation can even be negative in the tails. This can be a serious difficulty when the main use to be made of the Edgeworth approximation is to find critical values of a test statistic. The technique of tilted Edgeworth expansions (which are also equivalent to Saddlepoint approximations) can be used to get approximations which are accurate in the tails. The idea of the technique is to shift the center of the distribution, do an Edgeworth approximation, and then compensate for the shifted center. An easily readable exposition is available in Barndorff-Nielsen and Cox (1989). While this technique can be substantially more accurate than

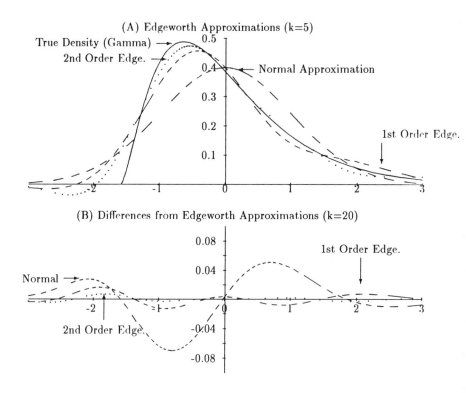

Figure 14.1: Edgeworth Approximations to Sums of Chi-Squares

the Edgeworth, it usually requires numerical calculations as well as accurate knowledge of the cumulant generating function. The Edgeworth expansion is more convenient theoretically since the formulas it yields are more manageable and have easy interpretations as corrections for skewness and kurtosis. It also demands less information as input.

We have given a two term Edgeworth expansion, involving corrections for skewness and kurtosis. Higher order expansions are possible, but rarely used. If one has accurate cumulants of high orders, it is usually possible to get the exact distribution, or else closer approximations via the saddlepoint technique. In typical applications we only have estimates or theoretical approximations to the higher cumulants. In this case, the additional error of approximation often outweighs any added benefits of higher order. A one term Edgeworth expansion is much simpler than the two term one, and is often adequate if the main concern is correction for skewness.

As discussed in the section below, it is possible to make Edgeworth expansions using the knowledge of the cumulants in very general situations. In such cases, it is important to ensure that the higher order cumulants are asymptotically small for the expansion to be valid. Finally, it is important to note that only the one term expansion is valid for discrete random variables such as the Binomial or Poisson. The second order term requires a continuity correction.

14.3 Related Expansions

The technique used to develop the Edgeworth expansion described in the previous section can be used in considerably more general situations. Three generalizations/extensions are of great importance in econometric applications. First we will give the Edgeworth-B variant of the expansion, which gives a genuine CDF as an approximation; recall that the original expansion does not integrate to one and can take negative values. Second we will give an example of the expansion for an i.n.i.d. case and also show how one can develop expansions around densities other than the normal one. The third extension is handled separately in Section 14.4, devoted to the Behrens-Fisher problem. There we show how to develop Edgeworth expansions for nonlinear functions of sums by using Taylor series to linearize them. See also exercise 5 for an illustration of this technique.

14.3.1 Edgeworth-B and Cornish-Fisher

Let X_i be a standardized i.i.d. sequence of random variables with κ_3, κ_4 being their common skewness and kurtosis. The expansion for the CDF of $Z_n = \sum_{j=1}^n X_j / \sqrt{n}$ given in Eq. 14.3 is known as the Edgeworth-A expansion. We will now give a modification, known as the Edgeworth-B expansion, which has the virtue of being a real CDF.

Theorem 14.3 *Another approximation to the CDF of Z_n which is accurate to order $\mathcal{O}(n^{-3/2})$ is given by*

$$F_B^{Z_n}(z) = \Phi\left(z - \frac{\kappa_3(z^2 - 1)}{6\sqrt{n}} + \frac{3\kappa_4(3z - z^3) + \kappa_3^2(8z^3 - 14z)}{72n}\right). \qquad (14.4)$$

Proof: A useful technique for discovering this and similar formulae is the following. Let a, b be unknown quantities and expand $P = \Phi(z + a/\sqrt{n} + b/n)$ in a Taylor series to get

$$
\begin{aligned}
P &\approx \Phi(z) + \left(\frac{a}{\sqrt{n}} + \frac{b}{n}\right)\phi(z) + \frac{1}{2}\left(\frac{a}{\sqrt{n}} + \frac{b}{n}\right)^2 \phi^{(1)}(z) \\
&= \Phi(z) + \frac{a}{\sqrt{n}}\phi(z) + \frac{b - za^2/2}{n}\phi(z) + \mathcal{O}(n^{-3/2})
\end{aligned}
$$

Matching terms of order $1/\sqrt{n}$ to the ones in Eq. (14.3 above, we see that $a = -\kappa_3 H_2(z)/6 = -(\kappa_3/6)(z^2 - 1)$. Matching terms of order $1/n$ and solving for b yields $b = (za^2/2) - (3\kappa_4 H_3(z) + \kappa_3^2 H_5(z))/72$. Substituting the explicit forms of a, H_3, H_5 yields the formula of the theorem.

In addition to providing a genuine CDF as a third order approximation, the Edgeworth-B expansion has two valuable by-products. The first is a normalizing transformation for the sum Z_n. Let $g(z) = z + a(z)/\sqrt{n} + b(z)/n$ so that $F^{Z_n}(z) = \Phi(g(z))$. Suppose that n is large enough so that g is monotonic on the domain of interest. Let $X \sim N(0, 1)$. Then the theorem says that to a close order of approximation, $\mathbf{P}(Z_n \le z) = \mathbf{P}(X \le g(z)) = \mathbf{P}(g^{-1}(X) \le z)$. Thus the distribution of Z_n is like the distribution of $g^{-1}(X)$, where X is standard normal. Equivalently $g(Z_n) \approx g(g^{-1}(X)) = X$ is approximately normal. Thus the Edgeworth-B formula gives a function which can be used to transform Z_n to approximate normality. This is known as the 'Cornish-Fisher' inversion.

As a second by-product, related to Cornish-Fisher, we can get a closer approximation to the critical values of test statistics. Given $\alpha \in (0, 1)$, let c_α satisfy $\Phi(c_\alpha) = \alpha$. For large n, $\mathbf{P}(Z_n < c_\alpha)$ will also be approximately α. Let $a(z), b(z)$ be as defined in the previous paragraphs and let $a'(z)$ be the first derivative of a. A more accurate approximation to the level α critical value of Z_n is given in the following corollary:

Corollary 14.3.1 *Define the adjusted critical value*

$$c_\alpha^* = c_\alpha - \frac{a(c_\alpha)}{\sqrt{n}} + \frac{a'(c_\alpha)a(c_\alpha) - b(c_\alpha)}{n}.$$

The $\mathbf{P}(Z_n < c_\alpha^*) - \alpha$ *goes to zero at rate* $\mathcal{O}(n^{-3/2})$.

Proof: We use the same technique as before. Define (*) $c_\alpha^* = c_\alpha + f/\sqrt{n} + g/n$ and find f and g which are required to make a good approximation as follows:

$$
\begin{aligned}
\mathbf{P}(Z_n < c_\alpha^*) &= \Phi(c_\alpha^* + \frac{a(c_\alpha^*)}{\sqrt{n}} + \frac{b(c_\alpha^*)}{n} \\
&= \Phi(c_\alpha + \frac{f + a(c_\alpha)}{\sqrt{n}} + \frac{g + b(c_\alpha) + f a'(c_\alpha)}{n}.
\end{aligned}
$$

The final equality comes by substituting for c_α^* from (*), and noting that $a(c_\alpha^*) \approx a(c_\alpha) + f a'(c_\alpha)/\sqrt{n}$ and $b(c_\alpha^*) \approx b(c_\alpha)$ to the required degree of accuracy. Since $\Phi(c_\alpha) = \alpha$, we must set the higher order terms to zero to get the desired accuracy. Doing so yield $f = -a(c_\alpha)$ and $g = a(c_\alpha)a'(c_\alpha) - b(c_\alpha)$, proving the corollary.

14.3.2   Weighted Sum of Chi-Squares

In this section we study Edgeworth approximation of a weighted sum of chi square random variables. There are three separate reasons to focus on this example. First, the weighted sum of chi squares arises in a myriad of applications and unfortunately, there is no convenient closed form expression for the density. While numerous numerical procedures have been devised to compute this density exactly (see, for example, Lye (1991), Ansley *et al.* (1992), and also Abadir (1992) for different techniques and further references), analytical approximations are more useful for theoretical purposes. Second, we will see some features of asymptotics in the i.n.i.d. case; in particular, it is frequently more natural to let some quantity other than the sample size n approach infinity in such cases. Third, we will illustrate how to approximate using a chi-square density as a leading term. We will need the following preliminary result giving the CGF and the cumulants of the Gamma density.

Lemma 14.2 *Suppose $X \sim \Gamma(p, \lambda)$, so that X has density $f^X(x) = \lambda^p x^{p-1} \exp(-\lambda x)/\Gamma(p)$ for $x > 0$. Then the CGF of X is $K(X, \theta) = -p \log(1 - \theta/\lambda)$, and the cumulants of X are $\kappa_j(X) = p\Gamma(j)/\lambda^j$.*

Proof: The MGF of X is easily computed to be $\mathcal{L}_X(\theta) = (1 - \theta/\lambda)^{-p}$. Taking logs yields the CGF. Expanding $-p \log(1 - \theta/\lambda)$ in a Taylor series gives

$$K(\theta, X) = \sum_{j=1}^{\infty} \frac{p\theta^j}{j\lambda^j}.$$

Matching coefficients with $K(\theta, X) = \sum_{j=1}^{\infty} \kappa_j \theta^j/j!$ yields the formula of the lemma.

Let $X_i \sim \chi^2_{f_i} \equiv \Gamma(f_i/2, 1/2)$ be independent random variables with chi-square densities. We wish to approximate the distribution of $S_n = \sum_{i=1}^{n} a_i X_i$ for large n. Define $\mu_n = \mathbb{E}S_n = \sum_{i=1}^{n} a_i f_i$ and $\sigma_n^2 = \mathbb{V}ar\, S_n = 2 \sum_{i=1}^{n} f_i a_i^2$. We will approximate the distribution of the standardized sum Z_n defined as

$$Z_n = \sum_{i=1}^{n} \frac{a_i}{\sigma_n}(X_i - f_i).$$

We need the cumulant of Z_n, which are easy to get from the cumulants of the chi-square. By standardization, $\kappa_1(Z_n) = 0$ and $\kappa_2(Z_n) = 1$. For $j \geq 2$, we use $\kappa_j(Z_n) = \sum_{i=1}^{n}(a_i/\sigma_n)^j \kappa_j(X_i)$ and the cumulants of the chi-square density given in Lemma 14.2 above to get

$$\kappa_j(Z_n) = 2^{j/2-1}\Gamma(j)\frac{\sum_{i=1}^{n} f_i a_i^j}{(\sum_{i=1}^{n} f_i a_i^2)^{j/2}} = 2^{j/2-1}\Gamma(j)\sum_{i=1}^{n} f_i(a_i/\sigma_n)^j. \quad (14.5)$$

The lemma gives the skewness and kurtosis of the standardized sum Z_n and hence the standard Edgeworth expansion around the normal density can be computed by plugging these cumulants into Eqs. 14.3 or 14.4. In order for the approximation to be meaningful, it is essential that the conditions for the CLT hold. Assume that $f_i \leq K$; this is without loss of generality since if $X \sim \chi^2_{2K}$ then $X = Y + Z$ where $Y, Z \sim \chi^2_K$. The assumptions for the CLT require that the quantity $\Delta^2_n \equiv \max_{\{i:1 \leq i \leq n\}} a^2_i/\sigma^2_n$ must go to zero. This implies that for $j \geq 3$ we have

$$\kappa_j = C \sum_{i=1}^{n}(a_i/\sigma_i)^j \leq C \sum_{i=1}^{n}(a_i/\sigma_i)^2 \Delta^{j/2-1}_n = C\Delta^{j/2-1}_n.$$

Thus the j-th cumulant goes to zero at rate $\mathcal{O}(\Delta^{j/2-1}_n)$, and this may differ from the $n^{j/2-1}$ which occurs in the i.i.d. case. This condition puts a real restriction on the weights and restricts the applicability of the approximations.

When the a_i are all the same, S_n has a chi square density. If the a_i vary little, it is reasonable to expect a chi-square density to be superior to a normal density in approximating the density of Z_n. We first give a second order approximation via a shifted Gamma density.

Lemma 14.3 *Suppose* $\lambda = 2/\kappa_3(Z_n)$, $p = \lambda^2$, *and* $\mu = \lambda$. *The density of* $Y - \mu$, *where* $Y \sim \Gamma(p, \lambda)$ *provides an approximation of order* $\mathcal{O}(\Delta^{-1/2}_n)$ *to the density of* Z_n.

Proof: Using Lemma 14.2 above, the CGF of $Y - \mu$ to four terms is

$$K(\theta, Y - \mu) = \frac{p\theta}{\lambda} - \mu + \frac{p\theta^2}{2\lambda^2} + \frac{2p\theta^3}{3!\lambda^3} + \frac{6p\theta^4}{4!\lambda^4} + \cdots.$$

To get an approximation accurate to $\mathcal{O}(\Delta^{-1/2}_n)$ it suffices to match the first three cumulants of $Y - \mu$ to those of Z_n. Equating the mean, variance, and skewness yields the three equations $p/\lambda - \mu = 0$, $p/\lambda^2 = 1$, and $2p/\lambda^3 = \kappa_3(Z_n)$. Solving for p, λ and μ gives the result of the lemma.

Getting a higher order approximation requires a bit more care than just matching moments. Define $\delta = \kappa_4(Z_n) - 6p/\lambda^4 = \kappa_4(Z_n) - (3/2)\kappa^2_3(Z_n)$ to be the discrepancy between the fourth cumulant of $Y - \mu$ and Z_n.

Lemma 14.4 *Let* $f(y)$ *be the density of* $Y - \mu$ *as given in the previous lemma. Then* $f(y) + \delta f^{(4)}(y)/4!$ *is an order* $\mathcal{O}(\Delta^{-1}_n)$ *approximation to the density of* Z_n.

Proof: This follows after noting that

$$
\begin{aligned}
K(\theta, Z_n) &= \frac{\theta^2}{2!} + \frac{\kappa_3(Z_n)\theta^3}{3!} + \frac{\kappa_4(Z_n)\theta^4}{4!} + \cdots \\
&= \frac{\theta^2}{2!} + \frac{\kappa_3(Y - \mu)\theta^3}{3!} + \frac{\kappa_4(Y - \mu)\theta^4}{4!} + \frac{\delta\theta^4}{4!} + \cdots.
\end{aligned}
$$

Taking exponents on both sides, we get

$$
\mathcal{L}_{Z_n}(\theta) \approx \mathcal{L}_f(\theta) \times \exp(\delta\theta^4/4!) \approx \mathcal{L}_f(\theta) \times \left(1 + \delta\theta^4/4!\right).
$$

Now applying Lemma 14.1 to invert the MGF yields the lemma.

14.4   The Behrens-Fisher Problem

Asymptotic expansions are used in a wide variety of problems. See Rothenberg (1984a) for a survey of econometric applications. Since that survey, many other applications have been made in the literature. The example to be presented is selected for two reasons. First, this is the simplest special case of hypothesis testing in regression problems with nonscalar covariance matrix. After understanding the tools in this simplest setting, the reader should be able to tackle Rothenberg (1984b), which studies a generalization with important applications. Higher order asymptotic theory reveals that it is relatively inexpensive to adapt our procedures to nonidentity covariance structures. The additional loss from the lack of knowledge of the exact variances is of order $o(n^{-1})$. On the other hand, if we assume that the covariance matrix is a scalar multiple of identity when it is not, our losses are much higher. This basic principle is supported by the detailed Monte Carlo study of Thursby (1992), who shows that the Chow test fails badly in the presence of heteroskedasticity, while several tests which take heteroskedasticity into account perform well even when heteroskedasticity is not present.

A second purpose in choosing this example is that it illustrates the basic principles in deriving asymptotic expansions for nonlinear functions. Given $T(S_n)$, where S_n is converging to a limit μ, the delta method uses the approximation $T(S_n - \mu + \mu) \approx T(\mu) + (S_n - \mu)T'(\mu)$

to obtain a linearization. Higher order terms may be retained for higher order approximations. This tool is of fundamental importance and can be used in a wide variety of situations to derive large sample approximations.

In its simplest form, the Behrens-Fisher problem can be stated as follows. Suppose we observe $X_1, \ldots, X_{N_1} \sim N(\mu_1, \sigma_1^2)$ and also $Y_1, \ldots, Y_{N_2} \sim N(\mu_2, \sigma_2^2)$ and we wish to test the null hypothesis $H_0 : \mu_1 = \mu_2$. The source of the difficulty is the unequal variances in the two samples. Somewhat surprisingly, this makes an exact solution along the classical lines impossible. An exact non-classical solution developed by Weerahandi is described in Section 10.4. Below we present an approximate solution developed independently by Welch and Aspin. For some discussion, references, and a second order optimality property of the Welch-Aspin solution, see Pfanzagl (1974).

It is convenient to define $n_i = N_i - 1$. Then sufficient statistics are the sample means $\overline{X} = (1/N_1) \sum_{i=1}^{N_1} X_i$, $\overline{Y} = (1/N_2) \sum_{i=1}^{N_2} Y_i$, and the (MVU) variance estimates $\hat{\sigma}_1^2 = (1/n_1) \sum_{i=1}^{N_1} (X_i - \overline{X})^2$, $\hat{\sigma}_2^2 = (1/n_2) \sum_{i=1}^{N_2} (Y_i - \overline{Y})^2$. All of these quantities depend on the sample sizes N_1, N_2 but this will be suppressed in the notation. A natural statistic for testing is the difference of means $\Delta = \overline{X} - \overline{Y} \sim N(\mu_1 - \mu_2, \sigma_1^2/N_1 + \sigma_2^2/N_2)$. Define $\sigma_\Delta^2 = \sigma_1^2/N_1 + \sigma_2^2/N_2$. Classical testing principles suggest that we should divide the estimate Δ by an estimate $\hat{\sigma}_\Delta$ of its standard deviation and test for a significant difference from zero. The MVUE estimator for σ_Δ^2 is $\hat{\sigma}_\Delta^2 = \hat{\sigma}_1^2/N_1 + \hat{\sigma}_2^2/N_2$.

It is easily shown that $T = \Delta/\hat{\sigma}_\Delta$ is asymptotically $N(0,1)$, and hence rejecting for $T \geq c_\alpha$, where c_α is the level α critical value of a $N(0,1)$ distribution, gives a test with asymptotic level equal to α. Unfortunately, the finite sample distribution of T is sensitive to the nuisance parameters σ_1^2 and σ_2^2 and the level of the test can deviate significantly from α, especially in small samples. By making a second order approximation to the distribution of T, we can adjust the critical value c_α and obtain a test which is more nearly of constant size α. This is done as follows.

Theorem 14.4 *Define* $c_\alpha^* = c_\alpha - (\sigma_D^2/8)(c_\alpha^3 + c_\alpha)$, *where* c_α *is the level* α *critical value of* $N(0,1)$ *distribution, and* $\hat{\sigma}_D^2$ *is defined below in Eq. (14.6). Then rejecting* H_0 *for* $T = |\Delta/\hat{\sigma}_\Delta| > c_\alpha^*$ *gives an approximation* of order $o(n^{-1})$ to a level α test in large samples.

Proof: The basic idea is to make an Edgeworth approximation to the distribution of T and then use a Cornish-Fisher-type adjustment to get the critical value. In previous examples, we were able to calculate the cumulants of the statistic of interest exactly. Since this is difficult to do for T, we will approximate T by a Taylor series for which the calculations are easier. This technique of approximating the statistic of interest to the required order is of wide applicability.

1:Approximate T   Note that $T = \Delta/\hat{\sigma}_\Delta = Z/(\hat{\sigma}_\Delta/\sigma_\Delta)$, where $Z = \Delta/\sigma_\Delta \sim N(0,1)$. In large samples the ratio $\hat{\sigma}_\Delta/\sigma_\Delta$ will be close to 1. It is convenient to define $D = \hat{\sigma}_\Delta^2/\hat{\sigma}^2 - 1$, and find an approximation to T suitable for small values of D. We claim that

$$T = Z\,(1+D)^{-1/2} \approx \frac{\Delta}{\sigma_\Delta}\left(1 - D/2 + 3D^2/8 + \mathcal{O}_p(n^{-3/2})\right),$$

where $\mathcal{O}_p(n^{-3/2})$ represent neglected terms which remain bounded even after being multiplied by $n^{3/2}$. A fine point, indicated by the subscript p on the $\mathcal{O}$, is that the neglected terms are random variables (instead of real numbers, as earlier). 'Bounded' refers to 'bounded in probability', for which a sufficient condition is: $X_n \in \mathcal{O}_p(c_n)$ if $\mathbb{V}ar(X_n/c_n)$ is bounded.

To prove the claim, define weights $w_i = (\sigma_i^2/N_i)/(\sigma_1^2/N_1 + \sigma_2^2/N_2)$ for $i = 1, 2$ and note that $\mathbb{E}D = 0$ and $\sigma_D^2 = \mathbb{V}ar(D)$ can be computed as follows:

$$
\begin{aligned}
\sigma_D^2 &= \mathbb{V}ar\,\frac{\hat{\sigma}_D^2}{\sigma_D^2} = \mathbb{V}ar\,\frac{\hat{\sigma}_1^2/N_1 + \hat{\sigma}_2^2/N_2}{\sigma_1^2/N_1 + \sigma_2^2/N_2} \qquad (14.6)\\
&= \left(w_1\frac{\hat{\sigma}_1^2}{\sigma_1^2} + w_2\frac{\hat{\sigma}_2^2}{\sigma_2^2}\right) = \frac{2w_1^2}{n_1} + \frac{2w_2^2}{n_2} = \mathcal{O}(n_1^{-1} + n_2^{-1}).
\end{aligned}
$$

The last equality uses $\mathbb{V}ar(\hat{\sigma}_i^2/\sigma_i^2) = 2/n_i$, which follows from the fact that $\mathbb{V}ar(\chi_f^2) = 2f$. It is convenient to let $n = n_1 + n_2$ and assume that both n_1 and n_2 go to infinity at the same rate as n. Then D is of order $\mathcal{O}_p(n^{-1/2})$ since $n^{1/2}D$ will have bounded variance asymptotically. To get the approximation claimed for T, note that $T = (\Delta/\sigma_\Delta) \times (D+1)^{-1/2}$ and make a Taylor series expansion of $(1+D)^{-1/2}$.

2:Approximate Cumulants of T   This representation makes it easy to compute the cumulants of T to the required order. Note that $Z = \Delta/\sigma_\Delta$ is $N(0,1)$ and independent of $\hat{\sigma}_i^2$. It follows immediately that $\kappa_1(T) = \kappa_3(T) = 0$. We claim that $\kappa_2(T) = 1 + \sigma_D^2 + \mathcal{O}(n^{-3/2})$ and $\kappa_4(T) = 3\sigma_D^2 + \mathcal{O}(n^{-3/2})$. For the second cumulant, note that

$$\mathbb{E}T^2 = \mathbb{E}Z^2 \times \mathbb{E}(1 - D/2 + 3D^2/8)^2.$$

The first expectation is 1, while $(1 - D/2 + 3D^2/8)^2 \approx 1 - D + D^2 + \cdots$, which has expectation $1 + \mathbb{E}D^2 = \sigma_D^2$. For the fourth moment, note that $\mathbb{E}(\Delta/\sigma_\Delta)^4 = 3$. Also,

$$\mathbb{E}(1 - D/2 + 3D^2/8)^4 \approx \mathbb{E}1 - 2D + 3D^2 + \cdots \approx 1 + 3\sigma_D^2.$$

Using exercise 2 and substituting zero for odd moments, we get

$$\kappa_4(T) = \mathbb{E}T^4 - 3(\mathbb{E}T^2)^2 = 3(1 + 2\sigma_D^2) - 3(1 + \sigma_D^2)^2 = 3\sigma_D^2 + \mathcal{O}_p(n^{-3/2}).$$

3:Approximate CGF   This gives the following cumulant generating function for T:

$$\log \mathbb{E}\exp(\theta T) = \kappa_2(T)\theta^2/2 + \kappa_4(T)\theta^4/4! = \theta^2/2 + \sigma_D^2\left(\theta^2/2 + \theta^4/8\right).$$

Inverting this, we get the following Edgeworth density for T:

$$f^T(t) = \phi(t) + (\sigma_D^2/2)\phi^{(2)}(t) + (\sigma_D^2/8)\phi^{(4)}(t),$$

where $\phi(x)$ is the standard normal density, and $\phi^{(j)}$ is the j-th derivative of ϕ. Integrating gives us the CDF

$$
\begin{aligned}
F^T(t) &= \Phi(t) - \phi(t)\sigma_D^2\left\{H_1(t)/2 + H_3(t)/8\right\} \\
&= \Phi(t) - \phi(t)(\sigma_D^2/8)\left\{4t + (t^3 - 3t)\right\} \\
&= \Phi(t) - \phi(t)(\sigma_D^2/8)(t^3 + t).
\end{aligned}
$$

Using the technique of Corollary 14.3.1, we get the adjusted critical value for the T test claimed above. Note that σ_D^2 involves unknown parameters σ_j^2 which must be replaced by estimates $\hat{\sigma}_j^2$; however, this does not affect the order of approximation.

14.5   Higher Order Accuracy of Bootstrap Estimates

Bootstrap estimates were introduced earlier. Applying the techniques of Edgeworth expansion sheds light on the properties of such estimate. In many situations, there are many possible methods one can use to implement bootstrapping. Standard asymptotic theory does not tell us which of various possible bootstrap methods are better, but the Edgeworth expansions reveal important differences. Roughly speaking, in many situations, bootstrapping can give us greater accuracy than the standard asymptotic normal approximation. However, achieving these gains depends on applying the bootstrap properly. The rest of this chapter is intended to provide enough examples of both the use and the misuse of bootstrapping to give a sense of how the bootstrap should be used in general situations.

14.5.1   Match Resample to Sample

We recall the basic concept of bootstrapping, as introduced in Section 12.5. Given an i.i.d. sample $X_1, \ldots, X_n$ with common distribution F, suppose we wish to estimate the distribution of some function $\theta(X_1, \ldots, X_n)$. If the distribution F is known, this is easily done by Monte Carlo — generate an i.i.d. sample $X_1', \ldots, X_1'$ from F, calculate $\theta(X_1', \ldots, X_n')$, and record the value as θ_1'. Repeat this process 1000 times to get $\theta_1', \ldots, \theta_{1000}'$. The empirical distribution of θ_j' will approximate the true distribution of $\theta(X_1, \ldots, X_n)$ with an arbitrarily high degree of accuracy as we increase the number of Monte Carlo repetitions. The bootstrapping technique is used when the distribution F is unknown. In this case, we replace F by $\hat{F}_n$, the empirical distribution function of the observed sample, and then proceed exactly as in the case of the Monte Carlo method. Note that the empirical distribution $\hat{F}_n$ assigns probability $1/n$ to each of the observed values $X_1, \ldots, X_n$. Thus an i.i.d. sample $X_1^*, \ldots, X_n^*$ from $\hat{F}_n$ will consist of the original observations, scrambled up, some repeated and some omitted, at random. This is the bootstrap sample, also called a 'resample' for obvious reasons. Since $\hat{F}_n$ converges F by the Glivenko-Cantelli theorem, bootstrapping should be nearly as good as Monte Carlo (which requires knowledge of the true distribution) asymptotically. In fact this simple description is not adequate, and we will find that complex situations require some care in formulating a correct version of the bootstrap.

The first lesson in using a bootstrap is that the resample should be as closely matched to the sample as possible. Any properties known to hold for the original sample should also hold for the resample. An example of bootstrap failure discussed earlier in Section 12.8 showed that if the original sample is heteroskedastic, the resample (which is i.i.d. and hence homoskedastic) does not succeed. In many situations, it is possible to transform the problem to make the elements i.i.d. or as close to it as possible. We must also be careful to impose known properties such as zero mean or symmetry on the residuals. By using some care in resampling, one can often succeed in making the bootstrap work in situations where a naive application fails. For example, a variation of the bootstrap suggested by Wu (1986) (see also Jeong and Maddala(1992)) succeeds in the heteroskedastic regression model where the naive version fails. This section provides a simpler example of how matching resampling to known properties of the sample improves the bootstrap.

Suppose $X_i \sim \chi_1^2$ are i.i.d. and let $Z_i = (X_i - 1)/\sqrt{2}$ be the standardized X_i. Let $S_n = (1/n)\sum_{i=1}^n Z_i$ be the average of the Z_i, and suppose we wish to approximate the distribution of the S_n. This is the same setup used to illustrate the Edgeworth approximation in Figure 14.1. We will see that the bootstrap competes favorably with the first order Edgeworth approximation, but not with the second order. A *naive* bootstrap procedure is the following: Let $Z_1^*, \ldots, Z_n^*$ be a resample of $Z_1, \ldots, Z_n$. Calculate $S_n^* = (1/n)\sum_{i=1}^n Z_i^*$. Repeated computations of S_n^* will generate a scatter of values which we could take to be the bootstrap estimate for the distribution of S_n. If we know that the Z_i variables have been standardized, this is a very poor estimator. Intuitively, this is because the known properties of Z_i after standardization, namely mean 0 and variance 1, have not been imposed on the resampling scheme. A theoretical analysis is given below.

The analysis of the performance of the bootstrap is straightforward. The Z_i^* are i.i.d. random variables with take values in the set of observations $(Z_1, \ldots, Z_n)$ and each observation has equal probability $1/n$.

We can compute all aspects of i.i.d observations on discrete variables
with great ease both theoretically and via Monte-Carlo. The moments
of the Z_i^* are easily seen to be

$$\mathbb{E}(Z_1^*)^\alpha = \frac{1}{n} \sum_{i=1}^n Z_i^\alpha.$$

In particular, $\mathbb{E}Z_i^* = (1/n)\sum_{i=1}^n Z_i = \overline{Z}_n$ and $\mathbb{V}ar\, Z_i^* = (1/n)\sum_{i=1}^n (Z_i - \overline{Z}_n)^2$. Note that the *theoretical* or *population* mean
and variance of bootstrap sample Z_i^* are the *sample* mean and variance
of the original sample Z_i^2. Since the S_n^* are averages of i.i.d. discrete
variables Z_i^*, we can easily analyze the distribution of S_n^*; to be more
precise, the conditional distribution of S_n^* given the observed samples
$Z_1, \ldots, Z_n$. Using the central limit theorem, we can conclude that for
large n, the distribution of $\sqrt{n}S_n^*$ will be approximately normal with
mean $\sqrt{n}\overline{Z}_n$ and variance $\hat{\sigma}_n^2 = (1/n)\sum_{j=1}^n (Z_i - \overline{Z}_n)^2$. Note that for
large n $\sqrt{n}\overline{Z}_n$ is $N(0,1)$. It follows that even asymptotically, the mean
of the naive bootstrap $\sqrt{n}S_n^*$ will be $N(0,1)$ instead of 0, which is the
mean of the $\sqrt{n}S_n$.

This problem arises because the bootstrap variables Z_i^*, which equal
one of the n original observations with equal probability, have expected
value $\overline{Z}_n$. However, we know that the original sample Z_i has mean 0. A
less naive bootstrap would be based on resampling from $Y_i = Z_i - \overline{Z}_n$;
this modifies the original sample to force it to have mean 0. Now
let Y_i^* be an i.i.d. resample of Y_i and define $S_n^{**} = (1/n)\sum_{i=1}^n Y_i^*$.
Consider the distribution of $\sqrt{n}S_n^{**}$ as a bootstrap approximation to the
distribution of $\sqrt{n}S_n$. Using the same line of reasoning as before, the
asymptotic distribution of $\sqrt{n}S_n^*$ will be $N(0, (1/n)\sum_{i=1}^n Y_i^2)$. Since
the sample variance estimate $(1/n)\sum_{i=1}^n Y_i^2$. converges to the true
variance 1, the asymptotic distribution of $\sqrt{n}S_n^*$ will be the same as that
of $\sqrt{n}S_n$. Thus the less naive bootstrap is successful to some extent,
while the naive bootstrap fails completely. Nonetheless, the less naive
bootstrap is quite poor as an estimator of the distribution of S_n. Its
variance depends on the sample variance which fluctuates substantially

---

[2]The population variance of Z_1^* is $(1/n)\sum_{i=1}^n (Z_i - \overline{Z}_n)^2$ whereas the sample
variance of Z_i is frequently defined to be $\sum_{i=1}^n (Z_i - \overline{Z}_n)^2/(n-1)$ so there is a small
difference, which we will ignore in the present analysis.

in small to moderate samples. If by luck the sample variance is close
to 1, then the bootstrap performs reasonably. In unlucky situations
(which are fairly common in small samples) it can perform very poorly.
We can put it more precisely by saying that the less naive bootstrap is
an $o(n^0)$ order approximation to the distribution of S_n, while the CLT
gives a better $\mathcal{O}(n^{-1/2})$ approximation.

For an illustration, see Figure 14.2 which plots the less naive bootstrap approximation to the density of S_n as a dotted line. In (A), (B),
and (C), the less naive bootstrap is completely off the mark, while in
(D), it performs adequately. The sample of size 20 used in (D) hap-
pened to satisfy the condition $\sum_{i=1}^2 0 Y_i^2 \approx 1$. However this event is of
low probability in small samples, and hence the less naive bootstrap
will be quite erratic in small samples. It is important to note that the
performance of the bootstrap depends on the sample at hand, and not
on the underlying probability distribution. For exactly the same probabilistic setup, the bootstrap will perform differently on different trials
as the random sample drawn from the distribution will vary from one
sample to the next. Thus, we can say that for most samples, the sample
estimates will be relatively far from the population values and the less
naive bootstrap will perform poorly. As the sample size increases, the
sample estimates will become closer to the true values and hence the
performance of the less naive bootstrap will improve. In practice, even
for samples of size 100 we found it to be very erratic.

The less naive bootstrap fails because it estimates the variance,
which is known to be one. If we define

$$Y_i = (Z_i - \overline{Z}_n)/\sqrt{(1/n)\sum(Z_i - \overline{Z})^2},$$

then the Y_i have sample mean 0 and sample variance 1. Now the
sample on which the resampling is to be based matches the population
in terms of known quantities. When the bootstrap is done in this
(correct) way, it displays superior properties. This can be seen in Figure
14.2. In all four cases displayed, which are representative, the bootstrap
approximation to the density of S_n displays skewness and is closer to
the true gamma density than the normal approximation obtained from
the CLT. Somewhat exceptionally, in Figure 14.2 (B), the bootstrap
density is almost the same as the normal; this particular sample of

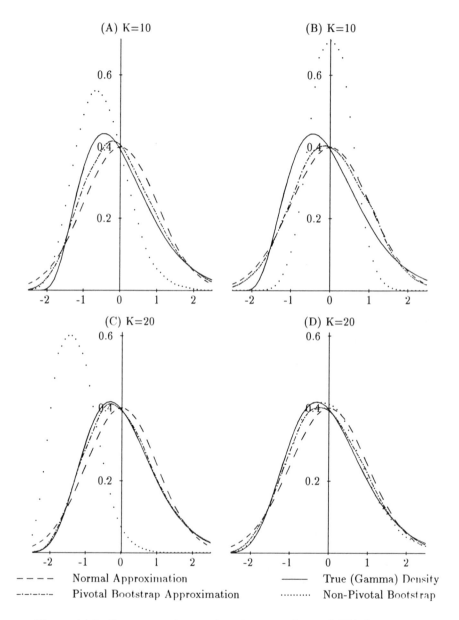

Figure 14.2: Bootstrap Approximation to a Sum of Chi Squares

10 observations happened to be nearly symmetric so that the sample measure of skewness was close to zero. As a result the bootstrap density was close to the normal approximation. It is theoretically possible that on a perverse sample, the observations happen to come out skewed in the opposite way from the true density, so that the bootstrap will be worse than the normal approximation. The probability of such events is low and becomes lower with larger sample sizes.

The diagrams show clearly that the bootstrap captures the skewness in the distribution of S_n and hence is more accurate than the symmetric normal asymptotic approximation. Navidi (1989) has shown that this phenomenon also holds in regression models. Depending on the behavior of the regressors, the bootstrap is as good as or better than the asymptotic normal approximation to the distribution of the OLS estimators. Used properly, the bootstrap provides additional accuracy over the standard asymptotic normal approximation. In other words, it is an automatic way of making a one-term Edgeworth correction. In fact, Hall (1992, Appendix 5) shows that the bootstrap has an edge over a one-term Edgeworth approximation, especially in the tails of a distribution.

Density Estimation: Figure 14.2 illustrates the performance of various types of bootstrap procedures by drawing the density of the bootstrap sample against the true density. To do this, we need to estimate a density from a given random sample. We drew bootstrap samples of size 10,000 and used one of the simplest density estimators to construct the figures. Given an i.i.d. sample $X_1, \ldots, X_n$, a kernel estimate for the common density $f(x)$ is

$$\hat{f}(x) = \frac{1}{nh_n} \sum_{i=1}^{n} \phi\left(\frac{x - X_i}{h_n}\right),$$

where $\phi(x) = \exp(-x^2/2)/\sqrt{2\pi}$ is the $N(0,1)$ density, and $h_n = 1.06\sigma n^{-1/5}$. This estimate is optimized to work best if the (unknown) estimated density is $N(0, \sigma^2)$. Since we were dealing with cases where the asymptotic density was close to $N(0,1)$, we used this formula with $\sigma = 1$. Another effective choice which is more robust to nonnormality and bimodality of the target density is to put $h_n = 0.9An^{-1/5}$, where

A = min(Sample Std. Deviation, (Sample Interquartile Range)/1.34). This second choice should be used when approximate normality of the target density cannot be assumed. See Silverman (1986, Section 3.4.1) for these and other more sophisticated density estimates.

14.5.2 Use of Theory to Guide the Bootstrap

Typically the bootstrap provides one additional term in an Edgeworth expansion and thus higher order accuracy. Several authors have suggested that if we use theory to guide the bootstrap, it can provide even higher accuracy. A version of this idea was formulated as a guideline for bootstrapping by Hall and Wilson (1991): they suggest that we should bootstrap pivotal quantities, or at least asymptotically pivotal quantities. Recall that a statistic T is (asymptotically) pivotal if its (asymptotic) distribution does not depend on any parameters. Heuristically, the reason for the superior performance of bootstrapping for pivotal quantities is quite simple. Suppose we bootstrap an asymptotically pivotal statistic T. Then the Edgeworth expansion for the distribution of T will have the form $\Phi(x) + n^{-1/2}\phi(x)A_1(x) + \cdots$. Because T is asymptotically pivotal, only A_1 and higher order terms in the expansion can depend on unknown parameters.[3] The Edgeworth expansion for the distribution of a bootstrap version of T^* will be of the form $\Phi(x) + n^{-1/2}\phi(x)\hat{A}_1(x) + \cdots$. The difference between $A_1(x)$ and $\hat{A}_1(x)$ is usually of order $\mathcal{O}(n^{-1/2})$ so that the difference between the Edgeworth expansion of T and that of T^* will be of order $\mathcal{O}(n^{-1})$.

If the distribution of T is not pivotal, the first term of its Edgeworth expansion will depend on unknown parameters — typically, it will have the form $N(0, \sigma^2)$. The Edgeworth expansion for the bootstrap version T^* will have leading term $N(0, \hat{\sigma}^2)$, where $\hat{\sigma}^2$ is the sample estimate of σ^2. The error of the bootstrap is then controlled by the error of $\hat{\sigma}^2$ which is usually of order $\mathcal{O}(n^{-1/2})$. This is of the same order as the error of the standard asymptotic normal distribution.[4] Sometimes it is possible

---

[3]If the first term depended on unknown parameters, the dependence would persist as $n \to \infty$.

[4]Thus use of pivotal quantities is helpful in improving the accuracy of the bootstrap, as discussed in detail by Hall (1992). In Section 3.10, he also discusses situations where this principle does not work well. In typical situations, one has to

in such cases to use a more accurate approximation to the variance of T derived from theory. Such an adjustment can substantially improve bootstrap performance. An example based on asymptotic expansions developed by Rothenberg (1984) is given by Rayner (1990).

We use asymptotic theory to obtain an asymptotically pivotal quantity — that is, one which has no unknown parameters in the first term of its asymptotic expansion. It is also possible to form statistics which have no unknown terms in the second order term. Bootstrapping such statistics can lead to even higher order accuracy. Several authors have suggested these kinds of extensions. See Rayner (1990) and also Ip and Phillips (1994) for examples of these techniques.[5] Another way to improve the bootstrap is to use transformations designed to reduce or eliminate skewness, or otherwise bring the target distribution closer to normality. Such transformations reduce the higher order terms and tend to thereby improve bootstrap performance.

14.5.3   Use the 'Analogy' Principle

Important insight into the behavior and use of the bootstrap method is obtained via the principle of analogy to be described in the present section. This is labeled the 'bootstrap principle' and used as the fundamental principle of bootstrap inference by Hall (1992). His exposition of the principle via the Russian 'matryoshka' dolls is delightfully clear and imaginative, and readers are urged to read it for themselves in the first chapter of Hall (1992). We will give a more prosaic (and also more limited in scope) description of the principle here.

Consider the first order autoregressive model, with $y_t = \rho y_{t-1} + \epsilon_{t-1}$, where ϵ_t are i.i.d. $N(0, \sigma^2)$ for $t = 1, 2, \ldots, T$. Suppose for simplicity that $y_0 = 0$. Suppose we wish to learn the distribution of $\hat{\rho} = (\sum_{t=1}^{T} y_t y_{t-1})/(\sum_{t=1}^{T} y_{t-1}^2)$, the OLS estimator of ρ. To be spe-

---

divide by an estimate of the variance to get a pivotal quantity (there are exceptions, where dependence upon unknown parameters can be removed by other techniques). If the variance estimate is very poor, then dividing by it can reduce the performance of the bootstrap.

[5]Note that if, as indicated by Hall(1992), the Edgeworth is worse than the bootstrap in the tails, doing a one term approximation by Edgeworth and a higher order approximation by the bootstrap may not produce much benefit in the tails.

cific, suppose we wish to estimate the bias of $\hat{\rho}$ — note the OLS is biased because of the stochastic regressors which are not ancillary here. Put $\hat{\rho}$ in place of the true parameter, and generate a bootstrap sample $e_1^*, \ldots, e_T^*$ *either* by (i) resampling the OLS residuals, *or* by (ii) generating an i.i.d. sample from a $N(0, \hat{\sigma}^2)$. It seems plausible that we should prefer method (ii) (which is called the parametric bootstrap) when the distribution of the errors is known to be normal, since this restriction is imposed on the resample. On the other hand, if there are doubts about normality, the nonparametric bootstrap of method (i) may be preferred since it gives the errors a chance to show there true distribution. Let $y_0^* = y_0 = 0$ and generate a resample of y's using $y_t^* = \hat{\rho} y_{t-1}^* + e_t^*$. Let $\rho^* = \sum_{t=1}^{T} y_t^* y_{t-1}^* / \sum_{t=1}^{T} y_{t-1}^{*2}$ be the OLS estimate based on the bootstrap sample. The asymptotic distribution of the OLS is known to be $\hat{\rho} \sim N(\rho, \sigma^2 / \sum_{t=1}^{T} y_{t-1}^2)$ for $\rho \neq 1$. It is known that the OLS is biased in finite samples. We propose to use bootstrapping to find out the extent of the bias. Abadir (1992,1993) has developed closed form formulae for asymptotic and some exact distributions related to the autoregressive regression model under study. Note however, that the problem of finding the distribution (and bias) of $\hat{\rho}$ is not solved by calculation of its exact or asymptotic distribution, since these typically depend on unknown parameters. It requires additional study to determine how well the exact distributions perform when the parameters required are replaced by estimates.

The analogy principle exploits the fact that the estimate ρ^* bears the same relationship to $\hat{\rho}$ as $\hat{\rho}$ itself bears to the true unknown parameter ρ. That is, the process used to generate $\hat{\rho}$ from ρ is *identical* to the process used to generate ρ^* from $\hat{\rho}$. Thus we should be able to make inference about the bias $\hat{\rho} - \rho$ by looking at $\rho^* - \hat{\rho}$. Consider the original sample as fixed, and generate many bootstrap samples of the residuals $e_1^*, \ldots, e_T^*$. For $b = 1, 2, \ldots, B$, each bootstrap sample generates an estimate ρ_b^*. The bias of ρ^* as an estimator of $\hat{\rho}$ is $\mathbb{E}(\rho^* - \hat{\rho} | y_0, \ldots y_T)$. This can easily be estimated by using $(1/B) \sum_{b=1}^{B} \rho_b^*$ as an estimate for $\mathbb{E}\rho^*$. One can correct for bias by adding the estimated bias to the original estimator.

Instead of studying $\rho^* - \hat{\rho}$, it is convenient to standardize and study $S = (\rho^* - \hat{\rho}) / \sum_{t=1}^{T} y_t^2$, which should, according to the analogy principle,

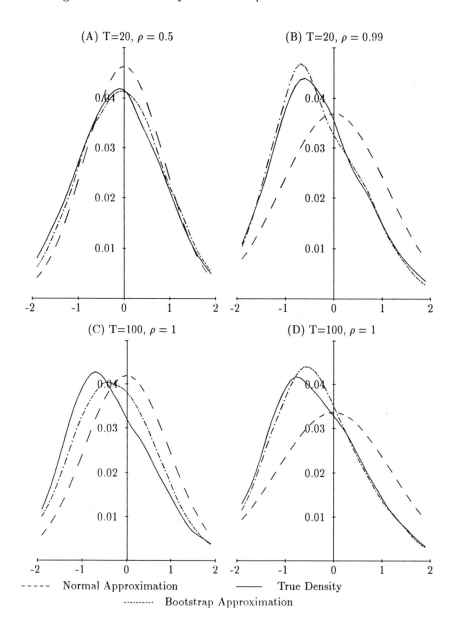

Figure 14.3: Approximations to OLS Density in AR-1 Model

be similar to the distribution of $(\hat{\rho} - \rho_0)/\sum_{t=1}^{T} y_t^2$. This second distribution is known to be $N(0, \sigma^2)$ asymptotically, but deviates from this substantially in finite samples, especially when ρ_0 is close to 1. Figure 14.3 studies the performance of the bootstrap estimator of the density of S with the true density (estimated by Monte Carlo, based on a sample size of 10,000), and also the asymptotic normal approximation.

Let $S_1^*, \ldots, S_B^*$ be samples of the statistic S, with T^p generated using the nonparametric bootstrap as described earlier. Figures 14.3 (A)-(D) show several important features of the bootstrap which we discuss in detail here. Before discussing the figures, it is important to understand that the performance of the bootstrap is conditional on the observations $y_1, \ldots, y_T$. For example, if we simply plot the distribution of ρ^* the (naive) bootstrap estimate, it will provide a very bad approximation to the distribution of $\hat{\rho}$ *except* when we are lucky enough to get a sample where $\hat{\rho}$ is close to the true value and also $\hat{\sigma}^2$ is also close to the true value. Thus the bootstrap will work well for some types of samples and poorly for others. In studying the behavior of the bootstrap, it is essential to do repeated experiments to determine how reliable the bootstrap is in a variety of samples. With this basic principle understood, we go on to discuss the many lessons we can learn from the figures.

Figure 14.3 (A) shows the behavior of bootstrap estimates when the errors are $N(0, 1)$, the sample size is $T = 20$, and $\rho = 0.5$. The bootstrap provides an excellent approximation to the true distribution. More precisely, the observed density of S^* resembles the density of $\hat{\rho}$ very closely. We found only slight differences between nonparametric and parametric bootstrap, both with normal and nonnormal errors, so we have just reported results for the nonparametric bootstrap. The uniformly good behavior (not shown in the figures) over many samples, a large range of values for sample size ($T = 20$ to $T = 200$), and a large range of ρ values ($\rho = 0.5$ to $\rho = 1.0$) was somewhat surprising. The only occasional difficulties arose for ρ close to 1, where the bootstrap distribution of T^p deviated somewhat (but not drastically) from the target distribution of the standardized density of $\hat{\rho}$; this took place for a minority of the samples. Figure 14.3 (B) gives a typical graph for $T = 20, \rho = 0.99$, a close-to-unit root case. As can be seen, the bootstrap works just fine.

Since the bootstrap is 'known' theoretically to fail in the unit root case (see Basawa *et al.* (1989)), the outcomes graphed in Figures 14.3 (C) and (D) for $T = 100$ and $\rho = 1.00$ were a bit surprising to the author. Although some evidence of 'failure' is evident in (C), the bootstrap is working well in (D), which was a typical case. In practice we found that the bootstrap provides approximations which are uniformly good in small samples. As the sample size increases, the probability of observing a sample where the bootstrap does not perform too well increases. Nonetheless, up to $T = 200$, the largest sample size we experimented with, with high probability the bootstrap works well, and when it fails, the failure is not drastic (as in (C) for example). Basawa *et al.* (1991) show how to modify the bootstrap so as to make it work in the case of unit roots. This modification, which sheds light on the Monte Carlo results above, shows that the discontinuity in the distribution of the OLS at $\rho = 1$ can be removed if we exclude a small portion of the sample. For any constant c, if we retain observations up to N satisfying $\sum_{t=1}^{N} y_t^2 < c\sigma^2$, the resulting OLS estimator is asymptotically normal even in the unit root case. The probability of this inequality holding (for fixed c) for the whole sample is high for small samples and decreases slowly; the bootstrap works fine for samples for which the inequality holds. This shows why the bootstrap never behaves very badly in unit root cases in our simulations, and also behaves excellently in some cases.

14.6   Bootstrap-Bartlett Corrections

As discussed in the previous section, it is frequently the case that asymptotic distributions of test statistics are poor approximations to the finite sample distribution of the test. We have discussed, in the Behrens-Fisher example, how asymptotic expansions can be used to get more accurate finite sample approximations, and to adjust the critical values for the tests. As an example of relevance in econometrics, Horowitz (1994) showed that critical values computed for White's Information Matrix test via bootstrap perform much better in obtaining a stable size in finite samples than the standard asymptotic theory. The above procedures leave the test statistic alone and work on changing the critical value to get greater accuracy. There is another technique used to achieve the same goal, where we adjust the test statistic itself so that its finite sample distribution is a closer match to its asymptotic distribution. This is known as the Bartlett correction. For a more detailed discussion and references, see Stuart and Ord (1991).

The Bartlett correction is based on a very simple idea, but in several examples it provides dramatic increases in accuracy for likelihood ratio tests. From general asymptotic theory we know that the $\lambda = -2\log(LR)$ is distributed asymptotically as a χ_q^2, where q is the number of restrictions imposed under the null hypothesis. Asymptotically λ/q has expected value unity. Find $c = \mathbb{E}\lambda/q$ and define $\lambda' = (\lambda/q)/c$. By forcing the finite sample expected value to equal one, we must improve the fit of the asymptotic distributon. Because of the complicated form of typical LR statistics, it is typically difficult to compute the expected value. It is usually much easier to find an asymptotic expansion of the expected value taking the form

$$\mathbb{E}\frac{\lambda}{q} = 1 + \frac{b_1}{n} + \mathcal{O}(n^{-3/2}). \tag{14.7}$$

Then if we define $\lambda' = \lambda/(1 + b_1/n)$, it is clear that $\mathbb{E}\lambda'$ will be closer to q than the original statistic λ. It is also likely, therefore, that the asymptotic distribution of λ' will be closer to χ_q^2 than that of λ. λ' is called a Bartlett correction of λ. For a simple correction, experience in a variety of problems shows that the Bartlett's correction performs remarkably well. Some theoretical explanation for this is given by Lawley (1956) who shows that the correction not only corrects the first moment but that *all* cumulants of the corrected statistic λ' match that of a χ_q^2 density to the order $\mathcal{O}(n^{-3/2})$.

According to standard methodology, in order to Bartlett-correct a likelihood ratio statistic, we require an asymptotic expansion of Eq. (14.7) for the expected value of the statistic. Typically the term b_1 occurring in the expansion will involve unknown parameters which then need to be estimated. Substituting a $\sqrt{n}$-consistent estimate of b_1 for b_1 still preserves the asymptotic accuracy of the Bartlett correction. In econometric problems of typical complexity, it can be a somewhat painful undertaking to calculate the required asymptotic expansion which enables us to make a Bartlett correction. Rocke (1989) observed that the bootstrap can save us from these calculations and demonstrated the effectiveness of the bootstrap-Bartlett correction in a seemingly unrelated regression model. Rayner (1990) shows theoretically that the bootstrap-Bartlett correction will provide accuracy to order $\mathcal{O}(n^{-3/2})$. Interestingly, in the case he studies, only the LR is improved by the Bartlett correction. Rayner shows that applying a similar correction to the Wald and LM tests fails to bring about the desired improvement.

Below, we give detailed instructions for applying the bootstrap-Bartlett correction to a frequently occurring and important hypothesis testing problem. Consider a regression model $y = X\beta + \epsilon$, where $\epsilon \sim N(0, \Sigma)$. Assume the covariance matrix Σ is parameterized by some small number of parameter θ. Consider testing the linear hypothesis H_0: $R\beta = r$, where R is a $q \times k$ matrix of rank q representing q linearly independent restrictions on the parameters. Let $\hat{\beta}, \hat{\theta}$ be the unconstrained ML estimates of the model parameters, and let $\tilde{\beta}, \tilde{\theta}$ be the ML estimates subject to the constraint imposed by the null hypothesis. Draw a bootstrap sample of errors $e^* \sim N(0, \Sigma(\hat{\theta}))$ from the estimated error distribution.[6] Since we are concerned with the null distribution, generate bootstrap values of the the dependent variables y from $y^* = X\hat{\beta} + e^*$. Now using the bootstrap sample (y^*, X) compute $\hat{\beta}^*, \hat{\theta}^*$ and $\tilde{\beta}^*, \tilde{\theta}^*$, the constrained and unconstrained bootstrap estimates. With likelihood function $f(\cdot)$ given by

$$f(y, X, \beta, \theta) = \det(\Sigma(\theta))^{-1/2} \exp\left\{-(1/2)(y - X\beta)'\Sigma^{-1}(\theta)(y - X\beta)\right\},$$

we get bootstrap LR statistics by defining

$$\lambda^* = -2\left(\log f(y^*, X, \hat{\beta}^*, \hat{\theta}^*) - \log f(y^*, X, \tilde{\beta}^*, \tilde{\theta}^*)\right).$$

Repeat this procedure 1000 times (for instance) and average the observed values $\lambda_1^*, \ldots, \lambda_{1000}^*$ to get the estimate $\overline{\lambda}^*$ of the average value

---

[6]To actually implement this, one can use a Cholesky decomposition $PP' = \Sigma(\hat{\theta})$ of Σ, where P is lower triangular. If $u \sim N(0, \mathbf{I})$ can be generated, Pu will be $N(0, PP')$ and hence have the desired distribution. The Cholesky decomposition is an easily computable 'square root' of Σ.

of the likelihood ratio. Now $q\lambda/\overline{\lambda}^*$ is the Bartlett corrected value of the likelihood ratio statistic.

14.7   Exercises

1. Using the definition of cumulants as the derivatives of the CGF evaluated at 0 $(\kappa_j(X) = K^{(j)}(0, X))$, show that $\kappa_j(aX) = a^j\kappa_j(X)$.

2. By expanding $\exp(\theta X)$ in a Taylor series, show that

$$K(\theta, X) = \log\left(1 + \sum_{j=1}^{\infty} \mathbb{E}X^j\frac{\theta^j}{j!}\right).$$

Next, show that $\log(1 + x) = x - x^2/2 + x^3/3 - \cdots$. Deduce that

$$K(\theta, X) = \sum_{k=1}^{\infty}(-1)^{k+1}\frac{1}{k}\left(\sum_{j=1}^{\infty}\mathbb{E}X^j\frac{\theta^j}{j!}\right)^k.$$

Show that the definition of cumulants ($\kappa_j = K^{(j)}(0, X)$) implies that the cumulant generating function has the following expansion:

$$K(\theta, X) = \sum_{j=1}^{\infty}\kappa_j\frac{\theta^j}{j!}.$$

Match the terms in the Taylor series to get the following equations for the first four cumulants in terms of the raw moments:

$$
\begin{aligned}
\kappa_1 &= \mathbb{E}X \\
\kappa_2 &= \mathbb{E}(X^2) - (\mathbb{E}X)^2 = \mathbb{Var}(X) \\
\kappa_3 &= \mathbb{E}(X^3) - 3\mathbb{E}(X^2)(\mathbb{E}X) + 2(\mathbb{E}X)^3 = \mathbb{E}(X - \mathbb{E}X)^3 \\
\kappa_4 &= \mathbb{E}X^4 - 4\mathbb{E}X^3\mathbb{E}X - 3(\mathbb{E}X^2)^2 + 12\mathbb{E}X^2(\mathbb{E}X)^2 - 6(\mathbb{E}X)^4 \\
&= \mathbb{E}(X - \mathbb{E}X)^4 - 3(\mathbb{Var}(X))^2.
\end{aligned}
$$

3. Let $Z_n = \sum_{i=1}^{n} a_iX_i$ where a_i are arbitrary real constants, and $X_i \sim \chi^2_{f_i}$ are independent chi-squares. We consider alternative approximations for $Z_n^* = (Z_n - \mathbb{E}Z_n)/\sqrt{\mathbb{Var}(Z_n)}$.

(a) Suppose $X \sim N(\mu, \mathbf{I}_k)$ and let $Y = \|X\|^2$. Then we say that Y has a noncentral chi-square density with k degrees of freedom and noncentrality parameter $\delta = \|\mu\|^2/2$. Show that the cumulant generating function of Y is

$$K(\theta, Y) = -\frac{k}{2}\log(1 - 2\theta) + \frac{2\delta\theta}{1 - 2\theta}.$$

(b) Consider approximating the distribution of Z_n^* by $\lambda Y + \mu$. We have four parameters λ, μ, δ, k in the distribution of $\lambda Y + \mu$. Expand the CGF $K(\theta, \lambda Y + \mu)$ to the fourth order. Match coefficients to get four equations. Unfortunately, these equations can only be solved for the four parameters by symbolic mathematics package such as Maple or Mathematica! This third order approximation is substantially more complicated than the one based on central chi-square, and hence not recommended.

(c) As discussed in the text, the approximation based on a shifted chi-square has support different from the support of Z_n^*. To solve this problem, fix $\mu = \mu_n/\sigma_n$, and consider the cumulant generating function $K(\theta, Y - \mu) + c_3\theta^3 + c_4\theta^4$, where $Y \sim \mathcal{G}(p/2, \lambda/2)$. By using μ to match the support of Z_n^*, we lose one order of approximation. To compensate, we need to add two constants c_3 and c_4 to get an equivalent order of approximation. Calculate the values of c_3 and c_4 needed to equate the cumulants. Invert the MGF $\exp(K(\theta, Y - \mu)) \times \exp(c_3\theta^3 + c_4\theta^4)$ to calculate the density which approximates Z_n^* and has the same support. Note that you will need to track the order of the terms in the expansion of $\exp(c_3\theta^3 + c_4\theta^4)$.

4. This exercise goes through the derivation of an Edgeworth approximation to the chi-square density, with enough details to permit replication of the Figure 14.1 via a computer. The basic setup is the following. We observe $X_1, X_2, \ldots, X_n$ independent and identically distributed as chi-squares with one degree of freedom (χ_1^2). Define $Z_i = (X_i - 1)/\sqrt{2}$. Then Z_i is standardized to

have mean 0 and variance 1. Define $S_n = (1/n)\sum_{i=1}^n Z_i$. According to the CLT, S_n is $N(0,1)$. We will study how to get better approximations to the distribution of S_n in this exercise.

(a) Show that the density of S_n is

$$f^{S_n}(s) = \frac{(n/2)^{n/4}}{\Gamma(n/2)}(s + \sqrt{n/2})^{n/2-1}\exp(-\sqrt{n}s/\sqrt{2}),$$

for $x \geq -\sqrt{n/2}$ and 0 if $x \leq \sqrt{n/2}$. Since the Gamma function blows up for large values, compute $f(s)$ as $\exp(\ln f(s))$ using a direct formula for the LNGAMMA(n/2)=$\log(\Gamma(n/2))$ — a formula for calculating LNGAMMA can be obtained from Abramovitz and Stegun (1965), a generally valuable resource for finding numerical formulae to compute commonly needed functions.

(b) If $Y \sim \mathcal{G}(p, \lambda)$, verify that the MGF is $\mathbb{E}\exp(\theta Y) = (1 - \theta/\lambda)^{-p}$, and hence the cumulant generating function is $K(\theta, Y) = -p\log(1 - \theta/\lambda)$. By differentiating and evaluating at $\theta = 0$, get the first four cumulants:

$$\kappa_1(Y) = p/\lambda, \ \kappa_2(Y) = p/\lambda^2, \ \kappa_3(Y) = 2p/\lambda^3, \ \kappa_4(Y) = 6p/\lambda^4$$

Conclude that the first four cumulants of $X_1 \sim \chi_1^2$ are 1,2,8, and 48. Conclude that the cumulants of $Z_i = (X_i - 1)/\sqrt{2}$ are $\kappa_j(Z_1) = 0, 1, 2\sqrt{2}, 12$ for $j = 1, 2, 3, 4$.

(c) Now compute the cumulant generating function of S_n, and use it to show that an approximation of order $n^{-3/2}$ to the MGF of S_n is $\mathcal{L}_\theta(S_n) \approx$

$$\exp(\theta^2/2) \times \left\{ 1 + \left(\frac{\theta^3}{\sqrt{9n/2}} + \frac{\theta^4}{2n}\right) + \frac{1}{2}\left(\frac{\theta^3}{\sqrt{9n/2}} + \frac{\theta^4}{2n}\right)^2 \right\}$$

$$\approx \exp(\theta^2/2) \times \left\{ 1 + \frac{(\sqrt{2}/3)\theta^3}{\sqrt{n}} + \frac{(1/2)\theta^4 + (9/2)\theta^6}{n} \right\}.$$

(d) Now invert this MGF to get the explicit form for the Edgeworth approximation to the density of S_n, $f^{S_n}(s) \approx$:

$$\phi(s) \left\{ 1 + \frac{\sqrt{2}/3}{\sqrt{n}} H_3(s) + \frac{1}{n} \left((1/2) H_4(x) + (9/2) H_6(x) \right) \right\}.$$

5. Let $X_1, \ldots X_n$ be an i.i.d sample from some unknown distribution F with mean μ and variance σ^2. According to the CLT, $\overline{X} = (1/n) \sum_{i=1}^{n} X_i$ can be expected to be approximately normal with mean μ and variance σ^2/n. Let h be a smooth function and suppose we wish to approximate the distribution of $h(\overline{X})$. Of course we can use the continuity principle to deduce that if $Z_n \sim N(\mu, \sigma^2/n)$ then $h(Z_n)$ should have the same distribution as $h(\overline{X})$. The following technique, sometimes called the *delta method*, is often more convenient. Since the difference between $\overline{X}$ and μ should be small in large samples, it is valid to make the following Taylor expansion:

$$h(\overline{X} = h(\mu) + (\overline{X} - \mu)h'(\mu) + \cdots.$$

Here $\overline{X} - \mu$ goes to zero at rate $n^{-1/2}$ and the omitted terms go to zero at rate n^{-1} or faster.

(a) Use this to derive a normal approximation to the asymptotic distribution of $h(\overline{X})$.

(b) Include one more term in the Taylor series, and derive an Edgeworth expansion for $h(\overline{X})$ which should be more accurate than the normal approximation of the previous part.

6. Consider the autoregressive process $y_t = \rho y_{t-1} + \epsilon_t$ for $t = 1, 2, \ldots, T$ and $y_0 \sim N(0, \sigma^2/(1 - \rho^2))$. Suppose that $\epsilon_t \sim N(0, \sigma^2)$. Let $\hat{\rho} = \sum_{t=1}^{T} y_t y_{t-1} / \sum_{t=1}^{T} y_t^2$. Each of the following methods of estimating the distribution of ρ is defective in some way. Explain.

(a) Use the asymptotic distribution $\hat{\rho} \sim N(\rho, \sigma^2 / \sum_{t=1}^{T} y_{t-1}^2)$.

(b) Let $y_0^*, \ldots, y_T^*$ be a resample of the observations $y_0, \ldots, y_T$. Get a new estimate $\rho^* = \sum_{t=1}^{T} y_t^* y_{t-1}^* / \sum_{t=1}^{T} y_{t-1}^{*2}$. Repeat this process 1000 times and consider the observed distribution of ρ^* to be the estimated distribution of $\hat{\rho}$.

(c) Let $e_j = y_j - \hat{\rho} y_{j-1}$ for $j = 1, 2, \ldots, T$. Let $e_1^*, \ldots, e_T^*$ be a resample of the residuals $e_1, \ldots, e_T$. Let y_0 be fixed.

7. *Eicker-White Heteroskedasticity Test.* Consider a standard linear regression model with $y_t = \sum_{j=1}^{K} \beta_j x_{tj} + \epsilon_t$ for $t = 1, 2, \ldots, T$. Suppose that $\epsilon_t \sim N(0, \sigma_t^2)$. As discussed in Section 1.4, OLS results are valid provided the fluctuation in σ_t^2 is uncorrelated with cross products of the regressors. An intuitive approach to testing this is to take the squared residuals $e_t^2 = y_t - x_t\hat{\beta}$ from an OLS fit and run a regression of these on all the (linearly independent) products of the regressors. If $x_1, \ldots, x_k$ are the original regressors, define $Z_{ij} = x_i \times x_j$ for all pairs i, j and then exclude any linearly dependent regressors. Regress e_t^2 on $Z_{ij,t}$. Then, under certain assumptions, the quantity TR^2, where R^2 is the standard coefficient of determination of this secondary regression, will be asymptotically χ_{P-1}^2, where P is the number of regressors in the secondary regression. Explain how you would do a bootstrap-Bartlett correction for this test for heteroskedasticity, and do a Monte Carlo study to compare the level of the bootstrap-Bartlett corrected test with the original White test.

Chapter 15

Asymptotic Optimality of ML and LR

The object of this chapter is to discuss the asymptotic optimality of various estimation procedures in some detail. The asymptotic optimality of tests is also treated briefly. This latter theory is not yet as fully developed as is the case for estimation theory.

15.1 Asymptotic Efficiency of the MLE

We first go over the basic result that asymptotically the ML is unbiased and its covariance matrix is the inverse of the information matrix. Then we show that the inverse of the information matrix is a lower bound for the covariance matrix of any unbiased estimator. Together these results suggest the the ML estimator is hard to beat in large samples. The Cramer-Rao Lower Bound is given below. Next we show that the Stein estimator exhibits 'superefficiency.' That is, it achieves an asymptotic variance below the CRLB. Such examples led to the development of techniques designed to establish the asymptotic superiority of the ML on more solid grounds. These attempts are reviewed in later sections.

15.1.1   Finite Sample Efficiency

Suppose observations $X_1, \ldots, X_n$ have density $f^X(x_1, \ldots, x_n, \theta)$, where $\theta \in \mathbf{R}^p$ is an unknown parameter. The Cramer-Rao lower bound provides a lower bound on the variance, or equivalently, an upper bound on the performance, of unbiased estimators. Unbiased estimators with variance equal to the CRLB are called 'efficient' — there is a sense in which such estimators extract the maximum possible information regarding the unknown parameters out of the observations. This finite sample efficiency result is not very satisfactory in the sense that comparisons are made only among unbiased estimators, and this class can be fairly small, or even empty. It turns out however that the CRLB provides an asymptotic bound for the performance of all estimators.

Theorem 15.1 (Cramer-Rao Lower Bound) *If* $X_1, \ldots, X_n$ *are random variables with density* $f^{(X_1, \ldots, X_n)}(x_1, \ldots, x_n\theta) = f^X(x, \theta)$ *and* $\hat{\theta}$ *is an unbiased estimator of* θ, *then* $\mathrm{Cov}(\hat{\theta}) \geq \mathcal{I}^{-1}(\theta)$.

Remark: The matrix inequality $A \geq B$ is to be interpreted in the sense that $A - B$ is a positive semidefinite matrix. To see that this is a sensible statistically, suppose that A is the covariance matrix of $\tilde{\theta}$ and B is the covariance matrix of $\hat{\theta}$, both of which are unbiased estimators of the $p \times 1$ vector of parameters θ. Let γ be any $p \times 1$ vector of constants. Then $\gamma'\tilde{\theta}$ and $\gamma'\hat{\theta}$ are both unbiased estimators of $\gamma'\theta$ and the difference between their respective variances is easily calculated to be $\gamma'(A - B)\gamma$ which must be positive. This shows that when the difference between the covariance matrices is positive definite, any linear function of the parameter is estimated with smaller variance by the corresponding linear function of $\hat{\theta}$. Before proceeding, we note two facts about this type of inequality. If $A \geq B$ and C is a positive definite matrix, then $CA \geq CB$ (and also $AC \geq BC$). In addition, if $A \geq B$ and A and B are positive definite, then $A^{-1} \leq B^{-1}$; see exercise 1.

Proof: Define the score function $S(X, \theta) = \nabla_\theta \log f^X(x, \theta)$ to be the $p \times 1$ (gradient) vector of first partial derivatives of the log likelihood. By definition $\mathcal{I}(\theta) = \mathrm{Cov}\, S(X, \theta)$. If $T(X)$ is any unbiased estimator of θ, we show below that $\mathrm{Cov}(S, T) = \mathbf{I}$. Now the Cramer-Rao lower bound follows from the observation that for any fixed matrix Σ

$$0 \leq \mathbb{C}ov(S+\Sigma T) = \mathbb{C}ov(S)+\mathbb{C}ov(S,T)\Sigma'+\Sigma \, \mathbb{C}ov(S,T)+\Sigma \, \mathbb{C}ov(T)\Sigma'.$$

On substituting $\Sigma = -(\mathbb{C}ov(T))^{-1}$, we get $\mathbb{C}ov(S) - (\mathbb{C}ov(T))^{-1} \geq 0$, which implies $\mathbb{C}ov(T) \geq (\mathbb{C}ov(S))^{-1} = \mathcal{I}^{-1}(\theta)$.

To complete the proof, it suffices to show that $\mathbb{C}ov(S,T) = \mathbf{I}$, an intermediate step in the reasoning above. Let $S_i(X,\theta) = \partial/\partial\theta_i \log f^X(X,\theta)$ and T_j the j-th coordinate of the vector T. Because T is unbiased, $\mathbf{E}T_i = \theta_i$. Since $\partial \log f/\partial\theta_i = (\partial f/\partial\theta_i)(1/f)$, we can compute $C_{i,j} = \mathbb{C}ov(S_i,T_j)$ as follows:

$$
\begin{aligned}
C_{i,j} &= \int T_j(X)\frac{\partial \log f^X(X,\theta)}{\partial\theta} f^X(x,\theta)\,dx = \int_X T_j(X)\frac{\partial f^X(x,\theta)}{\partial\theta_i}\,dx \\
&= \frac{\partial}{\partial\theta_i}\int T_j(X)f^X(x,\theta)\,dx = \frac{\partial}{\partial\theta_i}\mathbf{E}T_i(X) = \frac{\partial}{\partial\theta_i}\theta_j.
\end{aligned}
$$

This proves the desired result.

15.1.2 Asymptotic Efficiency

It turns out that in large samples the CRLB provides bounds on the performance of all estimators (not just unbiased ones). Thus estimators which asymptotically achieve the CRLB are, more or less, best possible, or asymptotically efficient. This result has been obtained under many different hypotheses and in many variant forms. The simplest form arises in the i.i.d. case. Since this forms the basis for sophisticated extensions, we review it first.

Suppose $X_1, \ldots, X_n$ are i.i.d. with common density $f(x,\theta)$. Since the joint density $f(X_1, \ldots, X_n, \theta)$ takes the form $\prod_{i=1}^n f(X_i, \theta)$, the information in a sample of size n is just n times the information in a sample of size 1:

$$\mathcal{I}_n(\theta) = \mathbb{C}ov \, \nabla_\theta \log f(X_1, \ldots, X_n, \theta) = \sum_{i=1}^n \mathbb{C}ov \log f(X_i, \theta) = n\mathcal{I}_1(\theta).$$

Thus, according to the CRLB, if $\hat{\theta}_n = \hat{\theta}_n(X_1, \ldots, X_n)$ is unbiased for θ then $\mathbb{C}ov(\hat{\theta}_n) \geq \mathcal{I}_n^{-1}(\theta) = (1/n)\mathcal{I}_1^{-1}(\theta)$. At the same time, Theorem

1.9 states that the ML estimator $\hat{\theta}_n^{ML}$ has the approximate distribution $\sqrt{n}(\hat{\theta}_n^{ML} - \theta) \to N(0, \mathcal{I}_1(\theta))$ for large n. This last is equivalent to the statement that $\hat{\theta}_n^{ML} \sim N(\theta, \mathcal{I}_n^{-1}(\theta))$ or that $\hat{\theta}^{ML}$ is approximately unbiased, with covariance matrix equal to the CRLB, in large samples.

Since even the simplest econometric examples involve heterogeneity, and dependent observations are also common, it is worthwhile sketching the extension of this basic optimality property to more general contexts. It was observed by Hájek, Le Cam, and a number of other authors that the key to this result was a second order approximation to the likelihood function. See Ibragimov and Has'minskii (1981) for a detailed discussion and references for the material below.

Definition 15.1 (The LAN condition) *Let $X^{(n)} = (X_1, \ldots, X_n)$ be random variables with joint density $f_n(\theta) \equiv f(X^{(n)}, \theta)$, where $\theta \in \Theta$, an open subset of $\mathbf{R}^p$. The family of densities is Locally Asymptotically Normal at θ if for some $p \times p$ matrices $B_n(\theta)$, the following approximation to the log likelihood ratio is valid. For $u \in \mathbb{R}^p$ define $Z_n(u) = \log f_n(\theta + B_n(\theta)u)/f_n(\theta)$. Suppose $Z_n(u)$ can be approximated by $\Delta_n(\theta)'u - (1/2)u'u$, where $\Delta_n(\theta)$ is a random variable converging to $N(0, \mathbf{I})$. To be more precise, we require that $\|Z_n(u) - \Delta_n(\theta)'u + (1/2)u'u\|$ should converge to 0 in probability as $n \to \infty$.*

Motivation: The LAN condition seems obscure. In fact it is the result of deep study of precisely what is required for the asymptotic normality and efficiency of the ML estimates. To see the relationship and motivate the condition, we give a brief discussion. Let $f_n(\theta) \equiv f_n(X_1, \ldots, X_n, \theta)$ be the density function for n observations (which may be heterogenous and dependent). If f_n is smooth function of $\theta \in \mathbb{R}^p$, B_n is a $p \times p$ matrix and $u \in \mathbb{R}^p$ is a fixed vector such that $B_n u$ is small, a Taylor series expansion of the log likelihood ratio shows that

$$\log \frac{f_n(\theta + B_n u)}{f_n(\theta)} \approx \nabla_\theta \log f_n(\theta)' B_n u + u' B_n' \nabla_\theta \nabla_\theta' \log f_n(\theta) B_n u.$$

Consider the effect of setting $B_n = \mathcal{I}_n^{-1/2}(\theta)$ in this approximation. Since $\mathcal{I}_n(\theta) = \mathbb{C}ov \, \nabla_\theta \log f_n(\theta) = \mathbb{C}ov \, S_n(\theta)$, it is clear that $B_n \nabla_\theta \log f_n$ has covariance identity. In the case of independent variables $S_n(X_1, \ldots, X_n, \theta) = \mathbb{E} \log \prod_{i=1}^n f(X_i, \theta) = \sum_{i=1}^n S(X_i, \theta)$, the score function is the sum of independent variables. It is reasonable to hope that $B_n S_n$ will converge to $N(0, \mathbf{I})$ in the general case (as it must in the i.i.d. case). The random variable $\Delta_n(\theta)$ in the LAN condition plays the role of $B_n S_n$ in the standard case. An important feature of the standard case is the relation between the matrix of second partial derivatives of the log likelihood and the information matrix; that is,

$$\mathbb{E}\nabla_\theta \nabla_\theta' \log f_n = -\mathcal{I}_n(\theta).$$

When the random variables are independent, it is easily seen using the law of large numbers that $B_n'(\nabla_\theta \nabla_\theta' \log f_n) B_n \approx \mathcal{I}_n^{-1/2}(-\mathcal{I}_n)\mathcal{I}_n^{-1/2} = -\mathbf{I}$. This accounts for the $u'B_n'(\nabla\nabla' f_n)B_n u = -u'u$ term in the LAN condition.

It can be proven that setting $B_n(\theta) = \mathcal{I}_n^{-1/2}(\theta)$ and $\Delta_n(\theta) = \mathcal{I}_\theta^{-1/2} S_n(X_1, \ldots, X_n, \theta)$ will fulfill the requirements of the LAN condition in typical cases. However, the LAN condition continues to hold even in irregular cases where the derivatives or second derivatives of the log likelihood function misbehave in some way. Under the LAN condition we get the following results which generalize the asymptotic efficiency of ML.

If a family of distributions satisfies the LAN conditions at θ, any estimator (sequence) for which $B_n^{-1}(\theta)(\hat{\theta}_n - \theta)$ converges asymptotically to $N(0, \mathbf{I})$ will be called *asymptotically efficient* at θ. This definition is justified by the results given below. Any estimator for which $B_n^{-1}(\hat{\theta}_n - \theta)$ converges weakly to some limit distribution, and this convergence is uniform in θ, will be called a *regular estimator*. The condition of regularity can be weakened (see Definition 9.1 in Chapter 2 of Ibragimov and Has'minskii (1981)), but it cannot be removed in the following theorem.

Theorem 15.2 *If* $X_1, \ldots, X_n$ *have densities* $f_n(x_1, \ldots, x_n, \theta)$ *satisfying the LAN condition at* θ *then for any sequence of regular estimators* $\hat{\theta}_n(X_1, \ldots, X_n)$

$$\lim_{n\to\infty} \mathbb{E}B_n^{-1}(\theta)(\hat{\theta}_n - \theta)(\hat{\theta}_n - \theta)'B_n'^{-1}(\theta) \geq \mathbf{I}.$$

This result shows that among regular estimators, asymptotically efficient estimators are best, since they achieve equality in the limit. This corresponds to classical results, since $B_n(\theta) = \mathcal{I}_n^{-1/2}(\theta)$ and convergence of $\mathcal{I}_n^{1/2}(\theta)(\hat{\theta}_n - \theta)$ to $N(0, \mathbf{I})$ is the classical condition for asymptotic efficiency. Since, under certain assumptions (see Theorems 1.1 and 1.2 Chapter 3 of Ibragimov and Has'minskii (1981)), we can prove that the ML estimator $\hat{\theta}_n^{ML}$ satisfies the condition that $B_n^{-1}(\theta)(\hat{\theta}_n^{ML} - \theta)$ converges weakly to $N(0, \mathbf{I})$ uniformly in θ, we can conclude from the theorem above that ML is asymptotically efficient. Several questions arise regarding this optimality result:

Q1 As a practical matter, it is not of much concern to us whether or not an estimator converges uniformly to a limit distribution. Thus there is a natural question of whether we can do better than the ML by considering 'irregular' estimators.

Q2 Consider an i.i.d. sample $X_1, X_2, \ldots \sim N(\theta, 1)$. Suppose we wish to estimate $1/\theta$. The ML estimator $1/\overline{X}_n$ has variance equal to $+\infty$ for all finite sample sizes n. In this situation, does it make sense to say that having a small asymptotic variance is a virtue?

Q3 Is the ML unique in being asymptotically efficient?

Q4 This last issue raises two related questions. How can we differentiate between asymptotically equivalent estimators? Also, how reliable are asymptotic properties as guides to finite sample behavior?

Regarding the second question, it can be shown that an estimator which is asymptotically efficient not only has small variance — a criterion which depends on the existence of moments and fails in the example cited — but also has high probabilities of concentration around the true parameter. Stronger forms of this result are given later. One form of it, under the restriction that competing estimators be regular, is given below. Let $L(x)$ be a loss function such that (a) $L(x) \geq 0$, (b) $L(x) = L(-x)$, and (c) the set $\{x : L(x) < c\}$ is convex for any $c > 0$.

Theorem 15.3 *Let Z be a $N(0, \mathbf{I})$ random variable. For any regular sequence of estimators $\hat{\theta}_n$, the following inequality holds:*

$$\liminf_{n \to \infty} \mathbb{E}^{f_n(\theta)} L(B_n^{-1}(\theta)(\hat{\theta}_n - \theta)) \geq \mathbb{E}L(Z).$$

Since asymptotically efficient estimators converge in distribution to Z, they automatically achieve this lower bound on risk. This shows that efficient estimators are superior relative to a large class of loss functions, and not just the variance measure.

15.1.3 Superefficiency

Fisher conjectured that all estimators have asymptotic variance larger than or equal to the CRLB. If true, this would imply that the ML is impossible to beat asymptotically. However, an example was produced by Hodges, as reported by Le Cam (1953), which showed that there exist estimators $\hat{\theta}_n$ which are consistent and have asymptotic variance less than or equal to the CRLB at all parameter points, with strict inequality at one point. Instead of the original example of Hodges, which is somewhat artificial, we show below that the Stein estimator, widely used in statistical applications, has variance equal to the CRLB at all points other that $\theta = 0$ and has variance less than the CRLB at $\theta = 0$.

Consider $X_1, \ldots, X_n \overset{iid}{\sim} N(\theta, \sigma^2 \mathbf{I}_p)$. Then $\overline{X} \sim N(\theta, \sigma^2 \mathbf{I}_p/n)$ is a sufficient statistic for the sample. The positive part Stein estimator can be written as

$$\hat{\theta}_{JS} = \left(1 - \frac{(p-2)\sigma^2/n}{\|\overline{X}\|^2}\right)^+ \overline{X} \equiv (1 - S)^+ \overline{X}.$$

Consider the asymptotic behavior of the quantity $Z = \sqrt{n}\overline{X}/\sigma \sim N(\sqrt{n}\theta/\sigma, \mathbf{I}_p)$. If $\theta = 0$ then $\|Z\|^2 \sim \chi_p^2$, while $\mathbf{E}\|Z\|^2 = p + n\|\theta\|^2/\sigma^2$ when $\theta \neq 0$. It follows that when $\theta \neq 0$, the quantity $S = (p-2)/\|Z\|^2$ converges to zero quite fast and so $\hat{\theta}_{JS} \approx \overline{X}$. Thus in this case the variance of $\hat{\theta}_{JS}$ approaches the variance of $\overline{X}$ which achieves the Cramer-Rao lower bound. When $\theta = 0$ on the other hand, $S \sim (p-2)/\|Z\|^2$, where $\|Z\|^2 \sim \chi_p^2$. It follows that $(1 - S)^+ \overline{X}$ will be strictly smaller than $\overline{X}$ and hence $\mathbf{E}\|\hat{\theta}_{JS}\|^2 < \mathbf{E}\|\overline{X}\|^2$. Thus $\hat{\theta}_{JS}$ beats the CRLB at 0.

Estimators having variance less than or equal to CRLB everywhere with strict inequality at some points in the parameter space are called *superefficient*. After the discovery that such estimators could exist, substantial research effort was put into learning the nature and significance of the phenomenon. This research, reviewed briefly in later sections, ultimately leads to the conclusions that we can safely ignore superefficient estimators, as they have unpleasant properties asymptotically.

15.2 Probabilities of Concentration

In an early battle, the ML had won the championship over the method of moments proposed by Pearson, when Fisher showed that it was asymptotically efficient (while MOM is not). The discovery of superefficiency threatened the longtime reigning champion, and counterattacks were immediately mounted to protect against this new threat. Le Cam (1953) extended the example of Hodges to obtain estimators which improved on the CRLB at an infinite number of points in the parameter space, but showed that this improvement could only occur on a set of Lebesgue measure 0. Thus it was immediately clear that at worst, the ML could be beaten at a *small* number of points. This was, however, not sufficient for the proponents of the MLE, and more vigorous defenses were devised. The results to be reviewed show that, in a sense to be made precise, the MLE has maximal probabilities of concentration in sets around the true parameter. Note that this characterization automatically takes care of the counterexample given in question Q2. By looking at concentration probabilities, we avoid the issues raised by infinite variance in finite samples.

15.2.1 Asymptotic Median Unbiasedness

One simple device which gives a clear victory to the ML forbids threatening groups of estimators from competing. It was established early that if attention was restricted to estimators which converge uniformly to their asymptotic distributions, the phenomenon of superefficiency could not arise. However, Pfanzagl (1970) — who gives a brief review and selected references — protested against this discrimination. He said that all estimators which do well in estimating the parameter of interest should be allowed to compete regardless of how, or whether or not, they converge to a limiting distribution. Instead he proposed a condition which is more appealing from an operational point of view.

For the rest of this section, we consider the case of a one-dimensional parameter.

Definition 15.2 *A sequence of estimators $\hat{\theta}_n$ of a real parameter θ is asymptotically median unbiased (AMU) if both the inequalities $\mathbf{P}(\hat{\theta}_n \geq \theta) \geq 1/2$ and $\mathbf{P}(\hat{\theta}_n \leq \theta) \geq 1/2$ hold in the limit as n goes to infinity.*

This is substantially less restrictive than conditions which had been placed on competitors to the ML earlier. In particular, every consistent and asymptotically normal estimator must be asymptotically median unbiased. Pfanzagl also showed that some very irregular estimators would be allowed to enter the competition under his rules, which only exclude those estimators which are not asymptotically median unbiased. He still obtained the desirable result that the ML has the highest probabilities of concentration in intervals around the true value:

Theorem 15.4 (Pfanzagl) *Let $X_1, X_2, \ldots$ be an i.i.d. sequence of random variables with common density $f(x, \theta_0)$. Let $\hat{\theta}_n^{ML}$ maximize the log likelihood $\ell_n(\theta) = \sum_{i=1}^n \log f(X_i, \theta)$ over all $\theta \in \Theta$, where Θ is an open subset of $\mathbf{R}$. If $\theta_n^*(X_1, \ldots, X_n)$ is any asymptotically median unbiased sequence of estimators, then for every pair of reals $a > 0$ and $b > 0$*

$$\mathbf{P}(-a < \sqrt{n}(\hat{\theta}_n^{ML} - \theta_0) < b) \geq \mathbf{P}(-a < \sqrt{n}(\theta_n^* - \theta_0) < b).$$

Sketch of Proof: This elegant result has an elegant proof, which we sketch here. For details of rigor consult Pfanzagl (1970) or Akahira and Takeuchi (1981). Given any estimator $\hat{\theta}_n$, define a test T_n of the hypothesis $\theta = \theta_0 + b/\sqrt{n}$ versus the alternative $\theta = \theta_0$, which rejects the null whenever $T_n \leq \theta_0 + b/\sqrt{n}$. Since T_n is median unbiased, the probability that $T_n \leq \theta_0 + b/\sqrt{n}$ under the null hypothesis that $\theta = \theta_0 + b/\sqrt{n}$ is asymptotically $1/2$. The power of the test is the asymptotic probability that $T_n \leq \theta_0 + b/\sqrt{n}$ or equivalently $\mathbf{P}\sqrt{n}(T_n - \theta_0) < b$ under the alternative hypothesis θ_0. This probability must be bounded above by the power of the Neyman-Pearson most powerful test of level $1/2$ for these hypotheses. This last is computed to be $\Phi(b\sqrt{\mathcal{I}(\theta_0)})$ in Lemma 15.1. A symmetric argument shows that for any estimator T_n, $\mathbf{P}(\sqrt{n}(T_n - \theta_0) > -a$ is bounded above by $1 - \Phi(-a\sqrt{\mathcal{I}(\theta_0)})$. Combining these results, we find that $\mathbf{P}(-a < \sqrt{n}(T_n - \theta_0) < b)$ is bounded above by $\Phi(b\sqrt{\mathcal{I}(\theta_0)}) - \Phi(-a\sqrt{\mathcal{I}(\theta_0)})$. That is, for all estimators T_n, the maximum probability of concentration in the interval $(-a/\sqrt{n}, b/\sqrt{n})$ around θ_0 is bounded by this value. However, using the fact that the asymptotic normal distribution of $\sqrt{n}(\hat{\theta}^{ML} - \theta_0)$ is $N(0, \mathcal{I}_1(\theta_0)^{-1})$, we find that the ML estimator achieves this upper bound. Hence it is asymptotically maximally concentrated around the true value in the class of all median unbiased estimators.

Lemma 15.1 *For large n, the power of the level 50 % Neyman-Pearson most powerful test of the null $H_0 : \theta_0 = \theta + b/\sqrt{n}$ versus the alternative $H_1 : \theta_1 = \theta$ is approximately $\Phi(b\sqrt{\mathcal{I}(\theta)})$.*

Proof: Define $Z_{ni} = \log f(X_i, \theta_0) - \log f(X_i, \theta_0 + b/\sqrt{n})$. Then the NP test rejects the null for large values of $T_n = \sum_{i=1}^n Z_{ni}$. To evaluate the behavior of the NP test for large n, expand $\log f(x, \theta_0 + b/\sqrt{n})$ around θ_0 in a Taylor series to get

$$Z_{ni} = -\frac{b}{\sqrt{n}}\frac{\partial}{\partial\theta}\log f(X_i, \theta_0) - \frac{b^2}{2n}\frac{\partial^2}{\partial\theta^2}f(X_i, \theta) + \mathcal{O}(n^{-2}).$$

Let $S(X_i, \theta) = \partial \log f(X_i, \theta)/\partial\theta$ be the score function, and $S'(X, \theta)$ be the first derivative of the score with respect to θ.

When the alternative hypothesis holds so that $\theta = \theta_0$, we can easily compute the mean and variance of T_n. The mean is

$$\mathbf{E}T_n = -\frac{b}{\sqrt{n}}\sum_{i=1}^n \mathbf{E}S(X_i, \theta_0) - \frac{b^2}{2n}\sum_{i=1}^n \mathbf{E}S'(X_i, \theta_0) = \frac{b^2}{2}\mathcal{I}(\theta_0).$$

For the variance, we calculate the second moment of T_n first:

$$\begin{aligned}
\mathbf{E}T_n^2 &= \frac{b^2}{n}\mathbf{E}\left(\sum_{i=1}^n S(X_i, \theta_0)\right)^2 + \frac{b^4}{4n^2}\mathbf{E}\left(\sum_{i=1}^n S'(X_i, \theta_0)\right)^2 \\
&\quad + \frac{b^3}{n^{3/2}}\mathbf{E}\left(\sum_{i=1}^n S(X_i, \theta_0)\right)\left(\sum_{i=1}^n S'(X_i, \theta_0)\right) \\
&= b^2\mathcal{I}(\theta_0) + \frac{b^2}{4n}\mathbf{E}(S'(X_1, \theta_0)^2 + \frac{b^3}{\sqrt{n}}\mathbf{E}S(X_1, \theta_0)S'(X_1, \theta_0).
\end{aligned}$$

Thus the leading term of the variance is $b^2\mathcal{I}(\theta_0)$. Since T_n is the sum of independent random variables, it will be approximately normal with mean $\mu_1 = (b^2/2)\mathcal{I}(\theta_0)$ and variance $\sigma_1^2 = b^2\mathcal{I}(\theta_0)$ under the alternative hypothesis.

An exactly analogous computation yields the mean and the variance of T_n under the null hypothesis that $\theta = \theta_0 + b/\sqrt{n}$. Write $\theta^* = \theta_0 + b/\sqrt{n}$ and express θ_0 as $\theta_0 = \theta^* - b/\sqrt{n}$. Now Taylor the expression for the Neyman-Pearson statistic T_n around θ^* and follow the procedure detailed above to see that the mean of T_n under the null is $\mu_0 = -(b^2/2)\mathcal{I}(\theta^*)$ and the variance $\sigma_0^2 = b^2\mathcal{I}(\theta^*)$. Now $\mathcal{I}(\theta^*)$ equals $\mathcal{I}(\theta_0)$ to the first order. Thus a first order approximation to the power of the Neyman-Pearson test for large n is the power of a level 50% test for testing between H_1: $Z \sim N(\mu_1, \sigma_1^2)$ and H_0: $Z \sim N(-\mu_1, \sigma_1^2)$. The level 50% test rejects the null for $Z > -\mu_0$ and hence has power

$$\beta = \mathbf{P}(Z > -\mu_1 | Z \sim N(\mu_1, \sigma_1^2)) = \mathbf{P}(-\frac{Z - \mu_1}{\sigma_1} < \frac{2\mu_1}{\sigma_1}) = \Phi(\frac{2\mu_1}{\sigma_1}).$$

Substituting $(b^2/2)\mathcal{I}(\theta_0)$ for μ_1 and $b^2\mathcal{I}(\theta_0)$ for σ_1^2 yields the result of the lemma.

Pfanzagl's result establishes that the ML is efficient in the class of all AMU estimators, in the sense that the probabilities of concentration in any interval around the true value are maximized by the ML. We discuss the significance, strengths, and weaknesses of this basic optimality result in this section.

First we note that this result is substantially stronger than the assertion of the CRLB, which merely claims minimal asymptotic variance. As discussed in connection with question Q2 raised in connection with the asymptotic optimality of the MLE earlier, it is not at all clear that having minimal asymptotic variance is meaningful, especially in situations where the ML has infinite variance in all finite samples. We can use this result to establish that the ML minimizes the expected loss for any loss function in a rather large class of functions. Let $L(\hat{\theta} - \theta)$ be any loss function such that $L(0) = 0$ and L is monotone nondecreasing on both sides of 0. Explicitly, for reals t_1, t_2, t_1', t_2' satisfying $t_2 < t_1 < 0 < t_1' < t_2'$, we must have $L(t_2) \leq L(t_1)$ and $L(t_1') \leq L(t_2')$. We will call such a function a *bowl shaped* loss function in the sequel. Then asymptotically the ML $\hat{\theta}^{ML}$ will have smaller risk than any other estimator $\hat{\theta}^*$ relative to such a loss function. That is

$$\liminf_{n\to\infty} \mathbf{E} L(\sqrt{n}(\hat{\theta}_n^* - \theta) - L(\sqrt{n}(\hat{\theta}_n^{ML} - \theta) \geq 0$$

This follows easily from Pfanzagl's result. For a proof, see Pfanzagl (1970).

In the class of asymptotically normal estimators, the condition of asymptotic median unbiasedness coincides with the more usual asymptotic mean unbiasedness. Outside this class of estimators, the two are different and the median unbiasedness appears more natural and attractive due to Pfanzagl's result. To be more precise, suppose we wish to maximize the probability of concentration of $\sqrt{n}(\hat{\theta}_n - \theta)$ in given set (a,b) As we have seen, this probability can be maximized simultaneously for all intervals (a,b) if we constrain estimators to be AMU. On the other hand, if we try to solve this maximization problem under the constraint that the estimator sequence be asymptotically mean unbiased, it turns out that there is no solution; see Akahira and Takeuchi (1981,p. 3) for an example.

15.2.2   Multidimensional Case

Unfortunately, the significance of Pfanzagl's result is limited by it failure to generalize nicely to the case of vector parameters. Let $\hat{\theta} = (\hat{\theta}_1, \ldots, \hat{\theta}_k)$ be an estimator of $\theta = (\theta_1, \ldots, \theta_k)$. Imposing the condition of coordinate wise median unbiasedness is not enough to find an estimator maximizing $\mathbf{P}(\sqrt{n}(\hat{\theta} - \theta) \in C)$ for symmetric convex sets C around the origin. If we put the stronger restriction on the class of estimators that $a'\hat{\theta}$ should be median unbiased for $a'\theta$ there generally does not exist an estimator maximizing probabilities of concentration in symmetric convex sets subject to this constraint. See Akahira and Takeuchi (1981, Chapter 1) for further discussion and examples. For the vector generalization we have to fall back on older and more classical techniques. We note however that if only one parameter is of interest, and the remaining parameters are nuisance parameters, then result of Pfanzagl does generalize.

Under the usual hypothesis of uniform convergence to a limit distribution, we can show that the ML maximizes probabilities of concentration only for certain special kinds of sets in the multidimensional case.

If C is a measurable, convex, and symmetric set about the origin in $\mathbf{R}^p$ and θ is a p-dimensional parameter, the ML $\hat{\theta}_n^{ML}$ will outperform any other estimator T_n in the following sense:

$$\lim_{n\to\infty} \mathbf{P}(\sqrt{n}(\hat{\theta}_n^{ML} - \theta) \in C) \geq \lim_{n\to\infty} \mathbf{P}(\sqrt{n}(T_n - \theta) \in C).$$

Under the hypothesis that $\sqrt{n}(T_n - \theta)$ converges to some limiting distribution uniformly in θ, this theorem was proven by Kaufman (1966, p.157, Theorem 2.1). As usual, this can be converted into a statement that the ML minimizes expected loss for a class of loss functions $\mathcal{L}$. This class includes all functions L which are nonnegative, bounded, symmetric around the origin, and such that $-L$ is unimodal. For details, see Pfanzagl and Wefelmeyer (1978).

15.3   Hajek-LeCam Local Asymptotic Minimaxity

Suppose that $\sqrt{n}(\hat{\theta}_n - \theta)$ converges to the asymptotic distribution $N(0, \sigma^2(\theta))$. If the convergence is uniform, then $\sigma^2(\theta)$ must be continuous in θ. Thus if $\sigma^2(\theta) < \mathcal{I}^{-1}(\theta)$, then this inequality must hold in an open neighborhood. However, it was demonstrated early by Le Cam (1953) that the phenomenon of superefficiency could only occur on sets of Lebesgue measure 0. To restrict the class of estimators to those which converge uniformly to their limit distributions more or less explicitly rules out superefficient estimators. Investigation of the probabilities of concentration also fails to give the desired evidence of superiority of the ML except perhaps in one dimension. The question remains as to whether a new champion would emerge if the superefficient estimators were allowed to compete against the ML. Convincing evidence in favor of the ML and against superefficiency was obtained via the local asymptotic minimax result of Hajek and Le Cam. The result to be described shows that relative to a minimax criterion, the ML can defeat all comers, superefficient or not. For details of the proof, variants, ramifications, and references, see Le Cam and Yang (1990) or Ibragimov and Has'minskii (1981). A more elementary and readable presentation is given in Lehmann (1983).

The Cramer-Rao lower bound looks at the behavior of an estimator at a single point, and assesses the estimator by its asymptotic variance. The local asymptotic minimax theorem looks at the behavior of an estimator on a small neighborhood of a point. Instead of looking at variances, we consider the expected loss, for loss functions $L(x)$ satisfying the conditions (a), (b), (c) as described before Theorem 15.3 as well as an additional boundedness condition (d): $L(x) \leq A \exp(a|x|^2)$ for some constants $A > 0, a > 0$. We will assume that the family of densities $f_n(x_1, \ldots, x_n, \theta)$ of the observations $X_1, \ldots, X_n$ satisfies the LAN condition 15.1 for some matrices $B_n(\theta)$. Recall that $B_n(\theta) = \mathcal{I}_n^{-1/2}(\theta)$ in the regular case. The following theorem is due to Hajek; closely related results were also obtained by Le Cam. For a proof and reference to original sources, see Theorem 2.12.1 of Ibragimov and Has'minskii (1981). For a more elementary and readable presentation of a one-dimensional and simplified version of this result, see Lehmann (1983).

Theorem 15.5 *Suppose the loss function L and the likelihood $f_n(\theta)$ of the observations $X_1, \ldots, X_n$ satisfies the hypotheses listed above. Also suppose that $\lim_{n\to\infty} \mathbf{tr}\, B_n(\theta) B_n(\theta)' = 0$. Let $Z \sim N(0, \mathbf{I})$ be standard multinormal random variable. For any $\delta > 0$ the maximum risk of any estimator $\hat{\theta}_n$ in a δ neighborhood of θ is asymptotically bounded below as follows:*

$$\liminf_{n\to\infty} \sup_{\{\psi:\|\psi-\theta\|<\delta\}} \mathbb{E}L(B_n^{-1}(\theta)(\hat{\theta}_n - \theta)) \geq \mathbb{E}L(Z).$$

Since in typical regular cases, the normalized ML $B_n^{-1}(\theta)(\hat{\theta}_n^{ML} - \theta)$ converges in distribution to Z, it achieved the lower bound given in the theorem. Thus its asymptotic risk is equal to the constant $\mathbb{E}L(Z)$ over the neighborhood, while competing estimators have higher maximum risk over this neighborhood. This is why this result is referred to as the local asymptotic minimaxity of the ML.

It can also be proven that any superefficient estimator is necessarily worse than the ML when evaluated with respect to their maximum risk in small neighborhoods of the true parameter. The gains at θ are purchased at the price of losses at nearby points. This is an unpleasant property and practically rules out the use of superefficient estimators in practice. To use one would give us good behavior at a particular point combined with bad behavior at arbitrarily close points. We would never be in a position of being able to make such fine judgements over the whole parameter space, especially since the estimated parameter is unknown. The local minimax result more or less guarantees that one cannot find a good estimator which will beat the ML asymptotically.

15.3.1 Implications for Empirical Bayes

As we have seen, superefficient estimators have unpleasant properties asymptotically. This casts shadows on the Stein estimator, widely used in certain types of statistical applications. The Stein estimator shares these unpleasant properties asymptotically. An intuitive understanding of the source of the misbehavior of Stein estimator comes from its representation as an empirical Bayes rule. Consider the canonical situation where θ has the prior distribution $N(0, \nu^2 \mathbf{I})$. The proper Bayes estimator for this prior has the form

$$\hat{\theta}_B = \left(1 - \frac{\sigma^2/n}{\sigma^2/n + \nu^2}\right) \overline{X}.$$

Note that the effect of the prior parameter ν^2 vanishes asymptotically, and the shrinkage factor $\nu^2/((\sigma^2/n) + \nu^2)$ converges to 1 as n goes to infinity. With respect to the marginal density of X which is $N(0, (\sigma^2 + \nu^2)\mathbf{I})$, we have

$$\frac{\sigma^2/n}{\sigma^2/n + \nu^2} = \mathbb{E}\frac{(p-2)\sigma^2/n}{\|\overline{X}\|^2}.$$

Define $Z_n = \sqrt{n}\overline{X}/\sigma$. The empirical Bayes estimator replaces $\sigma^2/(\sigma^2 + \nu^2)$ in the Bayes rule by the unbiased estimate $(p-2)/Z_n$. Dropping the expectation, and taking reciprocals on both sides of the previous equation shows that we are estimating ν^2 by

$$1 + \frac{\hat{\nu}^2}{\sigma^2/n} = \frac{Z_n}{p-2}.$$

If X has nonzero mean θ, Z_n has mean converging to infinity. Thus the (estimated) prior variance $\hat{\nu}^2$ is very large compared to the data variance σ^2/n, so that the rule behaves like a standard Bayes rule.

The effect of the prior washes out asymptotically. For $\theta = 0$, Z_n has a limiting distribution, showing that $\hat{\nu}^2$ is of exactly the same order as σ^2/n and hence remains influential even as the sample size goes to infinity. This is clearly undesirable since for large sample sizes, we should let the data determine the estimate. Violating this rule is the source of asymptotic misbehavior which both causes superefficiency at $\theta = 0$ and poor behavior at neighboring points.

How does this problem effect practice? First note that using hierarchical Bayes procedure will eliminate this and other undesirable features of the empirical Bayes rules (such as inadmissibility). In practice the hierarchical Bayes rules are typically close to empirical Bayes. Thus, as long as the empirical Bayes rules are viewed as approximations to the hierarchical Bayes rules, and the sample sizes in question are small, no difficulties should arise. The empirical Bayesian should announce that he is lazy and intends to mend his ways as he gets larger and larger sample sizes. This will rectify his asymptotics and permit him to enter both Bayesian and frequentist heaven asymptotically.

15.4   Higher Order Asymptotics

While the superefficiency rebellion was successfully put down, other contenders for the crown have had greater success. In particular, it has been shown that large classes of Bayes rules share the asymptotic efficiency properties of the ML — in fact, ML's good asymptotic properties can be explained as the result of its being asymptotically close to a Bayes rule. See Lehmann (1983) for an elementary exposition of the asymptotic theory of Bayes rules, and Ibragimov and Has'minskii (1981) for an advanced and complete discussion. In addition to the class of Bayes rules, there are various other ad-hoc techniques which can be coaxed into performing well asymptotically.

Higher order asymptotic analysis attempts to answer two related questions. One is to discriminate among the crowd of competitors on the basis of the second term in the asymptotic expansion. It was Fisher's idea, borne out to some extent by subsequent analysis, that the ML will be superior to its competitors when the higher order terms are analyzed. In addition to this, it is well known that the usual asymptotic theory is highly inaccurate in many statistical problems. Obviously comparisons made on the basis of inaccurate asymptotics are also going to be inaccurate. Again higher order asymptotics may achieve greater accuracy in smaller samples and thereby assist in valid comparisons between estimators.

15.4.1   Second and Third Order Efficiency of ML

An estimator is first order efficient (FOE) if it has an asymptotic normal distribution with variance equal to the Cramer-Rao lower bound. The results discussed earlier show that such estimators are more or less unbeatable in large samples. Unfortunately, as Pfanzagl has put poetically, 'this is an elegant description of a desolate situation' — meaning that the CLT is rarely reliable in sample sizes which occur in practice. Analysis of the higher order terms in an asymptotic expansion for the ML and related estimators leads to the following results, which may be more reliable in smaller samples than required for the FOE results. In this section, we summarize results which are obtained by using Edgeworth expansions to study higher order properties of estimators.

Even though the nature of the results remains the same in more complex situations, we consider an i.i.d. sequence $X_1, X_2, \ldots$ with common density $f(x, \theta)$ for ease of presentation. Assume $\theta \in \Theta$ and the parameter space Θ is a subset of $\mathbb{R}^p$. Following the approach of Pfanzagl and Wefelmeyer (1978), let us restrict attention to the class $\mathcal{T}$ of estimator sequences $T_n = T_n(X_1, \ldots, X_n)$ which possess stochastic expansions of the following type:

$$\sqrt{n}(T_n - \theta) = \tilde{\lambda} + n^{-1/2} Q_1(\tilde{\lambda}, \tilde{f}, \cdot) + n^{-1} Q_2(\tilde{\lambda}, \tilde{f}, \tilde{g}, \cdot) + n^{-3/2} R_n.$$

The functions Q_1 and Q_2 (terms in the Edgeworth expansion of T_n), and the remainder R_n is required to satisfy some regularity conditions which we do not specify here. The first term $\tilde{\lambda}$ is a random variable which is the target of FOE considerations. Estimators belonging to $\mathcal{T}$ for which $\tilde{\lambda}$ achieves the CRLB will be called FOE, or first order efficient.

Motivation for studying SOE, or second order efficiency is provided by the fact that the ML and large classes of Bayes estimators, typically differing substantially in finite samples, all achieve FOE. Perhaps these estimators can be distinguished in finite samples on the basis of the second order term in the stochastic expansion. Second order investigations are complicated by the fact that efficient estimator sequences are typically biased to order $1/n$. For example, the ML estimator typically satisfies $\mathbb{E}(\hat{\theta}_n^{ML} - \theta) = B(\theta)/n + o(n^{-1})$ for some bias function $B(\theta)$. This bias has a big impact on second order properties, and is not, unfortunately, comparable in any natural way across estimators. A natural idea is to eliminate the bias by defining $\tilde{\theta}_n^{ML} = \hat{\theta}_n^{ML} - B(\theta)/n$. This 'bias-adjusted' ML make the ML unbiased to second order terms. One can remove the bias of any FOE estimator by following a similar procedure. Once this is done, it becomes meaningful to examine the second term of the stochastic expansions of (bias-adjusted) FOE estimators. Here a surprise appears — after bias adjustment, the second order terms in the Edgeworth expansion are exactly the same for all FOE estimators. This result is referred to in the literature as 'First Order efficiency implies Second Order Efficiency', and has the implications that the second order term does not help us to discriminate between FOE estimators.

Two caveats may be attached to this result. The first is that the regularity conditions we skimmed over lightly are really needed for this result. Examples can be constructed of estimators which are FOE but do not become SOE after bias adjustment when the regularity conditions do not hold. This may not be of great practical importance as large classes of estimators used in practice do satisfy the required conditions. The second difficulty is much more serious. It seems natural to remove second order bias and compare estimators after bias correction; however, it turns out that if we allow second order biased estimators to compete with the unbiased ones, estimators with lower mean square error may be found. We replicate the simple but instructive example (6.3.5) of Lehmann (1983):

Example Consider the estimation of σ^2 on the basis of an i.i.d. sample $X_1, \ldots, X_n \sim N(0, \sigma^2)$. The ML $\hat{\sigma}^2 = (1/n)\sum_{i=1}^n X_i^2$ is also MVU and unbiased to all orders. As long as we restrict competition to regular estimators which are second order unbiased, the Pfanzagl and Wefelmeyer (1978) result assures us that the ML is second order efficient. Consider however the second order biased estimator $\delta_n = ((1/n) + (a/n^2)) \sum_{i=1}^{n} X_i^2$. Elementary calculations show that

$$\mathbb{E}(\delta_n - \sigma^2)^2 = \frac{2\sigma^4}{n} + \frac{(4a + a^2)\sigma^4}{n^2} + \mathcal{O}(n^{-3}).$$

Thus, by setting $a = -2$ we can improve on the mean square error of the ML to the second order.

Examples like this show that it is not at all clear whether second order unbiased estimators should be preferred to second order biased ones. The 'FOE implies SOE' result has been reformulated to take care of the possibility that we may prefer an estimator with nonzero second order bias. Given FOE estimators $\hat{\theta}^a$ and $\hat{\theta}^b$ with second order bias functions $B_a(\theta)$ and $B_b(\theta)$, we can bias-adjust either estimator to match the other. For example, defining $\hat{\theta}'^b = \hat{\theta}^b - B_b(\theta)/n + B_a(\theta)/n$ will cause both estimators to have second order bias function $B_a(\theta)$. After equalizing the bias, we find that the second order terms in the Edgeworth expansion of both estimators are identical, giving no basis for preference. To summarize, the 'FOE implies SOE' result shows that choice to second order among estimators can only be based on their respective second order bias functions. The present author knows of no practical or theoretical guidelines regarding how one might select a suitable second order bias function in applied work. Further work appears to be needed in this direction.

Assuming that we can somehow fix a suitable second order bias structure and confine attention to estimators sharing this common bias structure, the question arises as to whether the third term in the Edgeworth expansion would be helpful in discriminating among such estimators. It turns out that this is so. In particular it turns out that the bias-adjusted ML is superior to alternative FOE estimators. To be more precise, let θ_n^* be any FOE and $\hat{\theta}_n^{ML}$ be the maximum likelihood estimator. Then $\hat{\theta}_n^{ML} + B_2(\theta_n^*, \theta) - B_2(\hat{\theta}_n^{ML}, \theta)$ is bias-adjusted to match the bias of θ^* to the n^{-1} order. The bias-adjusted ML is then at least as good, and may be better than θ^* on the basis of the third term in the Edgeworth expansion. The superiority of the MLE over alternative FOE estimators can be quantified in terms of larger probabilities of concentration in symmetric convex sets, or alternatively, smaller risk relative to a large set of loss functions. This phenomenon is referred to as the third order efficiency (TOE) of the MLE.

As a special case of this result, suppose we define $\tilde{\theta}_n^{ML} \equiv \hat{\theta}_n^{ML} - B_2(\hat{\theta}_n^{ML}, \theta)$ so that $\tilde{\theta}_n^{ML}$ is unbiased to order n^{-1}. Then among all estimators which are unbiased to order n^{-1}, $\tilde{\theta}^{ML}$ has maximal asymptotic probabilities of concentration in symmetric convex sets (and also minimal asymptotic expected loss for a large class of loss functions).

It might be thought that the TOE property restores the lead to the MLE. However, many classes of competitors also achieve third-order efficiency. In particular, Bayes estimators relative to suitable priors share the same property — after adjustment for bias, they become third order efficient. Another class of importance in practice arises naturally as approximations to the ML. This is discussed in the following section. Nonetheless, third order efficiency does show that some care is needed in obtaining an estimator which is efficient; that is, which extracts all of the information available from the data. Akahira and Takeuchi (1981) discuss the higher order efficiency of many types of estimators used in simultaneous equations models and show that some are, while others are not, efficient to the third order.

15.4.2   Approximate MLE

In problems where solving for the MLE is difficult, a common procedure is to start with some consistent estimator and iterate towards the maximum using Newton-Raphson or some similar procedure. To illustrate in the context of a simple example, suppose $\hat{\theta}^{ML}$ is the root of the equation:

$$0 = S(X, \hat{\theta}^{ML}) = \sum_{i=1}^{n} \frac{\partial}{\partial \theta} \log(f(X_i, \theta)).$$

If this equation cannot be solved for the ML, it is frequently unclear how to construct an efficient estimator. In such situations the following procedure is helpful.

Let $\tilde{\theta}$ be any $\sqrt{n}$ consistent estimator; this means that $\tilde{\theta} - \theta$ should be of order $O(n^{-1/2})$. Many ad-hoc methods can be used to construct $\sqrt{n}$-consistent estimators. For example, the method of moments and variants will typically work. Our estimator is consistent but typically inefficient. We can attempt to improve the efficiency of estimation by moving closer to the ML via an iterative procedure. We have

$$S(X, \tilde{\theta}) \approx S(X, \hat{\theta}^{ML}) + (\tilde{\theta} - \hat{\theta}^{ML})S'(X, \hat{\theta})^{ML}).$$

It follows that

$$\hat{\theta}^{ML} \approx \tilde{\theta} - \frac{S(X, \tilde{\theta})}{S'(X, \hat{\theta}^{ML})}.$$

The quantity $S'(X, \hat{\theta}^{ML})$ is not known if $\hat{\theta}^{ML}$ is not known, but it can be estimated by either $S'(X, \tilde{\theta})$ or by $-\mathcal{I}(\tilde{\theta} = \mathbf{E}S'(X, \tilde{\theta})$. In this case the LHS of the equation above gives a closer approximation to the ML than the original estimator $\tilde{\theta}$. It is frequently possible by iterating this procedure to converge to the ML estimator. However, it was noticed early that if asymptotic efficiency is the sole concern, only one iteration is enough to achieve this. Define

$$\theta_1 = \tilde{\theta} - \frac{S(X, \tilde{\theta})}{S'(X, \tilde{\theta})}.$$

Then it is easily shown by a Taylor's expansion that $\sqrt{n}(\theta_1 - \hat{\theta}^{ML})$ converges to 0, so that the two estimators have the same asymptotic distribution.

One has the intuitive feeling that further gains should result from further iterative steps, and this intuition is borne out by the higher order asymptotic theory. In order to get an approximation to the ML valid to order $o(n^{-1})$ we must iterate at least two times. Doing so will yield an estimator sharing the third order efficiency property of the ML. It not clear how one would do bias correction on such an estimator, however.

15.5   Comparison of Tests

In Part II of this text, we studied finite sample techniques for the comparison of tests. As discussed there, the finite sample techniques can be successfully used to find optimal tests in only a small minority of cases. In applied problems of any complexity, the finite sample techniques fail to provide us tools necessary to make comparisons. Nonetheless, it is essential to study the class of situations where finite sample comparisons are possible. Asymptotic theory, which succeeds for a much larger class of problems, rests on the finite sample theory. That is, it finds large sample approximations which reduce complex problems to equivalent simpler comparison problems. These are then solved by the finite sample techniques.

There are more approaches to the problem of comparison of tests than there are to estimator comparison. A clear and readable exposition of six different methods of comparison is given by Serfling (1980, Chapter 10), but this by no means exhausts the criteria which have been used. Methods used can be classified into two broad categories – local, and nonlocal. Local methods compare tests in a shrinking neighborhood of the null hypothesis, while nonlocal methods compare test performance at a fixed element in the null and the alternative. The comparison of tests is complicated by the fact that we have an extra degree of freedom asymptotically. That is, we can control the level (or the power) of the test asymptotically. Different methods of doing so lead to different criteria for comparison.

Since fixed level testing is the prevailing paradigm, two of the most well-known methods of comparison use a sequence of tests with fixed level asymptotically. In the case of non-local comparisons, the power goes to unity for a fixed level. Bahadur efficiency looks at the rate at which the power goes to one to compare different tests. Bahadur (1966) and later authors have shown that, more or less, only the LR test is optimal with respect to this measure. In the case of local comparisons with fixed level, we move the alternative closer and closer to the null. This makes the problem harder and harder so that the power is prevented from going to one. The limit of the power is then a measure of the efficiency of tests which is closely related to Pitman's efficiency measure for the comparison of tests. Duals to this pair of measures fix the power of the test asymptotically and study the convergence of size to zero. Yet other measures are obtained following the plausible idea that asymptotically we would like both probabilities of type I and type II errors to go to zero. In this sense, criteria which fix either the size or the power are 'unbalanced.'

The most popular and widely used criteria in econometrics is the

Pitman efficiency measure for comparison of tests. Given the framework in which we are working, it can be most naturally formulated in the following terms. As we did in the course of the proof of Pfanzagl's Theorem 15.4 on the asymptotic efficiency of median unbiased estimates, we can calculate the asymptotic power of the Neyman-Pearson test. In that proof we calculated the first order power and this gives a first order local approximation to the power envelope, which is the maximum power achievable locally by any test. It can be shown that, under the LAN assumption, the popular trio of Lagrange multiplier, likelihood ratio, and Wald tests are all three Uniformly Most Powerful to the first order in one-dimensional one-sided hypothesis testing problems. This is what is meant by the local asymptotic equivalence of these three tests. Our derivation of the Lagrange multiplier and Wald tests as local approximations to the likelihood ratio test shows that they are equivalent to order $\mathcal{O}(n^{-1/2})$ and hence the local asymptotic equivalence is relatively easy to show.

One can pursue this approach further by taking into consideration higher order terms. A very clear summary of results and further references are available in Rothenberg (1984a). Here we will summarize the summary. By taking additional terms in the Taylor series expansion for the Neyman-Pearson test, we can get a third order approximation[1] to the asymptotic power envelope. Before making comparisons among tests, it is essential to make a correction for size. This can be done using the Cornish-Fisher technique based on the Edgeworth-B expansions discussed earlier. After correcting for size, the third order expansions for the power can be compared with the third order power envelope. An analog of the result that FOE implies TOE also holds in the case of testing. This means that that the $n^{-1/2}$ term of the Edgeworth expansion fails to discriminate between the three FOE tests (i.e.,LM, LR, and Wald) — the second order power for the three is the same as the asymptotic power envelope to the second order. To make comparisons we have to go to the third order term. Here the three tests show differences, although none is uniformly superior. The LM test is tangent to

---

[1] Rothenberg, in common with many other writers, considers the $n^{-1/2}$ term to be the first order term in the asymptotic expansion. Hence, these are second order results in his writing.

the third order (i.e.,including the n^{-1} term in the expansion) asymptotic power envelope at 0, the LR is tangent around $1/2$,while the Wald test is tangent at around $1 - \alpha$. If we evaluate on the basis of shortcoming, as recommended in Chapter 6, we would prefer the LR test. This recommendation must be tempered by the realization that third order approximations may still fail to be accurate in realistic sample sizes. Another possibility is that third order approximations are accurate, but the third order improvement of the LR is too small to be of practical importance. Heuristics suggest that the importance of the third order term will be governed by the 'statistical curvature' of the problem, as defined by Efron (1975). With high curvature, we would expect third order consideration to be important. Since curvature goes to zero asymptotically, we may expect the three tests to be equivalent in suitably large samples, but the issue of what is 'large enough' for equivalence will vary from problem to problem.

For two-sided testing of a one-dimensional parameter, the LR and Wald continue to have the characteristics described for the one-sided case. That is, the LR is optimal in the midrange of power, while the Wald is optimal at the higher power range. The locally unbiased Lagrange multiplier test becomes dominated and drops out of the competition in this case. The end result is that the LR wins in terms of stringency. For the multidimensional case, the three tests continue to be first order equivalent. The higher order theory is still being developed. It seems reasonable to conjecture that similar results will be obtained. Subject to the qualifications at the end of the previous paragraph, the local approach to hypothesis testing suggests that we should prefer the LR. Interestingly, nonlocal approaches also lead to similar conclusions. Results of Hoeffding (1965) for the multinomial, generalized to continuous distributions by Brown (1973), show more or less that Bahadur efficiency is restricted to the LR test, and does not hold for any test which is significantly different from the LR (such as Wald or LM). With regards to Chernoff Efficiency, which is a 'balanced approach' of the nonlocal variety, Kallenberg (1982) establishes the superiority of the LR to other types of test.

As far as the author knows, the LR either wins or ties for first place in every race, except in pathological cases. Thus there seem to be strong reasons to prefer the (size-corrected) LR to other methods of hypothesis testing, just as there are reasons to prefer the (bias-adjusted) ML estimator. Of course it must be remembered that there are alternatives, principally Bayesian estimators and tests for suitable priors, which enjoy the same asymptotic properties.

15.6 Exercises

1. For positive definite matrices A and B, let $A \geq B$ be notation for the statement that $A - B$ is a positive definite matrix.

   (a) If $A \geq B$ and C is positive definite, then $CA \geq CB$ and also $AC \geq BC$. Hint: the product of positive definite matrices must be positive definite.

   (b) If $A \geq B$ then $A^{-1} \leq B^{-1}$. Hint: use the previous fact, and that inverses of positive definite matrices are positive definite

2. Let X have density $f(x, \theta)$ indexed by a real parameter θ.

   (a) Suppose $T(X)$ is an estimator of θ and $g(\theta) = ET(X)$. $T(X)$ is not assumed unbiased, so $g(\theta)$ need not equal θ. Show that
   $$\mathbb{V}ar(T) \geq \frac{(g'(\theta))^2}{\mathcal{I}(\theta)}.$$

   (b) Introduce a new parameter $\psi = h(\theta)$, and suppose $T(X)$ is unbiased for ψ. According to the CRLB $\mathbb{V}ar(T) \geq \mathcal{I}^{-1}(\psi)$. Find the relationship between $\mathcal{I}(\psi)$ and $\mathcal{I}(\theta)$.

3. Let $X_1, X_2, \ldots$ be an i.i.d. sequence with density $N(\theta, 1)$. Consider the sequence of estimators $T_n(X_1, \ldots, X_n)$ defined by $T_n = \overline{X}_{(n)} = (1/n) \sum_{i=1}^n X_i$ for $|\overline{X}_{(n)}| > n^{-1/4}$, and $T_n = 0$ otherwise. Show that this sequence is superefficient at 0 and achieves the CRLB at all points.

4. Let $X_1, X_2, \ldots$ be i.i.d. with a Cauchy density: $f(x) = \{\pi(1 + (X - \theta))\}^{-1}$. Let $\tilde{\theta}$ be the median of the X_i. Define
$$S(\theta) = \sum_{i=1}^n \frac{X_i - \theta}{1 + (X_i - \theta)^2}.$$

(a) Calculate the CRLB for unbiased estimators of θ. You may use numerical integration to get the answer.

(b) Use Monte Carlo to assess the variance of $\tilde{\theta}$ and compare it with the CRLB for $n = 10$.

(c) Define $\theta^* = \tilde{\theta} - S(\tilde{\theta})/S'(\tilde{\theta})$. Argue that θ^* is asymptotically efficient.

(d) Estimate the variance of θ^* by Monte Carlo and compare it to that of $\tilde{\theta}$.

Part IV

Empirical Bayes: Applications

In the court of King Solomon, two women both laid claim to the same child. None of the evidence presented by either side was definitive, and the wise King pondered for a while. Finally, Solomon the Just came to a decision. 'Let each woman have half of the child,' he said, and took his sword out of it's scabbard. One of the women immediately abandoned her claim. Having recognized the true mother by his strategem, Solomon awarded her the child.

Chapter 16

Simple Examples

This last part of the text presents several applications of empirical Bayes methods, which are currently vastly underutilized in econometrics. There exist a number of misunderstandings/confusions regarding the technique which have been propagated in the literature. This chapter is devoted to examples which are 'simple' in the sense that the data involved are counts of various sorts, and genuine regression concepts are not involved. It is essential to build intuition for the method in a simple context. The theory behind the method has been exposited in Chapter 3. There are also links with Stein estimation, as discussed in Chapter 4. An excellent treatment of the theory, dealing in much greater detail with several aspects ignored here, is available in Berger (1985). An elementary account for general audiences is given in Efron and Morris (1977). Our goal here is to develop an intuitive understanding of the method, clear up several misconceptions, and to provide all formulas necessary to implement the method in practice.

Although all examples are taken from real life, they have been chosen to illustrate different, and successively more complex aspects of techniques involved in using empirical Bayes methods. The Stein estimate works only in the case of equal variances, a case difficult to find in practice. Efron and Morris constructed an ingenious example using Baseball batting averages where the assumption of equal variances is feasible. The Stein estimator is compared to two different empirical Bayes estimators. All calculations are based on normal approximations. The next example concerns seasonality in business cycles. It is chosen to illustrate Beta-Binomial calculations which are useful in small samples sizes where normal approximations are not valid. The example due to Carter and Rolph dealing with assessing the probability of a serious fire illustrates several issues which arise in the case of unequal variances. The final example of quality control and A.T. & T. develops several refinements of EB estimators which may be required in serious applications.

16.1   Play Ball!

We start with an illuminating baseball example from Efron and Morris (1975). Table 16.1 gives batting averages for selected major league baseball players. A reasonable model for the data can be constructed as follows. Define $H_j(i) = 1$ if on his j-th turn at bat, player i gets a hit, and $H_j(i) = 0$ otherwise. It is plausible to assume that $\mathbb{P}(H_j(i) = 1) = \theta_i$, where θ_i reflects the 'hitting ability' of player i. Also, it seems reasonable to take $H_j(i)$ for $j = 1, 2, \ldots$ to be i.i.d. The column labeled 'Mid-Season Average' is the average number of times each player hit the ball for his first 45 at-bats in the season. At mid-season, a natural estimate of the ability θ_i is just the mid-season average, $\hat{\theta}_i = (1/45) \sum_{j=1}^{45} H_j(i)$. Since the end-season average is based on a substantially larger number of trials (the total at-bats for the season, different for each player, are listed in the last column), it is reasonable to take it as a proxy for the true ability θ_i. On many grounds, the mid-season average $\hat{\theta}_i$ provides an optimal estimate of the end season average. For example, it is the ML and MVU estimate. The Stein estimator suggests that we can improve on this forecast by utilizing information on other players.

Player Name	Mid-Season Average	End-Season Average	Total At-Bats
1 Clemente (Pitts, NL)	.400	.346	367
2 F. Robinson (Balt, AL)	.378	.298	426
3 R. Howard (Wash, AL)	.356	.276	521
4 Johnstone (Cal, AL)	.333	.222	275
5 Berry (Chi, AL)	.311	.273	418
6 Spencer (Cal, AL)	.311	.270	466
7 Kessinger (Chi, AL)	.289	.263	586
8 L. Alvarado (Bos, AL)	.267	.210	138
9 Santo (Chi, NL)	.244	.269	510
10 Swoboda (NY, AL)	.244	.230	200
11 Unser (Wash, AL)	.222	.264	277
12 Williams (Chi, AL)	.222	.256	270
13 Scott (Bos, AL)	.222	.303	435
14 Petrocelli (Bos, AL)	.222	.264	538
15 E. Rodriguez (KC, AL)	.222	.226	186
16 Campaneris (Oak, AL)	.200	.285	558
17 Munson (NY, AL)	.178	.316	408
18 Alvis (Mil, NL)	.156	.200	70

Table 16.1: 1970 Major League Baseball Batting Averages

A transparent interpretation of the source of the gain is obtained from the empirical Bayes formulation of the Stein estimator. Our goal is to estimate the players' abilities θ_i. In this (and subsequent more complex) models, the argument hinges on finding a plausible model for the parameters being estimated. Once the problem is posed this way, we can think of many factors which would influence the abilities θ_i. As we shall see later, if data are available on relevant factors, it can be taken into account to improve forecasts. In the present case, even though no other data are available, a simple model suggests itself as plausible. We may assume that there is a common parameter θ_0 representing average human ability, and individual abilities are random variations around this common ability, so that

$$\theta_i = \theta_0 + \epsilon_i, \quad \text{where} \quad \epsilon_i \stackrel{iid}{\sim} N(0, \nu^2). \tag{16.1}$$

Now the batting averages are random departures from the batting abilities θ_i, which themselves cluster around a common mean θ_0. Common sense suggests that improved estimates of θ_i will result from shrinking the batting averages towards a common mean.

It is worth emphasizing that the plausibility of the model for parameters has everything to do with the success or failure of the empirical Bayes method in practice (though not in theory). If the parameters θ_i are unrelated to each other, the second stage model $\theta_i = \theta_0 + \epsilon_i$ is implausible, and in practice we will not obtain any gains from applying Stein-type shrinkage — even though the theory of Stein estimation works equally well in both cases. Intuitively, improving on the ML estimator requires addition of valid prior information; this takes the form of a plausible model for the parameters in the empirical/hierarchical Bayes case. No significant gains can be expected from mechanical applications of Stein estimator, as was discovered by several early econometric studies.

Interpreting the model (Eq. 16.1) as a prior on θ_i (namely $\theta_i \sim N(\theta_0, \nu^2)$, standard Bayes estimators for θ_i are

$$\hat{\theta}_i^B = \frac{\nu^2}{\sigma_i^2 + \nu^2}\hat{\theta}_i + \frac{\sigma_i^2}{\sigma_i^2 + \nu^2}\theta_0. \qquad (16.2)$$

It is convenient to use the term 'hyperparameters' for θ_0 and ν^2 occuring in the prior density, to differentiate these from the θ_i and σ_i^2 which are parameters of the data density. As already discussed in Section 3.4, the classical Bayesian prescription of selecting the hyperparameters on subjective grounds does not work well in practice. We will discuss below several alternative ways to estimate the hyperparameters and thereby arrive at a Bayesian estimator of $\hat{\theta}_i$ — as we will see, such estimators improve substantially on the crude ML.

16.1.1   A Simple Empirical Bayes Estimator

To present the argument, it is convenient to make the crude approximation that $\hat{\theta}_i \sim N(\theta_i, \sigma^2)$, where σ^2 is assumed known. It is convenient to postpone the discussion of how σ^2 should be fixed to make this approximation reasonable; more sophisticated approaches are presented in later sections. We can rewrite our assumption in the form of a model for the data $\hat{\theta}_i$ as follows:

$$\hat{\theta}_i = \theta_i + e_i, \quad \text{where} \quad e_i \overset{iid}{\sim} N(0, \sigma^2). \tag{16.3}$$

By combining substituting for θ_i using Eq. 16.1 above, we get $\hat{\theta}_i = \theta_0 + e_i + \epsilon_i$, so that $\hat{\theta}_i$ furnish valuable information about the parameter θ_0. In such a model (which is the simplest possible regression model, with the constant term being the only regressor), the usual estimate for θ_0 is $\hat{\theta}_0 = (1/n)\sum_{i=1}^{n} \hat{\theta}_i$, where $n = 18$ is the number of players. Let $v^2 = \sigma^2 + \nu^2$ be the common variance of the error $e_i + \epsilon_i$. The usual estimate for the error variance is $\hat{v}^2 = \sum_{i=1}^{n}(\hat{\theta}_i - \hat{\theta}_0)^2/(n-1)$. Under our simplifying assumption that the variance of e_t is a known quantity σ^2, we can estimate the parameter ν^2 by $\hat{\nu}^2 = \hat{v}^2 - \sigma^2$, provided this is positive; the case where this is negative will be discussed in greater detail later. With these estimates in hand, we can obtain an empirical Bayes estimator for the abilities θ_i by substituting these estimates in Eq. (16.2):

$$\hat{\theta}_i^{EB} = \frac{\hat{\nu}^2}{\sigma^2 + \hat{\nu}^2}\hat{\theta}_i + \frac{\sigma^2}{\sigma^2 + \hat{\nu}^2}\hat{\theta}_0. \tag{16.4}$$

To make this formula operational, we must specify how σ^2, the assumed common variance of the estimates $\hat{\theta}_i$, is to be chosen. To approximate the Binomial by a Normal, we should take $\hat{\theta}_i \sim N(\theta_i, \sigma_i^2)$ where $\sigma_i^2 = \theta_i(1 - \theta_i)/n_i$. The number of trials n_i is 45 for all players — in fact, the players were chosen so that they had the same number of at-bats in mid-season. The θ_i are different for all of them, but to a crude approximation, we can estimate θ_i by its average value θ_0, which is estimated by $\hat{\theta}_0$ as defined above. Then the right choice for σ^2 is (see Remark 3 below) $\hat{\sigma}^2 \approx \hat{\theta}_0(1 - \hat{\theta}_0)/45$. With this formula for σ^2, Eq. (16.4) provides a first operational empirical Bayes estimate of the end-season batting averages on the basis of the mid-season averages. Several remarks on these estimates are given below.

Remark 1: Figure 16.1 below compares the usual ML estimates (which are the mid-season scores) and the EB estimates of Eq. (16.4) to the true end-season scores. For 14 players out of 18, the EB estimates are closer to the true end-season score than the ML estimates. The mean absolute error of the EB estimates is about one-half of the mean absolute error for ML. Thus EB estimates are substantially superior to ML.

Remark 2: In empirical Bayes terminology, we would say that the *marginal distribution of the data*, $\hat{\theta}_i$, obtained after integrating out the prior distribution for θ_i, is $\hat{\theta}_i \sim N(\theta_0, \sigma^2 + \nu^2)$. We have expressed this in terms of Eqs. (16.1) and (16.3) since these forms are more familiar to econometricians.

Remark 3: The EB estimate was computed after estimating the variance σ_i^2 by $\hat{\theta}_0(1 - \hat{\theta}_0)/n_i$. A plausible alternative appears to be the estimate $\tilde{\sigma}_i^2 = \hat{\theta}_i(1 - \hat{\theta}_i)/n_i$. In fact, the estimates $\hat{\theta}_i$ are erratic, being based on a much smaller sample than $\hat{\theta}_0$ and the resulting estimate is quite inferior.

Remark 4: The inherent plausibility of the empirical Bayes hypothesis represented by Eq. (16.1) in the present case can also be seen as follows. Repugnant as it may be to baseball fans, consider the hypothesis that all players have exactly the same ability, so that the θ_i are identical. This would correspond to the case where ν^2, the variance of ϵ_i was 0. Then the best prediction for the end-season score for each player would just be the grand mean of all the mid-season scores. In Figure 16.1 arrows are drawn from the mid-season average to the end-season average; these arrows represent the movement of the average over the remaining season. Nearly all of these arrows point towards the grand mean, showing that the original fluctuations away from the grand mean can be attributed mostly to chance. The errors of various estimates are also listed in Figure 16.1, showing that this grand mean also improves substantially on the ML — in fact, it is only slightly worse than the EB estimate. Thus the Philistine who considers hits purely random and all players identical would do better at forecasting, at least on this data set, than the fan who bases estimates on individual abilities. The EB estimate is an 'optimal' compromise between the grand mean — corresponding to the case of equal abilities, and the ML — which represent the case where each player is an individual and

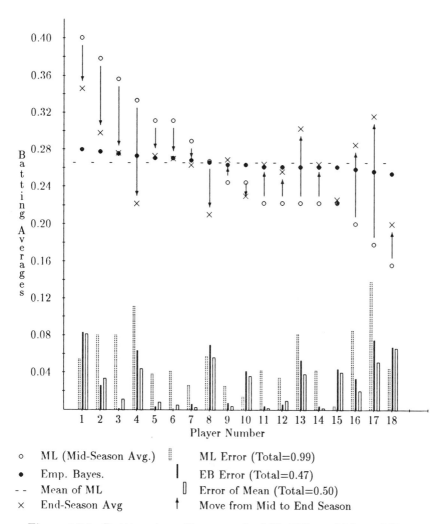

Figure 16.1: Batting Avg. Forecasts by ML, EB, and Mean ML

$\nu^2 = +\infty$, so that there is no advantage at all to shrinking towards any common mean.

Remark 5:   In the context of a general discussion of methodology for estimation, recognizing that subjective choice for prior parameters (θ_0, in the present case) advocated by classical Bayesians leads to somewhat arbitrary estimates, Leamer (1982) suggested the use of 'extreme bounds analysis.' As a type of sensitivity analysis he suggests moving the prior and studying how the estimator changes. Typically the bounds produced by extreme bounds analysis of this type are too large to be of use in practical situations. This has been taken to mean that the data are not informative. Empirical Bayes sheds light on the situation, in that it shows that the data also contain information about prior parameters, and all of the prior parameters are not equally plausible. Since extreme bounds analysis allows for values of prior parameters which are quite implausible, it does not use data in an efficient way, and produces loose inference.

16.1.2   Variance Stabilizing Transformation

In the previous analysis, we forced the variances σ_i^2 of the $\hat{\theta}_i$ to be the same (see Remark 3 above), even though they are known to be different. A technique known as *variance stabilizing transformation* allows us to handle the problem in a superior way. A general formulation of the technique is given below.

Lemma 16.1 (Variance Stabilizing Transformations) *Suppose X_n is a sequence of random variables with mean $\theta = \mathbb{E}X_n$ and variance $\mathbb{V}ar(X_n) = v(\theta)/n$. Let $h(\theta)$ be a function such that $h'(\theta) = 1/\sqrt{v(\theta)}$. Then $h(X_n)$ has mean $h(\theta)$ and variance $1/n$ asymptotically.*

Proof: This follows from a Taylor expansion, noting that

$$h(X_n) = h(X_n - \theta + \theta) = h(\theta) + (X_n - \theta)h'(\theta) + \cdots.$$

The remainder is of small order asymptotically. It is clear that the mean of the LHS is $h(\theta)$ and the variance is $v(\theta)(h'(\theta))^2$, which is enough to prove the lemma.

For the baseball averages, $\mathbb{E}\hat{\theta}_n = \theta$ and $\mathbb{V}\mathrm{ar}(\hat{\theta}_n) = \theta(1-\theta)/n$. It is easily checked that if $h(\theta) = 2\arcsin(\sqrt{\theta})$ then $h'(\theta) = \{\theta(1-\theta)\}^{-1/2}$, so that $h(\hat{\theta}_n)$ is a variance stabilizing transformation. Define $Z_i = 2\sqrt{45}\arcsin(\sqrt{\hat{\theta}_i})$. Then it is approximately true that $Z_i \sim N(\mu_i, 1)$, where $\mu_i = 2\sqrt{45}\arcsin(\sqrt{\theta_i})$. We can now apply the empirical Bayes procedure to estimate the μ_i, and re-transform to get estimates of θ_i. We go through the steps briefly.

It seems reasonable to assume that μ_i are i.i.d. $N(\mu_0, \nu^2)$. Natural estimates for the hyperparameters are $\hat{\mu}_0 = (1/18)\sum_{i=1}^{18} Z_i$, and $\hat{\nu}^2 = (1/17)\sum_{i=1}^{n}(Z_i - \hat{\mu}_0)^2$. Empirical Bayes estimates for μ_i are therefore:

$$\hat{\mu}_i^{EB} = \frac{\hat{\nu}^2}{1 + \hat{\nu}^2}Z_i + \frac{1}{1 + \hat{\nu}^2}\hat{\mu}_0.$$

To get estimates for θ_i, we use the inverse transform, getting $\hat{\theta}_i^{EB} = \sin^2(\hat{\mu}_i^{EB}/2\sqrt{45})$.

Following this theoretically superior route leads to slightly inferior results, a not uncommon outcome. The empirical Bayes estimates derived by this route have mean absolute error about 20% higher than those derived by the crude method of the previous section. In fact, these estimates are about 10% worse than the grand mean, which is twice as good as the naive ML method. The crude empirical Bayes estimates derived earlier are about 10% better than the grand mean. It is not clear what, if any, conclusion can be derived from this. Unless there is a larger amount of systematic evidence, we would be inclined to take it as a fluke. In any case, these estimates are still substantially superior to the ML estimates.

16.1.3 Stein Estimates

After the arcsin transformation, we get a vector $Z \sim N(\mu, \mathbf{I}_n)$, where $n = 18$ in the present example. Instead of using the empirical Bayes approach, we can use a Stein estimator for μ. In fact, the two are closely related, as we shall now see. The simplest Stein estimator is

$$\delta_{JS}(Z) = \left\{1 - \frac{n-2}{\|Z\|^2}\right\} Z.$$

This is not suitable, and does not work well in the present case, since it is implicitly based on a prior assumption that μ_i are i.i.d. $N(0, \nu^2)$ One can alter the estimator to shrink towards an arbitrary vector, say **m** as follows: define $Z' = Z - \mathbf{m}$ and apply shrinkage to Z'. After retransforming back to orign

$$\delta_{JS}(Z) = \mathbf{m} + \left\{ 1 - \frac{n-2}{\|Z - \mathbf{m}\|^2} \right\} (Z - \mathbf{m}).$$

Let $\mathbb{1}$ be the vector of ones. In the present case, we want to shrink the μ_i towards a common mean m_0. This corresponds to shrinking the vector μ towards $m_0 \mathbb{1}$. When m_0 is estimated by $\hat{m}_0 = \sum_{i=1}^{n} Z_i/n = \overline{Z}$, then it can be shown that the optimal constant become $n - 3$ instead of $n - 2$, so that the Stein estimator takes the following form:

$$\delta_{JS}^*(Z) = \overline{Z}\mathbb{1} + \left\{ 1 - \frac{n-3}{\sum_{i=1}^{n}(Z - \overline{Z})^2} \right\} (Z - \overline{Z}\mathbb{1}).$$

When using the Stein estimator in applications, it is important to use the positive part version, where the factor $(1 - c/\|z\|^2)$ is replaced by 0 if it is ever negative. This is noticeably superior to the original Stein estimator. We note that the positive part estimator is also dominated in terms of mean squared error, but the dominating estimators are substantially more complex to calculate without offering significant gains in mean square error. Another fact of interest is that the constant $n - 2$ ($n - 3$ in the case of estimated common mean) is optimal for the original Stein estimator, but there is no optimal constant when we use the positive part version.[1]

In the present context, there is a theoretical reason to prefer the Stein estimator to the empirical Bayes estimator developed earlier. To see this, note that the Z_i are i.n.i.d. $N(\mu_i, 1)$, and let the prior for μ_i be $N(m_0, \nu^2)$ as in the previous sections. The Bayes estimator of Eq.

---

[1] For the original Stein estimator $(1 - c/\|Z\|^2)Z$, the estimator with $c = n - 2$ has a mean squared error strictly smaller than that of estimators for other values of c. When the positive part is taken, the mean squared error functions for different values of c cross; thus no one value of c is uniformly superior. Nonetheless, we will continue to use $c = n - 2$ (or $c = n - 3$ is the case of estimated mean).

(16.2) can be rewritten in the following form:

$$\hat{\mu}_i^B = m_0 + \left(1 - \frac{1}{1 + \nu^2}\right)(Z_i - m_0).$$

Now the empirical Bayes estimators that we have developed simply take the estimates $\hat{\nu}$ and $\hat{m}_0$ and plug them into the formula above. This would be suitable if the estimates equalled the true values with certainty. Otherwise, it does not properly take into account the uncertainty associated with these estimates. To see that the Stein estimator *does*, consider the marginal distribution of Z_i, which is $Z_i \sim N(m_0, 1 + \nu^2)$. Since m_0 occurs linearly in the formula for the Bayes estimator, it is reasonable to replace it by the unbiased estimator $\overline{Z}$ which we get from the marginal distribution of Z. Next we need to find an estimator of $1/(1 + \nu^2)$. Now $S = \sum_{i=1}^n (Z_i - \overline{Z})^2$ has the distribution $(1 + \nu^2)\chi^2_{n-1}$. It follows that $\mathbb{E}(n - 3)/S = 1/(1 + \nu^2)$. Substituting these unbiased estimators into $\hat{\mu}_i^B$ yields the Stein estimator. In contrast to this, if we simply put $\hat{\nu}^2 = (S/(n-1)) - 1$ into the formula for the Bayes estimator as we did earlier, we get $(n-1)/S$ as an estimator for $1/(1 + \nu^2)$. This is a biased estimator, since the process of plugging in the estimator did not take into account the nonlinear way in which ν^2 appears.

This analysis shows that the empirical Bayes estimator is close to the Stein estimator. They differ in that the Stein estimator uses $n - 3$ while the empirical Bayes uses $n - 1$ in the shrinkage factor. The Stein estimator shrinks more heavily, and is based on a more careful analysis of the form of the estimator, and the form of the uncertainty surrounding the estimated hyperparameters, while the empirical Bayes just plugs in the estimate into the formula. In this case, the superior theory pays off. When the Stein estimator is applied to the arcsin transform and the resulting estimates are retransformed, we get a substantial improvement over the empirical Bayes estimators of the previous section. This set of estimates is about 5 % better than the grand mean, and about 15 % better than the EB estimators described earlier. Thus superior theory pays off in the form of superior estimates at least in this instance. Note that this is still a bit worse than the original crude EB estimates applied to the crude approximation. However, this is probably a fluke.

There is a serious limitation to Stein estimation in that it is restricted to the case of equal variances. The theory discussed here does not generalize easily to the case of unequal variances, which is most common in practice. Natural extensions of Stein estimates to unequal variances cases are the minimax estimates, which are qualitatively quite different from empirical Bayes, as discussed in Chapter 3. At present it appears that it pays to abandon minimaxity and use empirical Bayes in the unequal variances. We will not discuss Stein estimation further in the text. However, the issue of taking into account the uncertainty in estimated hyperparameters raised by Stein estimation is important to the theory. We will see that hierarchical Bayes estimates can handle this uncertainty in a suitable fashion later.

16.2   Santa Claus and Economic Recovery

NBER has compiled a long series of data which for each month tells us whether the economy was in a recession or an expansion. A summary given in Table 16.2 shows that the maximum number of recessions ended in the month of December. We propose to investigate whether there really is a 'Santa Claus' effect on the economy, as suggested by this data. Ghysels (1994) detects such a monthly effect in this data — that is, the probability of an expansion ending or a recession ending appears to depend on the month. We propose to analyze this data set using empirical Bayes methods.

In the baseball example discussed before, we have approximated the Binomial distribution by a normal distribution. Since the number of trials is 45, and the probabilities being estimated are around 0.25, the central limit theorem is very accurate, and the impact of the approximation is quite small. In the present example, the probabilities being estimated are low, making the normal approximation unsafe. The concept of empirical Bayes can also be directly applied to the Binomial density, as we will now show. Let R_i be the number of recessions which ended in month i, and let N_i be the total number of recession in month i. Then R_i are $Bi(N_i, p_i)$, where p_i is the probability of a recession end-

Table 16.2: Recessions and Expansions 1854 to 1990

Month	Recession ended	Total Recessions	Expansion ended	Total Expansions
January	1	44	5	93
February	1	48	1	89
March	4	48	2	89
April	2	46	2	91
May	3	46	3	91
June	5	46	3	91
July	3	44	4	93
August	1	45	3	92
September	0	47	1	90
October	2	48	3	89
November	3	49	2	88
December	6	48	2	89

ing in month i. A prior density which is computationally convenient to use is the Beta density. For our EB analysis, let us assume that $p_i \overset{iid}{\sim} \mathcal{B}e(a, b)$.

We have the density of the data R_i conditional on the parameters p_i, and also the prior, which is the marginal density of p_i. The first step in an EB analysis is to get the posterior, which the conditional density of p_i given R_i and also the marginal density of R_i. This merely involves factorizing the joint density appropriately, since for any X, Y, the joint density $f(X, Y)$ can be factorized in two ways: $f(X, Y) = f(X|Y)m(Y) = f(Y|X)m(X)$. In the present case we have for the density of $R = (R_1, \ldots, R_{12})$ and $p = (p_1, \ldots, p_{12})$

$$
\begin{aligned}
f(R, p) &= \prod_{i=1}^{12} f(R_i|p_i) \times m(p_i) \\
&= \prod_{i=1}^{12} \left(\binom{N_i}{R_i} p_i^{R_i}(1 - p_i)^{N_i - R_i} \right) \times \frac{\Gamma(a + b)}{\Gamma(a)\Gamma(b)} p_i^{a-1}(1 - p_i)^{b-1}
\end{aligned}
$$

$$= \prod_{i=1}^{12} \left(\binom{N_i}{R_i} \frac{\Gamma(a+b)\Gamma(a+R_i)\Gamma(b+N_i-R_i)}{\Gamma(a)\Gamma(b)\Gamma(a+b+N_i)} \right)$$

$$\times \left(\frac{\Gamma(a+b+N_i)}{\Gamma(a+R_i)\Gamma(b+N_i-R_i)} p_i^{a+R_i-1} (1-p_i)^{b+(N_i-R_i)-1} \right)$$

$$= \prod_{i=1}^{12} m(R_i) f(p_i|R_i).$$

This refactorization shows that the density of p_i given R_i is just $\mathcal{B}e(a+R_i, b+N_i-R_i)$, and gives the marginal density of R_i in an explicit way. Since the mean of a $\mathcal{B}e(\alpha, \beta)$ density is $\alpha/(\alpha+\beta)$, the Bayes estimator for p_i, which is just the posterior mean, can be written as

$$\hat{p}_i^B = \frac{R_i+a}{N_i+a+b} = \frac{N_i}{N_i+a+b} \hat{p}_i^{ML} + \frac{a+b}{N_i+a+b} \mu_0,$$

where $\hat{p}_i^{ML} = R_i/N_i$ is the usual ML estimate of p_i, and $\mu_0 = a/(a+b)$ is the common prior mean of all the p_i. This shows that the Bayes estimator is a weighted average of the prior mean and the data mean. It takes some calculation to see that the weights are proportional to the precisions (that is, reciprocals of the variances), exactly as in the normal case.

To see this, note that for any random variables X and Y, the marginal variance $\mathbb{V}ar(X)$ can be decomposed as $\mathbb{V}ar(X) = \mathbb{E}(\mathbb{V}ar(X|Y)) + \mathbb{V}ar(\mathbb{E}(X|Y))$; see exercise 2 on page 66. The marginal variance of the ML estimator can therefore be written as

$$\mathbb{V}ar(\hat{p}_i^{ML}) = \mathbb{E}\,\mathbb{V}ar(\hat{p}_i|p_i) + \mathbb{V}ar(\mathbb{E}\hat{p}_i|p_i). \qquad (16.5)$$

Now $\mathbb{V}ar(\hat{p}_i|p_i) = p_i(1-p_i)/n_i$. Using the Beta prior density of p_i, it is easily calculated that

$$\mathbb{E}\,\mathbb{V}ar(\hat{p}_i|p_i) = \mathbb{E}p_i(1-p_i)/n_i = \frac{ab}{n_i(a+b)(a+b+1)}.$$

Since $\mathbb{E}(\hat{p}_i|p_i) = p_i$, the second term in Eq. (16.5) is

$$\mathbb{V}ar\, p_i = \frac{a}{a+b} \left(\frac{a+1}{a+b+1} - \frac{a}{a+b} \right) = \frac{ab}{(a+b)^2(a+b+1)}.$$

Let $c = ab/[(a + b)(a + b + 1)]$. Then the averaged value of the data variance is c/n_i and the variance of p_i is $c/(a+b)$, so that data precision is proportional to n_i and the prior precision is proportional to $(a + b)$.

To proceed with an EB analysis, we need to estimate the prior mean $\mu_0 = a/(a+b)$ and the parameter $(a+b)$ which occurs in the formula for the Bayes estimator, and then substitute. The prior mean is easy; an obvious estimator is the average of all 12 monthly probabilities: $\hat{\mu}_0 = (1/12)\sum_{i=1}^{12} \hat{p}_i^{ML}$. This can be justified by noting that $E\hat{p}_i^{ML}|p_i = p_i$ and $Ep_i = \mu_0$ so that the unconditional mean of $\hat{p}_i^{ML}$ is just the prior mean μ_0. The estimation of $(a + b)$ is more complex. We can use either the method of moments or maximum likelihood. The MOM estimates are discussed in exercise (5). For ML, we use the marginal density of R_i calculated eariler. We must maximize

$$m(R) = \prod_{i=1}^{12} \binom{N_i}{R_i} \frac{\Gamma(a + b)\Gamma(a + R_i)\Gamma(b + N_i - R_i)}{\Gamma(a)\Gamma(b)\Gamma(a + b + N_i)},$$

with respect to the parameters a, b. Instead of jointly maximizing over a, b, it is convenient to fix $a/(a + b) = \hat{\mu}_0$, since this is our estimate of the prior mean. When we maximize we find that the likelihood function is increasing in $(a + b)$ so that there is no maximum!

To understand what the data are trying to tell us, it is useful to note that the variance of the $\mathcal{B}e(a, b)$ density is

$$\mathbb{V}\mathit{ar}(p_i) = \mu_0 \left(\frac{\mu_0 + (a + b)^{-1}}{1 + (a + b)^{-1}} - \mu_0 \right).$$

Fixing μ_0 and sending $(a+b)$ to infinity corresponds to fixing the mean of the prior and sending its variance to zero. Thus, the EB estimator would be one for which $\hat{p}_i = \hat{\mu}_0$ for all i. This is also what we get if we plug in $(a + b) = +\infty$ in the formula for the Bayes estimator above. Thus EB estimates suggest that, contrary to the finding of Ghysel (1994), the probabilities of recession do not depend on the season. Alas, as we eventually find out, there is no Santa Claus.

Because estimates of the hyperparameters are variable, and this variance is not formally taken into account in a crude analysis, we cannot rely on EB estimates alone for such a finding. It is essential to supplement this by a formal hypothesis test. A likelihood ratio test confirms that the data do not provide any evidence against the null hypothesis that the probabilities of ending of a recession do not depend on the month of the year. Ghysels (1994) is misled into finding a significant difference because he uses an inappropriate asymptotic distribution for his test statistics. Making proper finite sample adjustments leads to the correct results. See Blough and Zaman (1995) for further details.

16.3   The Bronx Is Burning!

We are sitting in a fire station in the Bronx when a fire alarm sounds from Box 245 in the area of Central Bronx. Immediately we must decide whether to make an all-out response and send several trucks and firemen, or to make a limited response and send out, for example, one truck. A great deal hangs on our decision since, if it is a serious fire, the delay caused by a limited response could run into big bucks of fire damage. On the other hand, if we make an all-out response, and the fire is not serious, maybe even a prank alarm, we might be caught with our pants down if a real alarm is sent in while all the trucks are out.

Prior to the Carter and Rolph (1974) study to be discussed, the traditional method for making the decision was as follows. We look at the box history and find, for example, that there have been 15 alarms rung in from box 245 in the past. Of these, 10 have been false alarms (or minor fires) while 5 alarms were for serious fires requiring an all-out response. Thus the box history-based estimate of the probability of a serious fire is $\hat{p} = 5/15 = 33.33\%$. We compare this estimate to a critical level p_0 (which is determined by budget, and other considerations), and if $\hat{p} \geq p_0$ then we make an all-out response. If not, we send out one truck to investigate.[2] If the number of previous alarms is small, the box history provides an unreliable estimate with high variance. In the extreme but not uncommon case where this is the first alarm from the given box, there is no box history. In such cases, it is usual to replace box history by the average probability for the entire neighbor-

---

[2]Under suitable assumptions, this can be shown to be an optimal decision rule. Our object here is not directly to assess the decision rule but to provide high quality estimates of the probability $\hat{p}$ using empirical Bayes estimates.

hood. This setup should ring a bell in the reader's mind because of its resemblance to the previous examples. We are concerned with finding a compromise between an estimate based on the highly precise (because it is based on a large sample) group average which may however fail to be representative for a particular box, and the imprecise but representative estimate based on box history.

16.3.1 The Model

All alarm boxes in the Bronx were grouped into 216 neighborhoods. Let i be the index of a box in a given neighborhood with a total of k boxes. Let n_i be the total number of alarms sounded in the neighborhood and Y_i the total number of structural fires reported by the i-th box. Let $N = \sum_{i=1}^{k} n_i$ be the total reported alarms for the neighborhood as a whole. The parameter of interest is p_i the probability of a structural fire given a report from the i-th box. It is reasonable to assume that $Y_i \sim Bi(n_i, p_i)$. Let $\hat{p}_i^{ML} = Y_i/n_i$; then it is approximately true by the central limit theorem that

$$\hat{p}_i^{ML} \sim N(p_i, p_i(1 - p_i)/n_i).$$

Since the n_i are unequal, even a variance stabilizing transformation will not reduce the problem to an equal variances case.

To proceed, we have two options, corresponding to the two types of empirical Bayes estimators discussed in the baseball example. The first step is to estimate the variances $V_i = p_i(1 - p_i)/n_i$. In subsequent steps we will ignore that fact that variances have been estimated — proper accounting for this uncertainty will be handled later in connection with hierarchical Bayes and the Gibbs sampler. It turns out that the simplest procedure of estimating V_i by $\hat{V}_i = \hat{p}_i^{ML}(1 - \hat{p}_i^{ML})/n_i$ does not work. The ML estimates are too noisy, and the resulting empirical Bayes estimates are of poor quality. Let $\hat{p}_0 = (1/N) \sum_{i=1}^{n_i} Y_i$ be the aggregated estimate of the probability of a big fire corresponsding to an alarm for the neighborhood as a whole. Using $\hat{V}_i = \hat{p}_0(1 - \hat{p}_0)/n_i$, corresponding to the first empirical Bayes procedure for baseball, results in reasonable estimates. The second procedure is to use the arcsin transformation, defining $X_i = 2\arcsin(\sqrt{\hat{p}_i^{ML}})$. Then it is approximately true that

$X_i \sim N(2\arcsin(\sqrt{p_i}), 1/n_i)$. These two possibilites were explored by Carter and Rolph in their study.

Since the sample sizes on several box histories were small, one can doubt the quality of the normal approximation. Also, variance stabilization is an asymptotic concept, and may not work well in small samples. It would thus be of interest to try the Beta-Binomial approach discussed in the context of business cycle data. However we ignore this possibility, since it was not raised or discussed by Carter and Rolph.

16.3.2   The Estimates

With either approach, we end up with a model of the type $X_i \sim N(\theta_i, D_i)$ independent for $i = 1, 2, \ldots, k$. Without data transformation, $\theta_i = p_i$ and $D_i = \hat{p}(1 - \hat{p})/n_i$, while with the arcsin transformation $\theta_i = 2\arcsin(\sqrt{p_i})$ and $D_i = 1/n_i$. This differs from the baseball example in that we have unequal variances. We now discuss how to obtain empirical Bayes estimators for this type of model.

As usual, the empirical Bayes approach requires us to posit a model for the parameters θ_i. The central tendency model adopted for baseball also seems suitable here.

$$\theta_i = \theta + \nu_i. \tag{16.6}$$

The θ represents the overall probability of a big fire corresponding to a fire alarm. Individual boxes differ from the main tendency by some random factor ν_i, which we assume i.i.d. $N(0, A)$ for ease of analysis. If the prior parameters θ and A are known then the Bayes estimates of θ_i are

$$\hat{\theta}_i^B = \frac{A}{A + D_i} X_i + \frac{D_i}{A + D_i} \theta. \tag{16.7}$$

Following the usual empirical Bayes approach, we must estimate the hyperparameters A and θ from the data and substitute into Eq. (16.7) to get the empirical Bayes estimators. To estimate θ and A from the data on X_i, we use the marginal distribution of X_i, which is $X_i \sim N(\theta, A + D_i)$. In terms of the models, we can represent this by saying that $X_i = \theta_i + e_i$, where $\mathbb{V}ar(e_i) = D_i$, and $\theta_i = \theta + \nu_i$.

Eliminating θ_i by substitution we get

$$X_i = \theta + e_i + \nu_i \equiv \theta + \epsilon_i, \qquad (16.8)$$

where ϵ_i are i.n.i.d. $N(0, A+D_i)$. This differs from the baseball example in that we have heteroskedasticity.

Two main approaches to getting estimates of hyperparameters out of the marginal densities are method of moments and maximum likelihood; there are also several variants based on modificitions and mixtures of the two ideas. Morris (1983), in an excellent though dated review of empirical Bayes methods, writes that the benefits of empirical Bayes depend on the validity of Eq. (16.6) and less critically on the procedure used to estimate the hyperparameters. The use of Bayesian estimators is also possible; this leads to the hierarchical Bayes model and will be discussed later. We discuss how to obtain the MOM and ML estimates here.

Estimators based on method of moments which prove useful in more general contexts can be developed as follows. Suppose for the moment that the prior variance A is known. Then the optimal estimator for θ is a weighted average of the X_i, with weights proportional to the precisions (reciprocals of variances):

$$\hat{\theta}(A) = \frac{\sum_{i=1}^{k}(A + D_i)^{-1}X_i}{\sum_{i=1}^{k}(A + D_i)^{-1}}. \qquad (16.9)$$

This is also the usual GLS estimate for θ in Eq. (16.8) when the variances $A + D_i$ are known. It is also the minimum variance unbiased estimator under the normality assumption.

Next note that if θ is known, it is easy to estimate the variance A. This is because $\mathbb{E}(X_i - \theta)^2 = A + D_i$ and so $(X_i - \theta)^2 - D_i$ is an unbiased estimator for A. This suggests using

$$\hat{A} = \frac{1}{k}\sum_{i=1}^{k}\left\{(X_i - \theta)^2 - D_i\right\}. \qquad (16.10)$$

The unknown θ must be replaced by $\hat{\theta}(A)$ to make this formula operational. In this case, k should also be replaced by $k - 1$. An alternative formula, also based on the fact that $\mathbb{E}(X_i - \theta)^2 = A + D_i$, makes the justification for using $k-1$ clearer. It is easily shown that if A is known,

$$\sum_{i=1}^{k} \mathbb{E}\frac{(X_i - \hat{\theta}(A))^2}{A + D_i} = k - 1.$$

Instead of Eq. (16.10), we can solve for A from

$$\sum_{i=1}^{k} \frac{(X_i - \hat{\theta}(A))^2}{A + D_i} = k - 1. \tag{16.11}$$

In either case, we have to solve a pair of Eqs. (16.9) and one of the two above (either Eq. (16.10) or Eq. (16.11)) for the two unknowns A and $\hat{\theta}(A)$. An intuitive approach which works well in practice is to start with an arbitrary estimate for A, and use Eq. (16.9) to solve for θ. Plug this θ into Eq. (16.10) or Eq. (16.11) and solve for A. Now iterate back and forth until convergence. This yields one set of estimators for θ and A which can be plugged into Eq. (16.7) to yield empirical Bayes estimators.

A second approach is to use maximum likelihood. The marginal distribution of the X_i is $N(\theta, D_i + A)$. Thus the log likelihood can be written as

$$\ell(X_1, X_2, \ldots, X_k) = C - \frac{1}{2}\sum_{i=1}^{k} \log(D_i + A) - \frac{1}{2}\sum_{i=1}^{k} \frac{(X_i - \theta)^2}{2(D_i + A)}.$$

By differentiating with respect to A and θ, we can obtain two first order conditions (FOCs). Setting the first derivative with respect to θ equal to zero yields, after a little manipulation, Eq. (16.9) above. Differentating with respect to A yields

$$\sum_{i=1}^{k} \frac{(X_i - \hat{\theta}^{ML})^2}{(A + D_i)^2} = \sum_{i=1}^{k} \frac{1}{A + D_i}. \tag{16.12}$$

It can be proven that iterating back and forth between Eqs. (16.9) and (16.12) yields the ML estimates. There is one caveat that if A ever becomes nonpositive, we stop iteration, set $A = 0$ and solve for $\hat{\theta}^{ML}$

from Eq. (16.9). This corresponds to the case where the model for parameters Eq. (16.6) is a perfect fit.

Neither of the estimates described above reduce to the Stein estimator in the special case that D_i are all equal. Since the Stein estimate has certain optimality properties, Carter and Rolph modified the estimation procedure to make it closer to the original empirical Bayes derivation of the James-Stein estimate. Their modified estimator coincides with the James-Stein estimator in the case of equal variances. The reader is referred to their original paper for details.

16.3.3 Performance of the Estimates

One of the main characteristics of the empirical Bayes estimates is their ability to compromise between the use of box history and neighborhood information. The form of the empirical Bayes estimator is

$$\hat{p}_i^{EB} = \frac{D_i}{\hat{A} + D_i}\hat{p}_0 + \frac{\hat{A}}{\hat{A} + D_i}\hat{p}_i^{ML}.$$

Thus the estimate is a weighted average of the box history estimate $\hat{p}_i^{ML}$ and the average for the neighborhood, which is $\hat{p}_0 = (1/k)\sum_{i=1}^{k}\hat{p}_i^{ML}$. As discussed earlier, traditional estimates use one of these two, but the EB estimator averages the two. The weights used in averaging are also of great interest. D_i is the variance of the $\hat{p}_i$, whereas $\hat{A}$ is a measure of how close the p_i are to each other, and hence the permissibility of using the group average to replace the individual estimate. As Carter and Rolph show, boxes which have long histories have D_i relatively small and hence the EB estimate collapses more or less to the box history. In such cases, EB does not improve much, if at all, on the box history. The EB estimator shines in situation where the box history is small to moderate, making the optimal compromise between the variable individual information of the box history and the reliable but possibly nonrepresentative group average.

In this particular example it was easy to assess the reliability of these estimation procedures. Carter and Rolph had data for the four years 1967 to 1970. They used the years 1967-1969 to compute their estimates and tested their estimates against conventional estimates on the year 1970. In general the empirical Bayes procedures provided more reliable estimates than the simple box history estimates. For boxes with large number of alarms, the box history and empirical Bayes rules were similar and performed similarly. For boxes with small numbers of alarms, the empirical Bayes rules were successfully able to supplement the box information with the neighborhood information to improve the estimate. Carter and Rolph performed various different kinds of exercises to demonstrate the overall superiority of their estimates to the box history. In one of them, the box history estimates incorrectly identified 183 fires as being major, while the empirical Bayes method only missed 130.

16.4   Goodness, Ma Bell!

Thousands of different products are produced by A.T. & T. Technologies (formerly called Western Electric). For quality control purposes, audit samples are taken for each product and tested for defects. The object of this section is to describe a procedure use to estimate the quality of the product from the sample, and to decide whether the product meets standard quality requirements or not. This procedure is based on a refined version of the empirical Bayes technique and was originally developed by Hoadley (1981). It was tested and compared with a more conventional technique for quality control for about three years, and finally adopted for use universally at A. T. & T.

   To illustrate just a few of the rough knocks received by theories in the wild world of applications, we tell our tale in the form of a parable.

16.4.1   The Pilgrim's Progress

The innovative management at A.T. & T. turn over the newly created quality control department to a Dr. Pilgrim, fresh out of grad school. Pilgrim learns that in every production period, a sample of each of the products was tested for defects. The number of defects D_t has the Poisson distribution with unknown parameter $\lambda_t > 0$, which is a measure of the quality:

$$\mathbf{Pr}(D_t = k) = \frac{\lambda_t^k}{k!} e^{-k\lambda_t} \quad k = 0, 1, 2, \ldots$$

Note that $\mathbf{E}D_t = \lambda_t$ so that small values of λ are good. It is desired to maintain production standards so that $\mathbf{E}D_t \leq e_t$; the average number of defects in the sample should be less than or equal to e_t the *expectancy*. The expectancy depends on the sample size and the quality standard it is desired to maintain.

Initially a bit nervous, Pilgrim feels quite happy after seeing the structure of the problem. "We learned that in first year," he says happily to himself. The MVUE of λ_t is simply $\hat{\lambda}_t^{ML} = D_t$. He announces that he will assess the quality as acceptable if the number of defects D_t is less than or equal to the expectancy e_t, and take steps to improve quality if $D_t > e_t$. Over a period of time, numerous complaints accumulate from engineers of being flagged as delinquent on the basis of having just one or two defectives above the expectancy e_t. They feel that the production quality of their unit is good, but the random sampling chanced to find the only few defectives in their whole output, for example. The management asks Pilgrim to look into the problem.

After some thought, the Pilgrim realizes that close to half of the complaints are probably justified! A bit embarrassed, he decides that he has taken the wrong approach, applying his course on estimation theory instead of hypothesis testing. He decides to test $H_0 : \lambda_t = e_t$ versus the alternative $\lambda_t > e_t$ for each product. An optimal test will reject for $D_t \geq c_t$, where c_t is chosen to achieve the desired significance level. Under the null hypothesis, $\mathbb{E}D_t = e_t$ and $\mathbb{V}ar\,D_t = e_t$, so $T = e_t - D_t/e_t$ should be approximately $N(0,1)$ under the null. A simple rule would be to reject the null for $T \leq 2$, as this indicates that with high probability the production quality is below standard. Note that $P(T < 2) \approx 97.7\%$ from the normal approximation. Thus Pilgrim lays down the foundation for the 'T-rate' system, which stays in effect at A. T. & T. as the basic quality control mechanism for over fifty years.

This new system successfully reduces the volume of complaints. However, an unforeseen difficulty arises. In the textbooks, when we do a 97.7% level test, this means we are reasonably sure of taking the right decision. In the quality control example, we are doing the same

procedure thousands of times. The law of large numbers says that we
will average 23 mistakes in a thousand decisions. If these mistakes
were inevitable, then one would resign oneself to this record. However,
preventable mistakes inevitably result in complaints. A vocal engineer
walks in and says that for the past 10 periods I have an excellent record.
Now I admit that the number of defectives in the current sample is too
high, but it is just a sampling error. My record speaks for itself. On
cross-examination, Pilgrim admits that his scheme has a type I error
probability, and it is possible that the engineer is right. To patch this
problem, Pilgrim modifies his scheme so that quality is flagged as defective when $T < -3$, to reduce type I errors. Values of T between 2
and 3 are not flagged if the past history is good, and flagged otherwise.
By combining intuition, pragmatism, and suitable bureaucratic compromises, Pilgrim does well, and retires with a comfortable salary. The
next chapter of our story is written by a revolutionary, with new ideas.

16.4.2   We All Want to Change the World

The actual system in effect at A.T.& T. prior to the implementation
of the EB algorithm labeled the QMP (Quality Measurement Plan),
was the result of over 50 years of adding patches to the basic scheme
described above. It had Ptolemaic complexity but worked reasonably
well. There were three problems addressed by the EB algorithm implemented by Hoadley.

1. The T-rate system used numerous patchwork rules to exploit in-
    formation in the past history of the production process. However,
    the information was not used in an efficient way. A situation
    where there is long past experience on the same or similar pa-
    rameters is ideally suited to the use of Bayesian methods.

2. When the quality of production is flagged as defective on the
    basis of a hypothesis test, we have no direct estimate of the actual
    quality level. To be specific, suppose the quality standard requires
    $\theta_t = 1$. If an estimate $\hat{\theta}_t = 0.99$ has standard deviation 0.003, it is
    three standard deviations away from the null, and we will reject
    the null. However, the dereliction does not appear quantitatively
    serious. This is quite different from a situation where $\hat{\theta}_t = 0.05$

and has standard deviation 0.05, and both are different from a case where $\hat{\theta}_t = 0.5$ and the standard deviation is 0.25. In the first case we know for sure that the quality is slightly below normal. In the second we know for sure that the quality is way below normal, and in the third, we have a lot of uncertainty as to the real quality. In all three cases we reject the null hypothesis, and no further information is provided. The EB scheme to be described provides direct estimates of quality (Thus reverting to the first, unsuccessful, 'estimation' technique of our hypothetical Pilgrim of the previous section).

3. When we say that we reject the null hypothesis of standard quality at the 95 % significance level, this does not mean that the null hypothesis is false with probability 95 %. The distinction is stressed in many statistics texts, and yet is quite subtle. In practice everyone interprets the statement in the 'wrong' way. See Berger (1985, Section 4.3.3) for interesting evidence that this is a misleading interpretation. In contrast, the EB scheme gives direct probabilities of whether or not the null hypothesis is true

16.4.3 The QMP

Recall that we observe defects $D_t \sim \mathcal{P}o(\lambda_t)$. It is convenient to define $\theta_t = \lambda_t/e_t$, where e_t is the expectancy (or the expected number of defects. Suppose the prior density for θ_t is a Gamma density, $\theta_t \sim \mathcal{G}(d, e)$. The form of the Gamma density is

$$X \sim \mathcal{G}(p, \psi) \Rightarrow f^X(x) = \frac{\psi^p}{\Gamma(p)} x^{p-1} e^{-\psi x}.$$

It is important to note that while it is mathematically somewhat more convenient to put an i.i.d. prior directly on λ_t, this would not be a reasonable prior, since the λ_t are heterogeneous — the expected defects depend on the sample size and on the standards set for the particular product. On the other hand, it appears more reasonable to expect the quality level as measured by θ_t to be similar across products, since all units wish to maintain $\theta_t \leq 1$.

Despite the different densities involved, the form of the empirical Bayes analyses is quite similar to the previous cases. For future reference note that if $X \sim \mathcal{G}(a,b)$ then $\mathbb{E}X = a/(a+b)$ and $\mathbb{V}ar(X) = a/(a+b)^2$. Given $D_t|\theta_t \sim \mathcal{P}o(\theta_t e_t)$ and $\theta_t \sim \mathcal{G}(p,\psi)$, then the posterior density of θ_t is also Gamma. The following calculation factorises the joint density of D_t and θ_t into the product of the marginal of D_t and the posterior of θ_t given D_t

$$
\begin{aligned}
f(D_t,\theta_t) &= f(D_t|\theta_t) \times m(\theta_t) \{= \mathcal{P}o(\theta_t e_t) \times \mathcal{G}(d,e)\} \\
&= \left\{ \frac{(\theta_t e_t)^{D_t}}{D_t!} \exp(-\theta_t e_t) \right\} \times \left\{ \frac{e^d}{\Gamma(d)} \theta_t^{d-1} \exp(-e\theta_t) \right\} \\
&= \left\{ \frac{e_t^{D_t} e^d \Gamma(D_t + d)}{D_t! \Gamma(d)(e+e_t)^{D_t+d}} \right\} \times \left\{ \frac{(e+e_t)^{D_t+d}}{\Gamma(D_t+d)} \theta_t^{D_t+d-1} e^{-(e+e_t)\theta_t} \right\} \\
&= m(D_t) \times f(\theta_t|D_t).
\end{aligned}
$$

The posterior density of θ_t given D_t is readily recognized as $\mathcal{G}(D_t + d, e_t + e)$. Let $\hat{\theta}_t^{ML} = D_t/e_t$ and let $\mu_0 = d/e$ be the mean of the prior. Then the Bayes estimator for θ_t, which is the posterior mean, can be written as a weighted average of the prior mean and the data mean:

$$
\mathbb{E}(\theta_t|D_t) = \frac{D_t + d}{e_t + e} = \frac{e}{e_t + e}\hat{\theta}_t + \frac{d}{e_t + e}\mu_t
$$

Furthermore the weights are proportional to the precisions as usual. To get empirical Bayes estimators, we have to replace the hyperparameters e and d by reasonable estimates based on the marginal density of D_t.

As usual in EB, we have two main choices for estimating the hyperparameters: MOM or the ML. For the MOM, note that

$$
\mathbb{E}\hat{\theta}_t = \mathbb{E}^{\theta_t}\left(\mathbb{E}^{\hat{\theta}_t^{ML}|\theta_t}(\hat{\theta}_t|\theta_t) \right) = \mathbb{E}^{\theta_t}\theta_t = \frac{d}{e} = \mu_0.
$$

All that the complex formulae say is that the conditional mean of $\hat{\theta}_t^{ML}$ is θ_t, which in turn has mean μ_0, forcing the marginal mean of $\hat{\theta}_t^{ML}$ to be μ_0. We can also compute the marginal variance of $\hat{\theta}^{ML}$ using the law of total variance (exercise 2, page 66) as follows:

$$
\begin{aligned}
\mathbb{V}ar(\hat{\theta}_t^{ML}) &= \mathbb{E}\,\mathbb{V}ar(\hat{\theta}_t^{ML}|\theta_t) + \mathbb{V}ar\,\mathbb{E}(\hat{\theta}_t|\theta_t) = \mathbb{E}\frac{\theta_t}{e_t} + \mathbb{V}ar(\theta_t) \\
&= \frac{\mu_0}{e_t} + \frac{d}{e^2} \equiv v_t.
\end{aligned}
$$

From these two equations for the marginal mean and variance of the ML estimates, we get two equations from which we can solve for the prior mean and variance. Since $\hat{\theta}_t$ has expected value μ_0, a natural estimate of the prior mean is a precision weighted average of the $\hat{\theta}_i$, or

$$\hat{\mu}_0 = \frac{\sum_{t=1}^{T} \hat{\theta}_t/v_t}{\sum_{t=1}^{T} \sum_{t=1}^{T}(1/v_t)} = \frac{\sum_{t=1}^{T}(eD_t)/(e+e_t)}{\sum_{t=1}^{t} e/(e+e_t)}. \tag{16.13}$$

Let $\sigma_0^2 = d/e^2$ be the prior variance. From the equation for the marginal variance of $\hat{\theta}_t^{ML}$ we have that $\mathbb{E}(\hat{\theta}_t^{ML}-\mu_0)^2-\mu_0/e_t = \sigma_0^2$. Thus a simple estimator for the prior variance is

$$\hat{\sigma}_0^2 = \frac{1}{T}\sum_{t=1}^{T}(\hat{\theta}_t^{ML}-\hat{\mu}_0)^2 - \frac{\hat{\mu}_0}{e_t}.$$

We must solve these two equations iteratively to get MOM estimators.

The MOM is not necessarily an efficient estimator, since the prior variance is estimated as a simple average of the estimates of variance from each $\hat{\theta}_t^{ML}$. It would be more efficient to use a weighted average, with weights being the precisions. To get these precisions, we will need the fourth moments of the marginal density of D_t. These are in fact derived and used by Hoadley (1981) in developing the formulae for the QMP. An alternative, perhaps simpler, route to getting efficient estimates is to use ML. For example, we could solve Eq. (16.13) for $\hat{\mu}_0$ and numerically maximize the marginal density $m(D_t)$ derived above (which is in fact a negative Binomial) with respect to the parameter e, after substituting $d = \hat{\mu}_0 e$. By this means we get a one parameter maximization problem which is typically substantially easier to solve than a two-dimensional one.

The actual scheme used by Hoadley begins with the formulae above, but with substantial developments of several types. It was a product of the necessity to sell the formula to the management, to the engineers, to be sufficiently close to the earlier quality management plan in effect as to be bureaucratically acceptable, and to be easy to compute — none of these factors can be adequately discussed in this type of a text. Another point to note is that Hoadley is interested in the posterior density, rather than just the estimates of the hyperparameters. The posterior density permits evaluation of the probability that the quality is acceptable directly, among other benefits. Given estimates $\hat{d}, \hat{e}$ of the hyperparameters, a crude estimate of the posterior density is $\theta_t|D_t \sim \mathcal{G}(D_t + \hat{d}, e_t + \hat{e})$. Hoadley develops more sophisticated estimates, which account for the variance in the estimated hyperparameters.

Table 16.3: Batting Average Data for Thurman Munson

Year	End of April			End of Season		
	At Bats	Hits	Avg.	At Bats	Hits	Avg.
1970	44	7	0.159	453	137	0.302
1971	37	6	0.162	451	113	0.251
1972	31	11	0.355	511	143	0.280
1973	56	20	0.357	519	156	0.301
1974	72	16	0.222	517	135	0.261
1975	48	18	0.375	597	190	0.318
1976	40	13	0.325	616	186	0.302
1977	42	9	0.214	595	183	0.308
1978	63	17	0.270	617	183	0.297

16.5   Exercises

1. Table 16.3 contains batting average data for Thurman Munson over the years 1970 to 1978.

   (a) Let $\theta_{70}, \ldots, \theta_{78}$ be the batting ability parameter in each year. If we make the Draconian assumption that the batting abilities remained the same in all nine seasons, then the best estimate of the common $\overline{\theta}$ is the average of the nine end season batting averages. Check that for each season the end season average is closer to $\overline{\theta}$ than the early season average. Why does this observation lend support to the underlying model?

   (b) Apply the James-Stein estimate directly to obtain a forecast of end season averages. Use the variant which shrinks the different coordinates towards a common mean.

(c) If θ is the probability of a hit for one time at bat, then the number of hits H in n at bats is $\text{Bi}(n, \theta)$, and the variance of H is $n\theta(1 - \theta)$. Thus the variance of the batting average H/n is $\theta(1 - \theta)/n$. Take a plausible value for θ, calculate the variances for each early season average, and check to see if the condition for the existence of a Stein-type estimate dominating the MLE holds in this case (see Chapter 4).

(d) In the above exercises, we have ignored the unequal variances. Taking these into consideration, derive an empirical Bayes estimates for the θ_i.

2. (Toxoplasmosis Prevalence) Table (16.4), taken from Efron and Morris (1975), gives the toxoplasmosis prevalence rate in 36 cities in El Salvador. The prevalence rate X_i is calculated as (observed-expected)/expected, where expected is a normalized expected incidence. From the sample size and binomial theory, it is possible to calculate the variance of X_i; the square root of this is listed as $\sqrt{d_i}$. It seems reasonable to model X_i as $N(\theta_i, d_i)$ and to assume that $\theta_i \sim N(0, A)$. The mean of zero is, theoretically, the expected prevalence rate.

(a) (Empirical Bayes Estimates) Calculate the MLE for A and use it to obtain empirical Bayes estimates for the prevalence rates. Efron and Morris used a variant estimator, and hence your estimates will not match theirs (given in the 4th and 8th column), but should be close.

(b) (Internal Validation) Without going out and actually measuring prevalence in each city (an expensive undertaking) it seem difficult to check whether the empirical Bayes estimates improve on the MLE. Efron and Morris suggest the following plausible strategy. The marginal density of the X_i is $X_i \sim N(0, d_i + A)$. Plot the likelihood of the data set above as a function of A. Take this likelihood function to be the probability density function of A; this looks very much like a normal density curve. Check that with high probability A satisfies the inequality $.004 \leq A \leq .037$. For values of

Table 16.4: Toxoplasmosis Prevalence Rates in El Salvador

City	X_i	$\sqrt{d_i}$	$\delta(X_i)$	City	X_i	$\sqrt{d_i}$	$\delta(X_i)$
1	.293	.304	.035	19	-.016	.128	-.007
2	.214	.039	.192	20	-.028	.091	-.017
3	.185	.047	.159	21	-.034	.073	-.024
4	.152	.115	.075	22	-.040	.049	-.034
5	.139	.081	.092	23	-.055	.058	-.044
6	.128	.061	.100	24	-.083	.070	-.060
7	.113	.061	.088	25	-.098	.068	-.072
8	.098	.087	.062	26	-.100	.049	-.085
9	.093	.049	.079	27	-.112	.059	-.089
10	.079	.041	.070	28	-.138	.063	-.106
11	.063	.071	.045	29	-.156	.077	-.107
12	.052	.048	.044	30	-.169	.073	-.120
13	.035	.056	.028	31	-.241	.106	-.128
14	.027	.040	.024	32	-.294	.179	-.083
15	.024	.049	.020	33	-.296	.064	-.225
16	.024	.039	.022	34	-.324	.152	-.114
17	.014	.043	.012	35	-.397	.158	-.133
18	.004	.085	.003	36	-.665	.216	-.140

A in this range calcuate the risk of the MLE and compare with the risk of the empirical Bayes estimate.

(c) Note that the ordering of the cities is changed by empirical Bayes. Why is this? Which is the city in which toxoplasmosis is most prevalent (i) according to the MLE, (ii) according to your empirical Bayes estimate, (iii) according to Efron and Morris. The standard deviation of X_i is $\sqrt{d_i + A}$. Suppose we limit the change allowed to two standard deviations, according to this measure. Which city has highest prevalence using this 'limited translation' estimator?

3. Suppose for $i = 1, 2, \ldots, k$, we observe independent random variables $X_i \sim N(\theta, D_i + A)$.

   (a) Let $p_i^2 = 1/(d_i + A)$ be the precision of X_i and let $P = \sum_{i=1}^k p_i^2$. Show that the MLE of θ is

$$\hat{\theta}_{ML} = \frac{1}{P} \sum_{i=1}^k p_i X_i.$$

   (b) Define $S = \sum_{i=1}^k p_i^2 (X_i - \hat{\theta}_{ML})^2$. Then $S \sim \chi_{k-1}^2$ so that $\mathbf{E}S = k-1$ and $\mathbf{E}(1/S) = 1/(k-3)$. Hint: introduce random variables $Z_i = p_i(X_i - \theta)$ and note that $Z = (Z_1, Z_2, \ldots, Z_k)$ has the standard normal density $Z \sim N(0, \mathbf{I}_k)$. Use the fact that if Q is an idempotent matrix, $Z'QZ$ has the chi-square density with degrees of freedom equal to the trace of Q. We can write the statistic S as follows.

$$S = \sum_{i=1}^k p_i^2 \left(\{X_i - \theta\} + \{\theta - \hat{\theta}_{ML}\} \right)^2.$$

   Expand the quadratic as $(X_i - \theta)^2 + 2(X_i - \theta)(\theta - \hat{\theta}) + (\theta - \hat{\theta})^2$. Combine the second and third terms above by noting that

$$\hat{\theta}_{ML} - \theta = \sum_{i=1}^k \frac{p_i^2}{P}(X_i - \theta).$$

   Let $\mathbf{p} = (p_1, p_2, \ldots, p_k)$ and show that $S = Z'Z + (1/P)(\mathbf{p}'Z)^2 = Z'MZ$ where $M = \mathbf{I}_k + (1/P)\mathbf{pp}'$. It is easily checked that M is idempotent with trace $k - 1$.

4. Suppose there were 40 alarms from a given box, and one of them corresponded to a structural fire. What is the ML estimate of the probability of a serious fire corresponding to the next alarm? How could you estimate the standard deviation for this estimate? What if there were zero structural fires associated with the 40 alarms?

5. Suppose for $i = 1, 2, \ldots, k$ $X_i|\theta_i \sim Bi(n_i, \theta_i)$ and $\theta_i \overset{iid}{\sim} Be(a, b)$. We show how to obtain MOM estimates for the hyperparameters a and b. It is convenient to introduce $\mu = a/(a+b)$ and $\lambda = (a+1)/(a+b+1)$.

(a) Show that $\mathbb{E}\theta_i = \mu$, $\mathbb{E}\theta_i(1 - \theta_i) = ab/(a+b)(a+b+1) = \mu(1 - \lambda)$, and $\text{Var}(\theta_i) = \mu(\lambda - \mu)$.

(b) Suppose $\hat{\theta}_i = X_i/n_i$. Show that the *marginal* variance of $\hat{\theta}_i$ is

$$\text{Var}\,\hat{\theta}_i = \mathbb{E}\,\text{Var}(\hat{\theta}_i|\theta_i) + \text{Var}(\mathbb{E}(\hat{\theta}_i|\theta_i))$$

$$= \mathbb{E}\frac{\theta_i(1 - \theta_i)}{n_i} + \text{Var}(\theta_i) = \frac{\mu(1 - \lambda)}{n_i} + \mu(\lambda - \mu).$$

(c) Define $\gamma \equiv \mu(\lambda - \mu)$. Based on the previous formulas, show that a reasonable MOM estimator for γ is

$$\hat{\gamma} = \frac{1}{1 - S}\left\{ \frac{1}{k-1}\sum_{i=1}^{k}\left((\hat{\theta}_i - \hat{\mu})^2 - \frac{\hat{\mu}(1 - \hat{\mu})}{n_i} \right) \right\},$$

where $\hat{\mu} = (\sum_{i=1}^{k} X_i)/\sum_{i=1}^{k} n_i$, and $S = (\sum_{i=1}^{k}(1/n_i))/k - 1$. Hint:$\mu(1 - \lambda) = \mu(1 - \mu) - \gamma$.

(d) An alternative is to define $Z_i = X_i/\sqrt{n_i} = \sqrt{n_i}\hat{\theta}_i$. Calculate the marginal variance of Z_i as a function of the parameters a, b and use to obtain an alternative MOM estimator for a, b.

6. (Insurance Pricing) In this exercise we consider a simple application of empirical Bayes to calculate the fair price for auto insurance. We observe Y_i which is the average claims cost in territory i per covered vehicle. We can model $Y_i \sim N(\mu + \alpha_i, \sigma^2/e_i)$. Here μ is the overall mean for all the territories combined, while α_i is the regional differential. The parameter e_i measures the number of claims submitted; the more claims, the less the variance in the average Y_i. What are the classical estimates for μ and α_i? Explain how we could calculate empirical Bayes estimates for μ and α_i? For what type of territories will the empirical Bayes methods tend to perform better than the classical procedures? Read

Morris and Van Slyke for evidence on actual performance of this procedure.

7. Data for the years 1977 and 1978 for defects in keys of telephone sets manufactured in Shreveport are presented in the Table 16.5, courtesy of Bruce Hoadley (private communication,1992) as an illustration of a real quality control data set. The quantities SS,D,E,I, and $\Pr(\theta_t > 1)$ represent audit Sample Size, Defects, Expected number of defects, the ratio $I = D/E$, and the estimated probability that the quality is below standard.

   (a) Show that if $\theta_t \sim \mathcal{G}(p, \psi)$ then $\mathbf{E}\theta_t = p/\psi \equiv \theta$ and Var $\theta_t = p/\psi^2 \equiv \gamma^2$. Show that the posterior density of θ_t is also Gamma: $\theta_t | D_t \sim \mathcal{G}(p + d_t, \psi + e_t)$.

   (b) Suppose that from previous data, it seems reasonable to assume $\theta = 1$ and $\psi^2 = 1/4$ for the prior Gamma density. Use the posterior to calculate $\mathbf{E}\theta_t$ and also $\mathbf{P}(\theta_t > 1)$ for each period. Compare your estimates to the empirical Bayes estimates given in the table.

   (c) Show that the marginal density of the data is the negative binomial density as claimed in the text. Use numerical maximization programs to evaluate the ML estimates for p and ψ in each period for the data above. Substitute these ML estimates into the proper Bayes procedure to get crude empirical Bayes estimates for the posterior and the associated probabilities.

   (d) Use the MOM estimators proposed by Hoadley to derive the empirical Bayes estimators for θ_t and the probability that the process is below standard quality 1.

Table 16.5: Defects in Keys of Telephone Sets - Shreveport

Period	SS	D_t	E_t	I_t	$\Pr(\theta_t > 1)$	$\hat{\theta}_t$
7701	691	821	1251	0.6763	.02	-
7702	777	2026	1368	1.7463	.99	1.17
7703	749	1318	1330	1.0699	.69	1.14
7704	866	1663	1569	1.2447	.93	1.16
7705	889	1331	1601	0.8320	.20	1.10
7706	603	849	1097	0.7739	.17	1.05
7707	786	806	1415	0.5979	.02	1.04
7708	724	1115	1281	0.8704	.17	0.91
7801	705	1141	1326	0.8680	.14	0.88
7802	740	1513	1378	1.1705	.47	0.86
7803	749	1045	1360	0.7684	.05	0.85
7804	741	1244	1377	0.9034	.15	0.87
7805	800	2420	1425	1.7144	.99	1.05
7806	748	1215	1340	0.9067	.34	1.05
7807	837	1172	1490	0.8268	.21	1.04
7808	742	1988	1347	1.6318	.99	1.11

*The gossip, who did not recognize Hoja Nasruddin, told
him several juicy tales about various people in town. Finally, he said, 'And do you know the latest? I have just
heard that Mrs. Nasruddin's husband fell off his donkey and died this morning.' On hearing this, Hoja be-
came greatly distressed and walked out, visibly agitated.
Friends asked him what the matter was, and he replied
that he had heard his wife had been widowed. They said
'How can that be, seeing that you are alive and well?' He
said 'I am puzzled at that myself, but the news is from a
most reliable source.'*

Chapter 17

Utilizing Information of Uncertain Validity

17.1   Introduction

Because the model and estimation techniques are new and relatively
unfamiliar, the empirical Bayes technique is vastly underutilized in
econometrics. The same situation prevailed in statistics, where these
techniques originated, until a series of articles by Efron and Morris
(1972,1973, 1975) were written to address what the the authors saw as
four main obstacles to the use of the newly developed Stein estimators
in practice:

1. The feeling that Stein's discovery was a mathematical trick of
   some sort, with no relevance to applications.

2. Any potential gains from the improved estimators were likely to be negligible in practice.

3. Difficulties in adapting Stein-type estimators to complications inevitably arising in applications.

4. Lack of practical experience with the use of the new estimators.

As an antidote to the first obstacle, the development of the Stein estimator as an empirical Bayes estimator has strong intuitive appeal and permits easy understanding of the source of the gains from the use of Stein-type techniques. The other objections were tackled through a series of examples, where Efron and Morris demonstrated that substantial gains were achievable in practice, and showed how to adapt the estimator to common complications. The efforts of Efron and Morris were successful and numerous successful applications of Stein type estimators have been made since. Nonetheless, Morris (1983) in an insightful and informative survey, suggests that these methods continue to be underutilized in statistics.

In addition to the four difficulties mentioned above, a number of widespread misunderstandings about empirical Bayes as well as additional complications in usage create even greater obstacles to their utilization in econometrics:

1. Early studies, such as Aigner and Judge (1977), failed to find noticeable improvements from the use of Stein estimators in several econometric applications. In fact, as discussed in Section 4.6.1, mechanical applications of Stein estimators are bound to fail; one must carefully match the implicit prior information embodied in the estimator to the problem at hand. The empirical Bayes technology provides a method for doing so.

2. In regression applications, the Stein estimators dominate OLS with respect to quadratic loss only under certain rather restrictive conditions on the design matrix. It appears that these conditions are rarely fulfilled in practice. As discussed in Section 4.6.2, this is a red herring, as quadratic loss is not a suitable criterion for regression models. Furthermore, as discussed in Section 4.6.3, minimaxity or dominating OLS is not necessarily desirable.

3. Practitioners also avoid Stein estimates because their distribution theory is not known, or just being developed. This problem is also sidestepped by the empirical Bayes development, which provides an approximate posterior distribution for the parameters. Posteriors are even more useful than distributions of estimators, and both empirical and theoretical studies have shown that they possess frequentist validity.

4. Adapting Stein-type estimators to regression-type models requires making significant adaptations and generalizations of the original framework.

A number of authors have adapted the empirical Bayes techniques and successfully applied them to regression models. However, each has done so in his own way, and the unity behind the diverse techniques is hard to recognize. We hope to address some of these difficulties in the present and following chapters. The first section below presents the empirical Bayes as a natural solution to a common econometric problem. It also develops a unified framework for applications of empirical Bayes and shows its superiority to several alternatives currently in wide use in econometric practice.

17.2 Improving on OLS

Consider a standard linear regression model with $y = X\beta + \epsilon$, where X is a $T \times k$ matrix of constants, $\epsilon \sim N(0, \sigma^2 \mathbf{I}_T)$ and β, σ^2 are $k \times 1$ and scalar parameters to be estimated. The usual OLS or MLE estimate for β is $\hat{\beta} = (X'X)^{-1}X'y \sim N(\beta, \sigma^2(X'X)^{-1})$. Our story requires that $\hat{\beta}$ be too imprecise for the purpose at hand. High correlations among the regressors frequently give rise to this situation by making $X'X$ nearly singular: this creates large entries in the diagonal of $(X'X)^{-1}$ and hence high variances for the estimated coefficients. For example, in dynamic models, high correlations among the lagged values of regressors typically prevent accurate estimates of the lag coefficients. Small sample sizes and/or large numbers of parameters have similar effects.

It is worth emphasizing that 'large' and 'small' variances are relative to the purpose for which the coefficients are to be used. If the OLS is accurate enough for a particular purpose, there is no need to improve it. On the other hand, what can we do if more accurate estimates are required? The obvious first response is to gather more data. Again our story requires that this be too expensive or otherwise not feasible. Even though additional data are not available, the only way to improve estimates, barring magic, is to introduce additional valid information. The only remaining source is prior information. The type of prior information demanded by classical Bayes techniques based on natural conjugate priors is very demanding. Most econometricians would be uncomfortable putting down a number as a estimated prior mean for some coefficient, especially when the resulting estimator would depend on such numbers chosen arbitrarily. Fortunately, the type of prior information demanded by empirical Bayes techniques is of a form very familiar to econometricians, is frequently available, and does not require making subjective judgements regarding magnitudes of unknown quantities.

Empirical Bayes techniques require that we provide a model for the regression parameters. That is, we must find a $k \times m$ matrix of regressors H such that

$$\beta = \beta_0 + H\alpha + e \qquad (17.1)$$

is a plausible model for the parameters β. The vector β_0 is a known vector of constants. The $m \times 1$ vector α and the parameters entering the covariance matrix of the error vector e are called the 'hyperparameters' in this setup. For the moment, we will simply assume that $e \sim N(0, \nu^2 I_k)$ so that the only hyperparameter associated with the error distribution is ν^2, the error variance. We will see later that applications frequently require more complex structures for the covariance matrix of e. While special cases of this type of model are frequently used in econometrics, the general structure and properties have not been widely explored. One special case which has received some attention is the random coefficients model, where $\beta_0 = 0$ and $H = I_k$. Empirical Bayes techniques provide new insights into this type of model as well.

While the structure of the second stage model may be unfamiliar to many econometricians, a reformulation is very familiar. Suppose it is suspected that the coefficient β may satisfy a linear constraint of the type $R\beta = r$, where R is $q \times k$ matrix of full rank and r is $q \times 1$. This can easily be reformulated into the desired form. We can find a $k \times (k - q)$ matrix H satisfying $RH = 0$. Fix some vector β_0 satisfying the constraints, so that $R\beta_0 = r$. Then the set of parameters $R\beta = r$ is equivalent to the set $\beta = \beta_0 + H\alpha$ where $\alpha \in \mathbb{R}^{(}k - q)$. This is equivalent to the model of the previous paragraph with the restriction that $\nu^2 = 0$.

17.2.1 Alternative Strategies

We have seen earlier in Chapter 9 some of the wide variety of examples covered by linear constraints on parameters in regression models. Some more examples will be presented to illustrate the empirical/hierarchical Bayes technique. Since the situation of plausible linear constraints on linear parameters is quite common in econometrics, a number of techniques have evolved to utilize such information. We review these briefly, in increasing order of sophistication, to show how empirical Bayes can be viewed as the culmination of a natural sequence of steps.

Ignoring the Information: The simplest technique is to ignore the information provided by the constraint (in view of its uncertainty). This corresponds to the use of OLS, or of setting $\nu^2 = +\infty$ in Eq. (4.1). When OLS is adequate for the purpose at hand, this technique is the recommended one.

Ignoring the Uncertainty: If OLS is not adequate, the simplest technique is to ignore the uncertainty of the constraints, and use RLS (restricted least squares). This corresponds to setting $\nu^2 = 0$ in Eq. (4.1).

Testing the Restrictions: Since RLS has very high mean squared error if the restrictions do not hold, and there typically exists uncertainty regarding the restrictions, the technique of 'pretesting' is widely used in econometric applications. We use the usual F-statistic to test the linear constraint $R\beta = r$, and use OLS if the null is rejected and

RLS otherwise. This corresponds to testing $\nu^2 = 0$, and using RLS if we cannot reject this, and setting $\nu^2 = +\infty$ otherwise.

Mixing OLS and RLS: In the light of Eq. (4.1), pretesting appears clearly inadequate, since it ignores the possibility that ν^2 may lie between the two extremes. It, and related strategies for model selection (see, for example, Judge *et al.* (1980) for a long list of alternatives), were conceived in a context that the restrictions either hold or do not hold. Classical Bayes procedures offer a richer menu of choices, namely a weighted average of the two estimators under consideration. Choosing the prior parameters is more or less equivalent to having to specify the weight given to the two, since the classical Bayes estimator is a precision weighted average of the prior and posterior mean:

$$\hat{\beta}^B = \left(X'X + \frac{\sigma^2}{\nu^2} \mathbf{I}_k \right)^{-1} \left\{ X'X\hat{\beta} + \frac{\sigma^2}{\nu^2}(\beta_0 + H\alpha) \right\}. \qquad (17.2)$$

The class of estimators covered by this general form is a large and appealing set of matrix-weighted averages of OLS and RLS. However, if we choose α and ν^2 a priori, only lucky choices will lead to good results. This intuitively obvious fact translates into technical defects in the classical Bayes procedure discussed in Section 3.4.

An 'Optimal' Mix: Since the crux of the matter is 'how reliable is the prior information,' the classical Bayes procedure does not really address the issue — it suggests that we decide this *a priori* by choosing weights, and does not offer guidance on how this can be done in a sensible way. The empirical Bayes technique is based on estimating the parameters α and ν^2. Thus it uses the data to assess the reliability of the prior information, and assigns weights to the prior information in accordance with its estimated reliability.

17.2.2   EB: Estimating 'Reliability'

We first present a basic framework for getting empirical Bayes estimators. In practice numerous variations and complications arise. We will illustrate some of them in the course of discussion of several applications of EB techniques.

Substituting Eq.(4.1) into the regression model $y = X\beta + \epsilon$ yields

$$y - X\beta_0 = X(H\alpha + e) + \epsilon = XH\alpha + Xe + \epsilon. \tag{17.3}$$

Thus this is the equation implied by our prior information combined with the model and data. Intuitively, the crux of the empirical Bayes procedure consists of assessing the reliability of the prior information according to the goodness of fit of this equation. If this equation is a bad fit, then our prior information conflicts with the data structure. This will result in high estimates of error variances in this equation, and low weight given to the prior information (and RLS) in the final estimation. Contrariwise, a good fit results in greater reliance being placed on the prior information and high weights for RLS. The technical details are as follows.

Defining $y^* = y - X\beta_0$ and $u = Xe + \epsilon$, we get a second regression model $y^* = XH\alpha + u$, but the covariance matrix $\Omega = \nu^2 XX' + \sigma^2 \mathbf{I}$ of u is not of the standard type. If Ω is known, an optimal estimator for α is

$$\hat{\alpha} = (H'X'\Omega^{-1}XH)^{-1}H'X'\Omega^{-1}y^*. \tag{17.4}$$

This is also a first order condition for maxizimizing the likelihood in the regression model (Eq. 4.3). The other first order condition is a bit complicated, and hence is frequently replaced in applications as follows. It is easily established that for known Ω, $Z = (y^* - XH\hat{\alpha})'\Omega^{-1}(y^* - XH\hat{\alpha})$ is distributed as χ^2_{T-m}. Thus $\mathbb{E}Z = T - m$, and this gives us a MOM-type equation:

$$(y^* - XH\hat{\alpha})'\Omega^{-1}(y^* - XH\hat{\alpha}) = T - m. \tag{17.5}$$

Fix $\sigma^2 = \hat{\sigma}^2 = \|y - X\hat{\beta}\|^2/(T-K)$ and ignore the uncertainty about this parameter. We can obtain estimates of the hyperparameters α and ν^2 by the following iterative procedure. Fix an arbitrary initial value $\nu^2 = \nu_0^2$, and calculate $\hat{\alpha}_0$ from (4.4). Substitute $\hat{\alpha}_0$ into Eq. (4.5) and solve for $\hat{\nu}_1^2$. Iterate this procedure until convergence, which is guaranteed by the E-M algorithm (Dempster *et al.* (1977)). By plugging the estimates $\hat{\alpha}$, $\hat{\nu}$ into Eq. (4.2), the formula for Bayes estimators based on known hyperparameters α and ν, we get one set of empirical Bayes estimates. Other ways to get similar estimators will be discussed later.

17.3   Estimates of Income for Small Places

We will review an article by R. E. Fay and R. A. Herriot (1979). Due to certain legislation passed in 1972, the Treasury was required to distribute funds to some 39,000 units of local governments within the States, on the basis of population and per capita income. At the time the 1970 Census was designed, the Census Bureau was unaware the estimates of Per Capita Income (PCI) would be demanded of it in this detail. Information about PCI was collected for a 20% sample. For large places, this provides a suitably precise estimate, but for places with fewer than 500 people (of which there are some 15,000), the Census Bureau estimates have significant sampling errors. Let Y_i be the Census estimate of the PCI for county i. Census estimates of the coefficient of variation $CV_i = \sqrt{\mathbb{V}ar(Y_i)}/Y_i$ show that $CV_i \approx 3/\sqrt{N}$. For instance, the coefficient of variation for a place with 100 people is about 30 %, while with 500 people it is 13 % . For all places with 500 people or less, the Census Bureau decided to use the more precisely measured (but potentially biased) county values for PCI in place of the (high variance) sample estimates.

Several features of the problem suggest that it is possible to improve this methodology. For one thing, the dividing line of 500 is arbitrary. Second, in many cases the county estimate was several standard deviations away from the (small) sample estimate. Third, some potentially valuable auxiliary information (tax returns, property values) was available. Fay and Herriot proposed an empirical Bayes estimator for the PCI which was eventually accepted by the Census Bureau both because of its internal logical consistency and because of independent empirical evidence validating the methodology.

17.3.1   Regression-Based EB estimates

Let Y be the $m \times 1$ vector of census estimates for each county. We may assume that $Y = \theta + \epsilon$. The estimate Y_i is an unbiased estimate of the true PCI θ_i, and $\epsilon_i \sim N(0, V_i)$. It is reasonable to assume that the ϵ_i are indepedent. If the coefficient of variation is $3/\sqrt{N_i}$, we can estimate the variance by $\hat{V}_i = 9Y_i^2/N$. Note that our basic regression model is the simplest possible; however, the error terms are heteroskedastic, making the setup slightly different from the one studied earlier. To use the information available on property values, tax returns, and county PCI, we hypothesize a linear regression relationship between these variables and the PCI (θ_i):

$$\theta_i = X_i\beta + \epsilon_i.$$

The 1×4 vector of regressors includes a constant term in addition to the independent variables mentioned. It is convenient and plausible to assume that ϵ_i are i.i.d. $N(0, A)$. Note that this is a considerable generalization of the original Census Bureau methodology, which corresponds to setting $\beta = (0, 0, 0, 1)$ (putting all weight on the county PCI), setting $\epsilon_i = 0$ and ignoring the data Y_i for all places for which the variance D_i exceed a certain value.

The essence of the empirical Bayes method is to estimate the parameters of the prior distribution, namely the regression coefficients β and A the common variance of the ϵ_i, from the marginal density of the data Y_i. This marginal density is $Y_i \sim N(X_i\beta, D_i + A)$. Let Y be the $m \times 1$ vector of PCI sample estimates, X the $m \times 4$ matrix with i-th row X_i, and V the $m \times m$ diagonal matrix with (i, i) entry $D_i + A$. The log likelihood function of the marginal density $l(Y, \beta, A)$ is

$$l(Y, \beta, A) = -\frac{m}{2}\ln(2\pi) - \frac{1}{2}\sum_{i=1}^{m}\ln(D_i + A) - \frac{1}{2}\sum_{i=1}^{m}\frac{(Y_i - X_i\beta)^2}{A + D_i}$$

$$= -\frac{m}{2}\ln(2\pi) - \frac{1}{2}\sum_{i=1}^{m}\ln(D_i + A) - \frac{1}{2}(Y - X\beta)'V^{-1}(Y - X\beta).$$

Taking the gradient with respect to β and solving the resulting first order condition, we get the equation for the ML estimate β^* of β:

$$\beta^* = (X'V^{-1}X)^{-1}X'V^{-1}Y. \tag{17.6}$$

The first order condition obtained by setting $\partial l/\partial A$ equal to zero is equivalent to

$$\sum_{i=1}^{m}\frac{(Y_i - X_i\beta)^2}{(A + D_i)^2} = \sum_{i=1}^{m}\frac{1}{A + D_i}. \tag{17.7}$$

We can get the MLE by iterating. Fix $A > 0$ and solve for β^* from Eq. (4.6). Plug β^* into Eq. (4.7) and solve for a new A. Iterate until convergence.

Fay and Herriot used an alternative estimator based on the method of moments. Note that $\mathbf{E}(Y_i - X_i\beta)^2 = A + D_i$ so that the expected value of the LHS in Eq. (4.7) equals the quantity on the RHS. Thus the equation Eq. (4.7) forces A to satisy an equality which holds exactly in expected value. Instead of Eq. (4.7), we could use a simpler equality which also holds in expected value:

$$\mathbb{E}(y - X\beta^*)'V^{-1}(y - X\beta) = m - 4.$$

Fay and Herriot suggest dropping the $\mathbb{E}$ from the above, and solving for A from

$$\sum_{i=1}^{m} \frac{(Y_i - X_i\beta)^2}{A + D_i} = m - 4$$

instead of Eq. (4.7). This corresponds to a method of moments approach. As we have discussed earlier, the relative merits of ML versus MOM for estimating hyperparameters have not been studied. Morris (1984) remarks that the merits of empirical Bayes come from the richer model for the parameters, and less importantly from the method of estimation used for the hyperparameters.

17.3.2   Validation

Fay and Herriot actually used two refinements not mentioned in the discussion above. First they used a log transform to stabilize the variance of the distribution ($Y_i = \log PCI$ instead of $Y_i = PCI$ as used above). This makes the normal a better approximation, and eliminates the term Y_i^2 from the variance. Second, after retransforming to get empirical Bayes estimates for the PCIs directly, they adjusted the PCI so that the sum of the place incomes within a county would match the (accurately estimated) county income.

How can the we assess the gains from using an empirical Bayes procedure of this type? A simple check is provided by the estimated A. Small values of A imply that the prior density is of low variance and so the gains from utilizing prior information will be high. Note also that A is the variance of the errors in the regression of Y_i on X_i and thus measures the goodness of fit in this regression. Using the fact that the coefficient of variation for the PCI is approximately $3/N^{1/2}$ it can be shown that the variance of $Y_i = \log PCI$ is approximately $9/N$ (see Section 16.1.2). This information can be used as follows. For the state of Kansas, A was estimated to be .02. If we solve $9/N = .02$ we get $N = 450$. This means that in Kansas, the regression provides a better estimate than the sample itself for all places of size 450 or less. Instead of using all the three regressors, one could follow the Fay and Herriot procedure with just one regressor, the county PCI. This would give us an idea of the accuracy of the county value as a proxy for the PCI in small places, the original Census Bureau procedure. For Kansas, doing this resulted in an estimated A of .064 indicating that the county value alone is considerably worse than the three regressors. Also note that $9/N = .064$ solves for approximately $N = 120$ so that the sample estimate PCI would be more accurate than the county value for places of size 120 or more (in contrast with the 500 originally used by the Census Bureau).

There are situations in which the estimated A's are the only means available to assess the performance of the empirical Bayes methods. In the present case, a special survey of per capita income was conducted by the Census Bureau in 1973. When values from this were compared with (i) sample estimate, (ii) county PCI, and (iii) the empirical Bayes estimates, the third set of estimates proved to be considerably more accurate overall. It should be clarified that while overall accuracy was improved, there were individual places for which (i) and/or (ii) were better estimates than (iii). Only improvement in average accuracy is promised by the shrinkage estimators and not improvement in each component separately.

17.4   EB Loses in Court

Freedman and Navidi (1986) and the discussion following their article are fascinating reading. In court battle over an application of EB methods, with expert statisticians on both sides, every assumption implicit in the model is laid bare and examined critically. Many statisticians objected to this airing of dirty laundry in public. This section, which provides a summary and review, should be read in conjunction with the original article.

The US Census Bureau attempts to count the entire population of the USA, broken down by area, every 10 years. The resulting counts are extremely important, as they determine allocation of Federal funds to the tune of $100 billion yearly in the early eighties. There is good evidence to suggest that about 2% of the population was missed (or, there was an *undercount* of 2%) in the 1970 census. If the undercount was uniform and random, this would not be a major issue. However, there is strong evidence to suggest that the undercount is larger in rural areas, big cities, and in certain segments of the population, such as blacks, Hispanics, and illegal immigrants. The Census Bureau made a much greater effort to eliminate the undercount in 1980, resulting in an overcount of about 1/4 of 1 % of the legal population. Nonetheless, the differential undercount problem persisted, and several local governments, including New York State, sued to get the Census Bureau to adjust its figures. Since New York State has a larger proportion of population which is harder to count (i.e., illegals and minorities), it appears clear that it would benefit from a revised estimate.

17.4.1   The Gory Details

We follow the notation of Freedman and Navidi (1986) (F-N in the sequel), and skip over some important issues not relevant to our goals (such as what is the true population anyway?). Let γ_i be the undercount in area i measured as a percentage. Thus $\gamma_i = (P_i - C_i)/P_i$, where P_i is the true population in area i and C_i is the Census Bureau count of the population in area i. Following the 1980 census, the Census Bureau used the PEP (postenumeration program) to estimate the undercount γ_i. The PEP used two types of studies to estimate the undercount. In one type, we collect a random sample of people (called the P sample), and attempt to assess the percentage which were counted by the census. In the second type, we take a random sample (called the E sample) of census records for 'counted' people, and attempt to estimate excess due to double counting, inclusion of fictitious persons, and errors arising from wrong addresses. As a result of these studies, we get an estimate y_i of the undercount. If the PEP estimates were exact, we would have $y_i = \gamma_i$ (estimated undercount equals true undercount), and no story to tell. Our story concerns the attempt to assess and improve the PEP measures of the undercount y_i as an estimate of the true undercount γ_i.

Suppose we assume that $y_i = \gamma_i + \epsilon_i$, for areas $i = 1, 2, \ldots, n$. At the same time that the PEP estimates y_i are computed, the Census Bureau also gets measures of the variability of these estimates: $\mathbb{V}ar(\epsilon_i) = K_{ii}$. We will assume for simplicity that $\epsilon \sim N(0, K)$, where K is a *known* diagonal matrix. If we can find a good model for the parameters γ_i, we should be able to use EB techniques exactly like those describe in the previous example to get improved estimates of the undercount. This was done by Erickson and Kadane (1985) (E-K in the sequel), who were the principal statisticians for New York. E-K model the undercount as $\gamma = X\beta + \delta$, where $\delta \sim N(0, \sigma^2 I)$. In addition to the constant term $X_{i,1} \equiv 1$, three regressors were chosen by E-K as explanatory variables for the undercount. These were $X_{i,2}$, the percentage of minority population, $X_{i,3}$, the crime rate, and $X_{i,3}$, the percentage of population conventionally enumerated.[1] This completes the specification of our empirical Bayes model.

As usual, we begin by getting estimates for the hyperparameters β and σ^2. Substituting for γ_i yields $y = X\beta + \delta + \epsilon$. Let $\eta = \delta + \epsilon$, and note that $\mathbb{C}ov(\eta) = K + \sigma^2 I \equiv \Sigma$. If σ^2 is known, the best estimate for β is the GLS $\hat{\beta} = (X'\Sigma^{-1}X)^{-1}X'\Sigma^{-1}y$. If β is known, the ML estimate for σ^2 satisfies the equation

$$\sum_{i=1}^{n} \frac{(y_i - X_{i,.}\beta)^2}{(K_{ii} + \sigma^2)^2} = \sum_{i=1}^{n} \frac{1}{K_{ii} + \sigma^2}.$$

---

[1] Conventional enumeration, done by face-to-face interviews mostly in rural areas, is considered more accurate than alternative methods using the mail.

One can get ML estimates for σ^2 and β by iterating between these two equations. The second equation can also be replaced by a simpler moment equation to simplify the procedure. If we let $\tilde{\beta}_{ML}$ and $\tilde{\sigma}^2_{ML}$ be the resulting estimators, an empirical Bayes estimator for the undercount can be written as follows. The prior mean for γ is $X\beta$ and the prior precision is $(1/\sigma^2)\mathbf{I}$. The data mean is just y and the data precision is K^{-1}. The EB estimator is the precision weighted average of the data and prior mean, with the hyperparameters β and σ^2 replaced by their ML estimates:

$$
\begin{aligned}
\hat{\gamma}^{EB} &= \left(K^{-1} + \frac{1}{\tilde{\sigma}^2_{ML}}\mathbf{I}\right)^{-1}\left\{K^{-1}y + \frac{1}{\tilde{\sigma}^2_{ML}}X\tilde{\beta}\right\} \\
&= X\tilde{\beta} + \tilde{\sigma}^2_{ML}\left(K + \tilde{\sigma}^2_{ML}\mathbf{I}\right)^{-1}(y - X\tilde{\beta}
\end{aligned}
$$

If we define $\Gamma = K^{-1} + (1/\tilde{\sigma}^2_{ML})\{\mathbf{I} - X(X'X)^{-1}X'\}$, an alternative simpler formula for $\hat{\gamma}^{EB}$ is $\hat{\gamma}^{EB} = \Gamma K^{-1}y$. This follows directly from Theorem 18.1 after setting $\Omega_2 = 0$.

Crucial to the argument for both sides is the issue of how precise the EB estimate is. Let $P = K^{-1} + \sigma^{-2}\mathbf{I}$. Under the assumption that β is known, the posterior distribution of γ given y is normal with mean $\hat{\gamma}^{EB} = (P^{-1}\{\sigma^{-2}X\beta + K^{-1}y\}$ and covariance P^{-1}. To get the covariance estimate of $\hat{\gamma}^{EB}$, we have to add to P^{-1} an estimate of the variance of $P^{-1}X\tilde{\beta}_{ML}/\sigma^2$. Since $\tilde{\beta}_{ML}$ has covariance matrix $(X'\Sigma^{-1}X)^{-1}$, we get the following formula for the covariance of $\hat{\gamma}^{EB}$:

$$
\mathbb{C}ov(\hat{\gamma}^{EB}) = P^{-1} + \frac{1}{\sigma^4}P^{-1}X(X'\Sigma^{-1}X)^{-1}X'P^{-1}.
$$

Using the matrix equality of exercise 1 (on page 502), it is easily shown that $\mathbb{C}ov(\hat{\gamma}^{EB}) = \Gamma = K^{-1} + \sigma^2\{\mathbf{I} - X(X'X)^{-1}X'\}$. Crude EB estimates of covariance are frequently imprecise and unreliable because they ignore the uncertainty in the estimated hyperparameters. The present estimate takes this fully into account, and hence it can be justified both in frequentist terms (see Lemma 7.2 of E-K) and in terms of a hierarchical Bayes (HB) model. As we will see in the next chapter, the HB models provide a natural way of accounting for uncertainty about the hyperparameters.

To all appearances, this is a perfectly standard and typical EB estimation problem. The undercount estimates y_i are more variable than desired, and we reduce their variance by shrinking them towards a central value. Instead of fixing a single number as a central value, we assume that the central value is explained by some regressors, and estimate the coefficient of these regressors first. In a devastating critique, F-N bring up for close examination every assumption that such an analysis rests on, and show that many fail in the case at hand. The EB world will never be the same.

17.4.2 The Attack

To begin with, we note that the Census Bureau decided not to adjust the census on the basis of the PEP studies because the variance of the PEP estimates was too high. We are trying to measure undercounts which are quite small — on the order of 1 % or less — using an instrument with not enough resolving power. The contention of E-K on behalf of New York is that we can improve the PEP estimates by shrinkage and reduce their variances to an acceptable level. F-N contend on the other hand that using E-K shrinkage estimates to adjust the census is like taking random numbers and adding them to the census counts — addition of pure noise will only make the counts less precise than before. Below we list reasons why F-N consider the shrinkage adjustments unreliable.

Bias An important assumption in the EB model is that the error ϵ in the model $y = \gamma + \epsilon$ is independent of the regressors X in the model for the parameter $\gamma = X\beta + \nu$. To the extent that ϵ also depends on the regressors in X, the EB estimates of γ will be biased. To see intuitively why this is the case, consider a simple situation, where the undercount exactly equals the crime rate (for example). That is, $\gamma_i = c_i$, where c_i is the crime rate in place i. Suppose now, as seems quite natural, that the deviations between PEP estimates of the undercount and the undercount are also affected by the crime rate, and for example $y_i = \gamma_i + c_i$. A regression of y_i on c_i will produces a very good fit and an estimate $\hat{a} = 2$ of the hyperparameter in the prior model $\gamma_i = ac_i + \eta_i$. Thus our adjustment procedure will take the crime rate and multiply by '2' (instead of the correct '1') to get an adjusted estimate of the undercount. The more complex situation which actually prevails in the situation at hand will almost certainly cause a bias, but of unknown size and direction, in the undercount estimates.

While this is a problem to watch for in all applications of EB, this bias is likely to be more serious in the present problem. This is because factors which cause differences between the census count and the true population are likely to be the same as the factors causing a difference by undercount estimates and the true undercount. In the discussion on E-K, Census Bureau statistician Wolters remarks that 'For those (with) ...experience ...of undercount studies, it is almost axiomatic that the greatest difficulties in measuring the undercount are encountered in the same areas where the census itself encounters great difficulties.' To see that this is not a typical problem, consider the formally similar model considered earlier, where we obtained estimates of average income for small counties by EB. The original estimates y_i differ from the true average incomes γ_i because of small sample sizes. The model for the county average income γ_i uses regressors of property values, tax returns data, and average income in a larger grouping. It seems unlikely that these three factors had any significant influence in the selection of size of sample by the Census Bureau. Even if there is some correlation, it would be plausible to assume that it is small.

Variations in PEP and Estimates of Bias   E-K and also Wolters attempt to quantify the bias in the EB estimates caused by this problem using several techniques. To describe one such technique, we note an important point heretofore not discussed. The Census Bureau has not one but 24 different PEP series, each estimating the same undercount. Why is there such a variety? The reasons, discussed in greater detail by E-K, are briefly as follows. In P samples, for a a rather large 8% of the sample, it could not be determined whether or not the person had been counted in the census or not. Similarly for the E sample of census records, it was not possible to determine for about 3% of the cases whether the record was correct or erroneous. The Census Bureau developed various rules for treating the uncertain cases in the P and E samples. A name like PEP 2/8 means that the uncertain cases in the P sample were treated using rule 2 and the uncertain cases in the E sample using rule 8. Each combination of rules leads to a new series and after some weeding, 12 series were retained as being reasonable. There was a very large variation among the 12 series — estimates of the undercount in New York ranged from 1 % to 8 %. Also estimates of the variances in the PEP series (referred to as K_{ii} in the previous section) could also be obtained and were quite high. Both of these factors played a role in the Census Bureau decision not to adjust the counts using the PEP series.

If the PEP series are all unbiased for the true undercounts γ, as assumed by EB, then the differences between any two series should be uncorrelated with the regressors in X. F-N show that the difference between two such series is highly correlated with the regressors in X, so that bias is likely to be a problem. Another way to show that the error $\epsilon_i = y_i - \gamma_i$ in the PEP estimates of undercount is related to the regressors is to use proxy variables for the error. One such proxy is the percentage of cases in the PEP samples for which rules had to be used to determine match status. This variable also turns out to be strongly related to the regressors in X. All in all, there seems to be conclusive evidence for the bias discussed earlier.

Standard Errors The claim of superiority for shrinkage estimators rests heavily on their smaller standard errors. There are several reasons why E-K estimates of the standard error are biased downwards. To begin with, the estimate σ^2 of error variance for the auxiliary regression model $\gamma = X\beta + \delta$ is treated as known, whereas it is estimated from the data. Similary the Census Bureau estimates K_{ii} of variances of the PEP estimates are treated as known constants, when they are potentially variable. In addition, there is substantial uncertainty as to the form of the auxiliary regression model. Instead of, or in addition to, the three regressor selected, a number of other regressors appear plausible. As shown by F-N and also Wolters, putting in other plausible regressors changes both undercount estimates and standard errors substantially. F-N also carry out a simulation study to establish the validity of their criticism. They take one of the PEP series as the true undercount, and generate random samples of PEP estimates and study the performance of the E-K procedure. The simulation study shows that the E-K procedure leads to overoptimistic estimates for the standard errors, and substantial variations occur in the regressors selected.

The Small Areas Problem   An internal inconsistency in the logic
of the E-K method is pointed out by F-N. Suppose that the model for
undercounts $y_i = \gamma_i + \epsilon_i$ is valid at the county level (of which there
are 39,000+). The undercount at the state level is a weighted average
of the undercounts in its counties. By the law of large numbers, if
the model is valid, then the PEP estimates must be more accurate
at any aggregated level than at the lower level. In particular, if the
E-K model holds for the county level, then the PEP estimates must
be more or less error free at the state level. Thus we must give some
reason to explain why the model is valid at the state level but not at
the county level. Subsequently, Cressie (1989) developed a more careful
EB analysis which is level-consistent (that is, the same model holds at
the micro and macro level), and also avoids the other criticisms of E-
K. Cressie's model shows that a regression model based on rates like
γ_i above (instead of counts) can be expected to be heteroskedastic, so
that the assumption that $Cov(\delta) = \sigma^2 \mathbf{I}$ is not realistic.

17.4.3   Political Feasibility

In the discussion, Erickson and Kadane do not dispute the points raised
by Freedman and Navidi. Their main argument is to attempt to put
their shrinkage estimates on par with the Census Bureau estimates. In
effect they say 'Fine, our estimates are biased and the standard errors
are higher than the ones we claim. What about the Census Bureau
estimates? Let's do a similar analysis of the bias and variance of these.
Only then can we make a comparison.'

In fact, the two sets of estimates are not on par. It is clear that
the Census Bureau estimates are free of political bias. The same cannot be said for the E-K estimates. Following the E-K methodology
with slight variations can lead to substantially different estimates of
the undercount across the USA. An interesting feature of the shrinkage
methodology is that it improves on OLS, at least theoretically, regardless of which point you shrink towards. Thus it is quite possible that a
large range of different shrinkage estimators would all be improvements
over the raw Census Bureau counts. However the improvement would

be on an average sense over all the states — one could not guarantee improvement in any one state. Unfortunately, the situation is such that each state is uniquely concerned with its individual counts. New York would not take a benign view of an adjustment which made the count worse for New York if it was assured that overall the other states were benefitted to a greater extent. Thus one essential ingredient for successful shrinkage, namely concern with the average loss, is missing. A second related problem is that the Census Bureau is politically constrained to follow objectively justifiable procedures. If it takes one out of a bundle of similar adjustment techniques and adjusts the counts on that basis, it will invite legitimate criticism from those who are hurt by that adjustment. They will propose an alternative technique of equal validity which helps them at the expense of others. There is no solution to such competing claims unless the procedure adopted is one which is objectively justifiable to all concerned parties.

17.5 Wages of Displaced Workers

While the EB technique has been widely used in many fields, there are relatively few serious econometric applications. One of these is Carrington and Zaman (1994) (C-Z in the sequel), which we now discuss. The type of model used is one of substantial generality, and is likely to be useful in a number of different contexts. The presentation given below is schematic, and does not exactly correspond to the paper in that some refinements of importance in practice are ignored.

17.5.1 The Model

The dependent variable y measures the ratio of the initial wage received by an unemployed worker to the last wage he/she received while on his/her last job. Explanatory variables of interest are Job Tenure, Experience, and Education which reflect individual characteristics, as well as certain other variables relating to local state-specific demand conditions. C-Z are interested in discovering the extent to which these unemployment shocks differ across industries. We posit a regression relationship $Y_i = X_i \beta_i + \epsilon_i$, where $i = 1, 2, \ldots, M = 50$ indexes the 50 two-digit industries on which we have separate data available. For each i, Y_i is a $N_i \times 1$ vector of observations on wage changes for individuals in industry i, X_i is the corresponding matrix of regressors, β_i is $K \times 1$ of industry-specific unknown parameters, and $\epsilon_i \sim N(0, \sigma_i^2 I)$ are the industry-specific errors.

We do not expect radically different parameters across industries, and many authors have estimated similar regressions aggregating across the industries, suggesting that the hypothesis that the β_i are all equal has some plausibility. Of course aggregation would defeat the purpose of discovering interindustry variation. The conventional alternative is to use OLS estimation for each industry separately. If sample sizes in each industry were sufficiently large, this procedure could be adequate. Note that even if the coefficients are exactly the same across industries, OLS will always produce different estimates for each industry. Shrinkage combats this tendency of OLS to scatter estimates of similar parameters. In the case at hand, while we have large samples for some of the industries, several industries have a small sample size, so that OLS estimates are imprecise. As a result, some of the OLS estimates are implausible in sign or size. This is to be expected when a large number of parameters are estimated by estimates of high variance.

A natural model for the parameters β_i is that $\beta_i = \beta_0 + e_i$, where e_i has mean zero, and a covariance matrix which will be specified later. For small errors, this corresponds to an aggregated model. For large errors, the estimates reduce to OLS separately in each model. However the model has substantially greater flexibility than either OLS or the aggregated model. The estimates combine the OLS and the aggregated estimates with weights which are data-dependent. Precise OLS estimates are left untouched, while cases where insufficient data lead to imprecise OLS estimates are shrunk towards the aggregated estimates. In fact this model is known in the econometrics literature as a random coefficients model. However, the estimation techniques for random coefficients are different. Also the focus in the random coefficients model is the estimation of the central tendency β_0 and the covariance matrix of the errors, whereas in the EB, we wish to estimate each β_i separately.

17.5.2 Estimation with d-Prior

It is convenient to reduce the data to the sufficient statistics. For each industry, $\hat{\beta}_i = (X_i'X_i)^{-1}X_i'y_i$ and $s_i^2 = \|y_i - X_i\hat{\beta}_i\|^2$ are sufficient statistics for the data. The conditional distribution of $\hat{\beta}_i$ given β_i is $\hat{\beta}_i|\beta_i \sim N(\beta_i, \sigma_i^2(X_i'X_i)^{-1})$. There are three computationally feasible choices for the prior density of β_i. One of them is the d-prior, where 'd' stands for diagonal. This involves assuming that $\beta_i \sim N(\beta_0, \Lambda)$, where Λ is a diagonal matrix. As a general rule, putting in too many hyperparameters leads to instability in empirical Bayes procedures, and it would be a bit risky to try to estimate a general covariance matrix unless the number of hyperparameters is small relative to the data set size. In the present case we have 13 regressors and 50 regimes is the size of the data set — too small to estimate a 13×13 covariance matrix. Thus it is essential to keep the Λ diagonal. It would not be reasonable to take $\Lambda = \sigma^2\mathbf{I}$, since different regressors are in different units, and there is no reason to expect the different parameters to be measurable on, and have standard deviations on, a common scale.

Since $\hat{\beta}_i|\beta_i \sim N(\beta_i, \sigma_i^2(X_i'X_i)^{-1})$ and the prior density of β_i is $N(\beta_0, \Lambda)$, the Bayes estimate for β_i is

$$\beta_i = \left(\frac{1}{\sigma_i^2}X_i'X_i + \Lambda^{-1}\right)^{-1}\left\{\frac{1}{\sigma_i^2}(X_i'X_i)\hat{\beta}_i + \Lambda^{-1}\beta_0\right\}.$$

To develop EB estimates, we must estimate the hyperparameters Λ and β_0. The marginal density of $\hat{\beta}_i$ is $N(\beta_0, \sigma_i^2(X_i'X_i)^{-1} + \Lambda)$. Thus the average of the estimates $\hat{\beta}_i$ is a good estimate of the aggregated central tendency for all industries β_0. Assuming Λ is known, the precision weighted average (which is both ML and MVU) is

$$\hat{\beta}_0 = \left(\sum_{i=1}^T \left(\sigma^2(X_i'X_i)^{-1} + \Lambda\right)^{-1}\right)^{-1}\left\{\sum_{i=1}^T \left(\sigma_i^2(X_i'X_i)^{-1} + \Lambda\right)^{-1}\hat{\beta}_i\right\}.$$

Of the several possible approaches to estimating Λ, the following was chosen by C-Z. Let $\hat{\beta}_{ij}$ be the coefficient of the j-th regressor in the i-th industry. For fixed j, and $i = 1, \ldots, T$, all $\hat{\beta}_{ij}$ cluster around the aggregate mean β_{0j}, the j-th coordinate of β_0. Let a_{ij} be the (j, j) diagonal entry in the matrix $(X_i'X_i)^{-1}$. Then the variance of $\hat{\beta}_{ij}$ around the mean β_{0j} is $\sigma_i^2 a_{ij} + \lambda_j$, where λ_j is the (j, j) diagonal entry of Λ. It follows that an unbiased estimator for λ_j is $(\hat{\beta}_{ij} - \beta_{0j})^2 - \sigma_i^2 a_{ij}$. This suggests the following estimator, which was used by C-Z.

$$\hat{\lambda}_j = \frac{1}{T-1} \sum_{i=1}^{T} \left((\hat{\beta}_{ij} - \hat{\beta}_{0j})^2 - \hat{\sigma}_i^2 a_{ij} \right)^+ .$$

The superscript '+' indicates that negative quantities were replaced by zero (since each entry estimates a variance, which is positive). Note that the variance estimate depends on having an estimate for β_0, while $\hat{\beta}_0$ depends on having an estimate for Λ. We can iterate back and forth until convergence (which is guaranteed by the E-M algorithm in the present instance). In practice this never took more than 10 iterations.

17.5.3   Estimation with the g-Prior

An alternative formulation for the prior leads to very tractable formulae. Suppose we assume that $\beta_i \sim N(\beta_0, \nu_i^2 (X_i' X_i)^{-1})$. Zellner (1986) and Ghosh et al. (1987) label this the g-prior and present some stories justifying the use of this prior. In fact the prior is plausible, but its main justification is computational convenience and the fact that it frequently leads to reasonable/good results in empirical applications.

With the g-prior, the posterior density for $\beta_i|\hat{\beta}_i$ is normal with covariance matrix $(\sigma_i^2 \nu_i^2)(X_i' X_i)^{-1}/(\sigma_i^2 + \nu_i^2)$, and the mean $\hat{\beta}_i^B$ given by:

$$\hat{\beta}_i^B = \frac{\nu_i^2}{\sigma_i^2 + \nu_i^2} \hat{\beta}_i + \frac{\sigma_i^2}{\sigma_i^2 + \nu_i^2} \beta_0 = \beta_0 + \left(1 - \frac{\sigma_i^2}{\sigma_i^2 + \nu_i^2} \right) (\hat{\beta}_i - \beta_0).$$

The form of the g-prior turns the matrix weighted average into a simple weighted average. In the d-prior formulation, the coefficient of each regressor gets assigned a shrinkage factor which is the same across all the industries. In the g-prior, every industry is assigned a single number which measures its conformity with the mean, and all coefficients of the regression for that industry are shrunk by a common factor towards the aggregate mean. It is also possible to develop more complex models which assign separate shrinkage factors to each industry and also to each regressor.

No iteration is needed to estimate the hyperparameter ν_i^2 in this model. Let L_i be any matrix such that $L_i(X_i'X_i)^{-1}L_i' = \mathbf{I}$. If $\hat{\theta}_i = L_i\hat{\beta}_i$, then $\hat{\theta}_i$ has the marginal density $N(L_i\beta_0, \sigma_i^2 + \nu_i^2)$. Thus $Z_i = \|L_i\hat{\beta}_i - L_i\beta_0\|^2$ is distributed as $(\sigma_i^2 + \nu_i^2)\chi_K^2$. This permits us to estimate ν_i^2 as $Z_i/T - \sigma_i^2$. However, it is also possible to develop a form closer to the original Stein estimate, which may be superior. When β_0 is known, $(K-2)\sigma_i^2/Z_i$ is unbiased for $\sigma_i^2/(\sigma_i^2 + \nu_i^2)$. This suggests that we estimate β_i by:

$$\hat{\beta}_i^g = \hat{\beta}_0 + \left(1 - \frac{(K-3)\hat{\sigma}_i^2}{(\hat{\beta}_i - \hat{\beta}_0)'X_i'X_i(\hat{\beta}_i - \hat{\beta}_0)}\right)^+ \left\{\hat{\beta}_i - \hat{\beta}_0\right\}.$$

17.5.4 Validation

Having obtained industry-specific estimates for the coefficients of the determinants of wage change after unemployment, how do we assess the performance of our estimates? There are several techniques which can be used. C-Z carried out studies of forecast performance, where the EB methods were applied on one part of the data, and they were used to forecast the remaining part. The EB estimated dramatically outperformed OLS, especially on the small industries, where the sample sizes were small. On the larger industries there was little difference between OLS performance and EB, since very little shrinkage towards mean took place. A second way to assess the performance is to look at signs and magnitudes of the coefficients for plausibility. Here it was clear that the EB estimates were substantially less variable (by construction) across industries and were generally much more plausible than OLS in sign and magnitude. Three types of EB estimates were used in the study: the *g*-prior, the *d*-prior, and also hierarchical Bayes using the Gibbs sampler — the technique to be described in the next chapter. The differences between the three were small, but the *d*-prior performed the best, and so these results were reported in the paper.

17.6   Exercises

1. In a typical regression model $y = X\beta + \epsilon$, why will application of Stein estimates to shrink the OLS estimates $\hat{\beta}$, say towards zero or towards a common mean, usually fail to produce noticeable improvement? Hint: using the empirical Bayes representation, what prior assumptions are reflected in the Stein estimates? Are these likely to hold in typical regression models?

2. Why is there no constant term in the Nebebe-Stroud Model?

3. The type of model structure estimated by Carrington-Zaman is very common and easy to find. For example, take any equation (micro, macro, or other type) at the US level and then break it down by state. Estimate the 50 state-by-state equations by the empirical Bayes method. As another example, take any macro relationship (such as the consumption function) and estimate it for all of the European Common Market countries using shrinkage. As they say, 'one applied regression analysis is worth a thousand pages of theory.'

A seeker of Truth travelled through the many stages of the Path — through the valley of despair, the mountain of hope, the plains of anxiety, the jungle of fear, and the fire of love. Still, it seemed to him that the path was long, the goal no nearer. In extreme despair, he cried out: 'You know that I am sincere. I seek You and only You. You know my condition, and You are able to elevate me to the goal I seek. Why then do You tantalize me, and keep away from, and try me with a thousand trials and difficulties?'

In his dream, he was answered thus; 'We have created Man in such a fashion that he does not value what he obtains easily. What he gets without effort, he also gives up without concern. Persist in your efforts and you will find treasures valuable beyond your wildest imagination.'

Chapter 18

Hierarchical Bayes & the Gibbs Sampler

In the previous chapters, we have described several applications of the empirical Bayes techniques. In each case, we have described how to estimate the hyperparameters using either ML or MOM. A third alternative for estimating the hyperparameters is to use a Bayes estimator. This leads to the hierarchical Bayes model, which will be discussed in detail in the present chapter. The HB methods enjoy several advantages over the EB methods discussed earlier. One of the main disadvantages of the EB method is that it does not properly account for the uncertainty caused by estimation of the hyperparameters. Several ad-hoc corrections have been suggested in the literature. However, both the model and the estimates are complex, and the task is difficult. A great advantage of the Bayes technique in this context is that it automatically accounts for the uncertainty in the hyperparameters and how it propagates through the various stages of the model. Thus we don't have to worry about the real possiblity that the several estimates necessitated by ad-hoc corrections to EB may conflict with each other in some subtle way.

Until recently the HB estimates were too computationally complex to be used routinely, and only EB estimates — regarded as approximations to the HB — could be used on a routine basis. The discovery of the Gibbs sampler has now made it possible to compute HB estimates in a large class of models. Does this mean that the EB estimates are obsolete? In the first place, the EB estimates are still substantially simpler to compute than the HB estimates. In many situations they provided adequate approximations to the HB estimates at substantially lower computational cost. It is possible that EB may be the only alternative available in large problems. In the second place, the logic and the formulas for the EB are almost transparent. Thus they are considerably easier to motivate and explain. In fact, the best approach to explaining the more complex HB methods is as a development of EB. This makes EB pedagogically valuable. In the third place, in several applications EB estimates outperform HB. Since HB estimates are theoretically superior, this is something of a surprise. The author suspects that further developments in HB techniques are needed to achieve dominance over EB. As of now, we cannot be sure that the theoretical superiority of HB will translate into a gain in practice, however.

18.1   Known Covariance Case

Consider the basic regression model $y = X\beta + \epsilon$ with y the $T \times 1$ matrix of dependent variable, X the $T \times K$ matrix of nonstochastic regressors, β the $K \times 1$ vector of unknown parameters, and $\epsilon \sim N(0, \Sigma)$ the unobserved error term. We allow for a general covariance structure as this is frequently needed in applications. For empirical Bayes applications, we need a second stage model for the parameters β, which we wrote as $\beta = \beta_0 + H\alpha + e$, with $e \sim N(0, \Omega)$. The EB methods are based on estimating the hyperparameter α and any parameters occuring in the covariance matrix Ω of the second stage model. Accounting properly for the uncertainty in the estimates of the hyperparameters causes difficulties for the EB method. For this reason, it is convenient to use a Bayes method for the third stage and directly model the uncertainty in the hyperparameters. Let us assume a third stage model for α, e.g. $\alpha = \alpha_0 + G\gamma + f$, where $f \sim N(0, \Omega_2)$. There are applications where one can think of suitable models for the hyperparameters; however, we will deal with situations where $\Omega_2^{-1} = 0$. This implies infinite uncertainty (variance) for the hyperparameter α, and is one representation of complete lack of knowledge about α. Use of such priors on the regression parameters β leads to classical OLS estimates and distribution theory, further evidence for the 'uninformativeness' of this prior.

Lindley and Smith (1972) explored the basic structure of this model, under the assumption that all the covariance matrices are known. With this simplification, the theory is manageable. Treating the case of unknown variances makes the methods much more complex, but is required by applications. We will see later that the Gibbs sampler can take these complications in stride. In the present section, we reproduce the formulas developed by Lindley and Smith for the known covariance case. The three stage model set out above can be summarized as follows. We will assume $\beta_0 = 0$ and $\alpha_0 = 0$ for convenience in the sequel. By transforming β to $\beta - \beta_0$ and α to $\alpha - \alpha_0$ the formulas for the general case can easily be derived. The distribution of Y given β is $N(X\beta, \Sigma)$, that of β given α is $N(H\alpha, \Omega)$. Finally, the marginal density of α is $\alpha \sim N(G\gamma, \Omega_2)$. The parameters γ and Ω_2 are assumed known. The theorem below gives results for the general case, but in applications, we will assume that $\Omega_2^{-1} = 0$, in which case the value $G\gamma$ does not enter the formulas.

Our object of interest is the posterior distribution of β given y. This summarizes the information available about the parameter β after aggregation information in all stages of the prior together with the data. So far, we have used Theorem 3.1 as the basis for Bayesian calculations in the normal case. That gives a representation of the posterior mean as a precision weighted average of the data mean and the prior mean,

which is easy to use and interpret. While the formulas can be derived
in that form, and are in fact algebraically equivalent, Theorem 18.1
below gives simpler forms which are easier to compute and manipulate
in the hierarchical Bayes case. We begin with a lemma which gives an
alternative to the formula of Theorem 3.1 for the regression case.

Lemma 18.1 *Suppose y given β is $N(X\beta, \Sigma)$ and $\beta \sim N(H\alpha, \Omega)$.
Then the marginal density of y is $N(XH\alpha, \Sigma + X\Omega X')$ and the posterior
density of β given y is $N(Bb, B)$, where $B^{-1} = \Omega^{-1} + X'\Sigma^{-1}X$ and
$b = X'\Sigma^{-1}y + \Omega^{-1}H\alpha$.*

Proof: The marginal density of y is most easily obtained by substituting
$\beta = H\alpha + e$ into $y = X\beta + \epsilon$ to get $y = XH\alpha + Xe + \epsilon$. This shows that
the mean of y is $XH\alpha$ and the covariance is $Cov(Xe + \epsilon) = X\Omega X' + \Sigma$,
as desired. To get the posterior, we proceed as follows. The joint
density of y and β, which is the product of the conditional density of y
given β and the marginal of β, can be written as $C \exp(-Q/2)$, where
the quadratic form Q is

$$
\begin{aligned}
Q &= (y - X\beta)'\Sigma^{-1}(y - X\beta) + (\beta - H\alpha)'\Omega^{-1}(\beta - H\alpha) \\
&= \beta'B\beta + 2\beta'b + \{y'\Sigma^{-1}y + \alpha'H'\Omega^{-1}H\alpha\} \\
&= (\beta - Bb)'B^{-1}(\beta - Bb) + \{y'\Sigma^{-1} - b'Bb + \alpha'H'\Omega^{-1}H\alpha\}.
\end{aligned}
$$

The factor in curly brackets ($\{,\}$) corresponds to the marginal density
of y while the term in β shows that the conditional density of β given
y is as claimed in the lemma. The results for the hierarchical model
follow as an easy consequence of the lemma.

Theorem 18.1 (Lindley-Smith) *Consider the hierarchical model
with $Y|\beta \sim N(X\beta, \Sigma)$, $\beta|\alpha \sim N(H\alpha, \Omega)$, and $\alpha|\gamma \sim N(G\gamma, \Omega_2)$, where
γ and Ω_2 are known constants. The posterior density of β given y is
$N(Dd, D)$, with D and d defined by:*

$$
\begin{aligned}
D^{-1} &= X'\Sigma^{-1}X + (\Omega + H\Omega_2 H')^{-1} \\
d &= X'\Sigma^{-1}y + (\Omega + H\Omega_2 H')^{-1}HG\gamma.
\end{aligned}
$$

Proof: From the previous lemma, the marginal density of β, obtained after integrating out α, is $N(HG\gamma, \Omega + H\Omega_2 H')$. Now use this as the prior density for β and apply the previous lemma to the the conditional of β given y to prove the theorem. Note that when $\Omega_2 = 0$, the posterior covariance is $D = (X'\Sigma^{-1}X + \Omega^{-1})^{-1}$ and the posterior mean is $DX'\Sigma^{-1}y$.

The known covariance case provides a basis for extensions to more general cases. The formulas developed here can also be used to get simpler formulas for empirical Bayes estimators than those we have used earlier. However, we have preferred the slightly more complex forms earlier because they have an easy and immediate interpretation. Since the main motivation for going to the complexity of the three stage HB model is to get a better grip on the variances, it is unsatisfactory to estimate variances and assume them known, as is necessary to use these formulas. We now go over one study which shows how to handle uncertainty in the hyperparameters properly.

18.2   Reading Levels and Test Scores

We now discuss a study by Nebebe and Stroud (1988) (NS hereafter) which makes use of the empirical and hierarchical Bayes methods. The basic model is as follows. NS study RGRADE, the difference in actual reading level and expected reading level based on age for children in school in different grades. This difference is to be explained by test scores on some 11 types of skills tested in the WISC-R battery of tests. Thus the model is of the type

$$RGRADE_i = \sum_{j=1}^{11} \beta_j T_{ij} + \epsilon_i$$

where $RGRADE_i$ is the dependent variable (reading ability) for the i-th child and T_{ij} is the score of the i-th child on the j-th component of the WISC-R test.

Given the data, the model can easily be estimated by least squares. NS were interested in exploring whether it was possible to improve on these estimates. In addition to OLS (and LAE), NS also calculated the empirical Bayes and hierarchical Bayes estimators for this model. Unlike most regression models, it seems reasonable to assume a priori that the β_i are exchangeable so that $\beta_j \overset{i.i.d.}{\sim} N(\mu, \tau^2)$, where τ and μ are the unknown hyperparameters. NS standardized the test scores by dividing each score by its standard deviation. This puts all the scores on par and makes the exchangeability more plausible. This model fits easily into the empirical Bayes model structure we have discussed in the previous chapter. It can be estimated by EB-type methods discussed there. NS estimate the model by hierarchical Bayes methods after introducing a diffuse (or uninformative) prior on the hyperparameters μ, ν^2. Since this fully accounts for the uncertainty about the hyperparameters, we use their study as an illustration of a direct application of hierarchical Bayes. Indirect methods using Gibbs sampler are discussed later.

18.2.1   An HB Analysis

Stack the dependent variable $RGRADE_j$ into a $T \times 1$ vector y, and form the $T \times k$ matrix X of test scores T_{ij}. The model has the standard form $y = X\beta + \epsilon$. For simplicity, we will follow NS and assume σ^2 is known, even though we will actually use the estimate s^2 for its value. The data density is $y|\beta \sim N(X\beta, \sigma^2 \mathbf{I}_T)$. Our prior assumption about β is that $\beta \sim N(\mathbb{1}\mu, \tau^2 \mathbf{I}_k)$, where $\mathbb{1}$ is the vector of 1's. To complete the HB model, we need to specify prior densities for the hyperparameters. NS assume that τ^2 and μ are independent, with prior densities $\pi(\mu) = 1$ and $\pi(\tau) = (\tau/\sigma^2 + \lambda_h)^{-\delta}$. In this last density, λ_h is the harmonic mean of the eigenvalues λ_i of $X'X$ and δ is a constant. They offer some justification for this prior, and also show that their analysis is not very sensitive to a range of values for δ.

We will deviate from the original analysis of NS by assuming that the prior density of μ is $N(0, \nu^2)$, where ν^2 is a fixed known constant. For large ν^2, our assumption provides a good approximation to NS and is computationally much more manageable.[1]. Our assumption con-

---

[1] For derivations of the original formulae of NS, see Nebebe and Stroud (1986). Normally taking limits as $\nu^2 \to \infty$ yields the diffuse prior analysis with $\pi(\mu) = 1$ as a special case. This is also true here, but the limit is a bit complex (see exercise 4) which is why our assumption here leads to a substantially simpler analysis than that of NS.

verts this to a three stage hierarchical model, amenable to analysis by standard methods described earlier. Analysis of hierarchical models proceeds in a sequence of similar steps, as we go up and down the hierarchy. Although each of the steps is simple, the overall effect can be baffling. We try to break down the analysis into digestible chunks below.

1: The first step is the simplest of all, since we proceed conditionally on the hyperparameters μ and τ^2. This conditioning reduces the problem to a standard Bayes model, where the hyperparameters are known constants. Lemma 18.1 above gives us the two densities of interest. The posterior density of β given y is $f(\beta|y, \tau^2, \mu) = N(Bb, B)$, where $B^{-1} = \tau^{-2}\mathbf{I} + \sigma^{-2}X'X$ and $b = \sigma^{-2}X'y + \tau^{-2}\mathbb{1}\mu$. The marginal density of y is $f(y|\mu, \tau^2) = N(X\mathbb{1}\mu, \sigma^2\mathbf{I}_T + \tau^2 XX')$.

2: The second step is to derive the posterior density for the hyperparameters μ and τ^2. This is a bit complex and will be broken down further into simpler steps. The density $f(y|\mu, \tau^2)$ derived above, relating the observed data y to the hyperparameters μ, τ^2, will be the focus of interest. In an EB analysis, this density is used to estimate the hyperparameters using ML, MOM, or some other ad-hoc method. In the HB analysis, instead of estimating μ, τ^2, we want the posterior density $f(\mu, \tau^2|y)$, which tells us what information the data contain about μ, τ^2.

2A: Conditioning on τ^2 makes it a constant, and reduces the problem to a standard Bayesian one. That is, $f(y|\mu, \tau^2) = N(X\mathbb{1}\mu, \Sigma)$, where $\Sigma = \sigma^2\mathbf{I}_T + \tau^2 XX'$. The prior density for μ is $N(0, \nu^2)$. From Lemma 18.1, we get that the posterior density of μ is $f(\mu|y, \tau^2) = N(Mm, M)$, where $M^{-1} = \nu^{-2} + \mathbb{1}'X'\Sigma^{-1}X\mathbb{1}$ and $m = \mathbb{1}'X'y$. We also get the marginal density of y, which is $f(y|\tau^2) = N(0, \Sigma + \nu^2 X\mathbb{1}\mathbb{1}'X')$.

2B: Linking y to τ^2 via $f(y|\tau^2)$ gets us to the bottom of the chain. Now we work back up, starting by obtaining $f(\tau^2|y)$, the posterior density for the bottom parameter. Unfortunately, as soon as we mess with the variance τ^2, the formulas become messy. The standard method for obtaining this posterior density is as follows:

$$f(\tau^2|y) = \frac{f(y|\tau^2)\pi(\tau^2)}{\int f(y|\tau^2)\pi(\tau^2)\, d\tau^2}.$$

The integral in the numerator cannot be evaluated analytically (to our knowledge) and this posterior density has no convenient form or name. However, we have derived $f(y|\tau^2)$ and specified $\pi(\tau^2)$ and computers have no difficulty with well-behaved one-dimensional integrals of this type, so we may consider that we have an explicit formula for the posterior $f(\tau^2|y)$.

2C: In step 3' below, we show that this step can by bypassed in the present problem. We include it for the sake of completeness, since it is necessary in alternative forms of the problem. From step 2A, we know that $f(\mu|\tau^2, y) = N(Mm, M)$ and we have just derived $f(\tau^2|y)$. The product of these two yields the posterior density of the hyperparameters:

$$f(\mu, \tau^2|y) = f(\mu|\tau^2, y) \times f(\tau^2|y).$$

3: In general, we could combine the density $f(\beta|\mu, \tau^2, y)$ of step 1 with $f(\mu, \tau^2|y)$ of step 2C to get the joint posterior density of β, μ, τ^2 as $f(\beta, \mu, \tau^2|y) = f(\beta|\mu, \tau^2, y) \times f(\mu, \tau^2|y)$. Now integrating out μ and τ^2 from this joint density would lead us to $f(\beta|y)$, the desired posterior density. This route involves integrating with respect to both μ and τ^2. We can bypass both step 2C and this double integration by following an alternative procedure detailed below.

3': This is an alternative to step 3 above which bypasses step 2C. Recall that $y \sim N(X\beta, \sigma^2 I_T)$, $\beta \sim N(\mathbb{1}\mu, \tau^2 I)$, and $\mu \sim N(0, \nu^2)$. This is a three stage hierarchical model. Treating τ^2 as known, the Lindley-Smith result provides us the posterior density $f(\beta|\tau^2, y) = N(Dd, D)$, where $d = \sigma^{-2} X'y$ and

$$D^{-1} = \sigma^{-2} X'X + \left\{ \tau^2 I_k + \nu^2 \mathbb{1}\mathbb{1}' \right\}^{-1} \approx \sigma^{-2} X'X + \tau^{-2} \left\{ I_k - \mathbb{1}\mathbb{1}' \right\},$$

where the approximation is valid for large ν^2. To get the object of our desire, the posterior density of β given y, we use $f(\beta|y) = f(\beta|\tau^2, y) \times f(\tau^2|y)$. The mean of the posterior gives the HB estimator. This can be computed by numerical integration — if we work through the formulas, we see that two univariate integrals need to be computed.

   Nebebe and Stroud attempted to compare the performance of (a variant of) the hierarchical Bayes estimator derived above with the OLS and LAE estimators in the following way. They split the sample in half, derived estimates for each half, and attempted to predict the

remaining half. This is known as cross-validation. For various group-
ings of the subjects, and for various criteria, it was universally true
that the empirical Bayes and the hierarchical Bayes methods achieved
higher accuracy at predicting the other half of the sample than OLS
and also the LAE.

18.3   The Gibbs Sampler

The hierarchical Bayes technique attempts to take into account properly the uncertainty created by the estimation of the hyperparameters.
Unfortunately, as the example of Nebebe and Stroud demonstrates, this
gets messy mathematically. Even though we obtained formulas for the
estimates which take uncertainty about the hyperparameters into account, we were forced to neglect the uncertainty in the estimation of
σ^2 and treated this as a known parameter even though we estimated it.
A proper treatment which sets up a prior on σ^2 as well gets even more
complicated and the resulting formulas involve bivariate integration.
Even with recent advances in computing, bivariate integration is not
easy to implement routinely at low cost. We now describe a computational technique which facilitates routine computation of hierarchical
Bayes estimates in very complex models which are difficult to treat by
any other method.

The Gibbs sampler is a technique for generating random variables
from a joint density which is hard to calculate, given the knowlege of
certain conditional densities associated with it. In the context of hierarchical Bayes models, the joint posterior density for the parameters
is frequently hard to calculate but the conditional densities required
by the Gibbs sampler are readily available. This accounts for the utility of the technique. A nice exposition of the Gibbs sampler, with
many elementary examples, is given by Casella and George (1992). See
also Chib and Greenberg (1994) for a good presentation of the related
Metropolis-Hastings algorithm. Below we treat the algorithm purely
as a tool, and outline the steps required to implement it in regression
models, without regards to its origin and logic.

Basic to the Gibbs sampler is the shift in focus from calculating
the desired density to obtaining a sample of random variables from

the desired density. Note that with a large enough sample we can recover any desired feature of the density to desired degree of accuracy. Typically we are interested in mean and variance of the target density, which are easily estimated from the sample mean and variance. Using density estimates, we can easily estimate the entire density if desired (see Section 14.5.1), but this requires substantially larger samples from the target density.

18.3.1   Generating Random Numbers

As a basic building block in the Gibbs sampler, we need to be able to generate a random variable X after we specify its density $f(x)$. Although this issue has a vast literature associated with it, we give a brief discussion adequate for our purposes.

Fundamental to (pseudo) random number generation is a mechanism for generating a uniform random variable on the interval $(0, 1)$. The 'multiplicative congruential' method is an old favorite and the uniform random number generators implemented in most programming languages typically use this method. Even though this method, like most, is not 'truly' random and in fact cycles, it passes most (but not all) tests for being random and uniform on $(0,1)$. As a practical matter, the randomness of the multiplicative congruential method is adequate for Monte Carlo-type studies provided that the sample size of the random numbers we generate is small relative to the period of the generator; this period is very large on 32-bit machines. See Marsaglia and Zaman (Arif) (1990) for a recently developed improved generator of uniform random numbers. See Marsaglia *et al.* (1993) for a battery of tests for uniformity to assess the quality of random number generators.

Theoretically, given density $f(x)$ and CDF $F(x) = \int_{-\infty}^{x} f(s) \, ds$ and U uniform on $(0, 1)$, $X = F^{-1}(U)$ will have CDF F and density f. Thus from uniform variables we can get all others. In practice, this method is not very efficient when the CDF is hard to compute. Fortunately, most programming languages of interest to econometricians also have built-in random $N(0, 1)$ generators, which suffice for most purposes. Suppose we wish to generate $X \sim N_k(\mu, \Sigma)$. Let T be a lower triangular matrix such that $TT' = \Sigma$; this is called a Cholesky decomposition of Σ and can be computed quickly and with high numerical accuracy. Generate $Z =$

$(Z_1, \ldots, Z_k)'$ as an i.i.d. sample from the built in $N(0,1)$ generator, and define $X = \mu + TZ$ to get the desired random variable. Similarly chi-square densities and related gamma densities can be obtained by adding squares of $N(0,1)$ variables, and perhaps multiplying by a suitable scale factor.

Finally suppose X has known density $f(x)$ which does not yield to the above methods. First transform X to Z so that the range of Z is finite; for example, $a \leq Z \leq b$. For instance, if X takes values in $(0, \infty)$ then $Z = X/(1 + X)$ has range $(0,1)$. Compute the density $f^Z(z)$ and also $M = \sup_{z \in (a,b)} f^Z(z)$. Now generate uniform random variables (U_1, U_2) where $a \leq U_1 \leq b$ and $0 \leq U_2 \leq M$. This is a random point in a rectangle inside of which the density $f^Z(z)$ can be inscribed. If the point lies under the curve of the density (that is $U_2 < f^Z(U_1)$), then we 'accept' the point and consider U_1 as a sample from the target density. Otherwise we 'reject' the point and draw a fresh pair (U_1, U_2). It is easily shown that the resulting sample has the desired density $f^Z(z)$ and inverting the original transform will get us back to the density of X. Because this simple *acceptance-rejection* method can be highly inefficient when the number of rejections is high (as will happen when the density has a sharp peak, for example), several sophisticated and more efficient variants of the method have been developed. A fast and effective algorithm for unimodal densities is given in Zaman (Arif) (1995).

18.3.2 Algorithm for Gibbs Sampler

Suppose we wish to generate a vector of random variables $X = (X_1, \ldots, X_k)$ from some density $f^X(x)$, but the density is not available in explicit form. Suppose however that we can calculate explicitly for each i, the density of X_i conditional on the remaining variables, denoted $X_i | X_1, \ldots, X_{i-1}, X_{i+1}, \ldots, X_k$. The 'Gibbs sampler' is the name of the following algorithm for obtaining a random sample $X_1, \ldots, X_k$ from the target density $f^X(x)$:

Fix $i = 0$ and arbitrary initial values $X_1^{(0)}, \ldots, X_k^{(0)}$.

LOOP : Generate the following random variables from the specified densities:

1: $X_1^{(i+1)}$ from the density of $X_1 | X_2, \ldots, X_k$.

2: $X_2^{(i+1)}$ from the density of $X_2 | X_1^{(i+1)}, X_3^{(i)}, \ldots, X_k^{(i)}$.

$\vdots$

j: $X_j^{(i+1)}$ from the density of $X_j | X_1^{(i+1)}, \ldots, X_{j-1}^{(i+1)}, X_{j+1}^{(i)}, \ldots, X_k^{(i)}$.

$\vdots$

k: $X_k^{(i+1)}$ from the density of $X_k | X_1^{(i+1)}, \ldots, X_{k-1}^{(i+1)}$.

Set $i = i + 1$ and GOTO LOOP.

This algorithm is an infinite loop. The theory of Gibbs sampling shows that for large i, the joint density of $X_1^{(i)}, \ldots, X_k^{(i)}$ will be approximately the same as the target joint density. Some mild regularity conditions on the conditionals and the joint density are needed, which are always satisfied in the type of models we discuss below. Roberts and Smith (1994) show how to verify the regularity conditions for a large class of models and provide further references to this literature. The question of how large i needs to be to get satisfactory approximations is also the subject of a body of research. Ritter and Tanner (1992) provide an interesting algorithm for making such decisions. We will discuss this further in the context of applications.

18.3.3   The Required Conditionals

Consider a standard regression model $y = X\beta + \epsilon$. We have uncertain prior information about β in the form of a linear restriction, which can be represented as $\beta = \beta_0 + H\alpha + \nu$. It will prove especially convenient to parametrize the models using precisions instead of variances, so we will assume that $\epsilon \sim N(0, s^{-1}\Sigma)$ and $\nu \sim N(0, t^{-1}\Omega)$. Thus ϵ_t are i.i.d. with mean 0 and variance $\mathbb{V}ar(\epsilon) = s^{-1}$, or $s = 1/\mathbb{V}ar(\epsilon)$, and similarly for ν. The parameters of the regression model are thus β and s. The hyperparameters are α and t. This much will normally be available from the application. We now reformulate the problem in HB terms and add elements necessary to complete the specification.

The data distribution is $y \sim N(X\beta, s^{-1}\Sigma)$. The parameter β has the prior density $\beta \sim N(\beta_0 + H\alpha, t^{-1}\Omega)$. We will assume that the prior density of s is $\pi(s) = s^a$ for some constant a; this is an improper prior, since it does not integrate to 1. Such a prior has the virtue of specifying minimal information (and hence letting the data decide on a suitable value of s) in addition to computational convenience. Thus we have specified priors for both parameters of the model. To complete the HB specification, we need to add a prior density for the hyperparameters α and t. Typically not much prior information is available about the hyperparameters, so the uninformative priors $\pi(\alpha) = 1$ and $\pi(t) = t^b$ are used. We will also work with this assumption, although it should become clear that a second stage prior on α of the usual type could be handled without any difficulty.

This completes the specification of the model. To implement the Gibbs sampler, we need four conditional densities, which are computed in the lemma below. Recall that $X \sim \mathcal{G}(p, \lambda)$ if $f^X(x) = \lambda^p x^{p-1} exp(-\lambda x)/\Gamma(p)$.

Lemma 18.2 *In the HB model specified above, the conditional density of*

1. *β given s, α, t, y is $N(Bb, B)$, where $B^{-1} = sX'\Sigma^{-1}X + t\Omega^{-1}$ and $b = sX'\Sigma^{-1}y + t\Omega^{-1}(\beta_0 + H\alpha)$.*

2. *α given s, β, t, y is $N(Aa, A)$, where $A^{-1} = tH'\Omega^{-1}H$ and $a = tH'\Omega^{-1}\beta$.*

3. *s given α, β, t, y is $\mathcal{G}((T/2) + a + 1, \|y - X\beta\|^2/2)$.*

4. *t given α, β, s, y is $\mathcal{G}(K/2 + b + 1, \|\beta - (\beta_0 + H\alpha)\|^2/2)$.*

Remark: We have already discussed how to generate samples from $N(\mu, \Sigma)$ densities. If a and b are half integers (i.e. 0,1/2,1,-1/2 etc.), the Gamma densities are easily generated as chi-squares. If m is an integer, a random variable Y having density $\mathcal{G}(m/2, \sigma/2)$ can be generated by first obtaining $X \sim \chi_m^2$ and then setting $Y = X/\sigma$.

Proof: Conditioning on everything in sight really simplifies the computations. The conditional density of β comes from classical Bayesian computations, as in Lemma 18.1.

The conditional density of α is slightly different, since α has been equipped with a diffuse prior $\pi(\alpha) = 1$, rather than a normal prior. Multiplying the conditional density $f(\beta|\alpha) = N(\beta_0 + H\alpha, t^{-1}\Omega)$ by the prior $\pi(\alpha) = 1$ gives the joint density of β, α, from which routine calculations permit the derivation of the conditional of α given β. Another route is to equip α with a normal prior, say $\alpha \sim N(0, v^2\mathbf{I})$. Now apply Lemma 18.1 to get $\alpha|\beta \sim N(Aa, A)$, where $A^{-1} = tH'\Omega^{-1}H + v^{-2}\mathbf{I}$, and $a = tH'\Omega^{-1}\beta$. Take limits as v^2 goes to infinity to get the desired result.

For the conditional density of s, write down the joint density of y, β, s, α, t as follows:

$$f(y, \beta, s, \alpha, t) = f(y|\beta, s) \times f(\beta|\alpha, t) \times \pi(s) \times \pi(\alpha) \times \pi(t).$$

The desired conditional density of s is proportional to the joint density, and hence:

$$f(s|y, \beta, \alpha, t) \propto (2\pi s^{-1})^{-T/2} \exp\left\{-(s/2)\|y - X\beta\|^2\right\} \times s^a,$$

where we have retained only terms involving s in the joint density. Letting $Q = \|y - X\beta\|^2/2$, we see that $f(s|y, \beta, \alpha, t) \propto s^{a+T/2}\exp(-Qs)$ from which we conclude that the desired conditional density is $\mathcal{G}(1 + a + T/2, Q)$.

A similar computation for the conditional density of t yields

$$f(t|\beta, \alpha, s, y) \propto t^{-K/2} \exp\left\{-(t/2)\|\beta - (\beta_0 + H\alpha)\|^2\right\},$$

from which the lemma follows.

18.3.4   Implementation

Using the conditional densities derived above, we can obtain samples from the target density, the posterior distribution of (β, α, s, t) given y. The posterior distribution combines the prior information and the data information and can provide the basis for all subsequent inference – estimation, testing, confidence intervals, etc.

One algorithm to get an i.i.d. sample from the posterior distribution is as follows. Arbitrarily choose initial values $\beta^0, \alpha^0, s^0, t^0$, and run the Gibbs sampler algorithm until $i = i^*$, where i^* is considered sufficiently large so that convergence has occured.[2] Record the observations at the i^* iteration as $\beta_1^*, \alpha_1^*, s_1^*, t_1^*$. Now repeat the entire process to get a second set of observations $\beta_2^*, \alpha_2^*, s_2^*, t_2^*$ from the target density. By repeating several times, we can get as large a sample as we want from the target density. This sample can then be used to estimate means, variances, and other statistics of interest of the posterior distribution. This original form of the algorithm was used by Blattberg and George (1991) in an application discussed in detail in a following section. It has the advantage of producing an i.i.d. sample from the target density. An alternative form of the algorithm, described below, has several advantages over this form.

One problem with the algorithm above is that as soon as the sampler 'heats up' and starts producing samples from the required density, we abandon it and start the process again from scratch. Instead, one might think of continuing to sample from the process, thereby producing more observations from the target density. The reason this is not done is that there is a very complicated dependence between the iterates $\beta^{(i+1)}$ and $\beta^{(i)}$. Thus if we keep on sampling from just one run of the Gibbs sampler, we do get a sample from the target density, but the sample is *dependent* but identically distributed (approximately), instead of being i.i.d. However, if we are only interested in the mean vector and the covariance matrix of the posterior distribution of β (which is frequently the case), the dependence of the sample does not really cause a problem. This is because if X_i are identically distributed then, as long the dependence declines suitably fast for distant indices i, j, it is still true that

$$\lim_{n \to \infty} \frac{1}{n} \sum_{i=1}^{n} X_i = \mathbb{E} X_1.$$

Thus if we take averages of a dependent (but identically distributed) sample, we will still converge to the mean of the underlying distribution.

---

[2]Note that this is hard to decide since the numbers will keep fluctuating (being a random sample); this is why theory has been developed to assess convergence. In simple problems with relatively few parameters, $i^* = 50$ has been suggested as a suitable value. The required i^* may be much higher if the model has a large number of parameters.

This suggests the following alternative algorithm.

As before, start with arbitrary $\beta^0, \alpha^0, s^0, t^0$ and run the Gibbs sampler to produce iterates $\beta^{(i)}, \alpha^{(i)}, s^{(i)}, t^{(i)}$. Suppose that our object of interest is the posterior mean and covariance of β — these correspond to a point estimator and estimates of variances in a classical analysis. Discard all initial observations up to some index i^*. After i^* keep track of just two quantities: $S_n = \sum_{i=i^*}^{n} \beta^{(i)}$ and $V_n = \sum_{i=i^*}^{n} \beta^{(i)}(\beta^{(i)})'$. Run the sampler until you reach some suitably large number n and then stop. Dividing S_n by $(n - i^*)$ provides us with an estimate of the posterior mean $E\beta|y$, and dividing V_n by $(n - i^*)$ is an estimate of $E\beta\beta'|y$, from which we can compute the posterior covariance matrix, since $Cov(\beta|y) = E(\beta\beta'|y) - (E(\beta|y))(E\beta|y)'$.

To implement this algorithm, we need to know how to choose i^* and n. Note that unlike the first algorithm described, errors in choosing i^* too low are not catastrophic — only the first few terms have distributions which are wrong (i.e., far from the target distribution), and the effect of these will become small for large n. In the first form of the algorithm, if i^* is too low, all observations will be far from the target density and seriously erroneous estimates can be obtained. Thus we suggest choosing some plausible value of i^*, such as 100 or 1000, without worrying too much about it.

In the i.i.d. case, it is easy to choose n according to the level of precision we desire for our estimates. The complicated dependence structure makes this impossible in the present context. Instead we recommend choosing some intuitively plausible value, for example $n = 50000$ in a moderately large problem. To assess whether our value of n is suitable, we should run the sampler two or three times and compare the final estimates for mean and covariance. If these match closely from one run to the next, that means that n is large enough to give us estimates to the desired degree of accuracy. If not, it should be adjusted upwards, and one might also modify i^* upwards. While this is somewhat ad-hoc method for chosing i^* and n, it should work just fine in applications. Note that we could also estimate skewness and kurtosis of the target density following this technique and thereby make an Edgeworth approximation to the target density if desired.

18.4 Estimates for Price and Promotional Elasticities

In this section, we will review an article by Blattberg and George (1991). This article, referred to as BG hereafter, applies many variants of the techniques we have described and reaches conclusions regarding their relative efficacy in the particular case under study.

18.4.1 A Model for Sales of Bathroom Tissue

BG study the sales of different brands of bathroom tissue in different chains of grocery stores. Let SL_t denote the logarithm of total sales in dollars for a particular brand in a particular chain during time period t (length of period was one week). This variable is to be explained by

- PR_t, *relative price* of the brand compared to a price index for all brands of bathroom tissue.

- DD_t, *deal discount* defined as the ratio of normal shelf price minus actual price to the normal shelf price.

- AD_t, *feature advertising*, the proportion of stores in chain using the ad.

- DP_t, *display*, the proportion of stores in chain displaying the brand.

- FL_t, *final* week, a dummy variable which is 1 if the time period is a last week of a multiperiod sale, and zero otherwise.

- CD_t, *competing discount*, the largest discount available on competing brands of bathroom tissue within the same chain.

Specifically, the regression model BG attempt to estimate is

$$SL_t = \beta_1 + \beta_2 PR_t + \beta_3 DD_t + \beta_4 DD_{t-1} + \beta_5 AD_t + \beta_6 DP_t$$
$$+ \beta_7 FL_t + \beta_8 CD_t + \beta_9 SL_{t-1} + \epsilon_t. \tag{18.1}$$

Justification for the choice of variables and the form of the equation is detailed in the article. Our concern here is the estimation technique used. BG remark that the possibility of constructing and studying such detailed models arises from the availability of scanner price data. Computerized weekly sales data were available to them for a period of over 100 weeks from three different chain stores and for four different brands of bathroom tissue.

18.4.2   Conventional Estimation Techniques

Conventional estimation techniques split the data by chain and by brand and then estimate some variant of the basic Eq. (18.1) by OLS. Here for example, we would estimate $12 = 3 \times 4$ equations, one for each chain-brand combination. BG discuss the OLS estimates obtained in this way and show that the results obtained are very reasonable overall. The adjusted R^2 ranges between .75 and .95 with a median value of .93. These high R^2 values come from the fact that the promotional variables (DD,AD, and DP in this model) account for a large proportion of the variation in the sales. Diagnostic checks for normality, linearity, robustness, and others reveal no abnormalities. There is nonetheless one important difficulty with the OLS results. This is that the estimated coefficients show considerable variation across chain-brands, and many have signs different from the theoretically expected ones.

Since it is reasonable to suppose that the coefficients should display some similarity across chains and across brands, the hierarchical Bayes technique suggests itself as a natural means of reducing the variability of the OLS estimates by shrinking them towards some group mean. For each $i = 1, 2, \ldots, 12$ let y^i denote observations on the dependent variable and let X^i denote the matrix of regressors in chain-brand i; the first column of each X_i will be the column of ones corresponding to the constant term. We will adopt the convention that $i = 1, 2, 3, 4$ are the observations on the four brands at chain 1, $i = 5, 6, 7, 8$ are the same brands in same sequence at chain 2, and $i = 9, 10, 11, 12$ are the brands in same sequence at chain 3. The OLS estimates are for the 12 regression models $y^i = X^i \beta^i + \epsilon^i$ for $i = 1, 2, \ldots, 12$. Data are available starting from the same week but somewhat different final week for each chain. Thus y^i, X^i, ϵ^i are $T_i \times 1$, $T_i \times K$, and $T_i \times 1$ respectively, where $K = 9$ is the number of regressors, and $T_i = 111$ for $i = 1, 2, 3, 4$, $T_i = 121$ for $i = 5, 6, 7, 8$, and $T_i = 109$ for $i = 9, 10, 11, 12$.

18.4.3 Alternative Prior Structures

In setting up a hierarchical Bayes model to exploit potential similarities between the β^i, BG assume that errors $\epsilon^i \sim N(0, \sigma^2 \mathbf{I}_{T_i})$ have the same variance for each i and are also independent for each i. Both of these assumptions are supported by their data. A version of the Bartlett's test for heteroskedasticity (adjusted for kurtosis, see Madansky (1988)) fails to reject the null hypothesis of equal variances, while regression residuals across chain-brands do not show significant correlations.

In order to link the coefficients across equations, we assume a prior density of the form $\beta^i \sim N(\theta^i, \Omega^i)$. We note that we have followed the notation of BG except in this one instance, where we use Ω^i for the prior covariance matrix instead of Σ_i used by BG. To reduce the number of parameters, we will, in all cases except for the g-prior discussed below, assume that $\Omega^i = \Omega$ is the same for all i. Various assumptions on the θ_i can be used to link the coefficients. For example, we could assume that θ^i are the same for all i, and thus shrink all coefficients towards the same mean. Or we could assume equality across brands but not chains, and also assume equality across chains but not brands. It is important to note that hypotheses of equality like this require that the variables in the models should be comparable across chain-brands. BG indicate that the variables AD and DP were defined in proportional terms so as to make them comparable. They also indicate that use of log sales for SL makes it possible to combine additively effects across different chain-brands even when the level of sales differs.

Each of the three constraints on θ^i considered by BG can be written in the form $\theta = W\gamma$, where $\theta' = (\theta'^1, \theta'^2, \ldots, \theta'^{12})$ is the $(12K \times 1)$ column vector obtained by stacking all 12 mean vectors. The constraint of overall equality is obtained by setting W to be 12 copies of $\mathbf{I}_K$ stacked on top of each other and letting γ be $K \times 1$. The constraint of equality across chains is obtained by stacking 3 copies of $\mathbf{I}_{4K}$ to get W and letting γ be $4K \times 1$ vector representing the separate regression coefficient for each brand. There is a similar representation for the constraint of equality across brands.

18.4.4   Empirical Bayes

By stacking y_i as y, writing X for a block diagonal matrix with X_i as the i-th diagonal block, and 0's on the off-diagonal blocks, we can combine all 12 subequations into one giant equation as $y = X\beta + \epsilon$. The β vector is $12K \times 1$, and satisfies a prior constraint $\beta \sim N(W\gamma, \Omega^*)$. The matrix W represents the restrictions placed on β and three different types are studied. Similarly Ω^* is the prior covariance matrix, and we will study different choices for it. In all cases, Ω is a block diagonal matrix with Ω^i on the i-th diagonal block, and 0's on off-diagonal blocks. In all cases except the g-prior to be discussed, the Ω^i are all equal to a common matrix Ω. This model, especially under the prior assumption that the coefficients β^i are the same for all i, is almost identical in structure to the model of Carrington-Zaman (1994) described earlier. In fact it is slightly simpler in that the variances are the same across the 12 different regimes. However, the assumptions that the coefficients are the same only across chains, or only across brands, add some novelty to the structure.

Classical Bayes: This setup describes a typical hierarchical Bayes model. If the hyperparameters γ and Ω are known, we are in the classical Bayes case, and the usual estimator for β^i will be

$$\mathbb{E}(\beta^i | y) = D_i^{-1}\left(\sigma^{-2} X'^i X^i \hat{\beta}^i + \Omega^{-1} \theta^i\right), \qquad (18.2)$$

where $D_i = \sigma^{-1} X'^i X^i + \Omega^{-1}$ is the posterior precision (inverse of covariance) matrix, and θ^i is the appropriate subvector of the known $W\gamma$.

BG recommend estimating σ^2 by

$$\hat{\sigma}^2 = \frac{1}{(\sum_i T_i) - 12K + 2} \sum_{i=1}^{12} \|y^i - X^i \hat{\beta}^i\|^2.$$

Note that this differs from the usual MVU estimate for σ^2 by the $+2$ in the denominator. This estimator can be proven to be the best scale invariant estimator and has been shown to yield desirable risk properties in shrinkage problems.

d-**Prior:**   Use of the empirical Bayes technique requires estimation of the hyperparameters γ and Ω. With only 12 observations on the $\hat{\beta}^i$, it is not possible to estimate a general covariance matrix Ω. Two prior specification which reduce the number of parameters in Ω have already been discussed in Section 4.5. One of these is the *d*-prior, which assumes that Ω is diagonal, with $\Omega = \text{diag}(a_1, a_2, \ldots, a_K)$. With this assumption, it is possible to estimate the hyperparameters as follows.

To develop estimates for β^i, let β be the $12K \times 1$ vector obtained by stacking the β^i and let $\hat{\beta}$ be the $12K \times 1$ vector of stacked OLS estimates. The marginal density of $\hat{\beta}$ is

$$\hat{\beta}|\gamma, \Omega \sim N(W\gamma, V), \tag{18.3}$$

where V is the $12K \times 12K$ block diagonal matrix

$$V = \text{diag}(\sigma^2(X'^1 X^1)^{-1} + \Omega, \ldots, \sigma^2(X'^{12} X^{12})^{-1} + \Omega).$$

With known σ^2 and Ω, the usual GLS estimator for γ would be

$$\hat{\gamma}_{GLS} = (W'V^{-1}W)^{-1} W'V^{-1}\hat{\beta}. \tag{18.4}$$

This immediately yields an estimate for θ defined by $\hat{\theta} = W\hat{\gamma}$. The $12K \times 1$ vector $\hat{\theta}$ shrinks the vector of OLS estimates $\hat{\beta}$ towards a common mean.

To develop an estimate for Ω under the *d*-prior assumption, we proceed, as in Section 4.5, as follows. Let $(X'^i X^i)^{-1}_{mm}$ denote the *m*-th diagonal entry of the indicated matrix. Then the density of the estimates $\hat{\beta}^i_m$ of the *m*-th coefficient in the *i*-th equation is, from Eq. (18.3)

$$\hat{\beta}^i_m \sim N(\hat{\theta}^i_m, \sigma^2(X'^i X^i)^{-1}_{mm} + a_m).$$

If $\hat{\theta}^i_m$ was known, then $(\hat{\beta}^i_m - \hat{\theta}^i_j)^2 - s^2(X'^i X^i)^{-1}_{mm}$ would be an unbiased estimate for a_m. Thus a reasonable estimate for a_m is given by (where $x^+ = \max(x, 0)$),

$$\hat{a_m} = \frac{1}{12 - q_m} \sum_{i=1}^{12} \left((\hat{\beta}^i_m - \hat{\theta}^i_m)^2 - s^2(X'^i X^i)^{-1}_{mm} \right)^+. \tag{18.5}$$

Here q_m is the degrees of freedom lost in estimating θ^i_m by $\hat{\theta}^i_m$. Note that $q_m = 1$ if all θ^i are considered equal, so that all the β^i_m have a

common mean. Also $q_m = 3$ if the regression coefficients are considered
equal across brands, for in this case there will be three separate means
to estimate, one for each chain. Similarly $q_m = 4$ in the case that
coefficients are considered equal across chains. This estimate $\hat{a}_m$ differs
slightly[3] from the definition given by BG, and should be somewhat
superior.

Now we can estimate θ using Eq. (18.4) given estimates of Ω and
we estimate Ω using Eq. (18.5) given estimates of θ. To estimate the
prior mean and variance simultaneously we use an iterative procedure,
setting $\Omega = \mathbf{I}$ initially and estimating θ from Eq. (18.4) and using this
estimate to update Ω using Eq. (18.5). We iterate back and forth until
convergence occurs. This procedure is guaranteed to converge. Label
the estimates obtained by this process as $\hat{\theta}_D^i$ and $\hat{\Omega}_D$. The empirical
Bayes estimates for the d-prior are obtained by plugging these estimates
into Eq. (18.2).

g-Prior: Instead of assuming the prior covariance matrices Ω^i are
all the same, the computationally convenient g-prior assumes that
$\Omega^i = \tau_i^2(X'^i X)^{-1}$. This assumption suggests that the prior covariances
between the true coefficients are of the same pattern as the covariances
between the OLS estimates. The virtue of g-prior lies in the simple
form taken by the marginal densities of $\hat{\beta}^i$ conditional on the hyperparameters:

$$\hat{\beta}^i|\theta^i, \tau_i^2 \sim N(\theta^i, (\sigma^2 + \tau_i^2)(X'^i X^i)^{-1}).$$

If θ^i is known then it is easily calculated that

$$\mathbb{E}\|X^i\hat{\beta}^i - X^i\theta^i\|^2 = K(\sigma^2 + \tau_i^2).$$

Taking into account a degree of freedom lost in estimating θ, we obtain
a method of moments estimate for τ_i^2 defined by

$$\hat{\tau}_i^2 = \frac{1}{K-1}\|X^i\hat{\beta}^i - X^i\hat{\theta}^i\|^2 - s^2.$$

---

[3]We take the positive part before averaging to get $\hat{a}_m$, while BG average first
and then take the positive part

Based on these estimates, the prior covariance matrix can be estimated by

$$\hat{V}_G = \text{diag}\left([s^2 + \hat{\tau}_1^2(X'^1 X)^{-1}], \ldots, [s^2 + \hat{\tau}_{12}^2(X'^{12} X^{12})^{-1}]\right)$$

Given an estimate of prior covariance, we can estimate the prior mean γ by GLS as

$$\hat{\gamma}_G = (W'\hat{V}_G^{-1}W)^{-1}W'\hat{V}_G^{-1}\hat{\beta}.$$

This of course yields an estimate for θ defined by $\hat{\theta}_G = W\hat{\gamma}_G$. We are now in a position to iterate back and forth until we converge to stable estimates for the hyperparameters. Once again, empirical Bayes estimates for β can be obtained by plugging in these estimates for the hyperparameters into the classical Bayes estimates given by Eq. (18.2). Because of the convenient structure of the g-prior, the formula reduces the estimate to simple weighted averages of the OLS $\hat{\beta}_i$ and shrunken estimate $\hat{\theta}_i$:

$$\hat{\beta}_g^i = (\hat{\sigma}^{-2} + \hat{\tau}_i^{-2})^{-2}\left\{\hat{\sigma}^{-2}\hat{\beta}^i + \hat{\tau}_i^{-2}\hat{\theta}^i\right\}.$$

18.4.5   Hierarchical Bayes

BG develop two Bayesian estimators for this problem, based on two slightly different prior specifications. Both of these are estimated by usings the Gibbs sampler. We describe the prior specifications below.

Common to both specifications is the prior for σ^2 which is $p(\sigma) = 1/\sigma$. This can also be written as $\sigma^2 \sim IG(0,0)$. Also for γ BG use the diffuse prior $p(\gamma) = 1$. In our previous discussions, prior covariance structures have been confined to be of the type $\Omega = \tau^2 I$ and ocassionally $\Omega = \tau^2(X'X)^{-1}$. BG introduce a more sophisticated model for the prior covariance. They assume that Ω^{-1} has a *Wishart* distribution with mean $(\rho R)^{-1}$ and degrees of freedom ρ. The Wishart distribution has the property of facilitating computations in this situation. It is defined as follows.

Definition 18.1 (Wishart Density) *If $Z_1, \ldots, Z_q$ are i.i.d. d-dimensional normal with $Z_i \sim N_d(0, \Omega)$ then the random variable*

$W = \sum_{i=1}^{q} Z_i Z_i'$ has the Wishart density with parameters Ω, q; we write $W \sim \mathcal{W}_d(\Omega, q)$. The Wishart density is, up to a constant, given by

$$f^W(W) \propto \det(W)^{(1/2)(q-d-1))} \exp\left(-\frac{1}{2}\, tr\, W\Omega^{-1}\right)$$

Remark: Note that the definition indicates how we can generate Wishart random variable, as required by the Gibbs sampler. Simply generate q random normal vectors $Z_1, \ldots, Z_q$ which are i.i.d. $N(0, \Omega)$ and set $W = \sum_{i=1}^{q} Z_i Z_i'$ to get a $\mathcal{W}(\Omega, q)$ random variable. Note also that the second parameter is an integer q which measures the number of normal observations used in forming W.

BG set $R = .0001 \times \mathrm{diag}(1, v_2, v_3, \ldots, v_{12})$, where v_j is the variance of the j-th regressor in each regression model. Inclusion of the v_i removes dependence on scale. The value of .0001 is chosen to make Ω^{-1} very small, and hence Ω very large — this will reduce the effect of the prior and give the data a greater influence in determining the estimates. The prior (B1) is obtained by setting $\rho = 2$, while (B2) sets $\rho = 12$. Note that we have essentially 12 OLS estimates $\hat{\beta}^i$ from which to estimate the prior covariance Ω. Setting $\rho = 2$ says that we have two prior observations reflecting the covariances specified in R, while $\rho = 12$ increases the weight on the prior.

In order to use the Gibbs sampler, we must calculate the required conditional densities $f_\theta, f_\sigma^2, f_{\Omega^{-1}}, f_\beta$, where each f_x is the conditional density of x given the other three and also the data $\hat{\beta}$. Except for the wrinkle introduced by the Wishart density, this calculation is similar to the one done previously.

Our model has $y \sim N(X\beta, \sigma^2 I$ and $\beta \sim N(W\gamma, \Omega)$. Set $s = 1/\sigma^2$. Assume the prior on s is $\pi(s) = s^a$, while $\pi(\Omega) \sim \mathcal{W}(\rho^{-1}R, \rho)$. We assume a diffuse prior for γ, or $\pi(\gamma) = 1$. The model parameters are β and s, while the hyperparameters are γ and Ω. The density of each of these, conditional on three of the others plus the data y, is given in the lemma below.

Suppose y given β is $N(X\beta, \sigma^2 I)$ and $\beta \sim N(H\alpha, \Omega)$. Then the marginal density of y is $N(XH\alpha, \Omega + X\Omega X')$ and the posterior density of β given y is $N(Bb, B)$, where $B^{-1} = \Omega^{-1} + X'\Omega^{-1}X$ and $b = X'\Omega^{-1}y + \Omega^{-1}H\alpha$. Let $\theta = W\gamma$ and define $S = \sum_{i=1}^{12}(\beta_i - \theta_i)(\beta_i - \theta_i)'$.

Lemma 18.3 *In the HB model above, the conditional density of*

1. β *given* s, γ, Ω, y *is* $N(Bb, B)$, *where* $B^{-1} = sX'X + \Omega^{-1}$ *and* $b = sX'y + \Omega^{-1}W\gamma$.

2. γ *given* s, β, Ω, y *is* $N(Aa, A)$, *where* $A^{-1} = W'\Omega^{-1}W$ *and* $a = W'\Omega^{-1}\beta$.

3. s *given* γ, β, Ω, y *is* $\mathcal{G}((T/2) + a + 1, \|y - X\beta\|^2/2)$.

4. Σ *given* β, γ, s, y *is* $\mathcal{W}(\rho R + S, 12 + \rho)$.

Proof: The densities for β, γ, and s are more or less exactly as derived in the previous case — the Wishart density does not come in when we condition on Ω.

To get the conditional density of Ω^{-1} in this specificition, note that $\beta^i | \gamma^i, \Omega^{-1} \sim N(\theta^i, \Omega)$ and hence $\beta^i - \theta^i \overset{i.i.d.}{\sim} N(0, \Omega)$. To get an expression which must be proportional to the posterior density for Ω^{-1} we multiply the data density (for β) by the prior density to get

$$f^{\Omega^{-1}}(\Omega^{-1}) \;\propto\; \left(\det(\Omega^{-1})^{(12/2)} \exp\left(-\frac{1}{2} \sum_{i=1}^{12} (\beta^i - \theta^i)^t \Omega^{-1} (\beta^i - \theta^i) \right) \right)$$
$$\times \left(\det(\Omega^{-1})^{(1/2)(\rho - K - 1)} \exp\left(-\frac{1}{2} \operatorname{tr} \Omega^{-1}(\rho R) \right) \right).$$

By noting that $\sum_i x_i^t \Omega^{-1} x_i = \operatorname{tr} \Omega^{-1} \sum_i x_i x_i^t$, we can combine the two densities and note that the density $f_{\Omega^{-1}}$ must be Wishart density with parameters as specified in the lemma.

Based on these conditional densities, we can run the Gibbs sampler as discussed earlier. After setting initial values for β and Ω, we can generate in sequence, randomly from the indicated conditional distributions, values for $\theta, \sigma^2, \Omega^{-1}, \beta^i$. After a suitable number of iterations (50 is suggested by BG), we may assume that the joint density of the newly generated sets will be approximately the joint posterior density. We can then average over subsequent observations to get estimates of moments of the posterior density required for estimation of the parameters.

18.4.6   Performance of the Estimators

In addition to the two empirical Bayes estimators based on d-prior and g-prior, and the two hierarchical Bayes estimators (B1) and (B2), BG also compared two pooled estimators (P1) and (P2) with OLS. The estimator (P1) is simply a restricted least squares estimators which constrains all coefficients to be equal — as usual, there are three versions: (i) equality in all 12 regressions, (ii) equality across chains, or (iii) equality across brands. This estimator performed very poorly in nearly all of the tests to which these estimators were subjected. However, estimator (P2) which constrains all coefficients *except the constant* to be equal had decent performance in many of the tests. Reasonable performance of the pooled estimator is a sure indication of potential for improvement by shrinkage.

One of the issues BG wished to explore was which of the three potential restrictions was more reasonable. Unfortunately, the estimates did not provide clear evidence on this issue. There does seem to be some slight superiority for the assumption that the coefficients are equal across the chains, but the other two restrictions (of equality across brands and of equality across all chains and brands) perform almost equally well.

Assessment of the estimates was based on two criteria. One was the sign and magnitude of the estimates. The second criteria employed 'hold-out' samples, which were not used in estimation, but were used to test forecasts from the various sets of estimates. In general all the shrinkage estimators reduced the variability of the estimates across equations as expected and had far fewer wrong signs. For example, the Bayes estimate (B2) shrinking towards the grand mean (equality across all chains and brands) gave the right signs for all the coefficients. For some coefficients, the OLS estimate had the right sign, but was of the wrong size, suggesting an elasticity much smaller or larger than reasonable. In such cases too, the shrinkage estimates were much more in line with intuitively plausible values for these elasticities.

With reference to forecast errors on the holdout samples, the shrinkage estimators also performed well, typically dominating OLS. The d-prior empirical Bayes estimator produced the best results, dominating OLS in all of its variants and with respect to three different

types of holdout samples. The *g*-prior frequently produced results very
close or identical to OLS, but did improve in forecasts when equality
across chains was assumed. The estimators (B1) and (B2) produced
even greater improvement in two of the three holdout samples but did
slightly worse than OLS in the third holdout sample. The pooling estimator (P1) performed very poorly in forecasts, producing errors five
times the size of OLS errors. Even though (P2) did worse than OLS,
it was only by margins ranging from 4% to 17%, and in one case it
improved on the OLS.

18.4.7   Conclusions

In order to assess which of three potential restrictions on coefficients
(equality across chains, brands, or both) was superior, BG performed
some formal tests like Akaike Information Criterion and others. All
these tests indicated that one should not pool — none of the restrictions
were valid. Since the formal criteria directly address the issue of the
whether restricted least squares would be superior to OLS, the poor
forecast performance of the estimators (P1) and (P2) is in accordance
with the failure of these criteria suggesting gains from pooling. Note
that even though restricted least squares does not produce gains, the
more flexible approach to incorporating this cross-equation information
(i.e., empirical Bayes) does produce substantial benefits. Some specific
issues deserve to be noted.

The unshrinkable constant: The performance of the pooled estimators clearly indicates that it is an error to shrink the constant —
assuming a common mean for the constants produces terrible results,
while assuming equality of coefficients does not. Nonetheless, all the
shrinkage estimators used by BG do assume equality for all coefficients
and hence shrink all the coefficients. As BG indicate, there is some
automatic safety against erroneous assumptions built into shrinkage estimators. For instance consider the *d*-prior empirical Bayes estimator.
The prior variance a_1 will be estimated by looking at the variance of the
OLS estimates. Since the constants fluctuate significantly across equations, a_1 will be estimated to be large, and hence the prior weight for
shrinkage of the constant towards its mean value will be low. This phenomena was in fact observed; the shrinkage estimators did not shrink

the constant very much in any of the equations. For this reason BG were not very concerned about makin an implausible restricition (there is no theoretical reason to assume equality for the constant term, while there is some plausiblity for assumption of equality, or similarity, of the elasticities).

It is possible that removing the restrictions from the constant will improve the performance of the shrinkage estimators studied by BG. Even a slight amount of wrong shrinkage will degrade the performance of the estimator. Also, due to the nonidentity covariance, the poor fit of the shrinkage estimator for the constant term will influence the estimation of other coefficients and further degrade performance. It seems plausible that the shrinkage estimators would perform better when the restrictions imposed in the W matrix do not include the constant term. It is possible that the relatively poor performance of the (G) estimates is due to this problem. The prior parameters τ_i^2 are estimated on the basis of the fit for *all* coefficients, including the constant term. One poor fit would be enough to make the (G) estimator discount all prior information. The closeness of the (G) prior to the OLS, and also its relatively good performance in the equality across chains case lends credence to this idea — equality across chains is precisely the case in which the estimator (P1), which restricts the constants to be equal, is least bad.

Another variation from the prior assumptions of BG would also be worth exploring. Even though the Wishart density is an elegant formulation, it is not clear whether this yields gains over the simpler formulation of a diagonal Ω. With $\Omega = \mathrm{diag}(a_1, \ldots, a_{12})$, the prior density $p(a_i) = |a_i|^{-c}$ leads to a manageable setup similar to the one analyzed earlier. This can also be estimated by the Gibbs sampler. It would be worth exploring if this makes any difference.

18.5   Exercises

1. Show that the formula of Lemma 18.1 for the posterior mean can be rewritten in the form of a precision weighted average of the data mean and the prior mean, as given in Corollary 3.1.1.

2. Obtain the marginal density of y as a continuation of the proof of Lemma 18.1. To get the desired quadratic form in y, you will need to prove the following matrix equality:

$$\Omega^{-1} - \Sigma^{-1} X \left(X'\Sigma^{-1} X + \Omega^{-1} \right)^{-1} X'\Sigma^{-1} = (\Sigma + X\Omega X')^{-1}.$$

3. Show that

$$\lim_{\nu^2 \to \infty} \left\{ \tau^2 \mathbf{I}_k + \nu^2 \mathbb{1}\mathbb{1}' \right\}^{-1} = \tau^{-2} \left\{ \mathbf{I}_k - \mathbb{1}\mathbb{1}' \right\}.$$

Hint: For scalar a and vector u, multiply $\mathbf{I} + a(uu')$ by $\mathbf{I} - b(uu')$ to discover the form of $(\mathbf{I} + auu')^{-1}$.

4. Suppose $X \sim N(\mu v, \mathbf{I}_k)$, where $v \in \mathbb{R}^k$ is a fixed vector and μ is a scalar constant.

   (a) Suppose the prior distribution of μ is $N(0, \nu^2)$. What is the posterior distribution of μ given X? What is the marginal distribution of X?

   (b) Suppose we equip μ with the diffuse prior $p(\mu) = 1$. First consider the case $k = 1$. Show that the marginal density of X is diffuse also — that is, $f(x) = $ constant. Since this is not a proper density, we have some trouble getting this result if use the prior density $f(\mu) = N(0, \nu^2)$ and taking limits as $\nu^2 \to \infty$. Show that there is no difficulty with the posterior distribution of μ — taking limits as $\nu^2 \to \infty$ leads to the right result.

   (c) Things get worse when $k > 1$. In this case the marginal density of X is improper in one dimension (spanned by the vector μ) and a proper normal distribution in the $k - 1$ dimensional subspace orthogonal to v. Derive these distributions and also the posterior distribution for μ. See if you can get these expressions by taking limits as $\nu^2 \to \infty$ using the prior $p(\mu) = N(0, \nu^2)$.

5. Assuming you did the last exercise of Chapter 17, take the same data set and apply hierarchical Bayes estimation to get a new set

of estimates. Compare with your earlier EB estimates as well as
OLS estimates. Attempt to ascertain which of the several sets of
estimates are 'best'. There are several techniques which can be
used. The simplest is 'intuitive'; we assess the plausibility (via
signs and magnitudes) of estimated coefficients. There are also
several variations of cross-validation analysis which can be carried
out.

Part V

Appendices

Appendix A

The Multivariate Normal Distribution

This appendix deals with the basic properties of the multivariate normal distribution. The first section introduces the Laplace transform, also known as the multivariate moment generating function. This is a convenient tool for obtaining the required properties.

A.1 Laplace Transform

For $x, \theta \in R^n$, we will use the standard notational conventions that $x = (x_1, x_2, \ldots, x_n)$, $\theta = (\theta_{1,2}, \ldots, \theta_n)$, $x'\theta = \sum_{i=1}^{n} x_i \theta_i$, and $\|x\|^2 = \sum_{i=1}^{n} x_i^2 = x'x$, where $\|\cdot\|$ stands for the Euclidean norm of x. For any function $f : \mathbf{R}^n \to \mathbf{R}$, we can define another function $g : \mathbf{R}^n \to \mathbf{R}$ as follows:

$$g(\theta) = \int_{-\infty}^{\infty} \cdots \int_{-\infty}^{\infty} f(x) \exp(\theta'x) \, dx_1 \, dx_2 \ldots dx_n.$$

The function g is called the Laplace transform of f. We also write $g(\theta) = \mathcal{L}_f(\theta)$ as shorthand for the equation above.

As an example of the Laplace transform, consider the function $f(x) = (2\pi)^{-(n/2)} \exp(-(1/2) \sum_{i=1}^{n} x_i^2)$. This is the standard n-dimensional normal density and satisfies

$$\int_{-\infty}^{\infty} \cdots \int_{-\infty}^{\infty} f(x) \, dx_1 \ldots dx_n = 1.$$

We will say that X is $N(0, \mathbf{I}_n)$ if the density of X is $f(x)$. The *moment generating function* of the X is defined to be $\mathbb{E}\exp(\theta'X)$ which is the same as the Laplace transform of the density f. This is derived below.

Lemma A.1 *If* $X \sim N(0, \mathbf{I}_k)$ *then* $\mathbb{E}\exp\theta'X = \exp(\theta'\theta/2)$.

Proof: The Laplace transform $g = \mathcal{L}_f$ is calculated as follows:

$$
\begin{aligned}
g(\theta) &= \int_{\mathbf{R}^n} \cdots \int f(x) e^{\theta'x} dx \\
&= \int_{\mathbf{R}^n} \cdots \int (2\pi)^{-(n/2)} \exp[-\frac{1}{2}(x'x - 2\theta'x)] dx \\
&= \int_{\mathbf{R}^n} \cdots \int (2\pi)^{-(n/2)} \exp[-\frac{1}{2}(x'x - 2\theta'x + \theta'\theta - \theta'\theta)] dx \\
&= \exp(\frac{1}{2}\theta'\theta) \int_{\mathbf{R}^n} \cdots \int (2\pi)^{-(n/2)} \exp(-\frac{1}{2}\|x - \theta\|^2) dx \\
&= \exp(\frac{1}{2}\theta'\theta).
\end{aligned}
$$

The last equality results from making the change of variables $z = x - \theta$ which reduces the integral to that of a standard n-dimensional normal density.

There are several properties of the Laplace transform which play an important role in statistical theory. Proving these requires advanced techniques, and so we will simply state the relevant properties below. An example given in exercise 1 shows that the Laplace transform may be well-defined for some but not all values of θ. For any function $f : \mathbf{R}^n \to \mathbf{R}$, let $D(\mathcal{L}_f)$ be the domain of the Laplace transform; that is, the set of values of θ for which the Laplace transform is well-defined.

Property 1: The domain of a Laplace transform is always a convex set. That is, if $\theta, \theta' \in D(\mathcal{L}_f)$ then for any $\lambda \in [0, 1]$, $\lambda\theta + (1 - \lambda)\theta' \in D(\mathcal{L}_f)$.

Property 2: If the domain contains an n-dimensional cube, then the Laplace transform is unique. Formally, suppose that for some $a = (a_1, \ldots, a_n)$, $b = (b_1, \ldots, b_n)$ with $a_i < b_i$ for all i, and for some $f :$

$\mathbf{R}^n \rightarrow \mathbf{R}$, we have $\mathcal{L}_f(\theta) = 0$ for all θ such that $a_i < \theta_i < b_i$ for all i. Then we must have $f(x) = 0$ for all x.

This property is equivalent to uniqueness. For $a, b \in \mathbf{R}^n$, let us write $a < b$ to indicate that the inequality holds in each coordinate. Suppose $a < b$ and suppose that the Laplace transform of two functions $f(x)$ and $g(x)$ is the same for all θ such that $a < \theta < b$. Define $z(x) = f(x) - g(x)$ and note that $\mathcal{L}_z(\theta) = 0$ for all θ such that $a < \theta < b$. Now the property above implies that $z(x) = 0$ for all x, and hence f is the same function as g.

Property 3: The Laplace transform $\mathcal{L}_f(\theta)$ is infinitely differentiable with respect to θ for all interior points of the domain. Furthermore, if X is a random variable with density f, the derivatives can be calculated as follows:

$$
\begin{aligned}
\frac{\partial}{\partial \theta_i} \int_{-\infty}^{\infty} \cdots \int_{-\infty}^{\infty} f(x) \exp(\theta' x) \, dx &= \int_{-\infty}^{\infty} \cdots \int_{-\infty}^{\infty} \frac{\partial}{\partial \theta_i} \exp(\theta' x) f(x) \, dx \\
&= \int_{-\infty}^{\infty} \cdots \int_{-\infty}^{\infty} x_i \exp(\theta' x) f(x) \, dx \\
&= \mathbb{E} X_i \exp(\theta' X).
\end{aligned}
$$

Note that evaluating the derivative at $\theta = 0$ will yield $\mathbb{E} X_i$. More generally by repeated application of this rule, we have

$$
\mathbb{E} X_1^{k_1} X_2^{k_2} \cdots X_n^{k_n} = \left(\frac{\partial}{\partial \theta_1} \right)^{k_1} \left(\frac{\partial}{\partial \theta_2} \right)^{k_2} \cdots \left(\frac{\partial}{\partial \theta_i} \right)^{k_n} \mathcal{L}_f(\theta) \bigg|_{\theta = 0} .
$$

A.2 Multinormal: Basic Properties

It is convenient to define the multivariate normal distribution via the moment generating function.

Definition A.1 *We say that $X = (X_1, X_2, \ldots, X_n)$ has the multivariate normal density with parameters $\mu \in \mathbb{R}^n$ and $n \times n$ positive semidefinite matrix Σ if*

$$
\mathbb{E} \exp \theta' X = \exp \left(\theta' \mu + \frac{1}{2} \theta' \Sigma \theta \right).
$$

We also write $X \sim N_n(\mu, \Sigma)$ to express the fact that the above equation holds.

We will establish that if Σ is nonsingular, then X has the standard multinormal density forming the basis for the usual definition. The added convenience of the MGF definition stems from the fact that we can allow for singular matrices Σ. The singular case arises with some frequency in application and the MGF permits us to handle it with ease. It will transpire that μ is the mean of X so that $\mathbb{E}X = \mu$ and also $\Sigma = \mathbb{C}ov\, X$, but we must establish these facts from the definition. The basic properties of the normal distribution fall out as easy consequences of the MGF, and are described in the following lemmas.

Lemma A.2 *If $X \sim N(\mu, \Sigma)$, and $Y = AX + b$ where A is an arbitrary $m \times n$ matrix and b is $m \times 1$ vector, then $Y \sim N(A\mu + b, A\Sigma A')$.*

Proof: We just compute the MGF of Y. Note that $\mathbb{E}\exp(\psi'Y) = \mathbb{E}\exp(\psi'AX)\exp(\theta'b)$, and introduce $\theta' = \psi'A$. Using the MGF of X, we conclude that $\mathbb{E}\psi'AX = \exp(\theta'\mu + (\theta'\Sigma\theta)/2) = \exp(\psi'A\mu + \psi'A\Sigma A'\psi)$ and hence $\mathbb{E}\psi'Y = \exp(\psi'(A\mu + b) + \psi'(A\Sigma A')\psi/2)$. Since this is the definition of $Y \sim N(A\mu + b, A\Sigma A')$, our proof is complete.

Lemma A.3 *If $X \sim N(\mu, \Sigma)$ then $\mathbb{E}X = \mu$ and $\mathbb{C}ov(X) = \Sigma$.*

Proof: Evaluate the gradient (i.e., the vector of first derivatives of the MGF at 0) to get $\nabla_\theta \exp(\theta'\mu + \theta'\Sigma\theta/2)|_{\theta=0} = \mu$. Define $Z = X - \mu$ and note that $Z \sim N(0, \Sigma)$ from the previous lemma. Since $\mathbb{C}ov(X) = \mathbb{C}ov(Z)$ it suffices to calculate the latter, which is simpler. Expanding e^x as $1 + x + x^2/2 + \cdots$ yields

$$\nabla_\theta \nabla'_\theta \exp(\theta'\Sigma\theta/2)|_{\theta=0} = \nabla_\theta \nabla'_\theta (1 + (1/2)\theta'\Sigma\theta + + \cdots)|_{\theta=0} = \Sigma.$$

Lemma A.4 *If $X \sim N_n(\mu, \Sigma)$ and Σ is nonsingular then X has the following density:*

$$f^X(x) = (2\pi)^{-n/2} \, \mathbf{det}(\Sigma)^{-1/2} \exp\left(-\frac{1}{2}(x - \mu)'\Sigma^{-1}(x - \mu)\right). \quad (A.1)$$

Proof: First consider the special case $\mu = 0$ and $\Sigma = \mathbf{I}$. Then the lemma follows immediately from the calculation in Lemma A.1. For general Σ find a matrix A such that $\Sigma = AA'$. Let $Z \sim N(0, \mathbf{I})$, and define $X = \mu + AZ$. The change of variables formula shows that the density of X must be as in Eq. (A.1) above. The MGF for linear transforms has already been computed and the two facts combine to yield the lemma.

A.2.1 Case of Singular Covariance

We take a deeper look at this case, to see the difference between singular normal distributions and nonsingular ones. Suppose $Y \sim N_n(\mu, \Sigma)$ where Σ, has rank $m < n$. This holds if and only if $\Sigma = AA'$ for some $n \times m$ matrix A of full rank m. If $X \sim N_m(0, \mathbf{I})$, then $Y' = \mu + AX$ will have the same density as Y. Thus every singular normal distribution is a linear transform of a nonsingular one.

It is important to clarify that even though $Y = AX + \mu$ fails to have a density when A is singular, it is a perfectly well-defined random variable. The simplest case is when A is just a matrix full of zeroes and hence of zero rank. In this case, Y is a (degenerate) random variable which takes value μ with probability one. Any random variable which takes values in some lower dimensional subspace of $\mathbf{R}^n$ will necessarily fail to have density on $\mathbf{R}^n$. An example we are already familiar with is that of the discrete random variables. These take values on finite or countable sets of points, which are 0-dimensional objects in the 1-dimensional space $\mathbf{R}^1$. Similarly, normal distributions with singular covariance matrices take values in a lower dimensional subspace of $\mathbf{R}^n$; there is nothing mysterious about them. As a concrete example, suppose $X \sim N(0, 1)$ and define $Y = AX$, where A is the 2×1 matrix $(1, 1)'$. Then $Y = (X, X)$ takes values only on the diagonal of $\mathbf{R}^2$. It should be clear that it is easy to compute the probability of Y lying in any subset of $\mathbf{R}^2$; we simply look at the intersection of that set with the diagonal, and do our computation in terms of X. However, Y does not have a multivariate normal density of the type given by Eq. (A.1). Application of previous results shows that the covariance matrix of Y is $AA' = \begin{pmatrix} 1 & 1 \\ 1 & 1 \end{pmatrix}$. This is singular with determinant zero, so density given in Eq. (A.1) is not well-defined.

A.3   Independence and Zero Covariance

An important property of the normal distribution is that it is only necessary to establish zero covariance to prove independence of jointly normal variables. Recall that covariance is a measure of <u>linear</u> association. If $Cov(X, Y)=0$, it follows that X and Y are not linearly related, but this is far from saying that they are independent. For example, if $X \sim N(0, 1)$ and $Y = X^2$, then $Cov(X, Y)=0$ (prove), but obviously X and Y are not independent. The nonlinear relationship of X and Y is not detected by the covariance. However, in the special case that X and Y are jointly normal, lack of a linear relation is enough to guarantee independence of X and Y. We now turn to a proof of this result. Suppose X is an $(m+n)$-dimensional random variable normally distributed as $N_{m+n}(\mu, \Sigma)$. Partition X, μ, Σ as follows:

$$X = \begin{pmatrix} X^1 \\ X^2 \end{pmatrix}, \mu = \begin{pmatrix} \mu^1 \\ \mu^2 \end{pmatrix}, \Sigma = \begin{pmatrix} \Sigma_{11} & \Sigma_{12} \\ \Sigma_{21} & \Sigma_{22} \end{pmatrix}$$

Using these notations we can state the basic result as follows:

Theorem A.1 *The random variables X^1 and X^2 are independent if and only if $\Sigma_{12} = 0$.*

Proof: It is easily checked that $\Sigma_{12} = Cov(X^1, X^2)$ so that the independence of X^1 and X^2 implies that $\Sigma_{12} = 0$. It is the converse which is a special property of the normal distribution. To prove it, we will show that the moment generating function of X factorizes into the moment generating functions of X^1 and X^2 when $\Sigma_{12} = 0_{m \times n}$. Partition θ and $\theta' = (\theta'_1, \theta'_2)$ where θ_1 is $m \times 1$ and θ_2 is $n \times 1$. If $\Sigma_{12} = 0$ then we can express the moment generating function of X as

$$\begin{aligned} \mathbb{E} \exp(\theta' X) &= \exp(\theta'\mu + (1/2)\theta'\Sigma\theta) \\ &= \exp\{\theta'_1\mu_1 + \theta'_2\mu_2 + (1/2)[\theta'_1\Sigma_{11}\theta_1 + \theta'_2\Sigma_{22}\theta_2]\} \\ &= \exp(\theta'_1\mu_1 + \theta'_1\Sigma_{11}\theta_1)\exp(\theta'_2\mu_2 + \theta_2\Sigma_{22}\theta_2) \\ &= \mathbb{E}(\theta'_1 X^1)\mathbb{E}(\theta'_2 X^2). \end{aligned}$$

If $X^1 \sim N(\mu_1, \Sigma_{11})$ and $X^2 \sim N(\mu_2, \Sigma_{22})$ were independent, then the MGF of $X = (X^1, X^2)$ would be the product of their separate MGFs, exactly as above. But the uniqueness of the MGFs guarantees that this must be the joint density of the variables X^1 and X^2, and hence their independence.

A corollary which is frequently used in econometrics is the following:

Corollary A.1.1 *Suppose $X \sim N(\mu, \sigma^2 I_p)$ and A_1, A_2 are $m_1 \times n$ and $m_2 \times n$ matrices of constants such that $A_1 A_2' = 0_{m_1 \times m_2}$. If $Y = A_1 X$ and $Z = A_2 X$, then Y and Z are independent.*

Proof: Define the matrix A by stacking A_1 on top of A_2

$$A = \begin{pmatrix} A_1 \\ A_2 \end{pmatrix}.$$

Define $W = AX$. Then $W' = (Y', Z')$ and Y and Z have a joint (singular) normal distribution by Lemma A.2 . To prove independence of Y and Z, it is enough to prove that their covariance matrix is 0. This is easy:

$$\begin{aligned} \mathbb{C}ov(Y, Z) &= \mathbb{E}[(A_1 X - \mathbb{E}A_1 X)(A_2 X - \mathbb{E}A_2 X)'] \\ &= A_1 \mathbb{E}[(X - \mathbb{E}X)(X - \mathbb{E}X)']A_2' \\ &= A_1 \mathbb{C}ov(X)A_2' = 0. \end{aligned}$$

The last equality results from substituting $\sigma^2 I_p$ for $\text{Cov}(X)$, and noting that $A_1 A_2' = 0$ by hypothesis.

It is important to note a subtle aspect of these results showing that zero covariance implies independence for jointly normal variables. They do NOT imply that if both X and Y are normal and $\text{Cov}(X, Y) = 0$ then X and Y are independent. The assumption that X and Y are jointly normal is considerably stronger than specifying that the marginal densities of X and Y are both normal. We illustrate the difference by an example.

Counterexample: We will construct random variables X and Y which are both normal, have zero covariance, but are not independent. Suppose $Z \sim N(0, 1)$ and B is independent of Z and takes values either $+1$ or -1 both with probability one-half. Define $X = BZ$ and $Y = Z$.

Now we claim that $X \sim N(0,1)$, $Y \sim N(0,1)$, $\text{Cov}(X,Y)=0$, but that X and Y are not independent. Obviously $Y = Z$ so that it has the same distribution, $N(0,1)$. To determine the marginal distribution of X note that $X = Z$ with probability one-half and $X = -Z$ with probability one-half. Thus the density of X is the average of the densities of Z and $-Z$. But Z is symmetric so that Z and $-Z$ have the same density $N(0,1)$, which must also be the density of X. To calculate the covariance of X and Y, note that

$$\text{Cov}(X,Y) = \mathbb{E}(BZ)(Z) - \mathbb{E}(BZ)\mathbb{E}Z = \mathbb{E}BZ^2 - 0.$$

Since B and Z are independent, $\mathbb{E}(BZ^2) = \mathbb{E}B\mathbb{E}Z^2 = 0$, proving that X and Y have zero covariance. It is obvious that X and Y are not independent since knowing the value of one of the two essentially determines the other (except for the sign).

A.4 Quadratic Forms in Normals

If $X \sim N_n(\mu, \Sigma)$, one frequently needs to know the distribution of quantities like $X'AX$ for some $n \times n$ matrix A. Rather surprisingly, this apparently simple problem has no closed form solution in the general case. Some numerical algorithms have been devised which can be use to accurately compute, or approximate the distribution. Ansley *et al.* (1992) show how to compute a special case which comes up frequently in applications. Lye (1991) describes available numerical methods and also provides references to the literature for the general case. Below we will discuss some special cases for which closed form solutions can be found.

Lemma A.5 *If $X \sim N(0, \mathbf{I})$ then $Y = X'X$ has a chi-square distribution with n degrees of freedom.*

Proof: To prove this, recall that $Z \sim \mathcal{G}(p, \lambda)$ is short for $f^Z(z) = \lambda^p z^{p-1} \exp(-\lambda z)/\Gamma(p)$. Also $Y \sim \chi_n^2$ is short for $Y \sim \mathcal{G}(n/2, 1/2)$. The MGF of Z is easily calculated:

$$\mathbb{E}\exp(\theta'Z) = \int_0^\infty \frac{\lambda^p}{\Gamma(p}x^{-1}\exp\{-(\lambda - \theta)z\}\,dz$$

$$= \frac{\lambda^p}{(\lambda - \theta)^p} \int \frac{(\lambda - \theta)^p}{\Gamma(p)} \exp\{-(\lambda - \theta)z\}\, dz.$$

The quantity inside the integral is the density of a $\mathcal{G}(p, \lambda - \theta)$ variable and must integrate to 1. Thus the MGF of Z is $\mathbb{E}\exp(\theta'Z) = \lambda^p/(\lambda - \theta)^p$. A direct calculation shows that if $X_1 \sim N(0,1)$ then $\mathbb{E}\exp(\theta X_1^2) = (1 - 2\theta)^{-1/2}$ which matches the MGF of a χ_1^2. If $X \sim N(0, \mathbf{I})$ then $X'X = X_1^2 + \cdots X_n^2$. Since the X_i are independent we have

$$\mathbb{E}\exp(\theta X'X) = \prod_{i=1}^{n} \mathbb{E}\exp(\theta X_i^2) = (1 - 2\theta)^{-n/2} = \frac{(1/2)^{n/2}}{(1/2 - \theta)^{n/2}}.$$

This matches the χ_n^2 density and establishes our result.

Next note that if A is symmetric, then $A = Q\Lambda Q'$ for some orthogonal matrix Q and diagonal matrix Λ. It follows that $X'AX = Z'\Lambda Z = \sum_{i=1}^{n} \lambda_i^2 Z_i^2$ with $Z = QX$. If $X \sim N(0, \mathbf{I})$, then $Z \sim N(0, \mathbf{I})$ since $QQ' = \mathbf{I}$. Furthermore, for arbitrary A, $X'AX = X'(A + A')X/2$ so that every quadratic form can be written in terms of a symmetric matrix. Thus quadratic forms can be reduces to weighted sums of squares of normals. Since idempotent matrices A have eigenvalues either 0 or 1, it follows that $X'AX$ must be chi-square for such matrices; see exercise 4 below.

The density of X^2 when $X \sim N(\mu, 1)$ is derived in Lemma B.1 below. The density of $\|X\|^2$ for $X \sim N(\mu, \mathbf{I})$ is derived in Lemma 7.3. These are the noncentral chi-square densities.

A.5   Exercises

1. Consider the function $f(x_1, x_2, \ldots, x_n) = \exp(-\sum_{i=1}^{n} |x_i|)$. Compute the Laplace transform $g = \mathcal{L}_f(\theta)$ and show that it is well-defined if and only if $|\theta_i| < 1$ for $i = 1, 2, \ldots, n$.

2. Some properties of the covariance matrix are listed below:

   (a) For any vector valued random variable X, prove that the covariance matrix $\text{Cov}(X)$ must be symmetric and positive semidefinite.

(b) For any X $\mathrm{Cov}(AX + b) = A\,\mathrm{Cov}(X)A'$.

(c) $\mathrm{Cov}(AX + b, CY + d) = A\,\mathrm{Cov}(X, Y)C'$.

3. Suppose $X \sim MVN_{m+n}(\mu, \Sigma)$ and we partition X as $X = (Y, Z)$ with Y being $m \times 1$ and Z as $nx1$. Partition μ and Σ conformably, with $\mu = (\mu_y, \mu_z)$ and

$$\Sigma = \begin{pmatrix} \Sigma_y & \Sigma_{y,z} \\ \Sigma_{z,y} & \Sigma_z \end{pmatrix}$$

(a) Show , by factorizing the density, that Y and Z are independent if $\Sigma_{y,z} = 0$. The MGF technique used in this section is more general because it works even when Σ is singular, so that the density is not defined. The next two questions deal with the case where $\Sigma_{y,z} \neq 0$.

(b) Find a matrix M such that $V = Y + MZ$ is independent of Z.

(c) Find the conditional density of Y given Z. Hint: $Y = V - MZ$. Conditional on Z, MZ is a constant vector, while V is independent of Z.

4. Show that if $X \sim N(0, \mathbf{I}_k)$ and A is a $k \times k$ matrix satisfying $A^2 = A$ then $X'AX$ is chi-suared with degrees of freedom equal to the trace of A. Hint: Diagonalize A as $Q'\Lambda Q$, show that $Z = QX \sim N(0, \mathbf{I}_k)$, and the eigenvalues of A, or diagonal entries of Λ, can only be 0 or 1.

Appendix B

Uniformly Most Powerful Tests

B.1   Introduction

The first section exposits the Neyman-Pearson theory of finding a most powerful test for a simple null hypothesis versus a simple alternative. The second section discusses several approaches to the choice of the significance level for the Neyman-Pearson test. Section B.4 discusses an example of a simple null versus a compound alternative where Neyman-Pearson tests are exactly the same for all alternatives. In this case the test is simultaneously most powerful for all alternatives and is called a uniformly most powerful test. The next section discusses the general situation. It show that the family of densities must possess the MLR (monotone likelihood ratio) property if a uniformly most powerful test exists. Some examples are given to illustrate how one can check for the MLR property. Section B.6 relates the MLR property to exponential families of densities.

B.2   Neyman-Pearson Tests

The general hypothesis testing problem can be described as follows. We have observations Y which come from a family of possible distributions, e.g., $f^Y(y, \theta)$, indexed by parameter $\theta \in \Theta$. The null hypothesis H_0

is that the parameter θ belongs to some subset Θ_0 of Θ, while the alternative is $H_1 \colon \theta \in \Theta_1$, where Θ_1 is a subset of Θ not intersecting with Θ_0. If the subsets Θ_i contain only one point, these are called *simple* hypotheses, while they are called *compound* hypotheses when the sets contain more than one point.

The Neyman-Pearson theory discussed in this section characterizes the set of all reasonable tests for the case where both hypotheses are simple. Every hypothesis test can be described as a function $\delta(y)$ taking values in the interval $[0, 1]$, with the interpretation that the value $\delta(y)$ is the probability with which we reject the null hypothesis. We will only consider tests for which $\delta(y)$ equals 0 or 1, so randomized tests will not be discussed. Recall that if we reject the null when it is true, this is called a type I error, while accepting the null when it is false is called a type II error. The probability of rejecting the null when the null is true (i.e., $\mathbb{P}(\delta(y) = 1|H_0)$) is called the *size* or *significance level* or simply *level* of the test. The probability of not making a type II error (i.e., $\mathbf{P}(\delta(y) = 1|H_1)$) is called the *power* of a test. In general there is a trade off between size and power. The smaller we make the probability of type I error the larger the probability of type II error, and hence the lower the power of the test. For any given size, the Neyman-Pearson theorem permits us to describe the test of maximum power. For use in the theorem below, define the likelihood ratio function

$$NP(y) = \frac{f^Y(y, \theta_1)}{f^Y(y, \theta_0)}.$$

Theorem B.1 (Neyman-Pearson) *Suppose that the densities of the observation Y under the null and alternative hypothesis are $H_0 \colon Y \sim f^Y(y, \theta_0)$ and $H_1 \colon Y \sim f^Y(y, \theta_1)$. Fix $\alpha > 0$. Then there exists a constant c_α such that*

$$\mathbb{P}(NP(Y) > c_\alpha|H_0) = \alpha.$$

Define hypothesis test $\delta^*(y)$ by $\delta^* = 1$ if $NP(y) > c_\alpha$ and $\delta^*(y) = 0$ if $NP(y) \leq c_\alpha$. Then this test has size α and if $\delta'(y)$ is a different test of smaller or equal size, then δ' has power less than the power of δ^*.

The proof of this theorem is given in the exercises. Except for the choice of level α or equivalently the constant c_α, the Neyman-Pearson theorem provides a complete solution to the hypothesis testing problem in the case of simple versus simple hypotheses. We now discuss four different approaches to the choice of constant. There has been some debate over which of these is 'correct.' Our point of view, detailed below, is that each is correct in its sphere of application. The validity of the approach depends on the application and also on the purpose for which the hypothesis test is being done.

B.3 Choice of Level in Neyman-Pearson Test

B.3.1 The Classical Approach

The original suggestion of Neyman and Pearson is to set α at a low level, say 10% or 5% or 1%. These are known as the conventional significance levels, and the choice among these is somewhat subjective. This approach is suitable in the contexts that Neyman and Pearson were discussing, namely where H_0 represents some scientific hypothesis. In this context, there are two possible consequences of the hypothesis test. Either we reject the null hypothesis (which amounts to saying that the data provide strong evidence against the null) or we fail to reject the null (which means that the data are not in strong contradiction to the null). These two outcomes are *not* symmetric. Our objective in testing in this kind of situation is to decide whether H_0 can be eliminated from consideration as a serious scientific hypothesis on the basis of the evidence provided by the data. If we fail to reject, this means we will continue to consider H_0 as a possible description of reality — it does not mean that the data in any sense prove the validity of H_0. Alternatively, a rejection means that the data are strongly in conflict with the null, so that H_0 does not deserve further serious consideration. In practice, in many such situations H_1 is just a dummy hypothesis; we do not really know what the correct theory should be if the null is not valid. Strong rejections (such as the rejection of Ptolemaic astronomy) lead us to search for good alternatives to the initial theory. Note that in the classical approach, the power of the test (or one minus the probability of type II error) is of lesser significance than the size. This is also

reasonable since when we fail to reject, this means we continue to regard
the scientific hypothesis as a workable one, but this does not commit
us to the hypothesis. Later on more convincing evidence could turn up
to show that the hypothesis is not valid. Thus failure to reject is not
as serious an error as rejecting a potentially valid hypothesis, which
means abandoning further research on this hypothesis.

B.3.2   The Bayesian Approach

A second approach is Bayesian. Suppose that we have a prior probability π_0 that the null hypothesis is true. Then $\pi_1 = 1 - \pi_0$ is the
probability that the alternative is true. It is a consequence of Bayes
rule that after observing the data Y, the probability of the null is updated to the posterior probability:

$$P(H_0|Y) = \frac{\pi_0 f^Y(Y,\theta_0)}{\pi_0 f^Y(Y,\theta_0) + \pi_1 f^Y(Y,\theta_1)}.$$

There is a similar expression for the posterior probability of the alterna-
tive hypothesis $P(H_1|Y)$. The Bayesian approach involves considering
the posterior odds ratio (POR) defined by

$$\text{POR} = \frac{P(H_1|Y)}{P(H_0|Y)} = \frac{\pi_0 f^Y(Y,\theta_0)}{\pi_1 f^Y(Y,\theta_1)}.$$

The odds ratio gives the strength of evidence in favor of H_1; if it is
greater than 1 then H_1 is more likely. From this, the constant c should
be set equal to π_1/π_0.

   A quantity useful in assessing the strength of evidence provided by
the data is the *Bayes Factor*, defined as:

$$BF = \frac{P(H_1|Y)/P(H_0|Y)}{P(H_1)/P(H_0)}$$

This is the ratio of posterior odds to prior odds. This eliminates some
of the arbitrariness involved in choosing prior probabilities. A Bayes
Factor of unity means that the prior odds and the posterior odds are
exactly the same, so the data provides no information. Values larger
than unity show that the data will cause an upward revision in the

probability of H_1 so that the alternative hypothesis is favored by the data. Similarly small values indicate that the data favor the null.

The Bayesian solution does not seem appropriate for the case of scientific hypothesis testing discussed by Neyman and Pearson for several reasons. For example, suppose the posterior odds ratio is 4, so that H_1 is four times as likely as H_0. This still gives 20 % posterior probability to H_0 and this is too large for us to say that the data permit us to rule out H_0 from consideration. The Neyman-Pearson problem is quite different from the problem of choosing the hypothesis more in conformity with the data, or the more likely hypothesis. Also, in typical situations, H_0 is a well-defined scientific theory, while H_1 is a vague alternative. For example, we may not know what type of theory will emerge if we decide to reject the Ptolemaic theory of astronomy. In such situations it seems implausible to have prior probabilities associated with the hypothesis.

B.3.3   Decision Theoretic Approaches

The last two approaches to the choice of constant come from decision theory, where we need to be able to assess the loss function. Suppose for example that the loss from a type I error is l_1 and the loss from a type II error is l_2. The loss is zero when no error occurs. There are two cases depending on whether we assume known prior probabilities or not. If the prior probabilities π_0 and π_1 are known then the expected loss from accepting the null is $l_2 \times \mathbb{P}(H_1|Y)$, while the expected loss from rejecting the null is $l_1 \times \mathbb{P}(H_0|Y)$. From this it follows that we should reject the null if $NP(Y) > l_1\pi_0/l_2\pi_1$. In other words, the constant should be set equal to $c = l_1\pi_0/l_2\pi_1$.

This is undoubtedly the best approach when the prior probabilities and losses can be determined reliably. The last method for choosing the constant is best for cases where the prior probabilities are not known. In this case, the minimax approach suggests that we should assess a hypothesis test according the higher of the following two values: (i) the expected loss given H_0 and (ii) the expected loss given H_1. Minimizing the maximum loss usually leads to a test with equal expected loss regardless of whether H_1 or H_0 is true. For example in the case where $l_1 = l_2$ so that the hypotheses are symmetric, the minimax test will

have equal probabilities of type I and type II error. Also any other test
will have higher error probabilities for either type I or type II error.

B.4   One-Sided Test of a Normal Mean

We consider a specific example of a simple null to be tested against a
compound alternative. Suppose $X \sim f(x, \theta)$ and we test $H_0 \colon \theta = 0$
versus the alternative $\theta > 0$. One possibility is to fix a particular
value θ_1 and use a Neyman-Pearson most powerful test for $\theta = 0$ versus
$\theta = \theta_1$. Typically this Neyman-Pearson test depends on the value of θ_1,
being different for different values. In this section we will study a case of
great theoretical importance where a single test can achieve maximum
power against all alternatives. This happens when the Neyman-Pearson
test is exactly the same for each alternative θ_1. When such a test
exists, it is called a uniformly most powerful test. This section gives an
example of such a case. In the next section we will study this issue in
greater generality.

Suppose we observe a sample $X_1, \ldots, X_n \overset{i.i.d.}{\sim} N(\theta, 1)$, and the null
hypothesis is $H_0 \colon \theta = 0$, while the alternative is $\theta = \theta_1 > 0$. The
Neyman-Pearson test will be based on the likelihood ratio

$$
\begin{aligned}
NP(X_1, \ldots, X_n) &= \frac{(2\pi)^{-n} \exp\left(-\frac{1}{2}\sum_{i=1}^{n}(x_i - \theta_1)^2\right)}{(2\pi)^{-n} \exp\left(-\frac{1}{2}\sum_{i=1}^{n} x_i^2\right)} \\
&= \exp\left(-\frac{1}{2}\sum_{i=1}^{n}(x_i - \theta_1)^2 - x_i^2\right) \\
&= \exp\left(\sum_{i=1}^{n} \theta_1 x_i - n\theta_1^2/2\right).
\end{aligned}
$$

Whenever $NP(y)$ is a monotonic function of $Z(y)$ for any statistic
Z, we say that the *Neyman-Pearson test can be based on* Z. Let ϕ
be a monotonic increasing function such that $NP(y) = \phi(Z)$. Then
the event that $NP(Y) > c_\alpha$ is equivalent to the event that $\phi(Z) > c_\alpha$,
which in turn is equivalent to $Z > \phi^{-1}(c_\alpha) \equiv c'_\alpha$. While it is typically
the case that the NP statistic itself has a complicated distribution, one
can usually find monotonic transforms with simpler distribution. This

facilitates the calculation of the constant. In the present case, define $Z = (1/\sqrt{n}) \sum_{i=1}^{n} X_i$. Then it is clear from the calculation above that $NP(X_1, \ldots, X_n) = \exp(\sqrt{n}\theta_1 Z - n\theta_1^2/2)$ so that NP is a monotone increasing function of Z (as long as $\theta_1 > 0$). Thus the event that $NP > c_\alpha$ is equivalent to $Z > c'_\alpha$ for some c'. But Z has a standard normal $(N(0,1))$ density under the null hypothesis. Given any α, we can find c'_α satisfying $\mathbb{P}(Z > c'_\alpha) = \alpha$ from tables of the standard normal density. Rejecting the null hypothesis for $Z > c'_\alpha$ gives the most powerful level α test for this problem. Note that, unlike the case of the previous section, the most powerful test does not depend on the value of θ_1. This means that the same test is simultaneously most powerful for all values of $\theta_1 > 0$.

B.5 Monotone Likelihood Ratio and UMP Tests

The previous section gives an example where the most powerful Neyman-Pearson level α test is the same for all alternatives $\theta_1 > 0$. This is the only situation where there exists a test which is simultaneously most powerful for all alternatives at the same time; in this case, we say that the test is uniformly most powerful. It is important to be able to describe situations where this property holds. This is the goal of the present section.

The fundamental Neyman-Pearson theorem furnishes both a necessary and sufficient condition for the existence of uniformly most powerful tests and a useful characterization. If a test is to be simultaneously most powerful at all the points in the alternative hypothesis, it must coincide with the Neyman-Pearson test at all points. Equivalently, the Neyman-Pearson level α test must be independent of the choice of the alternative θ_1. It is easily seen that this is the case when the family has *monotone likelihood ratio* defined as follows:

Definition B.1 *A family of densities $f^X(x, \theta)$ has monotone likelihood ratio at θ_0 if there exists a statistic $T(X)$ such that for all $\theta_1 > \theta_0$, there is a monotone increasing function ϕ satisfying*

$$\phi(T(x)) = f^X(x, \theta_1)/f^X(x, \theta_0) = LR(x, \theta_1, \theta_0).$$

This criterion will be very helpful in recognizing densities for which we can find uniformly most powerful tests.

Theorem B.2 *Suppose the family of densities of $X \sim f^X(x, \theta)$ has monotone likelihood ratio at $\theta = \theta_0$. Then the Neyman-Pearson tests coincide for all $\theta_1 > \theta_0$, and any one of them is uniformly most powerful for H_1: $\theta > \theta_0$.*

The proof of this result follows almost immediately from the Neyman-Pearson lemma. The Neyman-Pearson test characterized there is clearly independent of the value of θ_1 in the case of monotone likelihood ratio. It follows that it is simultaneously most powerful for all values of θ_1. We present some examples which will also be useful later.

An example of this type was given in the previous section, where the Neyman-Pearson test for a normal mean with H_0: $\theta = 0$ versus the alternative $\theta = \theta_1 > 0$ turns out to be the same regardless of the value of θ_1. It is important to note that if H_1 is changed to the two-sided alternative H_1': $\theta \neq 0$, then no UMP test exists. The function $LR(X)$ is not monotone increasing if θ^* is negative. UMP tests do exist if the alternative is H_1'': $\theta < 0$, and these tests reject H_0 for $t(X) < C''$ (with C'' chosen appropriately). Since the UMP tests differ for $\theta^* < 0$ and for $\theta^* > 0$, no test can be simultaneously most powerful for alternatives on both sides of zero (because if it is most powerful for one side, it is necessarily different from the most powerful test for the other). This shows that no UMP test can exist.

As another example, suppose we observe s^2 satisfying $s^2/\sigma^2 \sim \chi_k^2$ and wish to test H_0: $\sigma^2 = 1$ versus H_1: $\sigma^2 > 1$. Fixing $\sigma^2 = \sigma_1^2 > 0$ under H_1, we can form the Neyman-Pearson test statistic

$$NP(s^2) = \frac{\sigma_1^{-k} s^{k-2} \exp(-s^2/\sigma_1^2)}{s^{k-2} \exp(-s^2)} = \sigma_1^{-k} \exp((1 - 1/\sigma_1^2)s^2).$$

Regardless of the values of σ_1 this is monotone increasing in s^2. It follows that a test can be based on s^2 and such a test will be uniformly most powerful for all alternatives $\sigma^2 > 0$. Note that if alternatives to the null on the other side, i.e., $\sigma^2 < 1$, are also allowed, then this argument would no longer be valid, since $NP(s^2)$ is monotone decreasing in s^2 for such values.

A final example will be very important to us in subsequent work. Suppose we observe $Z = X^2$, where $X \sim N(\theta, 1)$ and wish to test $H_0 \colon \theta = 0$ versus $\theta > 0$. We must determine whether the family of densities of Z has monotone likelihood ratio or not, to find out whether or not there exists a uniformly most powerful test. The following lemma will be needed later.

Lemma B.1 *Let J be Poisson with parameter $\theta^2/2$, and suppose Y conditional on J has a chi-square density with $2J+1$ degrees of freedom. Then the marginal density of Y is the same as that of Z (where $Z = X^2$) and $X \sim N(0, 1)$.*

Proof: To prove equivalence, it is convenient to use the moment generating function. If $Y|J \sim \chi^2_{2J+1}$ then $\mathbf{E}^{Y|J} \exp(\psi Y) = (1 + 2\psi)^{-(J+1/2)}$ so the moment generating function of Y is

$$
\begin{aligned}
\mathbf{E}\exp(Y) &= \mathbf{E}^J \mathbf{E} Y|J \exp(\psi Y) \\
&= \sum_{j=0}^{\infty} \left(\frac{(\psi^2/j)^j}{j!} \exp(-\psi^2/2) \right) (1 + 2\psi)^{-(j+(1/2))}.
\end{aligned}
$$

The moment generating function of $Z = X^2$ is also easily calculated. Defining $\lambda = (1 - 2\psi)^{-1}$, we have

$$
\begin{aligned}
\mathbf{E}\exp(\psi X^2) &= \int \frac{1}{\sqrt{2\pi}} \exp\left(\psi X^2 - \frac{1}{2}(X - \theta)^2 \right) \\
&= \int \frac{\lambda^{1/2}}{\sqrt{2\pi\lambda}} \exp\left(-\frac{1}{2\lambda}(x^2 - 2\theta\lambda x + \lambda\theta^2) \right) \\
&= \sqrt{\lambda}\exp(\lambda\theta^2 - \theta^2/2) \int \frac{1}{\sqrt{2\pi\lambda}} \exp\left(-\frac{1}{2}(x - \lambda\theta)^2 \right) \\
&= \exp(-\theta^2/2) \sum_{j=0}^{\infty} \frac{(\theta^2/2)^j}{j!} \lambda^{j+1/2}.
\end{aligned}
$$

After substituting for λ this matches the moment generating function of Y derived above.

It is now easily verified that the family of densities has monotone likelihood ratio. Using the lemma, we can write out the density of Z as a (Poisson) weighted sum of central chi-square densities:

$$f^Z(z, \theta) = \sum_{j=0}^{\infty} f^Y(y|J = j) \mathbb{P}(J = j)$$

$$= \sum_{j=0}^{\infty} \left(\frac{(\theta^2/j)^j}{j!} \exp(-\theta^2/2) \right) \frac{2^{-(j+1/2)}}{\Gamma(j+1/2)} y^{j-1/2} e^{-y/2}. \quad \text{(B.1)}$$

For the case $\theta = 0$, density reduces to a χ_1^2, or just the first term of the series above. Thus the likelihood ratio is

$$LR(Z, \theta, 0) = \frac{f^Z(Z, \theta)}{f^Z(z, 0)}$$

$$= \sum_{j=0}^{\infty} \left(\frac{(\theta^2/j)^j}{j!} \exp(-\theta^2/2) \right) \frac{2^{-j}\Gamma(1/2)}{\Gamma(j+1/2)} Z^j.$$

This is obviously monotone increasing in Z. Hence a UMP test exists and rejects H_0: $\theta = 0$ for large values of Z. The constant is chosen according to the null distribution of Z which is just χ_1^2. We note for future reference that the alternative distribution of Z is noncentral chi-square with one degree of freedom and noncentrality parameter $\theta^2/2$; this is denoted $\chi_1^2(\theta^2/2)$.

B.6   UMP tests for Exponential Families

It is a general phenomenon that UMP tests fail to exist when the alternative is two sided. It is also true that UMP tests fail to exist when more than one parameter is involved. Thus UMP tests can only exist for one-sided hypotheses regarding one parameter; even here, they may not exist. This places severe limitations on the use of this type of test. Chapter 6 discusses what should be done in situations where a UMP test fails to exist. Below we discuss the relationship between UMP tests and exponential families of densities.

A family of densities $f(x, \theta)$ indexed by the parameter θ is called an exponential family if it can be written in the following form:

$$f(x, \theta) = A(x)B(\theta)\exp\left(C(\theta)T(x)\right).$$

The functions B and C must be monotonic and one-to-one. Suppose that $X_1, X_2, \ldots, X_n$ is an i.i.d. sample from an exponential family. It is easily verified that such a family has MLR and hence we can find UMP tests for one-sided hypotheses $H_0 \colon \theta = \theta_0$ versus $H_1 \colon \theta > \theta_0$. Fixing $\theta_1 > \theta_0$ the likelihood ratio can be written as

$$
\begin{aligned}
NP(X_1, \ldots, X_n) &= \frac{\prod_{i=1}^{n} A(X_i)B(\theta_1)\exp\left(C(\theta_1)T(X_i)\right)}{\prod_{i=1}^{n} A(X_i)B(\theta_0)\exp\left(C(\theta_0)T(X_i)\right)} \\
&= \frac{B(\theta_1)^n}{B(\theta_0)^n}\exp\left(\{C(\theta_1) - C(\theta_0)\}\sum_{i=1}^{n} T(X_i)\right).
\end{aligned}
$$

From this we see that the Neyman-Pearson likelihood ratio is monotone in the statistic $S(X_1, \ldots, X_n) = \sum_{i=1}^{n} T(X_i)$. If $C(\theta)$ is monotone increasing, then the test for $H_0 \colon \theta = \theta_0$ versus $H_1 \colon \theta > \theta_0$ can be based on $S(X)$ for all $\theta_1 > \theta_0$. It follows that the test rejecting H_0 for large values of $S(X)$ is uniformly most powerful.

This situation is of great interest since a result due to Pfanzagl (1968) shows that this is essentially the only one in which uniformly most powerful tests exist.

Theorem B.3 (Pfanzagl) *If for all levels there exists a uniformly most powerful test for $H_0 \colon \theta = \theta_0$ versus $H_1 \colon \theta > \theta_0$ then the family of densities $f(x, \theta)$ must be an exponential family.*

We illustrate this result by several examples. Note that the normal density $N(\theta, 1)$ can be written as

$$f(x, \theta) = A(x)B(\theta)\exp(\theta x),$$

where $A(x) = exp(-x^2/2)/(\sqrt{2\pi}$ and $B(\theta) = exp(-\theta^2/2)$. This shows that $N(\theta, 1)$ is an exponential family and hence a UMP test for $H_0 \colon \theta = \theta_0$ versus $H_1 \colon \theta > \theta_0$ exists and can be based on the statistic $S(x_1, \ldots, x_n) = \sum_{i=1}^{n} x_i$.

Similarly, consider the Gamma density $\Gamma(p, \alpha)$, which can be written as

$$f(x, \alpha) = \Gamma(p)^{-1} \alpha^p x^{p-1} exp(-\alpha x).$$

It is clear that this is also an exponential family and test for H_0: $\alpha = \alpha_0$ versus H_1: $\alpha > \alpha_0$ can be based on $S(x_1, \ldots, x_n) = \sum_{i=1}^{n} x_i$. Since the chi-square density is a special case of the Gamma, it follows that the test for the scale parameter in chi-square density (derived explicitly earlier) is uniformly most powerful.

As a final example, consider the Binomial density $X \sim Bi(n, \theta)$, which can be written as

$$f(x, \theta) = \left(\begin{array}{c} n \\ x \end{array} \right) \theta^x (1 - \theta)^{(n-x)}.$$

Define $A(x) = \left(\begin{array}{c} n \\ x \end{array} \right)$ and $B(\theta) = (1 - \theta)^n$. Note that $\theta^x(1-\theta)^{-x}$ can be written as $exp\{x \log(\theta) - x \log(1 - \theta)\}$. Thus with $C(\theta) = \log(\theta) - \log(1 - \theta)$, we have $f(x, \theta) = A(x) B(\theta) \exp(C(\theta)x)$, which shows that this is an exponential family. It can be verified that $C(\theta)$ is one-to-one and monotone increasing in θ so that a UMP test for $\theta = \theta_0$ versus $\theta > \theta_0$ exists and can be based on $S(x_1, \ldots, x_n) = \sum_{i=1}^{n} x_i$.

B.7   Exercises

1. **Proof of Neyman-Pearson** Let R be the rejection region of some alternative δ' to the Neyman-Pearson test. The alternative must be of equal or smaller size, so that

$$\mathbb{P}(y \in R | H_0) \leq \mathbb{P}(NP(y) > c_\alpha | H_0). \qquad (B.2)$$

Let A be the set of all points for which δ' rejects but $NP(y) \leq c_\alpha$ so that δ^* does not reject. Let B be the set of points not in R for which $NP(y) > c_\alpha$; on this set δ^* rejects but δ' does not.

(a) Prove that $\mathbb{P}(y \in A | H_0) \leq \mathbb{P}(y \in B | H_0)$ — this is merely a restatement of Eq. (B.2) above.

(b) Prove that (*) $\mathbb{P}(y \in A|H_1) < \mathbb{P}(y \in B|H_1)$. Show that this implies that the power of δ' is less than the power of δ^*. Hint: to prove (*) note that over the set B $f(y, \theta_1) > c_\alpha f(y, \theta_0)$, while over A the reverse inequality must hold (why?). Replace the probabilities by the integrals $\int_A f_i(y) dy$, use the inequalities to substitute in the integral and get (*) by applying Eq. (B.2). To show (*) implies that the Neyman-Pearson test has greater power, note that the sets A and B are the only values of Y for which the two tests behave differently. Above inequalities show that A and B have equal size but B has more power.

(c) Show that the function $p(c) = \mathbb{P}(NP(y) > c|H_0)$ is continuous and monotone decreasing from the value of 1 at $c = 0$ to 0 at $c = +\infty$. Apply the intermediate value theorem to conclude that we can find a c_α satisfying the condition of the theorem.

2. You are on the jury of a robbery case. The judge instructs you that the prosecutor must establish guilt beyond reasonable doubt. 'Beyond reasonable doubt' is interpreted to be 99 % probability. The prosecution presents two witnesses who testify that they saw the accused youth commit the robbery. The defense presents a large catalog of similar cases and establishes that witnesses make mistakes in identification about 10 % of the time.

(a) What is the likelihood ratio test? Hint: let $W_i = 1$ or 0 depending on the testimony of the i-th witness. Let H_0 be the null hypothesis of innocence, then $\mathbb{P}(W_i = 1|H_0) = 0.1$ while $\mathbb{P}(W_i = 1|H_1) = 0.9$. Note that this is an oversimplified model; we are assuming that the two witnesses were the only ones present and they could have testified either for or against.

(b) Which of the four theories of determining constant seems appropriate here? What action do you take?

(c) The prosecution establishes further that in the neighborhood where the young man lives, among people of his age group

and background, the incidence of robbery is about 25 %.
Using this as the prior probability for H_0 calculate the posterior probability of H_0. Suppose the prior information is
valid; is it fair to use this information to convict?

(d) Suppose the youth is from an upper class family and the defense establishes that the incidence of robbery among people
of his class is only one in a thousand. Based on the prior
suppose the posterior probability of guilt becomes very low.
Would it be correct to use the Bayesian argument to prevent conviction of someone who would on other evidence be
convicted?

(e) Instead of the 'reasonable doubt' criterion, suppose the judge
says that it is better to let 100 guilty people escape rather
than convict one innocent man. This implicitly defines a loss
function. Find the minimax test with respect to this loss.

(f) Suppose the crime in question is murder rather than robbery
and lives are at stake. Discuss whether the two possible
errors, convicting and hanging an innocent man, or letting a
murderer go free, are commensurable. Is it possible to assign
losses to the outcomes? Does it seem reasonable to use prior
probabilities, even when they can be found? Which of the
approaches to determining constant seems reasonable here?

3. The quality control unit at a paint factory tests each case of
paint, consisting of 10 cans. If any of the cans is defective, the
case is considered defective. Erroneously approving a defective
case results in a loss of $100. Thoroughly testing a single can
costs $10; the test determines with complete accuracy whether or
not the can is defective.

(a) How many cans in each case should be tested? What will
the average cost of quality control be?

(b) The answer to the previous question, in which no prior information is assumed, must be based on the minimax approach: equalize the losses from testing and the losses from
not detecting a defective. Suppose now that previous testing results from thousands of cases show that on the average

a percentage d of the cans is defective. What is the dollar value of this prior information for (a) $d=.01$, (b) $d=0.1$ (c) $d=0.25$?

4. We say that X is Cauchy if X has density

$$f^X(x) = \frac{1}{\pi \left(1 + (X - \theta)^2\right)}. \tag{B.3}$$

(a) Find the Neyman-Pearson test of $H_0 : \theta = 0$ versus $H_1 : \theta = 1$ at levels 10%, 5%, and 1%.

(b) Answer the previous question when H_1 is changed to H_1': $\theta = 2$.

(c) Find the minimax test of H_0 versus H_1.

(d) Let $R(a)$ be the set of values of X such that the level 10% Neyman-Pearson test of H_0: $\theta = 0$ versus H_1: $\theta = a > 0$ rejects H_0 if $X \in R(a)$. Find the *locally most powerful* test at level 10% by finding the limit of $R(a)$ as a goes to zero.

(e) Calculate the power of the locally most powerful test for H_1: $\theta = \theta_1 > 0$ in the limit as θ_1 goes to infinity.

5. Suppose we wish to test the null hypothesis that $X \sim \mathcal{U}(-1.5, 1.5)$ versus H_1: $X \sim N(0, \sigma^2)$. For which value of σ^2 is this most difficult (as gauged by the minimax error probability)? This density could be considered to be the closest normal density to the uniform one given. To answer the question proceed as follows:

(a) Check that the Neyman-Pearson tests reject H_0 for $|X| < c$ for some value of the constant c.

(b) Show that the probability of type I error as a function of c is $t_1(c) = c/1.5$. Show that the probability of type II error is $t_2(c) = 2(1 - \Phi(c/\sigma))$, where $\Phi(c) = \mathbb{P}(X < c)$ for $X \sim N(0, 1)$.

(c) Fix σ and use numerical methods to find the value of c for which type I and type II error probabilities are the same. Record this value, which is the error probability of the minimax test.

(d) Plot the minimax error probability versus σ. The value of σ for which this is the highest is the desired value.

6. Among the distributions $N(0, \sigma^2)$ find the one closest to the Cauchy distribution given in Eq. B.3 above. Use the method outlined in the preceding problem.

Appendix C

A Review of Decision Theory

C.1   Risk Functions of Decision Rules

Formally, an estimation problem is described in decision theoretic terminology as follows. The set Θ is a *parameter space*, consisting of the set of all possible values for all the unknown parameters in a given statistical model. We observe a sample X taking values in *the sample space* $\mathcal{X}$ according to the density $f(x, \theta)$. An *estimator* of θ is any function $\delta : \mathcal{X} \to \Theta$ (also written $\hat{\theta}(X)$ sometimes). Choosing an estimator $\delta(X)$ means that we decide to estimate θ by $\delta(x)$ when we observe $X = x$. Decision theory is concerned with the evaluation of estimators, also called decision rules. For a quantitative assessment of decision rules, we will need to introduce *a loss function*, $L(\theta, \hat{\theta})$; this measures the loss accruing to us from using an estimate $\hat{\theta}$ when the true value is θ. An important principle of decision theory is that decision rules are to be evaluated solely on the basis of their *expected loss* (also called *risk*). Note that since X is random, the loss $L(\theta, \delta(X))$ is also a random variable, and hence it is reasonable to talk about its expected value. Intuitively it appears plausible to be concerned about more than just the average loss; for example, a rule which produces a loss which is highly variable might appear unappealing. One can show that, under certain plausible hypotheses, we can always reformulate the loss func-

533

tion so that its expected value captures all our concerns. This is what justifies our evaluating decision rules solely on the basis of their risk, formally defined as

$$R(\theta, \delta) \equiv \mathbb{E}L(\theta, \delta(X)) = \int_{\mathcal{X}} L(\theta, \delta(x)) f(x, \theta) dx.$$

Two decision rules which have the same risk function are considered *equivalent* from the point of view of decision theory. Also, if decision rule $\delta_1(X)$ has smaller risk than $\delta_2(X)$ for all $\theta \in \Theta$, we say that δ_1 is *as good as* δ_2: formally if $R(\theta, \delta_1) \leq R(\theta, \delta_2)$ for all θ, then δ_1 is as good as δ_2. If both δ_1 is as good as δ_2 and δ_2 is as good as δ_1 then the two rules must be equivalent. On the other hand, if $R(\theta, \delta_1) \leq R(\theta, \delta_2)$ for all θ, and the inequality is strict for some θ, then we say that δ_1 is *better* than δ_2. In this case, we also say that δ_2 is *inadmissible* relative to δ_1, and that δ_1 *dominates* δ_2. The idea is that δ_1 always (i.e. for all possible parameter values) yields lower average loss than δ_2, so that δ_2 should not be used (is 'inadmissible'). If a given rule $\delta(X)$ is inadmissible relative to some $\delta'(X)$, we may simply say that δ is inadmissible. This terminology is useful as situations arise where we can show (by indirect arguments) that a given rule is inadmissible without being able to produce an alternative rule that is better. Finally, we say that a rule is *admissible* if it is not inadmissible.

C.2 Properties of Bayes Rules

We begin by reviewing basic facts about Bayes rules. Suppose a random variable X has a density of the type $f^X(x, \theta)$, where θ belongs to the set Θ (the parameter space). Let $\pi(\theta)$ be a probability density on the parameter space. For Bayesian calculations it is convenient to regard the density $f^X(x, \theta)$ as the conditional density of X given θ, and $\pi(\theta)$ as the marginal density of θ. The *posterior density of θ given $X = x$* is defined to be the conditional density of θ given the observation $X = x$. Formally,

$$f^{\theta|X=x}(\theta|X = x) = \frac{f^{(X,\theta)}(x, \theta)}{f^X(x)}.$$

The joint density of X and θ in the numerator is obtained by multiplying the conditional $f^{X|\theta}(x|\theta)$ by the marginal density of θ, i.e., $\pi(\theta)$. The marginal density of X in the denominator is obtained by integrating θ out of the joint density. The Bayes rule for the prior density $\pi(\theta)$ is defined to be a rule $\delta_\pi(x)$ which minimizes 'posterior expected loss'. If we fix a particular value of the random variable X, at $X = x$ for example, this yields a posterior density of θ. Now $\delta_\pi(x)$ must minimize over all δ the quantity $\mathbb{E}^{(\theta|X=x)}(L(\theta,\delta)$ (which is the posterior expected loss given $X = x$). More explicitly,

$$\int L(\theta, \delta_\pi(x)) f^{\theta|X=x}(\theta|X = x) \, d\theta \leq \int L(\theta, \delta) f^{\theta|X=x}(\theta|X = x) \, d\theta$$

for all x and δ. After fixing $X = x$, the Bayes rule for $\pi(\theta)$ is obtained by minimizing the integral on the RHS over δ; this is frequently possible using calculus. For example, if we have squared error loss, the first order conditions for a minimum over δ are

$$\frac{\partial}{\partial \delta} \int L(\theta, \delta) f(\theta|X = x) \, d\theta = 0.$$

Under certain regularity conditions (which hold for exponential families, as discussed earlier), one can interchange the order of integration and differentiation and write the first order condition as $0 = \mathbb{E}^{(\theta|X=x)} \frac{\partial}{\partial \delta}(\theta - \delta)^2 = -2\mathbb{E}^{(\theta|X=x)}(\theta - \delta)$. Rearranging, we get

$$\delta = \mathbb{E}^{(\theta|X=x)}\theta = \int \theta f^{(\theta|X=x)}(\theta|X = x) \, d\theta.$$

This proves that with quadratic loss the Bayes estimator is just the mean of the posterior distribution.

We will now prove an important property of Bayes rules which is closely related to their admissibility properties. Given a prior $\pi(\theta)$, let $\delta_\pi(x)$ be the Bayes rule for π and let $\delta'(x)$ be any other decision rule. Then we claim that

$$\int R(\theta, \delta_\pi)\pi(\theta) \, d\theta \leq \int R(\theta, \delta')\pi(\theta) \, d\theta. \tag{C.1}$$

The integral of $R(\theta, \delta)$ with respect to the prior density π written above is called the Bayes risk of the decision rule δ with respect to the prior

$\pi(\cdot)$ and denoted $B(\pi, \delta)$. Thus the above equation shows that among all decision rules, the Bayes rule for π minimizes the Bayes risk. The equation is easily proven by writing out the risk $R(\theta, \delta)$ as an integral with respect to x, and then changing the order of integration. This is done as follows:

$$
\begin{aligned}
B(\pi, \delta) &\equiv \int R(\theta, \delta)\pi(\theta)\, d\theta \\
&= \int \left[\int L(\theta, \delta(x)) f^{(X|\theta)}(x|\theta)\, dx \right] \pi(\theta)\, d\theta \\
&= \int \int L(\theta, \delta(x)) f^{(X,\theta)}(x, \theta)\, dx\, d\theta \\
&= \int \left[\int L(\theta, \delta(x)) f^{(\theta|X=x)}(\theta|X=x)\, d\theta \right] f^X(x)\, dx.
\end{aligned}
$$

The expression inside the square brackets is minimized for each value of x by $\delta_\pi(x)$, so that $B(\pi, \delta)$ must also be minimized by setting $\delta = \delta_\pi$.

C.3 Admissibility of Bayes Rules

The fact that Bayes rules minimize Bayes risk strongly suggests (but does not prove) that they are admissible. If δ_π is a Bayes rule for the prior π and δ' is a rule which dominates δ_π, the inequality $R(\theta, \delta') \leq R(\theta, \delta_\pi)$ immediately implies that $B(\pi, \delta') \leq B(\pi, \delta_\pi)$. We have just proven that the reverse inequality must also hold. Thus any rule which dominated δ_π must have equal Bayes risk to δ_π and hence also be Bayes for π. If Bayes rules are unique (as is the case in most problems) this is a contradiction. Formally we can state this admissibility property as follows:

Theorem C.1 *Suppose δ_π is a unique Bayes rule for the prior $\pi(d\theta)$. That is, if δ' is also Bayes for π then δ' is equivalent to δ_π in the sense that $R(\theta, \delta_\pi) = R(\theta, \delta')$ for all θ. Then δ_π is an admissible decision rule*

Intuitively speaking, Bayes rules can fail to be admissible only when the prior density puts zero mass on certain parts of the parameter space. In such situations, for observations which come from that part of the parameter space, we can do anything we like without affecting the Bayes risk — we don't have to optimize in this case because the Bayesian believes those observations to be impossible. This cause non-uniqueness of the Bayes rule (since you can do anything for certain observations and still have minimal Bayes risk) and also causes lack of admissibility. This is because if one of the parameter values considered impossible happens to be the true value, the Bayesian rule has not optimized for this possibility and may do something which could be dominated.

C.4 Complete Classes of Rules

Any set of decision rules which includes all admissible rules is called a *complete class* of decision rules: if any rule is outside the complete class, we can find a better rule inside it. Whenever the parameter set is finite, it is easily shown that the set of all Bayes rules forms a complete class. This no longer remains true with infinite parameter spaces that usually arise in estimation problems. In such cases it is possible to show that a complete class of estimators is obtained by joining to the proper Bayes rules certain limits of Bayes rules called 'extended Bayes rules.' This concept is illustrated by an example in the next section.

C.5 Extended Bayes Rules

Given an arbitrary decision rule δ and an arbitrary prior π, a measure of how close δ is to being a Bayes rule for π is the following number:

$$EB(\pi, \delta) = B(\pi, \delta) - B(\pi, \delta_\pi).$$

This is called the *excess Bayes risk* of δ for π. δ_π gives us the smallest possible Bayes risk for the prior π. If we use δ instead of δ_π the Bayes risk will be larger; $EB(\pi, \delta)$ measures just how much larger.

Definition C.1 *A decision rule δ is called an* extended *Bayes rule if for any $\epsilon > 0$, there exists a prior π such that δ has excess Bayes risk less than ϵ relative to π:*

$$EB(\pi, \delta) = B(\pi, \delta) - B(\pi, \delta_\pi) < \epsilon.$$

Thus, an extended Bayes rule comes arbitrarily close to being a Bayes rule. One of the fundamental results in decision theory due to Wald is that all admissible rules must be extended Bayes rules. In decision theoretic problems with infinite parameter spaces, the set of Bayes rules is a proper subset of the set of all admissible rules which in turn is a proper subset of the set of all extended Bayes rules. Thus while every Bayes rule is admissible, and every admissible rule is extended Bayes, there exist extended Bayes rules which are not admissible, and admissible rules which are not Bayes. These results, which we will not prove, show the close connection between admissibility and Bayes rules.

We will now illustrate these concepts by an example. Suppose we observe a random variable[1] X distributed as a univariate normal with mean θ and variance σ^2. We will assume that σ^2 is known, and set $\sigma^2 = 1$ for convenience. For mathematical simplicity assume squared error loss: $L(\theta, \delta) = (\theta - \delta)^2$. Let $\pi_\alpha(\theta)$ be a normal (prior) density on θ with mean zero and variance α^2. Then it is easy to show that the posterior density of θ given $X = x$ is normal with mean $\mathbb{E}(\theta|X = x) = (\alpha^2/(1 + \alpha^2))x$ and variance $\mathbb{V}ar(\theta|X = x) = \alpha^2/(1 + \alpha^2)$. It follows that the Bayes estimator of θ for the prior π_α is simply $\delta_{\pi_\alpha}(x) = (\alpha^2/(1 + \alpha^2))x$.

Note that the Bayes rule is a linear estimator of the form $\delta_a(x) = ax$ with $0 < a < 1$. This proves (more or less) the admissibility of such rules. The maximum likelihood estimator is $\delta_1(x) = x$, and is *not* a Bayes rule for this or any other set of prior densities. It is however extended Bayes and also admissible. To prove that it is extended Bayes, we calculate the excess Bayes risk of δ_1 with respect to the prior densities $\pi_\alpha(\theta) = N(0, \alpha^2)$. Note that $R(\theta, \delta_1) = 1$ for all θ so that $B(\pi, \delta_1) = \int R(\theta, \delta_1)\pi(\theta)\,d\theta = \int \pi(\theta)\,d\theta = 1$. The straightforward route to calculating the Bayes risk of δ_a with $a = \alpha/(1 + \alpha^2)$ for the prior π_α is to integrate its risk function (calculated earlier) with respect to this prior density. The student is encouraged to use this route as an exercise. A shortcut is to interchange the order of integration with

---

[1]The more general case, where we observe $X_1, \ldots, X_n$ i.i.d. $N(\theta, \nu^2)$ can be reduced to the case of a single observation by defining the average $A = \sum_{i=1}^{n} X_i$. Since the average is a sufficient statistic, we can assume that we just observe $A \sim N(\theta, \nu^2/n)$, without loss of generality

respect to x and θ, as follows:

$$B(\pi_\alpha, \delta_a) = \mathbb{E}^\theta[\mathbb{E}^{X|\theta} L(\theta, \delta_a)] = \mathbb{E}^{(X,\theta)} L(\theta, \delta_a) = \mathbb{E}^X[\mathbb{E}^{\theta|X}(\theta - \delta_a)^2].$$

Here the first expression is simply the expected value of the risk function (written out in expanded form inside the square brackets) with respect to the marginal density of θ. This iterated expectation can also be expressed as the expected loss over the jointly random variables X and θ. The final expression writes the Bayes risk as the expected value with respect to the marginal density of X, of the posterior expected loss. Note that δ_a is just the mean of the posterior density of $\theta|X$ so that $\mathbb{E}^{\theta|X}(\theta - \delta_a(X))^2 = \mathbb{E}^{\theta|X}(\theta - \mathbb{E}^{\theta|X}\theta)^2 = \mathbb{V}\mathrm{ar}(\theta|X)$. The posterior expected loss is just the variance of the posterior density, which is just $\alpha^2/(1 + \alpha^2)$ as discussed earlier. This variance does not depend on X, and hence taking expectation with respect to X does not change anything. It follows the $B(\pi_\alpha, \delta_a) = \alpha^2/(1+\alpha^2)$. Thus the excess Bayes risk of δ_1 is just:

$$EB(\pi_\alpha, \delta_1) = B(\pi_\alpha, \delta_1) - B(\pi_\alpha, \delta_a) = 1 - \frac{\alpha^2}{1 + \alpha^2} = \frac{1}{1 + \alpha^2}.$$

This proves that as α goes to ∞, the excess Bayes risk of δ_1 goes to zero; δ_1 comes closer and closer to being the Bayes rule for π_α. Thus δ_1 is an extended Bayes rule.

In general the set of all Bayes rules is a proper subset of the set of all admissible rules, which in turn is a proper subset of the set of all extended Bayes rules. Thus, extended Bayes rules need not necessarily be admissible, although every admissible rule is necessarily extended Bayes. As a concrete illustration of these ideas, note the following facts. It is possible to show that $\delta_1(x)$ cannot be a Bayes rule. It is, nonetheless, extended Bayes and also admissible (this is shown in Chapter 4). The arguments given above generalize easily to the multidimensional case. If we observe $X \sim N(\theta, \mathbf{I}_k)$, the ML $\delta(X) = X$ is not Bayes, but continues to be extended Bayes. However, for $k > 3$, these extended Bayes rules are not admissible. This illustrates that extended Bayes rules may be inadmissible. For further details, see Chapter 4.

C.6   Exercises

1. Prove the inadmissibility of the rules $\delta_a(x)$ for $a \in (-1, 0)$.

2. Suppose $X \sim Bi(n, \theta)$. Consider the class of decision rules $\delta_{a,b}(X) = (X + a)/(n + b)$.

   (a) Calculate the risk of these rules with squared error loss.

   (b) Show that $a < 0$ or $b < 0$ leads to inadmissible rules by finding dominating rules.

   (c) Suppose θ has the Beta prior density $Be(\alpha, \beta)$. Calculate the posterior density.

   (d) Show that the rules $\delta_{a,b}$ are proper Bayes for $a, b > 0$. Prove that these are admissible rules.

   (e) Prove that the rule $\delta_{0,0}(X) = X/n$ is an extended Bayes rule. Is it also admissible ?

   (f) Find values of a, b such that $\delta_{a,b}$ has constant risk.

   (g) Calculate the maximum risk of each decision rule $\delta_{a,b}$. Find the rule with the smallest maximum risk.

3. Show that if $X \sim N(\theta, 1)$, then $\delta_{a,b}(X) = aX + b$ is an admissible estimator of θ for $a \in (0, 1)$ with squared error loss. Hint: find a prior for which this is a proper Bayes rule.

4. Suppose $X_1, X_2, \ldots, X_n \overset{i.i.d.}{\sim} \mathcal{G}(p, \theta)$.

   (a) Calculate the maximum likelihood estimate of θ.

   (b) Calculate the Bayes estimate relative to the prior density $\pi(\theta) \sim \mathcal{G}(k, \lambda)$.

   (c) Calculate the excess Bayes risk for the MLE relative to these priors. Is the MLE extended Bayes?

   (d) Is the MLE admissible?

This bibliography doubles as an author index. Numbers in italics at the end of each bibliography entry indicate pages on which the work is cited. The bibliography is alphabetical by first author's last name. Second and subsequent authors have been cross-referenced. Also, the following abbreviations have been used for journal names:

AS	*Annals of Statistics*
AMS	*Annals of Mathematical Statistics*
Ecta	*Econometrica*
JoE	*Journal of Econometrics*
JASA	*Journal of the American Statistical Association*

Bibliography

Abadir, K. M. (1993), 'On the asymptotic power of unit root tests', *Econometric Theory* **9**, 189–221. *42, 377*

Abadir, K. M. (1992), 'A distribution generating equation for unit root statistics', *Oxford Bulletin of Economics and Statistics* **54**, 305–323. *42,364, 377*

Abramowitz, M. & Stegun, I. A. (1965), *Handbook of Mathematical Functions*, Dover, New York. *387*

Aigner, D. J. & Judge, G. G. (1977), 'Applications of pre-test and Stein estimators to economic data', *Ecta*, **45**, 1279–1288. *86,452*

Akahira, M. & Takeuchi, K. (1981), *Asymptotic Efficiency of Statistical Estimators: Concepts and Higher Order Asymptotic Efficiency*, Springer-Verlag, New York. *397, 400, 408*

Amemiya, T. (1985), *Advanced Econometrics*, Harvard University Press, Cambridge, Massachusetts. *xviii,13, 269, 287*

Andrews, D. W. K. (1986), 'Complete consistency: A testing analogue of estimator consistency', *Review of Economic Studies* **53**, 263–269. *271*

Andrews, D. W. K. (1992), 'Generic uniform convergence', *Econometric Theory* **8**, 241–257. *302*

Andrews, D. W. K. (1993), 'Tests for parameter instability and structural change with unknown changepoint', *Ecta* **61**(4), 241–257. *242*

Andrews, D. W. K. (1994), 'Applications of functional central limit theorem in econometrics: A survey', *Handbook of Econometrics*, North Holland. *324*

Andrews, D. W. K. & McDermott, C. J. (1993), Nonlinear econometric models with deterministically trending variables, Technical Report 1053, Cowles Foundation, Yale University, Box 2125 Yale Station, New Haven, CT 06520. *308*

Ansley, C. F., Kohn, R. & Shively, T. S. (1992), 'Computing p-values for the generalized Durbin-Watson and other invariant test statistics', *JoE* **54**, 277–300. *364,514*

Ashley, R. (1992), A statistical inference engine for small dependent samples with applications to postsample model validation, model selection, and granger-causality analysis, Technical Report E92-24, Virginia Polytechnic Institute, Economics Dept., Blacksburg, VA 24061. *228*

Ashley, R. (1994), Postsample model validation and inference made feasible, Technical Report E94-15, Virginia Polytechnic Institute, Economics Dept., Blacksburg, VA 24061. *228*

Atkinson, A. C. (1986), 'Masking unmasked', *Biometrika* **73**, 533–541. *113*

Bahadur, R. R. (1966), 'An optimal property of the likelihood ratio statistic', *Proceedings of the Fifth Berkeley Symposium on Mathematical Statistics and Probability* **1**, 13–26. *409*

Banerjee, A., Dolado, J., Galbraith, J. W. & Hendry, D. F. (1992), *Co-Integration, Error Correction and The Econometric Analysis of Nonstationary Data*, Oxford University Press, Oxford. *223*

Barndorff-Nielsen, O. E. & Cox, D. R. (1989), *Asymptotic Techniques for Use in Statistics*, Chapman and Hall, London and New York. *359*

Başçı, S. (1995), Detecting structural change when the changepoint is unknown, Master's thesis, Economics Dept., Bilkent University, 06533 Ankara, Turkey. *243,245*

Başçı, S., Mukhophadhyay, C. & Zaman, A. (1995), Bayesian tests for changepoints, Technical Report 95-15, Economics Department, Bilkent University, 06533 Ankara, Turkey. *243,245*

Basawa, I. V., Mallik, A. K., McCormick, W. P. & Taylor, R. L. (1989), 'Bootstrapping explosive autoregressive processes', *AS* **17**, 1479–1486. *381*

Basawa, I. V., Mallik, A. K., McCormick, W. P. & Taylor, R. L. (1991), 'Bootstrap test of significance and sequential bootstrap estimation for unstable first order autoregressive processes', *Communications in Statistics: Theory and Methods* **20**, 1015–1026. *381*

Belsley, D. A., Kuh, E. & Welsch, R. E. (1980), *Regression Diagnostics: Identifying Influential Data and Sources of Collinearity*, Wiley, New York. *103*

Bera, A. K. *see* Jarque, C. M.

Berenblutt, I. & Webb, G. I. (1973), 'A new test for autocorrelated errors in the linear regression model', *Journal of the Royal Statistical Society, Series B* **34**(1), 33–50. *137*

Berger, J. O. (1982), 'Selecting a minimax estimator of the normal mean', *AS* **10**(1), 81–92. *72*

Berger, J. O. (1985), *Statistical Decision Theory and Bayesian Analysis*, second edition, Springer-Verlag, New York. *89,91,244,417,441*

Berndt, E. R. (1991), *The Practice of Econometrics: Classic and Contemporary*, Addison-Wesley, Reading, Massachussetts. *xviii,225,226*

Bhattacharya, R. N. & Rao, R. R. (1976), *Normal Approximation and Asymptotic Expansion*, Wiley, Springer-Verlag, New York. *354*

Blattberg, R. C. & George, E. I. (1991), 'Shrinkage estimation of price and promotional elasticities: Seemingly unrelated equations', *JASA* **86**, 304–315. *489,491*

Blough, S. & Zaman, A. (1995), '(Lack of) periodic structure in business cycles', submitted to *Journal of Business and Economic Statistics.*

Bondar, J. V. & Milnes, P. (1981), 'Amenability: A survey for statistical applications of Hunt-Stein and related conditions on groups', *Zeitschrift fur Wahrscheinlichtkeitstheorie und verwandte Gebiete* **57**, 103–128. *173*

Boos, D. D. & Brownie, C. (1989), 'Bootstrap methods for testing homogeneity of variances', *Technometrics* **31**(1), 69–82. *191,192*

Brooks, R. O. & King, M. L. (1994), Hypothesis testing of varying coefficient regression models: Procedures and applications, Economics Dept., Monash University, Clayton, Victoria 3168, Australia. *241*

Brown, L. D. (1966), 'On the admissibility of invariant estimators of one or more locations parameters', *The AMS* **37**, 1087–1136. *72*

Brown, L. D. (1973), 'Estimation with incompletely specified loss functions' *JASA* **70**, 417–427. *82*

Brownie, C. *see* Boos, D. D.

Brownstone, D. (1990), How to data-mine if you must: Bootstrapping pretest and Stein-rule estimators, Technical Report MBS 90-08, University of California at Irvine, Research Unit in Mathematical Behavioral Science. *229*

Carrington, W. & Zaman, A. (1994), 'Interindustry variation in the costs of job displacement', *Journal of Labor Economics* **12**(2), 243-275. *469,494*

Carter, G. & Rolph, J. (1974), 'Empirical Bayes methods applied to estimating fire alarm probabilities', *JASA* **69**, 880–885. *432*

Casella, G. & George, E. (1994), 'Explaining the Gibbs sampler', *American Statistician* **46**, 167. *483*

Cenk, T. (1995), The bootstrap: three applications to regression models Master's thesis, Economics Dept., Bilkent University, 06533 Ankara, Turkey. *201*

Chen, C. *see* Diebold, F. X.

Chesher, A. (1984), 'Testing for neglected heterogeneity', *Ecta* **52**(4), 865–872. *185*

Chib, S. & Greenberg, E. (1994), Understanding the Metropolis-Hastings algorithm, Technical Report Working Paper 188, Washington University, One Brookings Dr., St Louis, MO 63130. *483*

Chung, K. L. (1974), *A Course in Probability Theory*, Academic Press, New York and London. *270,271,282,350*

Coakley, C. W. & Hettmansperger, T. P. (1993), 'A bounded influence high breakdown efficient estimator', *JASA* **88**(43), 872. *116*

Cox, D. R. *see* Barndorff-Nielsen, O. E..

Cox, D. R. (1990), 'Role of models in statistical analysis', *Statistical Science* **5**(2), 169–174. *xx*

Cox, D. R. & Hinkley, D. V. (1974), *Theoretical Statistics*, Chapman and Hall, London and New York. *xx,31,246,247*

Cressie, N. (1989), 'Empirical Bayes estimation of undercount in the decennial census', *JASA* **84**(408), 1033–1044. *468*

Darnell, A. C. & Evans, J. L. (1990), *The Limits of Econometrics*, Edward Elgar Publishing, Ltd., Hants, England. *xvii,xxiii, 31,226*

Davies, R. B. (1969), 'Beta-optimal tests and an application to the summary evaluation of experiments', *Journal of the Royal Statistical Society, Series B* **31**, 524–538. *142*

Deb, P. & Sefton, M. (1994), The distribution of a Lagrange multiplier test of normality, Technical Report ES250, The University of Manchester, School of Economic Studies, Oxford Road, Manchester M13 9PL. *181*

Dempster, A. N., Laird, N. & Rubin, D. (1977), 'Maximum likelihood estimation from incomplete data via the EM algorithm', *Journal of the Royal Statistical Society, Series B* **39**, 1–38. *132, 457*

Diebold, F. X. (1989), 'Forecast combination and encompassing: Reconciling two divergent literatures', *International Journal of Forecasting* **5**, 589–592. *228*

Diebold, F. X. & Chen, C. (1995), 'Testing structural stability with endogenous break point: A size comparison of analytic and bootstrap procedures', *JoE* **57**. *243*

Dolado, J. *see* Banerjee, A

Domowitz, I. & White, H. (1982), 'Misspecified models with dependent observations', *JoE* **20**, 35–58 *296*

Draper, N. & Smith, H. (1981), *Applied Regression Analysis, Second Edition*, Wiley, New York. *226*

Dufour, J.-M. (1980), 'Dummy variables and predictive tests for structural change', *Economics Letters* **6**(3), 241–247. *238*

Dufour, J.-M. (1982), 'Generalized chow tests for structural change: A coordinate-free approach', *International Economic Review* **23**(3), 565–575. *236*

Durbin, J. & Watson, G. S. (1950), 'Testing for serial correlation in least squares regression I', *Biometrika* **37**, 409–428. *134*

Dutta, J. & Zaman, A. (1990), What do tests for heteroskedasticity detect?, Technical Report 9022, Center for Operations Research and Econometrics, Universite Catholique de Louvain, Belgium. *xxvi,185, 192,276*

Dyer, D. D. & Keating, J. P. (1980), 'On the determination of critical values for Bartlett's test', *JASA* **75**(370), 313–319. *192*

Efron, B. (1975), 'Defining the curvature of a statistical problem', *AS* **3**(6), 1189–1242. *xix,140, 345, 412*

Efron, B. (1982), *The Jackknife, the Bootstrap, and Other Resampling Plans*, SIAM, Philadelphia, Pennsylvania. *318*

Efron, B. (1983), 'Estimating the error rate of a prediction rule: Improvement on cross-validation', *JASA* **78**(382), 316–331. *221*

Efron, B. & Morris, C. (1971), 'Limiting the risk of Bayes and empirical Bayes estimators — Part I: The Bayes case', *JASA* **66**, 807–815. *91*

Efron, B. & Morris, C. (1972), 'Limiting the risk of Bayes and empirical Bayes estimators — Part II: The empirical Bayes case', *JASA* **67**, 130–139. *91, 451*

Efron, B. & Morris, C. (1973), 'Stein's estimation rule and it's competitors — An empirical Bayes approach', *JASA* **68**(341), 117–130. *418,445, 451*

Efron, B. & Morris, C. (1975), 'Data analysis using Stein's estimator and its generalizations', *JASA* **70**, 311–319.

Efron, B. & Morris, C. (1977), 'Stein's paradox in statistics', *Scientific American* **246**(5), 119–27. *417*

Eicker, F. (1966*a*), 'Limit theorems for regression with unequal and dependent errors', *Proceedings of the Fifth Berkeley Symposium on Mathematical Statistics and Probability* **1**, 59–82. *276, 278*

Eicker, F. (1966*b*), 'A multivariate central limit theorem for random linear vector forms', *AMS* **37**(6), 1825–1828. *331*

Engle,R. F., Hendry, D F. & Richard, J.-F. (1983) 'Exogeneity', *Ecta* **51**, 77–304. *31*

Erickson, E. P. & Kadane, J. B. (1985), 'Adjusting the 1980 Census of population and housing', *JASA* **84**, 927–944. *463*

Evans, J. L. *see* Darnell, A. C.

Evans, M. A. & King, M. L. (1985), 'A point optimal test for heteroscedastic disturbances', *JoE* **27**, 163–178. *139*

Fay, R. E. & Herriot, R. A. (1979), 'Estimates of income for small places: An application of James-Stein procedures to Census data', *JASA* **74**(366), 269–277. *458*

Fraser, D. A. S., Guttman,I. & Styan, G. P. H. (1976), 'Serial correlation and distributions on a sphere', *Communications in Statistics A* **5**, 97–118. *139*

Freedman, D. A. & Navidi, W. C. (1986), 'Regression models for adjusting the 1980 Census', *Statistical Science* **1**, 3–39. *462*

Friedman, J. H. (1991), 'Multivariate adaptive regression splines', *AS* **19**(1), 1–67. *242*

Galbraith, J. W. *see* Banerjee, A.

George, E. I. *see* Blattberg, R. C., Casella, G.

Ghosh, M., Saleh, A. K. M. E. & Sen, P. K. (1988), 'Empirical Bayes subset estimation in regression models', *Statistics and Decisions. 58,472*

Ghysels, M. (1994), 'On the periodic structure of business cycles', *Journal of Business and Economic Statistics.* **12**(3), 289–297 *428*

Gnedenko, B. V. & Kolmogorov, A. N. (1954), *Limit Distributions for Sums of Independent Random Variables*, Addison-Wesley, Reading, Mass. Translation by K.L. Chung from Russian. *282,326, 330*

Goldfeld, S. & Quandt, R. (1965), 'Some tests for homoskedasticity', *JASA* **60**, 539–547. *193*

Gourieroux, C., Maurel, F. & Monfort, A. (1987), Regression and non-stationarity, Technical Report 8708, Institut National de la Statistique et des Etude Economiques (INSEE), Unite de Recherche, INSEE, Paris. *287*

Greenberg, E. *see* Chib, S.

Greenberg, E. & Webster, C. E. (1983), *Advanced Econometrics: A Bridge to the Literature*, Wiley, New York. *42,42, 86*

Greene, W. H. (1990), *Econometric Analysis*, Macmillan Publishing Co., New York. *xviii, 185*

Griffiths, W. E. *see* Judge, G. G.

Guttman,I. *see* Fraser, D. A. S.

Hall, P. (1992), *The Bootstrap and Edgeworth Expansion*, Springer Series in Statistics, Springer-Verlag, New York. *357, 375, 377, 377*

Hall, P. & Wilson, S. R. (1991), 'Two guidelines for bootstrap hypothesis testing', *Biometrics* **47**, 757–762. *376*

Hampel, F. R., Ronchetti, E. M., Rousseeuw, P. J. & Stahel, W. A. (1986), *Robust Statistics: The Approach Based on Influence Functions*, Wiley, New York. *114*

Hansen, L. (1982), 'Large sample properties of generalized method of moments estimators', *Ecta* **50**, 1029–1054. *306*

Härdle, W. & Linton, O. (1993), Applied nonparametric methods, Discussion Paper 9312, Institut für Statistik und Ökonometrie, Humboldt-Universität zu Berlin, Spandauerstr.1 Berlin, Germany. *121*

Harville, D. (1980), 'Predictions for national football league games via linear model methodology', *JASA* **75**(371), 516–524.

Has'minskii, R. Z. *see* Ibragimov, I. A.

Hawkins, D. M. (1994), 'The feasible solution algorithm for least trimmed squares regression',*Computational Statistics and Data Analysis* **17**, 185–196. *106*

Hendry, D. F. *see* Banerjee, A., Engle,R. F.

Hendry, D. F. (with Duo Qin & C. Favero) (1991), *Lectures on Econometric Methodology*, Nuffield College, Unpublished, Oxford. *xvii,xviii,xxii,31,178,183,184, 227*

Herriot, R. A. *see* Fay, R. E.

Hettmansperger, T. P. *see* Coakley, C. W.

Hildreth, C. & Houck, J. P. (1968), 'Some estimators for a linear model with random coefficients', *JASA* **63**, 584–595. *240*

Hill, R. C. *see* Judge, G. G.

Hillier, G. H. (1991), 'On multiple diagnostic procedures for the linear model', *JoE* **47**, 47–66.

Hinkley, D. V. *see* Cox, D. R.

Hinkley, D. & Schechtman, E. (1987), 'Conditional bootstrap methods in the mean shift model', *Biometrika* **74**, 85–93. *243*

Hoadley, B. (1981), 'The quality measurement plan (QMP)', *The Bell System Technical Journal* **60**(2), 215–273. *62, 64,438,443*

Hoeffding, W. (1965), 'Asymptotically optimal tests for multinomial distributions', *AMS* **36**, 369–401. *412*

Horowitz, J. L. (1994), 'Bootstrap-based critical values for the information matrix test', *JoE* **61**, 395–411. *381*

Houck, J. P. *see* Hildreth, C.

Hunt, G. & Stein, C. (1946), Most stringent tests of statistical hypotheses, (unpublished manuscript). *130,155*

Ibragimov, I. A. & Has'minskii, R. Z. (1981), *Statistical Estimation: Asymptotic Theory*, Springer-Verlag, New York. Translation by Samuel Kotz from Russian. *304,307, 392, 394,401,404*

Ip, W. C. & Phillips, G. D. A. (1994), Second order correction by the bootstrap, Technical Report 93-346, Economics Dept., University of Bristol, 8 Woodland Rd, Bristol BS8 1TN. *377*

Jarque, C. M. & Bera, A. K. (1987), 'A test for normality of observations and regression residuals', *International Statistical Review* **55**, 163–172. *181,327*

Jayatissa, W. A. (1977), 'Tests of equality between sets of coefficients in two linear regressions when the disturbance variances are unequal', *Ecta* **45**, 1291–1292. *246*

Jeong, J. & Maddala, G. S. (1992), 'A perspective on applications of bootstrap methods in econometrics', in *Handbook of Statistics Volume 11: Econometrics. 42,371*

Judge, G. G. *see* Aigner, D. J.

Judge, G. G. *et al.* (1985), *The Theory and Practice of Econometrics*, second edition, Wiley, New York.xviii, *179,192,223,235, 456*

Judge, G. G. *et al.* (1988), *Introduction to the Theory and Practice of Econometrics*, second edition, Wiley, New York. *224*

Kadane, J. B. *see* Erickson, E. P.

Kallenberg, W. C. M. (1982), 'Chernoff efficiency and deficiency', *AS* **10**, 583–594. *412*

Kaufman, S. (1966), 'Asymptotic efficiency of the maximum likelihood estimator', *Annals of the Institute of Statistical Mathematics* **18**, 155–178. *401*

Keating, J. P. *see* Dyer, D. D.

Kelezoglu, H. (1995), Improved Estimates of Investment Functions, unpublished manuscript, submitted to *Journal of Forecasting*. *225*

King, M. *see* Stuart, A.

King, M. L. *see* Brook, R. O., Evans, M. A.

King, M. L. (1987-1988), 'Towards a theory of point optimal testing', *Econometric Reviews* **6**(2), 169–218. *145*

Kipnis, V. (1989*a*), Evaluating the impact of exploratory procedures in regression prediction: A pseudosample approach, Tech. Report MRG Working Paper M 8909, University of Southern California, University Park, Los Angeles, CA 90089-0035. *228*

Kipnis, V. (1989*b*), Model selection and prediction assessment in regression analysis, Tech. Report MRG Working Paper M 8910, University of Southern California, University Park, Los Angeles, CA 90089-0035. *228*

Kiviet, J. F. *see* van Giersbergen

Klein, L. R. (1990), 'The concept of exogeneity in econometrics', in R. A. L. Carter, J. Dutta & A. Ullah, eds, *Contributions to Econometric Theory and Application: Essays in Honour of A. L. Nagar*, Springer-Verlag, New York. *39*

Kohn, R. *see* Ansley, C. F.

Kolmogorov, A. N. *see* Gnedenko, B. V.

Koschat, M. A. & Weerahandi, S. (1992), 'Chow-type tests under heteroscedasticity', *Journal of Business and Economic Statistics* **10**(2), 221–228. *250*

Krasker, W. S. & Welsch, R. E. (1982), 'Efficient bounded influence regression estimation', *JASA* **77**, 595–604. *114*

Kruskal, W. (1968), 'When are Gauss-Markov and least squares identical? a coordinate-free approach', *AMS* **39**, 70–75. *14*

Kuh, E. *see* Belsley, D. A.

LaFontaine, F. & White, K. J. (1986), 'Obtaining any Wald statistic you want', *Economics Letters* **21**, 35–40. *223*

Lai, T. L. & Wei, C. Z. (1982a), 'Asymptotic properties of projections with applications to stochastic regression problems', *Journal of Multivariate Analysis* **12**, 346–370. *284*

Lai, T. L. & Wei, C. Z. (1982b), 'Least squares estimates in stochastic regression models with applications to identification and control of dynamic systems', *AS* **10**(1), 154–166. *284, 286*

Lai, T., L. Robbins, H. & Wei, C. Z. (1978), 'Strong consistency of least squares estimates in multiple regression', *Proceedings of the National Academy of Sciences, USA* **75**, 3034–36. *283*

Laird, N. *see* Dempster, A. N.

Lawley, D. N. (1956), 'A general method for approximation to the distribution of the likelihood ratio criteria', *Biometrika* **43**, 295–303. *381*

Leamer, E. E. (1978), *Specification Searches*, Wiley, New York. *xvii,272*

Leamer, E. E. (1982), 'Sets of posterior means with bounded variance priors', *Ecta* **50**, 725–36. *xvii,424*

Leamer, E. E. (1983), 'Let's take the con out of econometrics', *American Economic Review* **73**, 31–43. *xvii*

LeCam, L. (1953), 'On some asymptotic properties of maximum likelihood and related Bayes estimates', *University of California Publications in Statistics* **1**, 277–330. *394,401*

LeCam, L. & Yang, G. L. (1990), *Asymptotics in Statistics: Some Basic Concepts*, Springer-Verlag., New York. *307,401*

Lee, T.-C. *see* Judge, G. G.

Lehmann, E. L. (1983), *Theory of Point Estimation*, Wiley, New York. *401,404,406*

Lehmann, E. L. (1986), *Testing Statistical Hypotheses*, second edition, Wiley, New York. *127,130,207,341*

Lehmann, E. L. (1990), 'Model specification: The views of Fisher and Neyman and later developments', *Statistical Science* **5**(2), 160–168. xx

Leroy, A. M. *see* Rousseeuw, P. J.

Lindley, D. V. & Smith, A. F. M. (1972), 'Bayes estimates for the linear model', *Journal of the Royal Statistical Society, Series B* **34**(1), 1–41. *477*

Linton, O. *see* Härdle, W.

Luenberger, D. C. (1969), *Optimization by Vector Space Methods*, Wiley, New York. *4,6*

Lye, J. N. (1991), 'The numerical evaluation of the distribution function of a bilinear form to a quadratic form with econometric examples', *Econometric Reviews* **10**(2), 217–234. *364,514*

Maasoumi, E. & Zaman, A. (1995), Tests for structural change, Technical Report, Economics Dept., Bilkent University, Ankara, Turkey. *257*

Madansky, A (1988), *Prescriptions for Working Statisticians*, Springer-Verlag, New York. *180,493*

Maddala, G. S. *see* Jeong, J.

Mallik, A. K. *see* Basawa, I. V.

Marsaglia, G. & Zaman, Arif (1993), 'Monkey tests for random number generators', *Computer & Mathematics With Applications* **26**(9), 1–10. *484*

Marsaglia, G., Narasimhan, B. & Zaman, Arif (1990), 'A random number generator for PCs', *Computer Physics Communications* **60**(3), 345–349. *484*

Maurel, F. *see* Gourieroux, C.

McCabe, B. & Tremayne, A. (1993), *Elements of Modern Asymptotic Theory with Statistical Applications*, Manchester University Press, Manchester and New York. *324*

McCormick, W. P. *see* Basawa, I. V.

McDermott, C. J. *see* Andrews, D. W. K.

Milnes, P. *see* Bondar, J. V.

Morgenthaler, S. & Tukey, J. W., eds. (1991), *Configural Polysampling: A Route to Practical Robustness*, Wiley, New York. *114*

Monfort, A. *see* Gourieroux, C.

Morris, C. *see* Efron, B.

Morris, C. N. (1983), 'Parametric empirical Bayes inference: Theory and applications', *JASA* **78**(381), 47–55. *435,452, 460*

Mukhophadhyay, C. *see* Başçı, S.

Nagar, A. *see* Ullah, A.

Nagarsenker, P. B. (1984), 'On Bartlett's test for homoegeneity of variances', *Biometrika* **71**(2), 405–7. *193*

Narasimhan, B. *see* Marsaglia, G.

Navidi, W. C. *see* Freedman, D. A.

Navidi, W. (1989), 'Edgeworth expansions for bootstrapping regression models', *AS* **17**(4), 1472–1478. *375*

Nebebe, F. & Stroud, T. W. F. (1986), 'Bayes and empirical Bayes shrinkage estimation of regression coefficients', *The Canadian Journal of Statistics* **14**(4), 267–280. *480*

Nebebe, F. & Stroud, T. W. F. (1988), 'Bayes and empirical Bayes shrinkage estimation of regression coefficients: A cross-validation study', *Journal of Educational Studies* **13**(4), 199–213. *63,479*

Ord, J. K. *see* Stuart, A.

Penm, J. H. W., Penm, J. H. & Terrell, R. D. (1992) 'Using the Bootstrap as an aid in choosing the approximate representation for vector time series', *Journal of Business and Economic Statistics* 10(2), 213-19. *229*

Peracchi, F. (1992), 'Bootstrap methods in econometrics', *Ricerche Economiche* 45(4), 609–622. *317*

Perlman, M. D. (1972), 'On the strong consistency of approximate maximum likelihood estimators', *Proceedings of the Sixth Berkeley Symposium on Mathematical Statistics and Probability* 1, 263–281. *300,304*

Pfanzagl, J. (1968), 'A characterization of the one-parameter exponential family by existence of uniformly most powerful tests', *Sankhyā A* 30, 147–156. *130,527*

Pfanzagl, J. (1970), 'On the asymptotic efficiency of median unbiased estimates', *AMS* 41(5), 1500–1509. *397,400*

Pfanzagl, J. (1974), 'On the Behrens-Fisher problem', *Biometrika* 61(1), 39–47. *212,246,247,366*

Pfanzagl, J. & Wefelmeyer, W. (1978), 'A third-order optimum property of the maximum likelihood estimator', *Journal of Multivariate Analysis* 8, 1–29. *401,405,406*

Phillips, G. D. A. *see* Ip, W. C.

Pitman, E. J. G. (1938),' Tests of hypotheses concerning location and scale parameters', *Biometrika*, 31, 200-231. *185,186*

Poirier, D. J. (1973), 'Piecewise regression using cubic splines', *JASA* 68(343), 515–524. *215*

Pollard, D. (1984), *Convergence of Stochastic Processes*, Springer-Verlag, New York. *326*

Pötscher, B. M. & Prucha, I. R. (1991a), 'Basic structure of the asymptotic theory in dynamic nonlinear econometric models part I: Consistency and approximation concepts', *Econometric Reviews* 10(2), 125–216. *296,301*

Pötscher, B. M. & Prucha, I. R. (1991*b*), 'Basic structure of the asymptotic theory in dynamic nonlinear econometric models part II: Asymptotic normality', *Econometric Reviews* **10**(3), 253–325. *335*

Pötscher, B. M. & Prucha, I. R. (1994), 'Generic uniform convergence and equicontinuity concepts for random functions', *JoE* **60**, 23–63. *301,302,338*

Pratt, J. W. & Schlaifer, R. (1984), 'On the nature and discovery of structure', *JASA* **79**(385), 9–23. *xvii*

Prucha, I. R. *see* Pötscher, B. M.

Quandt, R. *see* Goldfeld, S.

Ramsey, J. B. (1974), 'Classical model selection through specification error tests', in P. Zarembka, ed., *Frontiers in Econometrics*, Academic Press, New York, pp. 13–47. *212*

Rao, R. R. *see* Bhattacharya, R. N.

Rayner, R. K. (1990), 'Bootstrap tests for generalized least squares regression models', *Economics Letters* **34**, 261–265. *247, 377, 383*

Richard, J.-F. *see* Engle, R. F.

Ritter, C. & Tanner, M. A. (1992), 'Facilitating the Gibbs sampler: The Gibbs stopper and the griddy-Gibbs sampler', *JASA* **87**(419), 861–868. *486*

Rivest, L.-P. (1986), 'Bartlett's, Cochran's and Hartley's tests on variances are liberal when the underlying distribution is long tailed', *JASA* **81**(393), 124–128. *193*

Robbins, H. *see* Lai, T., L.

Roberts, G. O. & Smith, A. F. M. (1994), 'Simple conditions for the convergence of the Gibbs sampler and Metropolis-Hastings algorithm', *Stochastic Processes and their Applications* **49**, 207–216. *486*

Rocke, D. M. (1989), 'Bootstrap Bartlett adjustment in seemingly unrelated regression', *JASA* **84**(406), 598–601. *383*

Rolph, J. *see* Carter, G.

Ronchetti, E. M., *see* Hampel, F. R

Rothenberg, T. J. (1984*a*), 'Chapter 15: Approximating the distributions of econometric estimators and test statistics', in Z. Griliches & M. D. Intriligator, eds, *Handbook of Econometrics: Volume II*, Elsevier Science Publishers, Amsterdam, pp. 881–935. *366, 410*

Rothenberg, T. J. (1984*b*), 'Hypothesis testing in linear models when the error covariance matrix is nonscalar', *Ecta* **52**(4), 827–842. *247,366, 377*

Rousseeuw, P. J. *see* Hampel, F. R

Rousseeuw, P. J. (1984), 'Least median of squares regression', *JASA* **79**, 871–880. *104*

Rousseeuw, P. J. & Leroy, A. M. (1987), *Robust Regression and Outlier Detection*, Wiley, New York. *97,99,105*

Rousseeuw, P. J. & van Zomeren, B. (1990), 'Unmasking multivariate outliers and leverage points', *JASA* **85**, 633–639. *113*

Rousseeuw, P. J. & Wagner, J. (1994), 'Robust regression with a distributed intercept using least median of squares', *Computational Statistics and Data Analysis* **17**, 65–76. *107*

Rousseeuw, P. J. & Yohai, V. J. (1984), 'Robust regression by means of s-estimators', in *Robust and Nonlinear Time Series*, Springer-Verlag, New York. *111,114*

Roy, K. P. (1957), 'A note on the asymptotic distribution of the likelihood ratio', *Calcutta Statistical Association Bulletin* **7**, 73–77. *343*

Rubin, D. *see* Dempster, A. N.

Saleh, A. K. Md. E. *see* Ghosh, M.

Sampson, P. D. (1979), 'Comment on 'splines and restricted least squares'', *JASA* **74**(366), 303–305. *216*

Sargan, J. D. (1980), 'Some tests of dynamic specification for a single equation', *Ecta* **48**, 879–897. *222*

Schechtman, E. *see* Hinkley, D.

Schlaifer, R. *see* Pratt, J. W.

Sefton, M. *see* Deb, P.

Sen, P. K. *see* Ghosh, M.

Serfling, R. J. (1980), *Approximation Theorems of Mathematical Statistics*, Wiley, New York. *410*

Shively, T. S. *see* Ansley, C. F.

Silverman, B. W. (1986), *Density Estimation for Statistics and Data Analysis*, Chapman and Hall, New York. *376*

Smith, A. F. M. *see* Lindley, D. V. Roberts, G. O.

Smith, H. *see* Draper, N.

Spanos, A. (1986), *Statistical Foundations of Econometric Modelling*, Cambridge University Press, Cambridge. *xvii,xviii*

Stahel, W. A. *see* Hampel, F. R.

Stegun, I. A. *see* Abramowitz, M.

Stein, C. *see* Hunt, G.

Strasser, H. (1981), 'Consistency of maximum likelihood and Bayes estimates', *AS* **9**(5), 1107–1113. *307*

Stroud, T. W. F. *see* Nebebe, F.

Stuart, A. & Ord, J. K. (1987), *Kendall's Advanced Theory of Statistics Volume 1*, fifth edition, Edward Arnold, London, Melbourne, Auckland. *181*

Stuart, A. & Ord, J. K. (1991), *Kendall's Advanced Theory of Statistics Volume 2: Classical Inference and Relationship*, fifth edition, Edward Arnold, London, Melbourne, Auckland. *192.343, 381*

Styan, G. P. H. *see* Fraser, D. A. S.

Takeuchi, K. *see* Akahira, M.

Tanner, M. A. *see* Ritter, C.

Taylor, R. L. *see* Basawa, I. V.

Taylor, R. L. (1978), *Stochastic Convergence of Weighted Sums of Random Elements in Linear Spaces*, Springer-Verlag, New York. *301*

Terrell, R. D. *see* Penm, J. H.

Thisted, R. A. (1976), Ridge regression, minimax estimation and empirical Bayes methods, Ph.D. thesis, Statistics Dept., Stanford University, CA. 94305. *89,89*

Thursby, J. (1992), 'A comparison of several exact and approximate tests for structural shift under heteroskedasticity', *JoE* **53**, 363–386. *213,246,247, 248,366*

Tire, C. (1995), Bootstrapping: Three applications to regression models, Master's thesis, Economics Dept., Bilkent University, 06533 Ankara, Turkey. *201,223*

Tomak, K. (1994), Evaluation of the Goldfeld-Quandt test and alternatives, Master's thesis, Economics Dept., Bilkent University, 06533 Ankara, Turkey. *192,255*

Tomak, K. & Zaman, A. (1995), A note on the Goldfeld-Quandt test, Technical report, Economics Dept., Bilkent University, 06533 Ankara, Turkey. *192, 194,255*

Tremayne, A. *see* McCabe, B.

Tukey, J. W., *see* Morgenthaler, S.

Ullah, A. *see* Zinde-Walsh, V.

Ullah, A. & Nagar, A. (1988), On the inverse moments of noncentral wishart matrix, Technical Report 6, Economics Department, University of Western Ontario, Canada. *42*

Ullah, A. & Zinde-Walsh, V. (1984), 'On the robustness of *LM*, *LR* and Wald tests in regression models', *Ecta*. *201*

Ullah, A. & Zinde-Walsh, V. (1985), 'Estimation and testing in a regression model with spherically symmetric errors', *Economic Letters* **17**, 127–132. *201*

van Giersbergen, N. P. A. & Kiviet, J. F. (1994), How to implement bootstrap hypothesis testing in static and dynamic regression models, Technical Report TI 94-130, Tinbergen Institute, Amsterdam-Rotterdam, Oostmaaslaan 950-952, 3063 DM Rotterdam, The Netherlands. *201*

van Zomeren, B. *see* Rousseeuw, P. J.

Wagner, J. *see* Rousseeuw, P. J.

Wald, A. (1949), 'A note on the consistency of the maximum likelihood estimate', *AMS* **20**, 595-601. *299*

Watson, G. S. *see* Durbin, J.

Webb, G. I. *see* Berenblutt, I.

Webster, C. E. *see* Greenberg, E.

Weerahandi, S. *see* Koschat, M. A.

Weerahandi, S. (1994), *Exact Statistical Methods for Data Analysis*, Springer-Verlag, New York. *234,248*

Wefelmeyer, W. *see* Pfanzagl, J.

Wegge, L. (1971), 'The finite sample distribution of least squares estimators with stochastic regressors', *Ecta* **38**, 241-251. *42*

Wei, C. Z. *see* Lai, T. L.

Welsch, R. E. *see* Belsley, D. A., Krasker, W. S.

White, H. *see* Domowitz, I.

White, H. (1980), 'A heteroskedasticity consistent covariance matrix estimator and a direct test for heteroskedasticity', *Ecta* **48**, 817–838. *185,335*

White, K. J. *see* LaFontaine, F.

Wilson, S. R. *see* Hall, P.

Wold, H. (1954), 'Causality and econometrics', *Ecta* **22**, 162–177. *39*

Wu, C.-F. (1981), 'Asymptotic theory of nonlinear least squares estimation', *AS* **9**(3), 501–513. *308, 309*

Wu, C.-F. (1986), 'Jackknife, bootstrap, and other resampling methods in regression analysis', *AS* **14** 1261–1295. *371*

Yang, G. L. *see* LeCam, L.

Yohai, V. J. *see* Rousseeuw, P. J.

Yohai, V. J. (1987), 'High breakdown point and high efficiency robust estimates for regression', *AS* **15**, 642–665. *114*

Yohai, V. J. & Zamar, R. H. (1988), 'High breakdown estimates of regression by means of the minimization of an efficient scale', *JASA* **83**, 406–413. *109, 116*

Zabel, J. E. (1993), 'A comparison of nonnested tests for misspecified models using the method of approximate slopes', *JoE* **57**, 205–232. *179*

Zaman, Arif *see* Marsaglia, G.

Zaman, Arif (1995), Cutting Corners: Efficient Random Number Generation from Unimodal Densities, Technical report, Lahor University of Management Science (LUMS), Lahore, Pakistan. *484*

Zaman, Asad *see* Başçı, S., Carrington, W., Dutta, J., Maasoumi, E. , Tomak, K.

Zaman, A. (1984), 'Avoiding model selection by the use of shrinkage techniques', *JoE* **25**, 239–246. *228*

Zaman, A. (1989), 'Consistency via type 2 inequalities: A generalization of Wu's theorem', *Econometric Theory* **5**, 272-286. *308, 309*

Zaman, A. (1991), 'Asymptotic suprema of averaged random functions', *AS* **19**(4), 2145–2159. *300,304*

Zaman, A. (1995), Implications of Bayesian rationality arguments, Technical report, Bilkent University, Economics Department, Bilkent University, Ankara, Turkey. *xxvi*

Zamar, R. H. *see* Yohai, V. J.

Zellner, A. (1986), 'On assessing prior distributions and Bayesian regression analysis with g-prior distributions', in P. Goel & A. Zellner, eds, *Bayesian Inference and Decision Techniques*, North-Holland Elsevier, New York. *58, 472*

Zinde-Walsh, V. *see* Ullah, A.

Zinde-Walsh, V. & Ullah, A. (1987), 'On the robustness of tests of linear restrictions in regression models with elliptical error distributions', in I. B. MacNeill & G. J. Umphrey, eds., *Time Series and Econometric Modelling*, D. Reidel Publishing Co., New York, pp. 235–251. *42,201*

Index